MW00513571

INTRODUCTION TO DESIGN AND TECHNOLOGY

INTRODUCTION TO DESIGN AND TECHNOLOGY

TODD, TODD, McCRORY

Ronald D. Todd, Ph.D.
School of Engineering ■ Department of Technological Studies ■ Trenton State College ■ Trenton, New Jersey

Karen Rohne Todd, Ph.D.
Department of Home Economics ■ Montclair State College ■ Montclair, New Jersey

David L. McCrory, Ph.D.
Technology Education ■ West Virginia University ■ Morgantown, West Virginia

Copyright © 1996
by Thomson Learning® Tools
an International Thomson Publishing Co.

IP

The ITP logo is a trademark under license.

For more information, contact:

Thomson Learning® Tools
5101 Madison Road
Cincinnati, OH 45227

International Thomson Publishing GmbH
Konigswinterer Str. 418
53227 Bonn
Germany

International Thomson Publishing
Berkshire House
168-173 High Holborn
London, WC1V7AA
England

International Thomson Publishing Asia
221 Henderson Bldg. #05-10
Singapore 0315

Thomas Nelson Australia
102 Dodds Street
South Melbourne 3205
Victoria, Australia

International Thomson Publishing Japan
Kyowa Building, 3F
2-2-1 Hirakawa-cho
Chiyoda-ku, Tokyo 102
Japan

Nelson Canada
120 Birchmont Road
Scarborough, Ontario
M1K 5G4, Canada

ALL RIGHTS RESERVED

The text of this publication, or any part thereof, may not be reproduced or transmitted in any form or by any means, electronic or mechanical, including photocopying, recording, storage in an information retrieval system, or otherwise, without the prior written permission of the publisher.

ISBN: 0-538-64465-6

1 2 3 4 5 6 7 8 9 K I 00 99 98 97 96 95

Printed in the United States of America

Library of Congress Cataloging-in-Publication Data
Todd, Ronald D.
    Introduction to design & technology / Ronald D. Todd, Karen
I. Todd, David L. McCrory.
        p. cm.
    Includes index.
    ISBN 0-538-64465-6 (hard cover)
    1. Engineering design. 2. Design, Industrial. I. Todd, Karen
I. II. McCrory, David L. III. Title.
TA174.T58 1995
620' .0042--dc20   95-20691
                                                     CIP

Publisher: Brian J. Taylor
Project Manager/Acquisitions Editor: Suzanne F. Knapic
Production Coordinator: Jean Findley
Editor: Carol Spencer
Associate Director/Photo Editing: Devore Nixon
Photo Editor: Kathryn Russell/Pix Inc.

Preface

TO THE STUDENT

Have you ever wondered what life was like for your great-grandparents when they were young? Maybe you have talked with them or heard family stories about them. Also, you probably have watched movies or read about life centuries ago. At that time (and even now in some parts of the world) it was critical that people understood their **natural environment**. They needed to protect themselves from natural elements and work with natural forces to obtain what they needed for survival and comfort. In the past, people made most of what they needed, or traded with their family and neighbors. For example, when they bought bread or a tool instead of making it themselves, they probably knew the baker or the blacksmith.

In the twentieth century, however, major changes occurred as a result of technology. We still need to know about the natural environment. However, in the United States and many parts of the world, many aspects of that natural environment have been modified and changed. It is now imperative that people today understand their **technological environment**.

Currently, most of what you and your family want and need is purchased from someone else, often someone unknown to you. In the past, those necessary items and services were provided locally, and events primarily affected people who lived close by, the neighbors and friends. Information used in making decisions was from local sources. Now, events that happen around the world have immediate and significant impact on the lives of us all. We have begun to recognize that decisions and actions made now, locally, have potentially far-reaching **consequences** globally, and for the next generations.

Much of your life will be spent in the twenty-first century. Many changes that you will experience go beyond what we can dream of today. These **technological changes** will influence the quality of your life and that of all living beings. We, the authors, hope that once you have read this book and done some of the activities, you will want to become an informed, active participant in these changes. We hope you will lend your resources to the development of a technological environment that enhances the lives of all, rather than impedes and destroys.

Most people do not know very much about technology. Many are confused and uncertain about its meaning. This book is intended to help you learn about technology by examining:

- its *processes*, the **design** approach to problem-solving,

- its *knowledge*, the understandings gained by people about the **design, resources, systems,** and **impacts** of technological activities,

- its *results*, the major technological **products, systems,** and **environments,** and

● ●

KEY TERMS

consequences
decisions and actions
design and technology
impacts of technology
natural environment
products
resources of technology
systems of technology
technological changes
technological environment
technologically literate
technology
wants and needs

- its *impacts*, some previously identified **consequences** and **projections** for future developments.

Figure 1 indicates how these major aspects of technology relate to each other. The book is organized to reflect this interaction.

Technology is a human endeavor, made by and for humans. For the purposes of this book, we use this **definition of technology**:

> **the use of our knowledge, tools, and skills to solve practical problems and to extend human capabilities.**

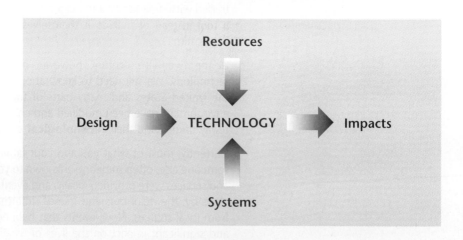

Figure 1 ■
The Major Aspects of Technology

Unit I of this book is **Design of Technology**. This unit presents an introduction to the problem-solving approach that is the "doing" of technology. You will learn the IDEATE model for designing, as well as drawing and documenting skills. You will create a portfolio and identify other methods for organizing data gathered as you carry out your design work. Developing and constructing models and prototypes for planning, implementing, and evaluating the outcomes of your design are explored.

Unit II, **The Resources of Technology**, identifies what is needed for doing any technological activity. Tools and machines, materials, energy, information, humans, and the resources of time, space, and capital are examined. Humans convert materials, energy, and information, through the use of tools and processes, in order to produce the outcomes they want. Management of all resources for effective and efficient use is explored. The application of the concepts and ideas to selecting and managing resources in your own design work is also addressed in this unit.

As technological activities have become more complex and interrelated, systems have been developed. We explore these major systems in Unit III, **The Systems of Technology**. Physical systems, including construction, transportation, and production, are presented. These systems are the source of many of the goods and services you use every day. Biotechnology has had a major impact on agriculture, health care, and many other aspects of our lives. These systems that use living materials as a resource are addressed. Communication systems provide the information needed for understanding and using technology. An exploration of these systems and the use of information to set up control systems, often to automate a design, concludes this unit.

Finally, in Unit IV, **The Impact of Technology**, we examine how to improve ideas and outcomes through investigation and development. A close look is taken at how to assess the outcomes of technological development. Ways to project intended and unintended outcomes, for the present and the future, are examined. Decision-making and the importance of weighing costs and benefits, using the best information, and deciding on acceptable risk levels are discussed. We conclude the book with an indication of how you can use your new skills and understandings to make a difference in your life and the lives of other people.

For centuries, technology has been used to serve both human **wants** and human **needs**. Sometimes it is used for the wants of one group of people, at the expense of other groups and other species of life. Technology itself is a tool. As with any other tool, it can be used by people for good or for evil, to enhance or to destroy.

All around the world, people strive to make a better life, to have maximum comfort and minimum pain. We all want to know more, live longer, have more fun. Because you live in a democracy, you have the ability to influence decisions that are made. These decisions, however, require that you be **technologically literate** and that you make informed decisions about technology.

In addition, the technological approach involves a "can-do" attitude. We hope you begin to think about and design better solutions for everyday issues facing you and the people around you. In that way, you can get a very real sense about how your decisions and actions can make a difference. You can begin your technology awareness by completing the following Enrichment Activity.

ENRICHMENT ACTIVITIES

Conduct a survey of at least four people who are not taking this course. Ask them to define "technology" and give some examples of technological products and activities. Also, ask them why people use technology.

Analyze the definitions. For each, circle yes (Y) or no (N) in the following chart. (Yes indicates the person included this concept in his/her definition. No indicates he/she did not.)

Definition	1	2	3	4
use of knowledge	Y/N	Y/N	Y/N	Y/N
use of tools	Y/N	Y/N	Y/N	Y/N
use of skills	Y/N	Y/N	Y/N	Y/N
solve practical problems	Y/N	Y/N	Y/N	Y/N
extend human capabilities	Y/N	Y/N	Y/N	Y/N

Did the people you surveyed understand the definition of technology used in this book?

How do your findings compare with those of the rest of the class?

From the overall results of the survey, would you conclude that "technology" is a well-understood term?

TO THE TEACHER

This is a time of great opportunity for making significant and fundamental changes in the nature of schooling. First, education must become more inclusive and the school must supply some of the support that was formerly provided by the family. Approaches to learning are needed that are appropriate for students from a variety of learning styles, from all racial, economic, and ethnic groups, including females as well as males. Family support that ensures that students come to school well-fed and healthy, ready to learn, and motivated to succeed is often lacking. Schools must create that motivation by engaging students in meaningful learning and often must attend to the physical and emotional needs of the learner as well.

Second, evidence about how people learn should be applied more systematically to the choice of experiences for students. Research clearly indicates that students are better able to retain and use what they learn if they are active participants and they can apply what they learn. Understanding requires that content be taught in an integrated, real manner in order for transfer to occur. All learners must construct their own sense of reality, and each makes the unique connections between prior learning and current instructional activities. Information is rapidly exploding and the teacher cannot be the primary source of relevant resources. Students need to work with each other, gain access to resources beyond the classroom, and come up with their own unique solutions.

Third, current requirements for informed citizenship and effective workers indicate that changes in schools are needed. Today, citizens must understand their technological environment, as well as the social and economic forces around them. Graduates of our school systems must be technologically and scientifically knowledgeable and responsible. The workplace has few openings for students who can do only unskilled labor and the wages they can earn will not support independent living. Work, consumer behavior, and citizenship requires the ability to communicate, work with others, critically think through potential solutions, and make tough decisions. Graduates must be ready to design, build, operate, and maintain the essential technological systems, both in our country and globally. They must recognize that everyday decisions have consequences, for themselves and others, currently and in the future.

Technology education addresses many of these problems. By its very nature of problem-solving and design, technology education demands integrated learning. The approach requires that students engage in collaborative learning and actively solve real problems. Technology education can be taught by teachers from several backgrounds, but ideally involves a team approach, crossing traditional discipline lines. Students, other teachers, and community and global resources are critical to the process. This book is intended to support students as they learn this approach by carrying out a design and problem-solving process that will:

(1) require them to use what they have learned and/or seek out the resources that are missing,

(2) create a deeper interest and commitment to improving the quality of life for themselves and others, and

(3) provide concrete, practical experiences to help them gain new conceptual insights and skills for making decisions.

The content of technology as identified in this book has evolved to consider the design process, resources, systems, and impacts of technology—all important in their own right. The book pays attention to both the process and content dimensions of technological understanding. We believe it is critical that students understand more about the world in which they live and the technological products and environments that affect them. We also want them to have skills they can use to make a difference, both in what they choose as consumers and what they change as designers.

Through their direct involvement in the activities and materials provided, we hope students will gain new understandings and capabilities in doing technology, as well as knowing about their technological environment.

Additional Features

A *Portfolio and Activities Resource* accompanies the text. This student resource book includes additional materials for design activities and design briefs and the resources necessary for student to keep a portfolio of their work.

The *Teacher's Resource Guide* provides an overview to design and technology, answers to text questions, information on how to teach a course on design and technology, as well as additional instructional aids.

Design Brief Manager Software (by Dan Engstrom and Larry Hatch), available in Macintosh™ and IBM® formats, is available with the text. The software is an electronic database that organizes, stores, retrieves, prints, and exports design briefs. The software allows searches by key words, grade level, and content organizers. Teachers can search for design briefs and copy and modify them to suit their classroom needs. All design briefs included in the text appear on the software with the addition of 50 new ones.

ABOUT THE AUTHORS

Ronald D. Todd is a research professor in the Department of Technological Studies at Trenton State College. He received his Ph.D. from Case Western Reserve University. He is cofounder of *TIES Magazine* and continues to serves as its Publisher. He has been active in the International Technology Education Association and has taught technology as a subject at both the high school and university levels. Dr. Todd has directed funded grants for more than twenty-five years and is currently directing two technology education related grants supported by the National Science Foundation. He has worked as a consultant in more than twelve states developing programs in technology education. His special interest is to integrate design and problem solving with proven technology education approaches to enhance the learning of science, mathematics, and other subjects for all students.

Karen Rohne Todd is a professor at Montclair State University in the Department of Home Economics. She received her Ph.D. from the University of Iowa. She is certified as a Family Life Educator and has a strong background in child development and family studies. In addition to teaching at the college level, she has taught nursery school, elementary, and junior high school. She has been active in curriculum development in those aspects of home economics related to family life education and technology education. Her special interest in technology is the impact of technological development on the well-being of individuals and families and the impact that individuals can make on the development of technology by actions taken as consumers and informed citizens.

David L. McCrory is a professor and Chairman of the Department of Technology Education, West Virginia University, where he teaches graduate courses in technology education curriculum development, innovation and invention, and multidisciplinary studies. He holds a Ph.D. from Case Western Reserve University. He has thirty years teaching and supervisory experience, including junior high school, senior high school, and university levels. He has directed numerous funded research and development projects in technology education. Dr. McCrory has published in state and national journals and professional yearbooks. He serves as a member of the Board of Examiners, National Council for Accreditation of Teacher Education, and is an active member of the International Technology Education Association and other professional organizations.

This team of authors wrote *Understanding and Using Technology*, published in 1985 as a ground-breaking effort to help introduce technology education into middle schools nationally. This book provided a comprehensive conceptual foundation for technology education as a newly emerging school subject.

Contributing Authors

John Wells, Assistant Professor, Department of Technology Education, West Virginia University, Morgantown, West Virginia

Alan Paul, Senior Lecturer, Design and Technology, The Nottingham Trent University, Nottingham, England

ACKNOWLEDGMENTS

This book, like most good design and technology activities, draws heavily on the work and ideas of others. We would like to thank the following individuals for their gracious assistance during the preparation of the book and the *Portfolio and Activities Resource*. They made significant contributions to the content, process, and activities of the book, thus enriching the learning experiences for the readers.

Peter Bartcherer, Director, Design and Imaging Studio, Drexel University, Philadelphia, PA

Tim Brotherhood, QLS-STEP, Staffordshire Technology Education Programme, Stafford, England

Marc Dabe, Freelance Art Director, Frisco Lakes, CA

Paul Devine, Chair, Department of Technology Education, Glasgow High School, Glasgow, DE

Leonard Finegold, Professor of Physics, Drexel University, Philadelphia, PA

John Hindhaugh, Teacher Advisor, QLS-STEP, Staffordshire Technology Education Programme, Stafford, England

Catherine Houghton, Teacher, Department of Biology, Glasgow High School, Glasgow, DE

Pat Hutchinson, Editor-in-Chief, TIES Magazine, Trenton State College, Trenton, NJ

Geoffrey Oliver, Senior Lecturer, In-Service Department Unit, Charlotte Mason COllege, Lancaster University, Cumbria, England

Edward Rockel, Professor of Biology, Trenton State College, Trenton, NJ

Phillip Forrest-Smith, Design Consultant, Nottingham, England

Jonathan Renaudon-Smith, Head of Science and Information Technology, Queenswood School, Hatfield, Herts, England

Wendy Snyder, Teacher, Technology Education, Shore Regional High School, West Long Branch, NJ

The authors would like to thank the many reviewers who offered insightful suggestions to improve this book. Their assistance was a constructive and necessary part of the creative writing process. We appreciate all they have done.

John Brousseau, Technology Education, Cherry Creek High School, CO

Bob Hunt, Technology Education, Johnson County Community College, KS

Phil Hamel, Technology Education Teacher

Additionally, we would like to thank the following for their help and interest:

Amy Banghart, Jeff Bell, Frank Capelle, Catherine Farrup, Keith Finkral, Ted Gill, Hazel Hannant, John Hutchinson, John Karsnitz, Jamie Mulligan, Gordon Pritchett, Ray Shackelford, Ben Sprouse, Melvin Sprouse, Barbara Stala, Roni Todd, Victor Vinci, Shelly Woodring, and Cindy Yayac.

Finally, we would like to thank the editors and art directors who play such an essential role in bringing this effort to a successful conclusion.

Sue Knapic, Carol Spencer, Kathy Russell, Sandy Clark, and Wendy Troeger.

Contents

Unit 1 · Design and Technology

Chapter 1 • Introduction to Designing and Problem Solving   **3**

Introduction—What Is Designing and Problem Solving?   3
The Design Process at Work   5
Revisiting the Design Loop   16

Chapter 2 • Designing, Documenting, and Drawing   **23**

Introduction—The Process of Designing   23
Design Considerations   25
Documenting Techniques   37
Drawing Techniques   39
Pictorial Drawings   40
Nonpictorial, Flat Drawings   44
Presentation Techniques   45
Presentation Portfolios   50

Chapter 3 • Models and Model-Building   **55**

Introduction—Translating Ideas into Real Objects   55
Types of Models   56
Model-Making   62

Unit 2 · The Resources of Technology

Chapter 4 • Tools, Mechanisms, and Machines   **77**

Introduction—Devices to Extend Our Abilities   77
Tools Extend Our Abilities   78

Tools and Mechanisms ... 79
Fluidic Mechanisms ... 90
Electromechanical Mechanisms ... 94
Machines ... 96

Chapter 5 • Materials and Materials Processing ... **105**

Introduction—The Basic Ingredients of Technology ... 105
Characteristics of Materials ... 106
The Structure of Materials ... 108
Functions of Materials ... 118
Conversion of Materials ... 124
Similarities in Materials Conversion Processes ... 136

Chapter 6 • Energy and Energy Processing ... **141**

Introduction—The Driving Force of Technology ... 141
Sources of Energy ... 142
Forms of Energy ... 150
Conversion of Energy ... 160

Chapter 7 • Information and Information Processing ... **187**

Introduction—Building Blocks for Communication and Control ... 187
Putting Information to Use ... 188
Signals: Information for Machines ... 190
Symbols: Information for Humans ... 202
Using Symbols in Design and Technology ... 208

Chapter 8 • Humans as Designers and Consumers ... **221**

Introduction—The Creators and Users of Technology ... 221
What Is a Person? ... 222
Designing for the Human Body ... 222
Designing for Human Users ... 224
Growth and Development ... 226
Humans As Material Users ... 227
Humans and Energy Use ... 232
Humans As Information Processors ... 233
Designing and Human Cognition ... 237
Emotions and Behavior ... 240

Human Wants and Needs   241
Humans As Decision-Makers   248

Chapter 9 • Time, Space, Capital, and Management   **253**

Introduction—Putting It All Together   253
A Sense of Time and Its Use   254
A Sense of Place and Space   258
Capital—Human and Monetary Resources   263
Management of Labor and Other Resources   270

Unit **3** **The Systems of Technology**

Chapter 10 • Physical Systems   **289**

Introduction—Systems of Constructing, Transporting, and Producing   289
Construction Systems   290
Transporting Systems   302
Pathways, Vehicles, Loads, and Purposes   303
Production Systems   309

Chapter 11 • Bio-related Technology Systems   **329**

Introduction—Harnessing Living Organisms   329
Stages of Development   332
Cells and Genetic Engineering   338
Bioprocessing   342
Biotechnology in Agriculture   349
Bioethics   356

Chapter 12 • Communication Systems   **361**

Introduction—Sending Messages to Humans and Machines   361
Senders and Receivers   362

Transmitting Through Media   366
Feedback   371
Forms of Communicating   374
Three-Dimensional Forms of Communicating   382

Chapter 13 • Control Technology Systems   **387**

Introduction—Directing and Regulating Systems   387
Control Systems   389
Design and Development of Control Systems   391
Controlling Technology Systems   397
Developing Smart Machines   402
Computers and Control Technology   408

Unit 4 — The Impact of Technology

Chapter 14 • Investigating, Developing, and Improving   **419**

Introduction—How Technology Grows   419
Investigating Through Exploring   420
Investigating Through Forecasting   426
Expanding Knowledge for Improving Future Investigations   437
Developing—Putting Knowledge to Work   438
Improving—Making Things Better   440

Chapter 15 • Consequences and Decisions   **449**

Introduction—Dealing with Change, Now and Beyond   449
Looking Toward the Future   450
Consequences—Results and Effects   453
Decisions—Making Tough Choices   466

Epilogue   **477**

Glossary   **481**

Index   **503**

UNIT ONE

Design and Technology

Introduction to Unit I

All around you are examples of attempts to improve on natural occurrences, as well as products that people have made. In this unit, you will explore the "how to do it" aspects of technology. Chapter 1 introduces the IDEATE model for design and problem solving. This formal approach can be useful to ensure that you cover all aspects of design. This approach includes reworking designs and continually looking for ways to improve them. We illustrate the use of the model as you read about how two students, April and Enrique, solved a problem experienced by Aunt Sarah.

In Chapter 2, you will learn some skills for drawing and documenting your work. The intent is not only to enhance your thinking, but also to move this thinking to a tangible form that can be shared with other people. The focus includes principles of design and the use of color to communicate information more effectively. Ways to organize and present information you produce and gather at each design step are illustrated. Portfolio development and its uses are included as well.

Chapter 3 explores types of models and their uses. Constructing models and prototypes for implementing, evaluating, and communicating about your design will be a necessary and important part of your work.

In this unit, you will use, and hopefully enhance, your skills for creating new and improved ideas. Next, you will critically analyze ideas and information against given design criteria. Then you will take the next step into the real world of application. You will take action and try to make tangible and useful solutions to practical problems. As you test out and analyze how well your solution works, you will reenter the design process and make something better next time. Thus, you will contribute to technology.

CHAPTER 1

Introduction to Designing and Problem Solving

Figure 1.1 ■
The Grande Arche at La Defense Center in Paris, France is an architectural wonder and the direct result of the design process at work. *(© J.A. Kraulis/Masterfile)*

INTRODUCTION—WHAT IS DESIGNING AND PROBLEM SOLVING?

This may be the first time you have been involved with designing something. It is quite appropriate to ask, "What is designing and problem solving?" **Designing**, in its simplest form, is the translation of ideas into something

KEY TERMS

alternatives
assessing
constraints
criteria
design brief
design limits
design loop
design process
designing
desired outcomes
environments
evaluation
fabrication
idea generation
IDEATE
improvements
information
intended outcomes
investigating
iteration
making
model
modifications
patent
planning
portfolio
primary sources
problem
products
prototype
reiteration
report
research
resources
secondary sources
specifications
systems
testing
timetable
undesired outcomes
unintended outcomes

tangible. Designing provides a plan, and includes how to make and improve a product. Sometimes the word *design* is used to relate to the **design process**. At other times, it refers to the product of the design process.

In technology, designers are people who set out to solve practical problems that arise from real situations. When you engage in designing, you take what you think is a good idea and turn it into something that will solve the problem you have identified. In this sense, you will be engaged in a form of technical problem solving often referred to as a **Design Loop**. The Design Loop presented in Figure 1.2 uses the word *IDEATE* as an acronym to help you remember the general process you can follow as you translate your ideas into an end result. The outcome you produce can be of several types—**products**, **systems**, and **environments**.

The **IDEATE** Design Loop represents a formal approach, but you may do problem solving without going through each step. However, there are three times in which using a more formal process can be helpful. First, when you want to communicate with others about your work, you can be clear about what occurred. Maybe the other person can suggest different approaches to improve your work. Maybe they want to repeat the process to get the same result. The second time the IDEATE Design Loop will be helpful is when you want to document the process you have gone through to prove that this is your work. This may be especially important if you want to be sure that you get the credit. Maybe you want to get a **patent** or copyright, or maybe you are being evaluated on the thinking that you did as well as the outcome. The third time that the IDEATE Design Loop can be very useful is when you want to **assess** how well you have done and if you have left out important steps. This is useful when you do not get the results you hoped for. Mistakes can be very valuable, especially if you can tell where in the process you made them.

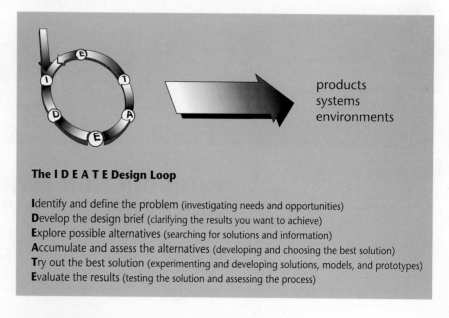

products
systems
environments

The I D E A T E Design Loop

Identify and define the problem (investigating needs and opportunities)
Develop the design brief (clarifying the results you want to achieve)
Explore possible alternatives (searching for solutions and information)
Accumulate and assess the alternatives (developing and choosing the best solution)
Try out the best solution (experimenting and developing solutions, models, and prototypes)
Evaluate the results (testing the solution and assessing the process)

■ **Figure 1.2**
The design process.

THE DESIGN PROCESS AT WORK

In the following pages, an example is presented of how two students, April and Enrique, implemented the design process. This case study is intended to help you understand the design process and how you might use the Design Loop in your own work.

Context and Situation

April asked Enrique to join her in solving a problem that is important to her. April's elderly Great Aunt Sarah is a very independent woman, but over the years she has developed a vision problem. Aunt Sarah is almost completely blind, and there is little hope that her sight can be restored. Aunt Sarah wants to live alone in her home, where she has lived more than 50 years.

Identify and define the problem.   Before April and Enrique try to solve the problem for Aunt Sarah, it is important that they analyze the situation and determine what is actually the problem. This aspect of the design process leads to identifying and defining the problem. April has some ideas about what is wrong and what to do, but her ideas may or may not be what Aunt Sarah wants. The problem with her vision affects Aunt Sarah in several ways. Only one problem can be tackled at a time, and some are too complicated for April and Enrique. Information must be gathered before choosing what to tackle first.

One Saturday, the teenagers visited Aunt Sarah in her home. They discussed what they wanted to do and some of their ideas. Aunt Sarah was pleased with their helpfulness and asked them to stay for lunch. April carefully observed how Aunt

Figure 1.3A ■
Good design and problem solving requires an analysis of the situation in which the problem occurs.

use with students

Problem Statement:
Storing and finding food is very difficult for someone who is visually impaired. The problem is even more difficult when the food is stored in cans. We think there is a need for a means of storing and retrieving canned goods so that they can be identified easily and safely.

yellow clipboards

Sarah moved around in her house. It seems that she knows where things are. She only has trouble if someone leaves things in unexpected places. Enrique noted that she was dressed neatly, yet there were some spots on her blouse that she may have missed in doing laundry or in choosing what to wear. He made a note to investigate that further. While they were there, Aunt Sarah got three phone calls and a neighbor from the third floor stopped by with some groceries.

Figure 1.3B ■
Aunt Sarah describes the problem she has in finding the food she wants.

It was when Aunt Sarah opened a can to make the tuna salad for sandwiches that April and Enrique caught each other's eye. Aunt Sarah opened a can of pineapple. She laughed and asked if they wanted pineapple for dessert. She then proceeded to open another can—pineapple, again! When she started to open a third can of what she thought was tuna, April asked her if she had trouble telling the difference. Aunt Sarah sighed and indicated that she did this often. When the cans were the same size and color, she could not tell just what she was going to eat for a meal. This seemed to be a **problem** worth **investigating**.

Develop the design brief. Once April and Enrique identified and understood the problem, they were ready to write a **design brief**. The design brief is a short statement that outlines the problem to be solved.

A design brief should be fairly specific, but should still allow the designer some flexibility. It is usually a short statement; that is why it is called a brief. April and Enrique might start with a design brief that states: Design and develop a kitchen storage system. This statement is rather vague and unclear as to what is to be done. In contrast, the design brief must not be so specific that it rules out the possibility and freedom of creativity. The teenagers worked on their design brief and agreed on the following statement.

Design Brief:
Design and develop a method for storing and retrieving canned goods in Aunt Sarah's kitchen. The system should allow Aunt Sarah to find specific cans in a safe and easy manner after she has stored them.

Figure 1.4 ■
Once the problem is clarified, it is time to write the design brief.

Explore possible alternatives. Once the design brief was developed, April and Enrique were ready to begin their investigation and research. In some instances, they could use their past experience and knowledge to solve a problem. In most cases, they must search for new information and knowledge related to the problem. This requires investigation and research.

Research may take the form of reading books, magazines, and other materials. It may involve interviewing, talking, and listening to people who have experience related to the problem. April and Enrique's research may involve writing and sending letters. They may make telephone calls to seek information from government, business, and industry. They need to know what systems now exist that aid people who are blind or have limited vision. Observation is also an important part of their work. Looking at products that are similar to what they have in mind, or looking at examples from nature may help. Finally, they can explore the community where they live. The shops, businesses, museums, and industries, as well as the people from the community, are possible sources of information and support.

Enrique made some brief sketches of Aunt Sarah's kitchen during their visit. April noted what happened after the neighbor gave Aunt Sarah the bag of groceries. They discussed the traffic patterns in the kitchen and the cooking area. They decided that these notes would do for now as they explored alternative storage patterns. They would go back to Aunt Sarah's after preliminary plans were made to see if they had forgotten anything.

The pair then talked to people who had encountered problems with their own relatives and friends who are visually impaired. April wondered whether helping Aunt Sarah continue to make her own meals would have some **undesired outcomes** as well as **desired outcomes**. Would she continue to go to the Community Center for lunch with other senior citizens? She might miss a chance to play games and socialize with friends if she did not. There is an exercise class Aunt Sarah might also miss. If they solved the problem of the cans, would there be not only **intended outcomes** but also **unintended outcomes** that were worse than Aunt Sarah's opening a can of green beans when she wanted chili?

Enrique reminded April that in the short time they visited, Aunt Sarah had three phone calls and a visit from a neighbor. Her social life and contacts seemed to be okay. April remembered that her aunt is a member of an exercise group. Physical activity may not be something they have to worry about. However, they both knew they should be alert to consequences while they think ahead for the future.

Figure 1.5 ■
April and Enrique begin their research by gathering information.

As their research progressed, April and Enrique began to understand some of the possible **alternatives** as they thought of their design. They became better prepared to clarify what their design should be able to do. This clarifying statement that goes with the design brief is called the **specifications**. The specifications for their project spell out the specific details of the design. This detailed description indicates what must be satisfied by the finished design.

As they continued their research and investigation, April and Enrique gained a better understanding of the problem and what might be accomplished—the **design limits**. As they developed specifications for their design, they began to see the limits within which they were working. For example, as they considered one plan rather than another, they were alert to costs. What was possible and practical? They predicted how much time the project would take. Similarly, what limits were placed on their design by the current skills they had?

Accumulate and assess the alternatives. This is the stage of their work where April and Enrique let their imagination have free rein. They used such processes as brainstorming and free association. Through brainstorming, they generated a lot of ideas. Free association helped them think of some radical and funny solutions, as well as ordinary and straightforward ones. They made many sketches and drawings of some of the most interesting ideas. Each idea

Figure 1.6A, B ■
A design brief includes a list of specifications (criteria) for the problem solutions. Now is the time to think creatively and come up with a number of possible solutions.

failure ok

full class Brainstorming good

Ease of use
" to Keep Clean

criteria for young children = good

represented a possible approach to the problem faced by Aunt Sarah, as identified in the brief.

April and Enrique were aware that it was very easy to settle on the first good idea they came up with. They had been warned that this can lead to some frustrating blind alleys and dead ends. This could be a problem, especially if they got bogged down with that one idea. To guard against this, they developed several ideas, spending only a short period of time on each. They were told three ideas were the minimum, but they generated more. Sketches, plans, and notes on each of several possible solutions were brief, but they were enough to identify pitfalls and to share with Aunt Sarah. These plans led to assessing which of the alternatives appeared to be the best possible solution for the problem.

Specifications:
a) The system must be very easy to use.
b) It must be easy to keep clean.
c) It must ensure that the contents of the cans match what the user has selected.
d) It must allow for the safe storage and retrieval of groceries.

e) It must not restrict the normal use of the kitchen.
f) The cost should be modest (not to exceed $200 for implementation and $20/year for maintenance).

April and Enrique now faced an important decision. They had to decide which of the possible solutions had the best potential for solving the problem. One way of determining this is to see which solution will fulfill the greater number of the specifications that have been spelled out.

Figure 1.7 ■
April and Enrique use the specifications for the design to help select what seems to be the best solution.

SPECIFICATIONS FOR THE DESIGN

	I	II	III	IV	V
	Alternatives*				
a) Very easy to use.	4	2	4	3	__
b) Easy to keep clean.	3	2	3	4	__
c) Ensures can contents match selection.	2	3	4	3	__
d) Allows for the safe storage and retrieval of groceries.	3	2	3	3	__
e) Does not restrict normal use of kitchen.	2	2	3	2	__
f) Maximum cost: $200 start-up, $20/yr. for maintenance.	3	4	4	4	__
	17	**15**	**21**	**19**	
Average	**2.8**	**2.5**	**3.5****	**3.1**	

* Rank alternatives on a scale of 1–5
(1= poor, 2 = fair, 3 = acceptable, 4 = good, 5 = excellent)

** Apparent best solution

[handwritten margin notes:] *Modeling System Lego. etc. Interaction of Mind & Hand Hazy to sharper scetch/writing *Christmas tree Model Doing thinking

Try out the best solution.   Now they were ready to try out their choice for the best solution—a combination of a storage rack and magnetic labels with raised letters. In trying out the best solution, April and Enrique engaged in two general activities—**planning** and making. Planning activities include developing necessary drawings to construct the device or system they selected. These drawings are created in detail to show as many aspects of the solution as possible. Drawings can be developed in a number of ways. Enrique and April may begin by making rough sketches with annotations, then developing more formal drawings with drafting tools, and then using a computer-based program to generate final drawings. Planning also includes listing resources they will need, including tools, materials, energy, information, time, money, and the like.

(A)

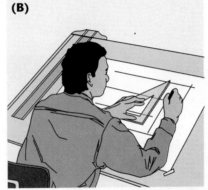

(B)

(C)

Figure 1.8A, B, C ■
Design and problem solving requires many drawings, from rough sketches to detailed illustrations.

April and Enrique found that planning is difficult work. However, they knew from previous experience that planning is very important to the success of their project. Good planning will help them identify how much time they need. Good planning will also help them get their work done on time. April and Enrique developed a **timetable** of events as they planned for their project. (In later chapters of this book, a number of techniques for the management of time are examined. These should be helpful in planning your own projects.)

Figure 1.9 ■
This stage of planning includes listing the required resources and developing needed timetables.

Planning was also important as April and Enrique determined what materials, tools, and other **resources** they needed for their project. Resource planning will ensure that they have the necessary materials and components on hand when those resources are needed. They do not want to start work and make a mess, only to find out that they must order a part and wait for it to be delivered.

Making is a simple way of saying April and Enrique will build something that represents their solution to the problem. In this instance, making involves the **fabrication** of a **model** of their solution. Models are usually smaller and simpler versions of the real thing. At this point, they do not worry too much about

Figure 1.10A ■
April develops a two-dimensional model of Aunt Sarah using her kitchen cabinets. She also develops a prototype of their storage device.

These large models of the raised labels will help improve our solution.

Figure 1.10B ■
After initial testing of the proposed solution, April and Enrique make a number of revisions. They found out that raised labels were definitely needed on the cans.

aesthetics or details. April and Enrique worked out the problems that became apparent through their model. Then they developed a **prototype** of their solution. A prototype is a full-size version of the product, system, or environment that represents their solution to the problem. Prototypes actually work and can be tested in real life. At this point in the development, aesthetics and attention to details become more important.

At several points along the way, April and Enrique **tested** different aspects of their design. They checked the strength of the materials they planned to use. They tested how well the dispensing aspect of their design actually worked. Most importantly, they asked Aunt Sarah to try it while they watched. They tested whether she was able to sort and find specific cans of food she is interested in using. The data collected from this testing helped to determine what modifications were needed in the design. A comparison of earlier tests with current tests was done. They made progress on improving the overall design, but there are still a number of things to think about and document.

Figure 1.11 ■
The revised solution is
now checked.

Evaluate the results. After making modifications and testing in Aunt Sarah's kitchen one more time, April and Enrique were ready for a more complete **evaluation**. Evaluating the results of their design work included both testing the solution and assessing the process. This means April and Enrique were concerned with both the product and process of their efforts. The products, systems, and environments are the most apparent results of design work. They tested the entire project to determine if it did what the design brief and specifications called for. They looked back at the specifications and checked each requirement carefully.

Figure 1.12 ■
Now is the time to evaluate the design against the specifications for the design.

As they assessed how well the product performed, April and Enrique wrote an evaluation of their work. This **report** outlined the strengths and weaknesses of their design. They noted where their project succeeded and failed to achieve the specifications.

Some of the specific questions they asked as they prepared the evaluation of their project and their own work included:

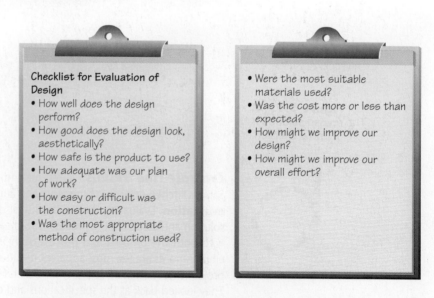

Checklist for Evaluation of Design
• How well does the design perform?
• How good does the design look, aesthetically?
• How safe is the product to use?
• How adequate was our plan of work?
• How easy or difficult was the construction?
• Was the most appropriate method of construction used?

• Were the most suitable materials used?
• Was the cost more or less than expected?
• How might we improve our design?
• How might we improve our overall effort?

Evaluating and testing the product and its design is an important part of the design approach. It is important to see how well the product met the design brief and solved the problem. Also, assessing the overall approach and process used in implementing the project helped April and Enrique profit from both their successes and failures. They can improve this project and use the information when they do other design projects in the future.

Figure 1.13 ■
This stage of evaluation involves the actual use of the solution by Aunt Sarah in her kitchen.

Developing the Portfolio

As the assessment phase of the project came to an end, April and Enrique turned more attention to developing their final report in the form of a **portfolio**. The notes and drawings they developed and kept were very helpful in preparing the document. (The form of the final portfolio and some of the skills that will help in developing the document will be treated in some depth in the next chapter.) In simple terms, the portfolio developed by April and Enrique recorded and presented the ideas they implemented. The report also documented some of the major steps that April and Enrique recycled again and again as they attempted to implement their ideas.

Aunt Sarah was delighted with her new canned food storage and retrieval system. She especially liked the idea of using raised labels that allowed her to "read" the contents by touch. She was also appreciative that April and Enrique had paid so much attention to helping her with an everyday problem. She bragged so much to her neighbors that several of them asked to have a system like hers. April and Enrique felt that some more testing and improvements were needed before other systems were produced. They were particularly concerned about some safety aspects, and they were concerned that someone other than Aunt Sarah had to place the labels on the cans and the cans on the shelves. There was more work to be done in improving the system they had designed.

Figure 1.14 ■
April and Enrique organize their sketches, notes, and other evidence of their research and development for their portfolio.

If April and Enrique only build a product for Aunt Sarah, the documentation of this design process can be done in a way that is useful for just themselves. But they might also want to use the documentation to get ideas from others, or communicate with other people who are trying to solve similar problems. Then documentation will need to be more complete. If they turn in the project for a grade in school, even more attention to the details and completeness of the documentation might be needed. If they want to patent the idea or start a business making their product, even more detailed records and carefully constructed drawings would be needed. If they want to produce the product, then questions concerning cost analysis and package design would need to be examined. (We will give more attention to the development of a **portfolio** to organize these documents in the next chapter.)

REVISITING THE DESIGN LOOP

If you had been working with April and Enrique on the problem for Aunt Sarah, you could now move through the Design Loop again with added insight. Often, the first time through the Design Loop on any problem helps focus on what the problem actually is. It is quite common to cycle through the Design Loop several times. Going through the same process several times is called **reiteration**. (Each cycle through the process is called an **iteration**.) The experiences you gain each time you go through the design process helps you ask better questions. Often, these questions become more meaningful and penetrating relative to the overall problem and its context as well as its parts.

The Guide for Design Questions is provided to help you ask and answer better questions in your design work.

GUIDE FOR DESIGN QUESTIONS

Identify and Define the Problem—Investigating Needs and Opportunities

- Did you identify a problem that represents a real need or potential opportunity?
- Did you set the problem within a larger context or situation?

Develop the Design Brief—Clarifying the Results You Want to Achieve

- Did you develop specifications and criteria such as:

time	cost	materials
size/shape	special function	user(s)
place of use		

Explore Possible Alternatives—Searching for Solutions and Information

- Did you conduct an adequate investigation?
- Did your research help clarify the needs, wants, and opportunities of the design?
- Did you generate a number of possible and diverse solutions?

Accumulate and Assess the Alternatives—Developing and Choosing the Best Solution

- Did you use a systematic approach to assessing the possible alternatives?
- Did you compare and rank the alternatives before choosing the best solution?

Try Out the Best Solution—Experimenting and Developing Solutions, Models, and Prototypes

- Did you conduct experiments and tests on your design and its operation?
- Did you engage in modeling and prototyping your proposed answer?

Evaluate the Results—Testing the Solution and Assessing the Process

- Did you conduct appropriate testing of your finished product?
- Did you collect, analyze, and report the essential data on the product?
- Did you assess its operation and your efforts in developing the product?

Design Questions

Identify and define the problem. Investigating needs and opportunities. *As you engaged in your design and technology work, did you identify a problem that represents a real need or potential opportunity? Did you set the problem within a larger context or situation?*

■ **Figure 1.15**
Wheelchair athletes require durable
wheelchairs specifically designed
for speed and precision. (© Robert
E. Daemmrich/Tony Stone Images)

All of us have needs. We have basic physical needs such as food, shelter, air, and water. Without them we could not survive. But people also have emotional needs. The most basic of emotional needs is to feel safe and secure. Additionally, all of us need to be wanted, to be a part of a larger community, and to feel a sense of achievement through what we can contribute. (Needs are explored in more depth in the chapter on Humans.)

The needs and wants that people express can be opportunities for designing. Many opportunities for design exist, if we can become sensitive to them. For example, most of us use many different products without paying much attention to how well the products work. In some instances, we may become annoyed because the product does not work very well. That state of being annoyed is an indication that the product may require redesigning.

Develop the design brief. **Clarifying the results you want to achieve.** *In describing the design brief, did you develop specifications and criteria such as time, cost, materials, size/shape, special function, user(s), and place of use?*

Generating a design brief requires attention to what you are trying to achieve. You need to give considerable thought to the goal or results you are trying to reach. Additionally, you need to consider the specifications and criteria for the problem. Such issues as time, cost, materials, size, shape, special function, use by a particular person, and use in a particular place are usually considered. Are there other issues that are not included in this list?

Explore possible alternatives. **Searching for solutions and information.** *As you pursued your design, did you conduct an adequate investigation? Did your research help clarify the needs, wants, and opportunities of the design? Did you generate a number of possible and diverse solutions?*

There are many ways of collecting **information**. You might use a questionnaire to conduct a survey. You might interview people. These approaches are examples of using **primary sources** for collecting data. They are called primary sources because you are directly involved and you collect the data from people who have firsthand experiences. In many cases, it is not possible for you to collect the data directly. You must rely on information collected by someone else. In such cases, you will be collecting data through **secondary sources** (the writer, the photographer, etc.). You need to examine the qualifications of the secondary source. Did they gather data and put it together accurately? Are they likely to have personal biases that slant their interpretations of the information they have reported? Did your primary sources include interviews, surveys, questionnaires, potential users, and experts? Did your secondary sources include the library, resource centers, computer databases, the media (television, radio, newspapers, magazines), museums, galleries, exhibits, books, journals, catalogs, brochures, and the like?

As you worked to identify possible alternatives, did you engage in a range of **idea generation** activities? Did you use brainstorming techniques, and develop mind maps and question webs? (In the chapter on Documentation, you will learn to make mind maps, question webs, and spider charts. These approaches can help you develop your ideas more fully.) Did you take time to think and play with the problem and activities related to your design?

Accumulate and assess the alternatives. **Developing and choosing the best solution.** *As you worked on your design, did you use a systematic approach in assessing the possible alternatives? Did you compare and rank the alternatives before choosing the best solution?*

Assessing alternatives includes collecting, analyzing, interpreting, and explaining information. This helps you identify the strengths and weaknesses of each design and to choose the design that best satisfies the specifications. You can present information by using graphics such as pie charts, bar graphs, and pictographs, as well as through individual and group presentations.

Try out the best solution. **Experimenting and developing solutions, models, and prototypes.** *In this phase of your design work, did you conduct experiments and tests on your design and its operation? Did you engage in modeling and prototyping of your potential answer?*

Trying out a solution requires planning. This includes considering the **constraints** of time, materials, and other resources. These show up on the timetable, flowcharts, and material lists. During this aspect of work, you will keep notes on the progress of your work and the tests you carried out. You will also keep notes on the problems you encountered in making the object and the decisions you made to deal with these problems. Reflecting on your efforts should result in identifying possible **modifications** and **improvements** to the design and to the overall effort.

Evaluate the results. **Testing the solution and assessing the process.** *Did you conduct appropriate testing of your finished product? Did you collect, analyze, and report the essential data on the product, its operation, and your efforts in developing the product?*

Conducting a final evaluation of your design project should include an assessment of the strengths and weaknesses of your design. This will include an assessment of the product, the materials used in its fabrication, and the means through which you developed it. This information will help to identify how successful you were in fulfilling the specifications. It will also identify where you were unsuccessful. Other aspects to be evaluated will include the economics, the social and environmental effects, and the changes you would make to improve the overall effort. The use of charts and graphs for collecting and presenting data are effective means of communicating how a solution performs.

In some instances, you will want to consider if the product has a potential market. If so, you may want to initiate plans for production. This aspect of design and technology will be treated in some depth in the chapter on Physical Systems.

SUMMARY
CHAPTER 1

Designing and problem solving involves an uncertain search for elusive answers to ambiguous problems. Nearly everything related to design and problem solving is tentative and vague. The design process, represented by the IDEATE Design Loop, can be very helpful in providing general guidelines. It can be used for identifying need and opportunities and generating design briefs. Steps involve conducting research and investigation, planning, and making objects (products, systems, and environments). These results are tested and your own work and the efforts of others are assessed. The entire process and product can then be used in improvement of the design.

In the chapters that follow in this unit, you will be introduced to a range of skills that will help you in your design work. In subsequent chapters, you will gain important knowledge and insights about technology and the role it plays in supporting designing and problem solving. As you accumulate new insights and skills, you will become more proficient and creative in design and technology activities.

■ **Figure 1.16**
The IDEATE Design Loop.

ENRICHMENT ACTIVITIES

There are several design and technology activities you can try out at this time. Suppose you and one of your classmates were April and Enrique's friends and you joined their team. Given the design brief and the specifications that April and Enrique developed, consider the following:

- Identify at least three alternative solutions you would suggest.

- What source could you use to get good information on each of these alternative solutions?

- Use the checklist from page 14 and assess each of your proposed solutions.

- Which solution to the problem seems to be best and worth exploring further?

- Make a sketch of your proposed solution and what it would look like.

- How would you go about testing your proposed solutions?

- What are some of the strengths and weaknesses of your solution?

- Write a short statement describing those strengths and weaknesses in comparison to April and Enrique's solution using magnetic labels.

REVIEW AND ASSESSMENT

- List and describe the general steps in the Design Loop.

- Identify a problem of how tools are displayed in your school facility.

- Develop a design brief to solve that problem for a specific tool.

- Develop specifications for the tool display device.

- Generate several alternative solutions to your problem.

- Assess which of the alternative solutions is potentially the best.

- Develop plans for how you would build the device or a model of it.

- Describe how you would evaluate the product of your work.

CHAPTER 2

Designing, Documenting, and Drawing

Figure 2.1 ■
Designers use a variety of tools to turn creative thoughts into tangible objects. (© G. & V. Chapman/The Image Bank)

INTRODUCTION—THE PROCESS OF DESIGNING

Designing is translating ideas into things and thought into action. Designing is the **process** part of technology. As such, it is different from other types of creative work. The intent of design work is to satisfy a human **want** or **need**.

KEY TERMS

aesthetics
anthropometrics
asymmetrical
auditory representation
balance
color
color wheel
comfortable
contrast
contrasting colors
crating
cutaway drawings
designing
dissonance
documenting
effectiveness
ergonomics
exploded view drawings
form
free form
functional
geometrical
harmonious colors
harmony
highlights
isometric
line
motif
natural
need
norm
oblique
orthographic projection
patterns
perspective
point
portfolios
primary color
process
proportion
prototypes
radial balance
secondary colors
sectional drawings
shading
shape

Often, this involves solving everyday, practical problems with an outcome or product that is useful. To be successful, a design must function properly and fulfill that identified want or need. In addition, since humans are affected by the form and appearance of a product as it is used, good design satisfies the senses as well.

As you develop ideas, it is helpful to get them out of your head and into a more tangible form. This may be doodling, sketching, writing a diary, or making up a song. Even if you do not share the ideas with someone else, getting them out in the open often helps clarify the idea or use up some emotional energy. If you have ever drawn an ugly picture of someone you dislike and then torn it up, you have experienced this. Or maybe you have a vague idea of what would be your dream bedroom. Making different sketches of what a "private sleeping space" might look like could help clarify what you want. Drawing can help you get a handle on what you feel and some ideas about what to do.

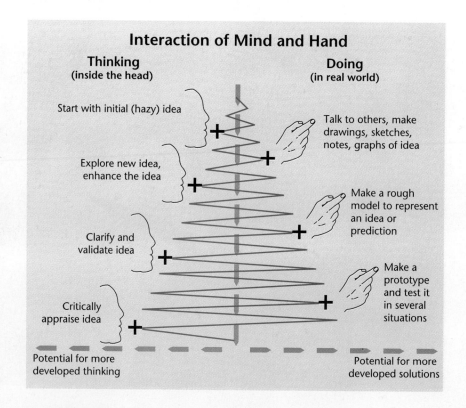

■ **Figure 2.2**
A model showing how the interaction of mind and hand leads to a higher level of thinking and creating.

Communicating ideas to others can be expanded and enhanced if you not only talk about the ideas, but also add some form of **visual** or **auditory representation.** Sharing your ideas involves listening to yourself as well. Designing includes organizing ideas with an audience in mind. The design process is valuable for creating, clarifying, and communicating ideas and for developing desired outcomes.

source of light
stability
stamina
standardization
symmetrical
tactile
tertiary color
tessellations
texture
three-dimensional
tonal range
tone
visual design
visual representation
want

DESIGN CONSIDERATIONS

Aesthetics and Appearance

It stands to reason that a design should function properly, but it also must attract the attention of the users. Attractiveness must go beyond looking good. We experience and appreciate a design through all our senses of sight, hearing, touch, taste, and smell. If a design has qualities that make it attractive or pleasant to experience, it is said to have **aesthetic** appeal. If you wish to convey something shocking and uncomfortable, you want to create that effect in your audience. You will be interested in aesthetic **effectiveness**. Most of us like to avoid unpleasant sensations, so this communication may be tricky.

In the chapter on Humans, we examine factors that influence people's perceptions and what influences their attentiveness. People differ considerably in what they think is appealing or shocking. Understanding what influences people is very helpful to you as you engage in designing something for someone else to use. This becomes especially important if that person has a different background, culture, and interest than you. The following aspects will be helpful as you try to design something that creates the effects you want. Some of the same principles apply whether you are concerned about designing space vehicles, presentation of food on a plate, public spaces in a city landscape, or a dance routine.

Elements of Visual Design

Visual design translates ideas into objects of effective appearance. The elements of visual design include **point, line, shape, form, texture,** and **color**. Examples of these elements are illustrated in Figure 2.3A. This can involve two- or three-dimensional presentations. Examining the elements of visual design helps you understand your own preferences about aesthetics and appearance while you develop skills of designing. You can also note the effects of placement, movement, and color on your own reactions. These principles apply when you are drawing, making a videotape, or taking a photograph.

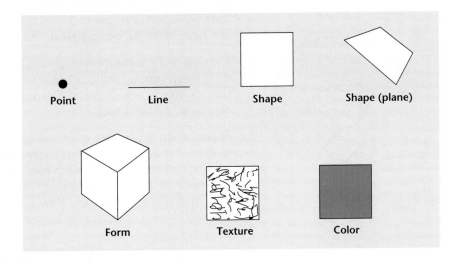

Figure 2.3A ■
The six basic visual elements.

Point. The most basic design element is the point. In visual design, you can use a point to create a dot, or many dots. Patterns of dots can create lines. Two points can be used to describe a straight line, while three points can be used to describe an arc or circle.

Technically, a point has no size—only position. In visual design, points are used to create lines. Lines are used to enclose areas and create shapes. Shapes and areas are then used to create three-dimensional forms.

Line. Of the different elements of visual design, line is perhaps the most essential. Lines can be used alone or they can be combined to create shapes and forms. Lines can also be used to provide information, apply decoration, and evoke feeling. (Refer to Figure 2.3B.) Lines take into consideration the likely movement of the eye as it explores the presentation. A design will attempt to create a point that attracts attention and guides the eye around and through the total composition.

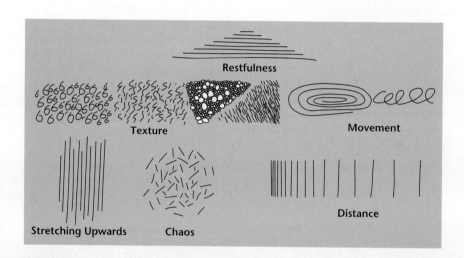

Figure 2.3B ■
Lines and their many uses.

Shape. When a line is used to enclose space, we call this **shape**. Shapes in print or drawings are two-dimensional in that they have no thickness. Although there are countless shapes in the natural and built world, they all fall into three categories: **natural**, **geometrical**, and **free-form** shapes.

The view of the audience is important in considering shape. Will all aspects be seen? When using a computer to assist in design, simulations of the view from all sides of objects can be useful. Will edges or points or connections between objects be of importance?

Form. Form is generated when a shape takes on thickness. Form indicates that an object has three dimensions. Designs are often created by the combination of different forms. Like shapes, forms may be geometric, natural, or free-form. The placement of forms in relationship to each other is important in **three-dimensional** representations. How much bulk and weight will the objects represent? How much space will be assigned to each in relation to the others?

■ **Figure 2.4**
A natural shape and form.

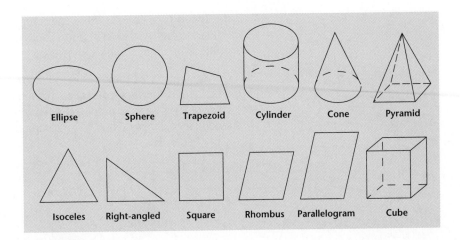

Figure 2.5A ■
Geometric shapes and forms.

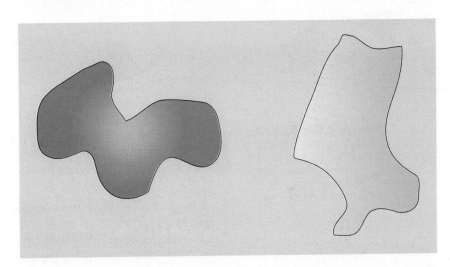

Figure 2.5B ■
Free-form shapes and forms.

Texture. There are two forms of texture—the tactile and the visual. The **tactile** aspect of texture means you can actually feel the surface finish through your sense of touch. The **visual** aspect refers to what you see that gives you the impression of texture.

Color in Visual Design

Color. Color is an important aspect of visual design. Shape, form, and texture provide the basics for the appearance of an object or arrangement, but it is color that provides the detail and difference of these characteristics. Color also evokes emotional reaction and is important for directing attention to various portions of the design in the manner you desire.

Color wheel. A useful aid in understanding the use of color is the **color wheel**. This model explains color from pigments, used in various materials. The first wheel, illustrated in Figure 2.6A, shows the three **primary colors** of red, yellow, and blue. These are called primary because all other colors are made from these three. The second wheel, illustrated in Figure 2.6B, shows the addition of the **secondary colors**. These colors are made by mixing equal amounts of two primary colors. As shown in Figure 2.6B, red and blue yields violet, red and yellow produces orange, and blue and yellow produces green.

Continuing to mix these colors will allow you to produce a third level of color, described as **tertiary colors,** as shown in Figure 2.6C. For example, the tertiary color of blue-green is produced when the primary color blue is mixed with the secondary color of green or a small amount of the primary color yellow.

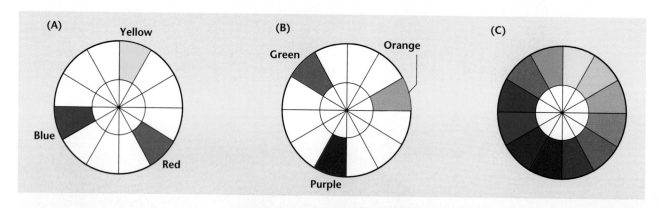

Figure 2.6A, B, C ■
Color wheels: (A) primary colors, (B) secondary colors, (C) full color wheel with tertiary colors. *(From: Hunter, The Art of Floral Design, © 1994 Delmar Publishers Inc.)*

PRODUCING COLOR BY MIXING PIGMENTS VERSUS COMBINING LIGHT

The process described above applies to inks used for printing and dyes used in clothing. Both are based on pigments. Color generated from light is different, however. White light can be separated into the primary colors and all the other colors of the spectrum. You can use filters to project light, red, blue, and green. Combining these three colors will result in the color white. Additional attention will be given to light in the chapter on Energy.

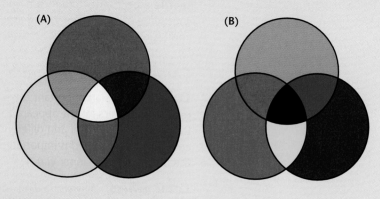

■ **Figure 2.7A, B**
Mixing colors: (A) from projected light, (B) from pigments.

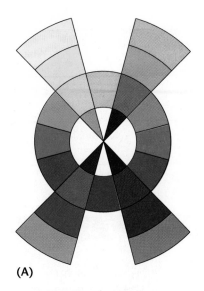

(A)

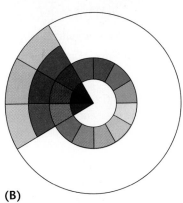

(B)

■ **Figure 2.8A, B**
Color wheels: (A) contrasting
colors, (B) harmonious colors.
*(From: Hunter, The Art of Floral
Design, ©1994 Delmar Publishers
Inc.)*

The development of two other color wheels should prove helpful in understanding the nature of contrasting and harmonious colors. Examples of contrasting colors are shown in Figure 2.8A. **Contrasting colors** are opposite from one another on the color wheel. Red contrasts with green, purple with yellow, and so on. Contrasting colors stand out and give emphasis when they are used together.

Harmonious colors are close to each other on the color wheel. The grouping of colors from a wheel of primary and secondary colors is shown in Figure 2.8B. Only a few of the three color groups are shown. Any set of three adjacent colors will make a new group.

Another important variation of pigments is the ability to darken them by adding black and lighten them by adding white. Adding black or white changes the **tone** of the color. In Figure 2.9A, the range of tones that can be achieved by mixing black and white with blue is shown. This instance is referred to as the **tonal range** of blue. If the black and white pigments were mixed without the blue, they would produce a tone chart for the two colors as illustrated in Figure 2.9B.

Color has some interesting effects on people when used in designs. Colors can portray a sense of warmth and coolness. Some colors appear to have a calming effect, while others have an exciting effect. Strong primary colors make objects look bulkier. Dark colors will tend to make something look smaller, while light colors make objects appear larger.

We also tend to make associations with different colors. White is associated with cleanliness, red with danger, and green with security, while purple may be seen as serene and regal. Many of us become confused if the expected associations are broken. We would have a problem eating blue peas and purple mashed potatoes. (Although some Dr. Seuss fans may have eaten "green eggs and ham.")

Primary colors in a playground may be fun, but not very peaceful in a library. As you add color to your designs, you will need to keep in mind the response that people may have to different colors and combinations of color.

CONTRASTS AND COMPARISONS

The principles of design used with color and light (harmony and contrast) can be applied to designing with sounds, tastes, and smells. **Contrast** gains attention and **harmony** is more pleasant. Just as bottom and top notes are needed to give a sense of balance in music, similar qualities are needed in flavoring and scents. For example, chili that uses only hot pepper for flavoring will not have an interesting taste. Adding garlic or other spices gives a balance (a low note) to the hot pepper (the high note). In general, considering contrast and harmony will help you develop and communicate a complete idea with yourself and with others.

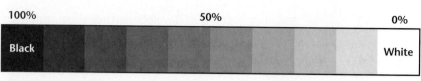

(A) Stepped Tones

Figure 2.9A, B ■
Tones of colors: (A) mixing
black and white with blue,
(B) tone chart from two colors.

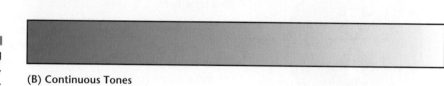

(B) Continuous Tones

■ **Figure 2.9C**
In many cases, perceptions are
influenced by color; therefore,
confusion can occur when
common color associations are
radically different. (© Philip
Habib/Tony Stone Images)

DESIGN ACTIVITIES

Identify a machine in your lab. By yourself or in a team, sketch out a design that uses one of the elements of visual design to warn users of potential danger areas on the machine.

Carry out several trials to determine which combinations of colors are best at drawing attention to these potentially dangerous areas. Determine if other people choose similar color combinations.

Date, sign, and keep these ideas as records of your work. Create a file folder in which you can accumulate your design and problem solving work.

Principles of Visual Design

Visual elements can be combined in many different ways. There are a number of principles that you can use in your design work. Those principles are **proportion**, **balance**, **harmony**, **contrast**, and **pattern**. Using the principles as a set of guides can help you develop designs that "look right" and are consistent with the appearance you want.

Proportion. In your design work, you will combine and arrange different elements to make a composition. If these parts seem to fit together, they are seen as being in proportion. If selected parts are too large while others are too small, they will be out of proportion. Arrangement of the parts can also enhance or detract from the feeling of proportion.

When objects are out of proportion to what we expect, **dissonance** is created. The eyes may continue to move back and forth as the brain tries to make these parts fit together in expected ways. It is distracting. Examples of good proportion are found in nature, and we may have learned our expectations from this source. If you want people to react strongly with aggravation, you may deliberately develop a design that is dissonant. Most of the time, however, you want the audience to find the proportion appealing; then they will attend to the message to be communicated.

Figure 2.10 ■
These objects show good and poor proportions. Which of these objects do you think is in good proportion?

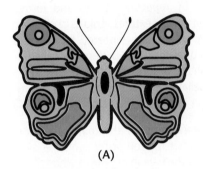

(A)

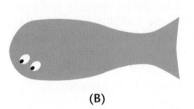

(B)

(C)

■ **Figure 2.11A, B, C**
Balance in design can be achieved in several ways: (A) symmetrical, (B) asymmetrical, and (C) radial balance. *(C: © Michael Dzaman Photography)*

Designs that show good proportion must also fit well within the environment for which they are intended. A large object may look fine if designed to be in a large room. It will be out of proportion if placed or used in a small room. Objects must also be in proportion to the people who are to use them. Fitting the object to its function and the user are also important.

Balance. Balance implies a sense of stability. Our sense of balance comes from our experience with gravity. Balance in a design can be accomplished by using the same treatment on both sides of the object. Balance can also be created by working around a center point.

One way that balance is achieved is when one side of the object is the mirror image of the other. Placing an imaginary mirror in the middle of an object creates a **symmetrical** image. The object shown in Figure 2.11A illustrates symmetry and provides a sense of balance and stability.

It is also possible to create visual balance in compositions that are **asymmetrical**. Asymmetrical objects are different on each side of an imaginary center line. The balance of the object can be enhanced using appropriate proportions of different parts. The object shown in Figure 2.11B is symmetrical in one view and asymmetrical in another view.

Stability of design can also be attained by the use of **radial balance**. Radial balance arranges the elements of the design in a uniform manner as they radiate out from a center point. An example of radial balance is illustrated in Figure 2.11C.

Harmony and contrast. Harmony implies that the parts of the whole design come together in a proportionate or orderly manner. The elements are in harmony when there is an agreement in the arrangement of color, size, shape, form, texture, and the like that has been used in the design. The object shown in Figure 2.12A has elements that are seen as harmonious.

Contrast is applied in a design when you want to emphasize or compare differences. This will often result in calling attention to some element or aspect. This in turn can add a sense of tension and liveliness to the design. Harmonious and contrasting elements are often used together to create interesting appearances and effects. The object in Figure 2.12B represents the use of contrast.

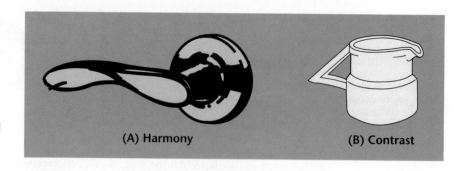

(A) Harmony   (B) Contrast

Figure 2.12A, B ■
Objects showing (A) harmony and (B) contrast.

(A)

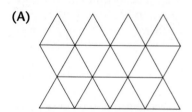

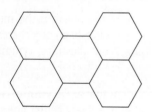

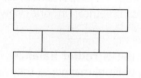

(B)

■ **Figure 2.13A, B**
Instances of (A) patterns and (B) tessellations.

Pattern. It is possible to create patterns by repeating shapes and forms. Patterns can also be generated by repeating lines, textures, colors, and other elements, as shown in Figure 2.13A. The elements are repeated to create a pattern, referred to as the **motif**. A unique form of pattern is called a **tessellation**. Tessellations are elements of shape or form that fit together in an interlocking manner. An example of a tessellated pattern is shown in Figure 2.13B.

Ergonomics. As you engage in design work, most of the things that you develop will be for human use. This can be either for you or for someone else. Part of a good design means that the object fits the person or people who will be users. As a designer, you will need to know what size things should be so they are **functional** and **comfortable** for the users. You will want to predict the effects on the person and the object as it is used over time. You may also consider how to take the object apart so that materials can be reclaimed when the person is ready to discard the product. This means you will need to know a fair amount about the dimensions of people and their movements as they use objects.

Ergonomics is an important area of study in designing things for people. Ergonomics refers to the study of ways in which objects, systems, and environments can be made safe, effective, and efficient for use by people.

As you design things for people, four main factors will need to be considered and investigated:

• the *size* of the part of the human body that will be involved in using the things being designed,

• the *movements* the humans will make while using the designs,

• the *energy* requirements (strength, stamina, and endurance) for use, and

• the *sensory* responses of the person to the design.

Size. When you are designing for humans, you must consider the measurements important to the safe and comfortable use of the product. If the design is to be used by one person, it seems obvious that you just take appropriate measurements of that person. If that person's size is expected to change over the period of use, however, you need to plan for expansion or contraction in appropriate places. For example, have you ever seen an adolescent riding a bike he/she rode as a child? Also, the dimensions of a tent when open for sleeping will be different than when it is folded up for carrying.

STATURE

Adult Male and Female Stature* in Inches and Centimeters by Age, Sex, and Selected Percentiles†

		18 to 79 (Total)		18 to 24 Years		25 to 34 Years		35 to 44 Years		45 to 54 Years		55 to 64 Years		65 to 74 Years		75 to 79 Years	
		in	cm	in	cm	in	cm	in	cm	in	cm	in	cm	in	cm	in	cm
99	MEN	74.6	189.5	74.8	190.0	76.0	190.0	74.1	188.2	74.0	188.0	73.5	186.7	72.0	182.9	72.6	184.4
	WOMEN	68.8	174.8	69.3	176.0	69.0	176.0	69.0	175.3	68.7	174.5	68.7	174.5	67.0	170.2	68.2	173.2
95	MEN	72.8	184.9	73.1	185.7	73.8	185.7	72.5	184.2	72.7	284.7	72.2	183.4	70.9	180.1	70.5	179.1
	WOMEN	67.1	170.4	67.9	172.5	67.3	172.5	67.2	170.7	67.2	170.7	66.6	169.2	65.5	166.4	64.9	164.8
90	MEN	71.8	182.4	72.4	183.9	72.7	183.9	71.7	182.1	71.7	182.1	71.0	180.3	70.2	178.3	69.5	176.5
	WOMEN	66.4	168.7	66.8	169.7	66.6	169.7	66.6	169.2	66.1	167.9	65.6	166.6	64.7	164.3	64.5	163.8
80	MEN	70.6	179.3	70.9	180.1	71.4	180.1	70.7	179.6	70.5	179.1	69.8	177.3	68.9	175.0	68.1	173.0
	WOMEN	65.1	165.4	65.9	167.4	65.7	167.4	65.5	166.4	64.8	164.6	64.3	163.3	63.7	161.8	63.6	161.5
70	MEN	69.7	177.0	70.1	178.1	70.5	178.1	70.0	177.8	69.5	176.5	68.8	174.8	68.3	173.5	67.0	170.2
	WOMEN	64.4	163.6	65.0	165.1	64.9	165.1	64.7	164.3	64.1	162.8	63.6	161.5	62.8	159.5	62.8	159.5
60	MEN	68.8	174.8	69.3	176.0	69.8	176.0	69.2	175.8	68.8	174.8	68.3	173.5	67.5	171.5	66.6	169.2
	WOMEN	63.7	161.8	64.5	163.8	64.4	163.8	64.1	162.8	63.4	161.0	62.9	159.8	62.1	157.7	62.3	158.2
50	MEN	68.3	173.5	68.6	174.2	69.0	174.2	68.6	174.2	68.3	173.5	67.6	171.7	66.8	169.7	66.2	168.1
	WOMEN	62.9	159.8	63.9	162.3	63.7	162.3	63.4	161.0	62.8	159.5	62.3	158.2	61.6	156.5	61.8	157.0
40	MEN	67.6	171.7	67.9	172.5	68.4	172.5	68.1	173.0	67.7	172.0	66.8	169.7	66.2	168.1	65.0	165.1
	WOMEN	62.4	158.5	63.0	160.0	62.9	160.0	62.3	158.2	62.3	158.2	61.8	157.0	61.1	155.2	61.3	155.7
30	MEN	66.8	169.7	67.1	170.4	67.7	170.4	67.3	170.9	66.9	169.9	66.0	167.6	65.5	166.4	64.2	163.1
	WOMEN	61.8	157.0	62.3	158.2	62.4	158.2	62.2	158.0	61.7	156.7	61.3	155.7	60.2	152.9	60.1	152.7
20	MEN	66.0	167.6	66.5	168.9	66.8	168.9	66.4	168.7	66.1	167.9	64.7	164.3	64.8	164.6	63.3	160.8
	WOMEN	61.1	155.2	61.6	156.5	61.8	156.5	61.4	156.0	60.9	154.7	60.6	153.9	59.5	151.1	59.0	149.9
10	MEN	64.5	163.8	65.4	166.1	65.5	166.1	65.2	165.6	64.8	164.6	63.7	161.8	64.1	162.8	62.0	157.5
	WOMEN	59.8	151.9	60.7	154.2	60.6	154.2	60.4	153.4	59.8	151.9	59.4	150.9	58.3	148.1	57.3	145.5
5	MEN	63.6	161.5	64.3	163.3	64.4	163.3	64.2	163.1	64.0	162.6	62.9	159.8	62.7	159.3	61.3	155.7
	WOMEN	59.0	149.9	60.0	152.4	59.7	152.4	59.6	151.4	59.1	150.1	58.4	148.3	57.5	146.1	55.3	140.5
1	MEN	61.7	156.7	62.6	159.0	62.6	159.0	62.3	158.2	62.3	158.2	61.2	155.4	60.8	154.4	57.7	146.6
	WOMEN	57.1	145.0	58.4	148.3	58.1	148.3	57.6	146.3	57.3	145.5	56.0	142.2	55.8	141.7	46.8	118.9

* Height, without shoes.
† Measurement below which the indicated percent of people in the given age group fall.

Figure 2.14A ■
Ergonomics: anthropometric charts showing height (stature) by gender, age, and percentile.
(Courtesy of Watson-Guptill Publications)

The study of human size and the science of measuring people is known as **anthropometrics**. This word has two roots—"anthropo," meaning "human," and "metric," meaning "measure." You may be surprised to know that many of the patterns for women's clothes are based on a study of the dimensions of United States women who volunteered for military service during World War II. Uniforms for these young women, who were in good physical shape from training, were constructed from these dimensions. Which women do you think these sizes will fit today?

When your design is to be appropriate for a group of people, you will have to make some decisions. For example, if you look at or measure real people, you find that no one may fit what is called "normal." This means that actual human beings vary widely from the **norm**. You must make decisions about which people you want to fit and which to leave out. Many products are made for a middle group. The largest people (those beyond the 95th percentile) and the smallest people (those below the 5th percentile) are left out. This decision often occurs with products, from computers to potato chips. For example, designers choose sizes and placement of elements on keyboards based on average finger and hand size. How much salt to put in packaged potato chips is decided from taste tests on people from different regions in the country. People who like a lot of salt or almost none have trouble finding products that suit their taste. **Standardization** is the term used to indicate designing based on the characteristics of an average.

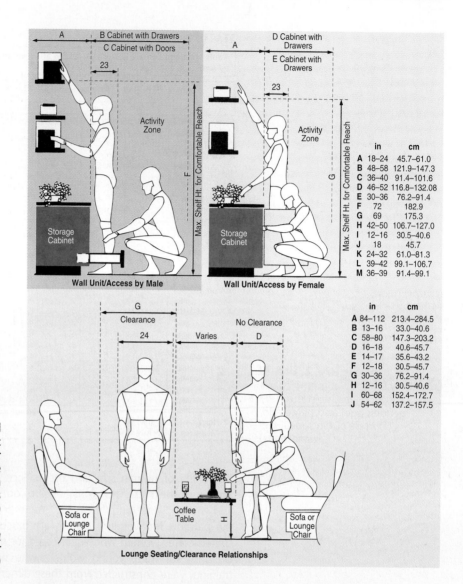

Figure 2.14B ■
Ergonomics: the use of ergonomic data for designing space for humans. (In this illustration, the larger figure is based on the 95th percentile data; the smallest is based on the 5th percentile data.) *(From: Hutchinson and Karsnitz, Design and Problem Solving in Technology, © 1994 Delmar Publishers, Inc.)*

Movement. Humans move in many intricate ways. The human body is a wonderful design and is able to make a wide range of complex movements. These movements allow us to complete tasks involved in work and play. If a product or device creates fatigue, pain, or discomfort when used in a normal way, it is considered to be a poor design. A plan for a work space should allow the person to move as needed and not bump into things. It should not require that the user spend extra energy making unnecessary movements.

Refer to the two stepladders shown in Figures 2.15A and B. The stepladder in Figure 2.15A is less **appropriate** in size. Although the user could climb either of the ladders, what do you think makes the ladder with the steps far apart, inappropriate? What makes the second ladder with the steps placed closer together more appropriate?

(A)

(B)

■ **Figure 2.15A, B**
Using knowledge of how muscles work results in designing better products for people.

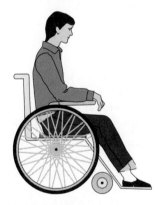

■ **Figure 2.16**
People in wheelchairs require stability as they sit. Also, they have special clothing needs. For example, when seated, their waistlines tend to drop in the back and bunch in the front. Hemlines in pants and skirts tend to ride up. *(Courtesy of M. Dolores Quinn and Renée Weiss-Chase, Drexel University.)*

Strength and stamina.   There are limits as to how much a person can lift or move. Although this differs considerably among individuals, there are accepted ranges of power that different muscle groups in the body can generate. Within these limits, we can expect that most humans (from the 5th percentile to the 95th percentile) will be able to use objects that do not exceed these strength requirements. In the example of the stepladder in Figure 2.15A, the large distance between the steps requires that the person use the weaker muscles in the front of the legs rather than using the stronger calf muscles in the back of the legs.

In addition to strength, humans exhibit **stamina**, or the ability to work at something over a period of time. Stamina is influenced directly by body fatigue. Humans are more effective and efficient in their work if they are comfortable. Designs, therefore, should take into account the movement and strength requirements of the operator. This will reduce excessive fatigue and muscle strain and increase the comfort factor for the user. The proposed design must also take into consideration the age and gender of the intended users, as these will have an effect on a person's strength and stamina.

It is also useful to analyze the way a person uses a product over time. Products can be designed to compensate for uneven use or potential for error. More energy may be applied at some location than others. The heels of boots may wear out on the outside first and need repair, even though the rest of the boot is still good. A child's pair of jeans may need reinforcement in the knees. More attention may be needed at one point of use than others.

Balance.   Although individuals may like the feeling of instability for short periods of time (as on a carnival ride), humans require a sense of **stability** and balance. This need for balance is seen in how people use a chair, a wheelchair, or how they ride a bike. All designs must ensure that the user can maintain balance; otherwise, they simply fall over. (Refer to Figure 2.16.) Expending energy to keep in balance causes muscle strain. Continue to walk in that pair of boots with the worn heels and you will feel the strain on your feet and legs.

Senses

Humans interact with designed objects through their senses. The objects we design come into contact with us through our senses of touch, smell, taste, sight, and hearing. A design can create pleasurable sensations or distracting and uncomfortable sensations. The following are some of the factors that will need to be considered in your designs:

Size, shape, form.   Size, shape, and form determine if your designs will fit the users. Does the design fit the body, or a specific part of the body, in a comfortable manner? If the device is hand-operated, it should fit the hands of the people for whom it was designed—toddlers, adults, or aged. Similarly, devices that come into contact with other parts of the body should fit well and not create pain or pressure points. (Refer to Figure 2.17.)

Heat-cold.   Humans can stand a relatively small range of temperatures. Objects that are too hot or cold will be uncomfortable to hold and use. Steps need to be taken in using handles and insulating materials, as appropriate.

Noise and vibration. There are levels of noise and vibration that are uncomfortable for humans to experience. Time spent at levels higher than recognized as safe can be damaging to an individual's hearing and health. Figure 2.18 includes specific information regarding the environmental tolerance zones.

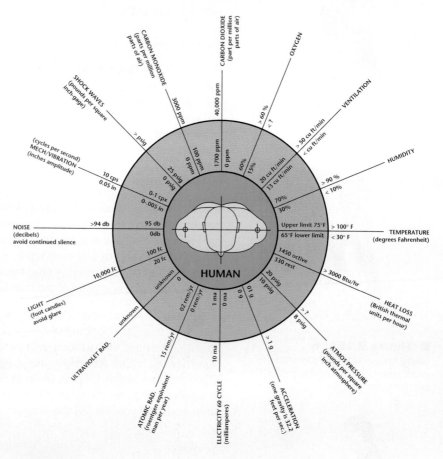

■ **Figure 2.18**
Environmental Tolerance Zones for Humans
The band between the circles indicates the zone from comfort to the tolerance limit. Outside this limit, great discomfort or physiological harm is encountered. Other factors not shown and to be considered are infrared radiation, ultra-sonic vibrations, noxious gases, dust, pollen, chemicals, and fungi.
(© 1966 Henry Dreyfuss)

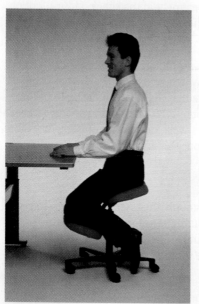

■ **Figure 2.17**
The design of a seat can vary significantly. All must provide for balance and comfort of the user.
(*Top: © Chris Noble/Tony Stone Images; Bottom: Courtesy of HAG, Inc.*)

Surface finish. Objects that will come into contact with the body must have satisfactory surface finishes. Edges and corners should not be too sharp. Appropriate textures and treatments are important if the object is to be held or operated by hand. Increasing or decreasing the friction on a handle can make operation easier.

Visual elements. As your design begins to take a final form, it will be important that the visual elements come together appropriately. Elements of the design that are to communicate information through symbols should be of appropriate size and color. They should be placed appropriately so that the symbols can be seen and read easily by the users. Attention must also be given to the balance and contrast of color, shape, and forms used in the design.

DESIGN ACTIVITIES

Carry out several activities in taking and comparing measurements of members of your class. Using an adjustable chair, determine the sitting height for each person (so that their thighs are parallel to the floor). Collect that data for the class population and calculate the mean (arithmetic average) for the class. Determine the variation (range) in the height of the chair seat for the class. What might this measurement mean if you were designing a chair to use at a work desk?

DOCUMENTING TECHNIQUES

As you become involved in design and technology activities, you will soon learn the importance and difficulty of keeping track of your ideas and work. A large part of designing is "thinking on paper." Keeping records of this thinking is an important part of being a designer, as shown in Figure 2.19. In the following sections of this chapter, you will be introduced to some important concepts and skills that you will need to develop.

Documenting is an important part of the design and technology process. In many cases, documenting helps to prevent the loss of good ideas. It is frustrating to come up with a good idea, only to forget what it was a few hours or days later.

In addition to documenting your ideas, it is also important to keep track of when you did the design work. Your record keeping should include the date to show when you made the notes and drawings. It is not easy to develop good habits of recording and dating your ideas, but it may be worth the effort. For example, there are many instances where the ownership of an idea is proven by the documentation that an individual kept. This is especially important when the idea or product can be patented or copyrighted. The originator gets the royalties and part of the money that comes from the production and sale of the product. Lacking good documentation has sometimes meant that the person who had the idea first could not prove it. Without such proof, someone else may get the credit and the profits.

Formats for Documenting

Formats for documentation may take many forms. The ones considered here are journals, logs, reports, and portfolios.

Journals.   Journals provide a general record of events for an individual or team. Entries may be made in a journal every day and may be similar to a diary. Your journal is a technical diary with entries made on a daily or periodic basis. Journals tend to be personal observations and records of the work in progress.

Logs.   Logs are somewhat more formal than journals. They serve as a daily record, like a ship's log. A project log details what happens each day in the conduct of a project.

Reports and portfolios.   Reports and portfolios are documents that are produced, when appropriate, to provide information on the progress or results of a project. (See Figure 2.20.) Although these can take a variety of forms, the format shown on page 50 is suggested for reporting your design and technology projects. This format is patterned after the flow of the IDEATE Design Loop presented in Chapter 1.

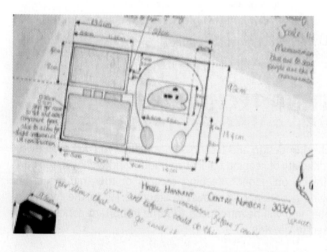
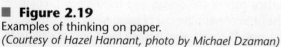

■ **Figure 2.19**
Examples of thinking on paper.
(Courtesy of Hazel Hannant, photo by Michael Dzaman)

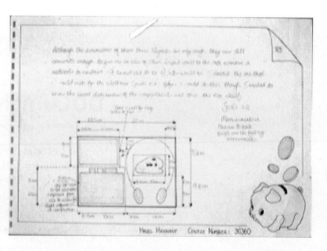

■ **Figure 2.20**
Sample page from a design and technology portfolio.
(Courtesy of Hazel Hannant, photo by Michael Dzaman)

Recording Ideas

Drawings.   Drawings are essential means of recording ideas that cannot be described in words alone. Drawings allow you to express ideas of what something will look like before it is actually made and help you think through how a product or device will work. Your drawings can help to isolate problems that might be encountered in an object's operation or in your making that object. Drawings are an indispensable part of design and technology work.

Notes.   A helpful way to communicate with others, or even with yourself, is by using notes. You will want to add notes to your journal and log entries. Often, you can combine these notes with sketches and drawings of the ideas to help you remember exactly what you had in mind. Notes relate to such things as how different parts might go together, how a material might be shaped or formed, how moving parts might interfere with each other, or where something might not be safe. (Refer to Figure 2.21.) Documentation without sketches and drawings tends to lose much of its value for sharing ideas. Sketches and drawings without explanatory notes also lose their value as a memory tool and for communicating your ideas to others.

Plans. Plans that you develop play an important part in helping you see what lies ahead in your work. Often, problems can be averted or solved through good planning. Many different types of plans are available. An example of time planning is shown as a Gantt timeline in Figure 2.22. These and other means of planning are discussed in more depth in Chapters 8 and 10.

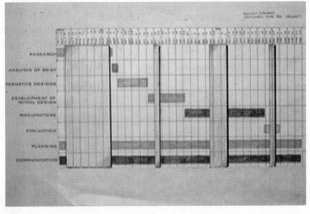

■ **Figure 2.21**
Notes and notations being used to record ideas.
(Courtesy of Hazel Hannant, photo by Michael Dzaman)

■ **Figure 2.22**
A timeline developed using a Gantt chart. *(Courtesy of Hazel Hannant, photo by Michael Dzaman)*

DESIGN ACTIVITIES

Using the drawing provided by your teacher, add notations that will help you understand and communicate to others how the device is to be used.

On a second copy of the drawing, add notations of how the device might be constructed.

DRAWING TECHNIQUES

Sketching and Drawing

Sketching is an important skill needed for your design and technology work. Freehand sketching is an easy way of recording your ideas and sharing them with others. Initial sketches can be "sketchy" and include only partial thoughts and ideas. Sketching is a rough means of producing drawings. With training and practice, sketching can become an important and usable skill, not only for your

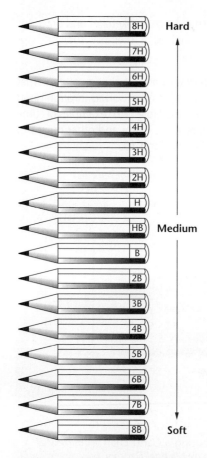

Hard

Medium

Soft

■ **Figure 2.23**
Drawing pencils range from very
soft to very hard.

design and technology work, but also for a wide range of communication tasks. A few basics of sketching and drawing help as a starting point for learning these skills. The concepts of line, shape, and form were introduced in Chapter 1. You use these ideas when you represent something through a drawing or sketch.

Lines.   Lines in sketches and drawings may range from being very thin and faint to lines that are thick and heavy. The drawing instruments illustrated in Figure 2.23 show a range of line thicknesses. For freehand drawing and sketching, the most useful pencil is an HB. An **HB** is similar to a #2 pencil and is halfway between soft and hard. It is considered a medium pencil.

Shape.   As you may recall, shape is generated when a line is used to enclose space. Shapes are two-dimensional in that they have no thickness. Although there are countless shapes in the natural and built world, they all fall into three categories of the natural, geometrical, and free-form shapes.

Form.   Form is generated when a shape takes on thickness. Form indicates that an object has three dimensions. Sketches and drawings take on the illusion of form by the use of shading, lines, and/or color. For example, a cube can be developed from the basic shape of a square, a cone is developed from a triangle, a sphere from a circle, and a cylinder from a circle and a rectangle. Each of these is shown in Figure 2.24.

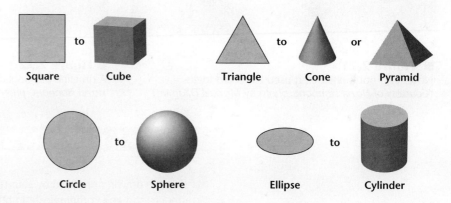

Figure 2.24 ■
Developing three-dimensional forms
from two-dimensional shapes.

Square   to   Cube   Triangle   to   Cone   or   Pyramid

Circle   to   Sphere   Ellipse   to   Cylinder

PICTORIAL DRAWINGS

• •

There are a number of ways that you can represent things pictorially. Some are easy to draw, but lack realism. More realistic drawings are possible, but these drawings usually require more advanced drawing skills. For purposes of design and technology, most sketching can be done by using one of four different kinds of drawings. These include **one-point perspective**, **two-point perspective**, **oblique**, and **isometric** drawings. **Crating** is another technique for doing freehand pictorial drawings.

Perspective Drawings

Perspectives are one form of pictorial drawing that allows us to draw realistic shapes quickly and easily. Shapes represented in perspective drawings closely resemble their actual appearance. Through perspective drawings, you can show what your design would look like when implemented.

There are a few key ideas to keep in mind as you look at objects drawn in perspective. In these drawings, you can see that:

(1) Parallel lines appear to come together (converge) as they extend away (recede) from you.

(2) Objects of the same or similar size appear to get smaller as they recede into the distance.

(3) Lines of equal length appear smaller as they recede into the distance.

(4) Perspective adds a sense of depth to the drawing.

One-point perspective. Drawing an object in one-point perspective is represented in the drawings in Figure 2.25.

(a)

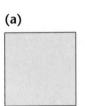

Use light construction lines to draw a flat view of one face of the cube. The flat view will be the true shape of the cube and can be drawn in scale or true size.

(b)

Draw a horizontal line representing the horizon. This will be at your eye level as you look at or visualize the object. If the line is above the flat view, you will see the top of the cube. If the line is below the flat view, you will see the bottom. If the line runsthrough the flat view, the cube will look as though it is on the same level as you are.

(c)

Place the vanishing point on the horizontal line. If the vanishing point is on the right of the flat view, you will see the right side of the cube. If it is on the left, you will see the left side.

(d)

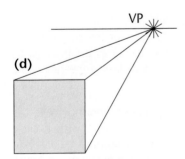

Sketch light lines that will extend from each corner of the flat view and connect with the vanishing point.

(e)

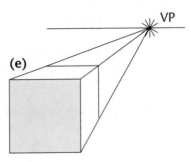

Determine the appropriate length of the lines, keeping in mind that the lines get shorter (are foreshortened) as they recede away into the distance.

(f)

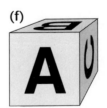

Complete the shape by drawing in the back of the cube. Since the shape is a cube, the back lines will be parallel with the sides of the front shape.

Figure 2.25 ■
Steps for drawing a form in one-point perspective.

Two-point perspective. The drawing of an object in two-point perspective is represented in the drawings in Figure 2.26.

(a)

Eye Level

Horizontal Line

As in one point perspective, draw a horizontal line representing the horizon. This will be at your eye level.

(b)

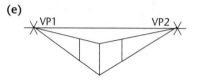

In a two-point perspective, all sides appear to be receding away from the viewer. This requires that two vanishing points be used—thus the title "two-point" perspective.

(c)

In a two-point perspective, only the closest vertical is of true size. It can be drawn to scale, if desired.

(d)

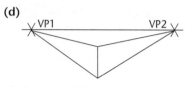

All other sides and lines are reduced in length as they recede away from you.

(e)

The outline of the cube is then completed with the vertical sides kept parallel with the first corner that was drawn.

(f)

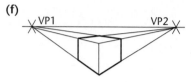

The top and bottom edges are drawn in to complete the shape of the cube. These lines are drawn so they point toward the vanishing points.

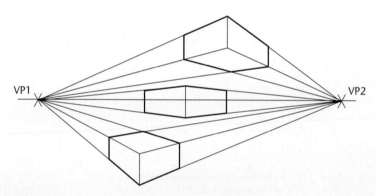

As in one-point perspective, we can alter the drawing by moving your point of view or your eye level.

Figure 2.26 ■
Steps for drawing a form in two-point perspective.

Oblique drawings. Oblique drawings are similar to one-point perspectives but that use parallel, slanted lines to provide a picture of the object. Oblique drawings are similar to one-point perspective. In both, the true shape of one side is drawn as the front view. The lines that form the sides of the shape do not converge, but are drawn at a 45° angle. This angle produces a fairly realistic pictorial drawing. In order to give the drawing a more realistic appearance, the sides are foreshortened.

Oblique presentations are relatively easy to draw. Oblique drawings have the advantage that circles drawn in the front view can be drawn with a circle template or a compass. Circles and curves on the angled sides must be drawn in freehand, as in perspective and isometric drawings. Oblique drawings are not as realistic as one-point perspectives. The lines in the front and back are the same size. Because the back lines are not drawn shorter, the feeling of perspective is reduced. Figure 2.27 demonstrates the steps involved in creating an oblique drawing.

Isometric drawings. Isometric drawings are similar to two-point perspectives but are drawn by using three equal angles. Using these angles makes

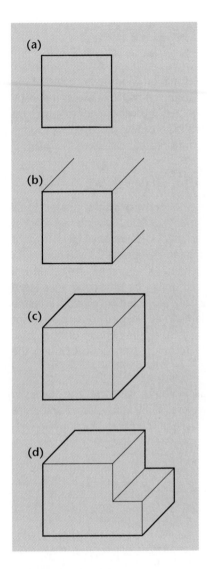

isometric drawings similar to two-point perspective drawings. In isometrics, the vertical lines remain vertical and all horizontal lines are drawn at 30° angles. All lines in an isometric are true size and all sides are drawn using their true measurements. (See Figure 2.28.)

Isometric drawings have the advantage that all lines can be drawn mechanically. They have the disadvantage that the drawings lack realism because of the distortion. Another disadvantage is the difficulty of drawing curves and circles. These usually have to be drawn freehand, but often they can be drawn with the aid of a template.

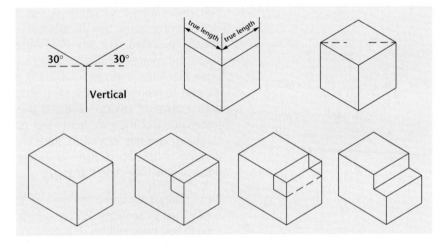

■ **Figure 2.28**
Developing an isometric drawing.

■ **Figure 2.27**
Developing an oblique drawing.

Crating. Pictorial drawings by freehand techniques are made considerably easier through the use of the transparent box, or crating technique. The crate helps you visualize the object you are drawing by placing it inside the transparent box. (See Figure 2.29.)

Crating allows you to think through what an object should look like before actually drawing it. After you are satisfied that the drawing represents faithfully the object of interest, you can add width and weight to the outline of the object.

Figure 2.29 ■
The use of crating in freehand drawing.

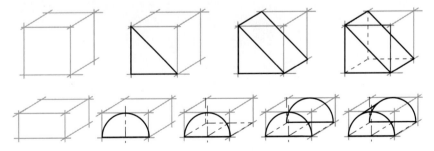

NONPICTORIAL, FLAT DRAWINGS

You have seen that by using angular lines, pictorial drawings help represent objects as you see them. The realism achieved by using angled lines makes it difficult to get true measurements from pictorial drawings. You will encounter problems if you try to use a pictorial drawing as a means of recording dimensions of something you plan to build. What is needed is a drawing that can serve as a plan.

Plan Drawings and Orthographic Projection

Plan drawings are sometimes given the name **orthographic projections**. Orthographic projection is different than many of the drawing approaches already presented. The word *orthographic* as used in geometry refers to right angles and perpendicular lines. Projection means to represent something on a plane much like a movie is projected on a screen. Orthographic projection is a method of representing the exact shape of an object by projecting lines (rays) onto a plane of vision. Three planes are usually projected to give an accurate presentation of the top, front, and side (sometimes referred to as plan, front elevation, and side elevation). Refer to Figure 2.30 to see how an orthographic drawing is developed. Refer to Figure 2.31 to see a plan drawing using the orthographic projection technique.

Plan drawings are important because they allow you to see edges and shapes in their actual dimensions. Plan drawings, therefore, allow you to draw objects as though you were looking straight at each surface of the object. By drawing with right angles and perpendicular lines, each view is seen in its true shape.

Figure 2.30 ■
The development of an orthographic (projection) drawing.

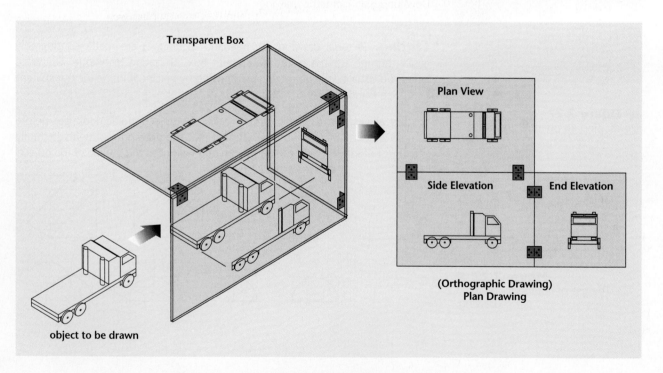

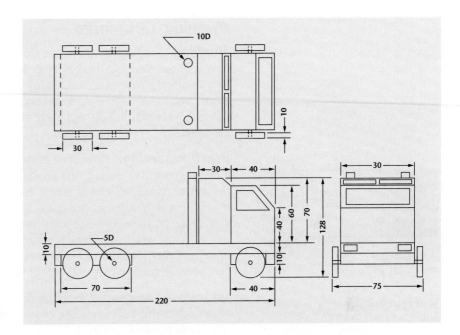

Figure 2.31 ■
A plan drawing using orthographic projection.

PRESENTATION TECHNIQUES

In order to communicate effectively through drawings, it is often helpful to use different presentation techniques. These techniques can help you add to the realism of your drawings or draw attention to a specific part of the drawing that is particularly important. Some examples are shown in Figures 2.32A-C and include **sectional drawings**, **exploded view drawings**, and **cutaway drawings**. The following will help you develop such presentation techniques.

(B)

(C)

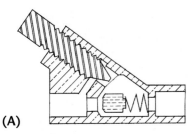

(A)

■ **Figure 2.32A, B, C**
Examples of other drawings:
(A) sectional, (B) exploded, and
(C) cutaway. *(Courtesy of Wendy Snyder)*

Drawing Techniques

Shading. Shading is a technique that helps to make the pictorial drawing of an object look more like the real thing. Shading of pictorial drawings requires you to understand how light is reflected from the object and where lighter and darker shades appear. In Figure 2.34, pyramids are shown as they appear to reflect different amounts of light. For each of the drawings, the **source of light** is identified.

Pencil rendering. One of the easiest ways of producing tones on a drawing is by using pencil rendering. The careful use of **shading** with a pencil allows you to produce realistic effects. As shown in Figure 2.33, many different tones can be made using only one medium grade (HB) pencil. These tones can be made by varying the pressure on the pencil. A very light pressure will produce a light tone that is almost white. A heavier pressure or many layers using light pressure will produce a dark, blackish tone. Pressures that fall in between the very light and very heavy produce a range of mid-tones of gray.

There are several standard forms used to create drawings of complex objects. These include the cube, cylinder, cone, pyramid, and sphere. Examples of these shapes are shown in Figure 2.24. Learning how to render these forms will help you combine and develop more complex forms of the objects you will be drawing. The illustrations that follow show how each of the shapes can be shaded in pencil to give them a more realistic appearance of form. (In a later section, you will see how color is used in a similar manner.)

Before shading a form, it is important to identify how the light falls on the different surfaces. This is determined by establishing the light source. The lamps shown in each of these drawings indicate that the light is coming from a specific direction. More light falls on the surface that is oriented toward the lamp and less light reaches the surfaces pointing away from the light source. Pencil shading can be used to represent these differences.

Cubes are drawn, starting with a square. The cube has three faces. With the light coming from the upper left, the top of the cube gets the most light and the right

■ **Figure 2.33**
Using a pencil to render continuous shading.

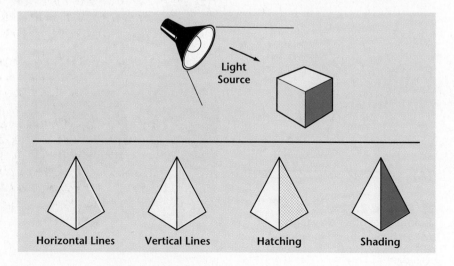

Figure 2.34 ■
Shading of a cube and pyramids showing darkest side away from light source and the lightest side closest to the light.

side, the least. Light falling on the right side will be mid-tone. The rendering of a cube, as in Figure 2.34, shows how shading the three surfaces can produce a more realistic form.

Prisms and pyramids are somewhat similar to cubes in that they are regular solids. Prisms have the same shape in their cross section, all along their length. Pyramids are similar to prisms but they come to a point, sometimes called an apex. Shading a prism or pyramid uses the same techniques as in shading a cube.

Cylinders are shaded by using some new ideas as well as some of the shading techniques described previously. As shown in Figure 2.35, part of the curved surface of the cylinder receives a lot of light. That part should receive no shading. The shading on the cylinder progressively gets darker as the surface moves farther away from the light source. The use of continuous shading becomes very important in rendering cylinders and other curved surfaces.

Cones are shaded in a manner very similar to the cylinder. The biggest difference is the shape of the lightest part of the shaded surface. The lightest part must have a triangular shape that reaches from the apex to the base. This shape is maintained by shading the rest of the surface at a slight angle. The basic triangular shape of the cone is matched to add realism to the drawing.

Spheres have a light shape that is circular, since the sphere is based on a circular shape. Shading is done in a circular manner, becoming darker as the surface of the sphere curves away from the source of light. Figure 2.36 shows the sphere in different stages of rendering.

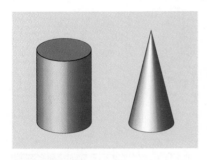

■ **Figure 2.35**
Shading of cylinders and cones.
Note: The highlights on the cylinders are rectangular and on the cones they are triangular.

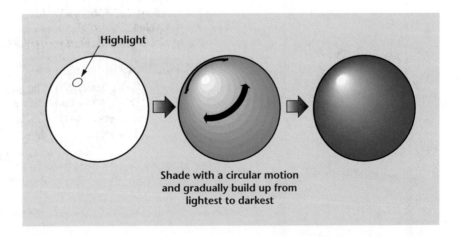

Figure 2.36 ■
Shading a sphere using circular motions, building gradually from lightest to darkest.
Note: The highlight is usually a circular shape on the sphere.

Highlight

Shade with a circular motion and gradually build up from lightest to darkest

Highlights. Highlights show different shades and light reflections. These are important for portraying form. Reflections are a natural part of nearly everything we see. That is how we see color, for example. Reflections also add realism to the forms we are drawing. As with previous drawings you examined, reflections are influenced by the shape of the object. This is especially noticeable in the cone, cylinder, and sphere. Highlights are somewhat exaggerated and represent a line or point of bright, reflected light. Examples of highlights are shown in the illustrations in Figure 2.37.

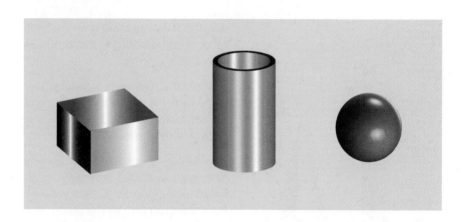

Figure 2.37 ■
Examples of objects showing highlights and reflections.

DESIGN ACTIVITIES

Using drawings provided in the *Student Portfolio Activity Resource*, try out the different drawing techniques of shading, rendering, and highlighting of cubes, pyramids, cylinders, cones, and spheres.

(A)

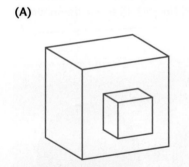

(B)

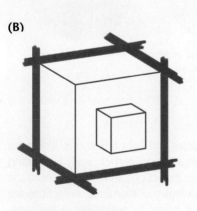

■ **Figure 2.38A, B**
Drawings or their parts can be emphasized with outlining.

Outlining. Outlining emphasizes objects and makes them stand out. One technique is to use thick lines for only the outside edges of the object, as illustrated in Figure 2.38A. Another technique, shown in Figure 2.38B, is to use a marker to create a background by outlining the object in heavier lines in black or in color. The use of color makes the product stand out and attract more attention.

Color Techniques

This chapter introduced a number of concepts related to color. These include primary, secondary, and tertiary colors; complementary and contrasting colors; and color wheels. In this section, you will apply these concepts as you use color to add to the realism and impact of your drawings and illustrations.

Color pencils. Color pencils utilize the same skills you developed in shading with regular pencils. Color pencils are an easy, inexpensive, and effective way to add color to a drawing and are useful in creating a range of colors. As shown in Figure 2.39, knowledge of the color wheel allows you to develop a full range of secondary and tertiary colors by using three pencils of primary colors.

Color markers. Color markers provide a wide range of colors that can be very helpful and effective in your color renderings. Using color markers requires practice, but the power and effect that color markers can make on drawings and other presentations will be well worth your time and effort. Figure 2.40 shows markers that come in a range of sizes and colors. Figure 2.41 shows how effective color markers can be in presenting a design.

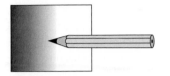

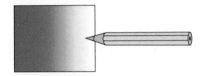

■ Figure 2.39
Shading with pencils can produce shades of one color or a merging of shades of multiple colors.

■ Figure 2.40
Examples of color markers.
(© Michael Dzaman Photography)

Figure 2.41 ■
The process of using color markers for illustrating a personal communicator. *(Courtesy of Philip Forrest Smith, photo by Michael Dzaman)*

Safety Note:
Use only color markers that are water-based rather than oil-based.

Color media. Color media include pastels, water paints, and other means of applying color to your presentations. A discussion with your art and design teacher or a visit to a local art supply store can help you select the approach that will achieve the effect you want.

Graphics

Graphics refers to the general approach of presenting information in a visual form. Specifically, graphics includes the presentation of data in charts, graphs, statistics, schematics, maps, sequence drawings, and the like. Examples of these presentation methods are illustrated in Figure 2.42. Each of these will be discussed in more detail in the chapter on Information.

You often find that photographs, tape recordings, videos, or other methods of recording and documenting work are important. Whether you are developing

Graphs

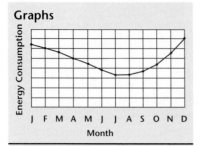

Bar Charts

School Activity Budget

Pictograms

Local Recycling Effort

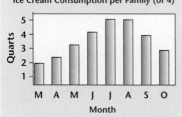

Histograms

Ice Cream Consumption per Family (of 4)

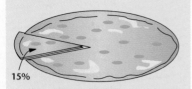

Pie Charts

Percentage of Take-Out Pizza

■ **Figure 2.42**
Examples of graphs, bar charts, pictograms, histograms, and pie charts.

rough sketches, planning the sequence of events, identifying the resources needed, or deciding on the placement of materials to make a product, you will develop a number and variety of documents. These documents represent your thought processes as they were put on paper. These preliminary ideas guide your work and should be saved for your presentation portfolio.

PRESENTATION PORTFOLIOS

Developing a portfolio to present a summary of your work will be easier if you follow a general outline from the very beginning. The **portfolio** is a compilation of the records that you keep in various forms throughout your design and problem solving process. A portfolio puts your work into a meaningful whole.

D E S I G N   B R I E F

Organization of a Portfolio—Using the IDEATE Model

Title Page

Introduction

Provide the background for your problems. (Describe the situation and context of the problem.)

Describe the Problem

Identify and define the problem. (Describe how you investigated the needs and opportunities.)

Design Brief and Specifications

Develop the design brief and specifications. (Describe how you clarified the results you wanted to achieve.)

Research and Investigation

Explore possible alternatives. (Portray how you conducted your search for information, resources, and solutions.)

Selected Solution

Accumulate and assess the alternatives. (Describe how the best solution was chosen and developed. Why is this the best?)

Implementation and Development Work

Try out the best solution. (Describe your experiments as you develop solutions, models, and prototypes.)

Final Testing and Overall Evaluation

Evaluate the results. (Present how you tested your solution. How did you assess the process used in solving the problem? What is your overall satisfaction with the results and the process?)

The portfolio is a formal method for you to use as you communicate with others about the project you implemented. It includes aspects of the work itself (the process) as well as of the outcome (the product). Your report or portfolio is also a form of evidence for your teacher and others of the work you have done. The document provides an indication of your skills in carrying out the steps of the IDEATE model, as well as in communicating your results.

Portfolio Development

The parts of a portfolio might differ, but the topics indicated in the design brief on page 50 represent the key aspects that you should include in your portfolio. It is important, however, that your portfolio be designed to be as effective as possible in communicating your ideas and efforts to others. In other words, apply the skills and principles discussed in this chapter and the preceding chapter to enhance your portfolio.

Other examples of portfolio work will be shown in later sections of the book. These examples should be helpful as you work on the overall design of your portfolio, the format you want to use, and the content you want to include. Figure 2.43 includes the components necessary to develop a portfolio. You will find portfolio work demanding, but worthwhile. The results of that work, and the insights you gain about your design, your project, and your efforts can be used to make even greater improvements in your future design work.

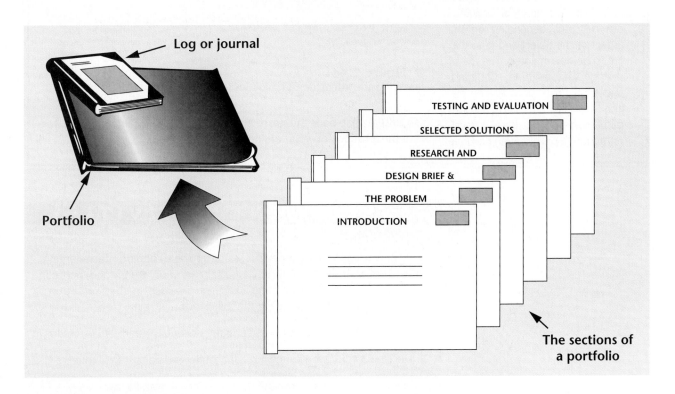

Figure 2.43 ■
Organization and development of a portfolio.

SUMMARY
CHAPTER 2

In this chapter, you explored some of the ways to document the process of design work. In addition, steps taken in developing and evaluating the product are recorded in order to keep track of ideas, indicate when the work is done, and show the thought processes at each step of the way. Journals, logs, sketches and drawings, notes, and plans are useful in developing thoughts and communicating ideas.

There are a variety of drawing techniques for presenting line, shape, and form, using a perspective or other pictorial approach that can enhance the presentation. Different types of drawings can be used, depending on the aspect of the object and plan that are to be documented. The use of shading, highlighting, outlining, and color can provide contrast and attract attention to various parts of the design. Careful development of a presentation portfolio

Interaction of Mind and Hand

Thinking
(inside the head)

Doing
(in real world)

Start with initial (hazy) idea

Talk to others, make drawings, sketches, notes, graphs of idea

Explore new idea, enhance the idea

Make a rough model to represent an idea or prediction

Clarify and validate idea

Make a prototype and test it in several situations

Critically appraise idea

Potential for more developed thinking

Potential for more developed solutions

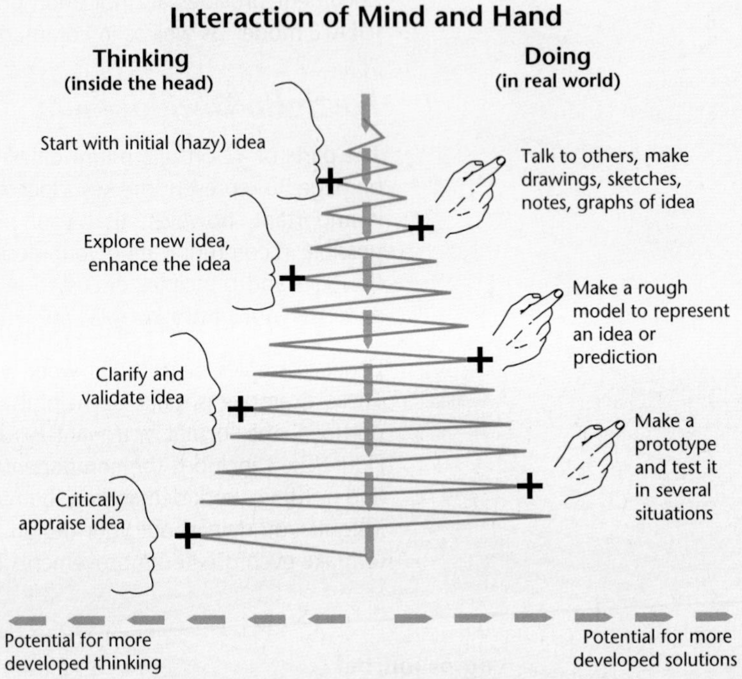

■ Figure 2.44
Revisiting the interaction of mind and hand model.

can ensure that both the process and the product of your design work is clearly communicated to others.

ENRICHMENT ACTIVITIES

Now is a good time to develop some of the designing, documenting, and drawing skills that will be important in your design and portfolio activities. Use the drawings and exercises provided in the *Portfolio and Activities Resource,* and complete the following:

- Develop a color wheel showing primary, secondary, and tertiary colors.
- Develop a color wheel showing contrasting and harmonious colors.
- Identify objects that show the use of balance, harmony, and contrast.
- Develop instances of simple and more complex tessellation patterns.

- Develop a one-point perspective of a simple tool that you have used.

- Develop a two-point perspective of a piece of furniture or a building.

- Complete the shading and rendering of a cube and a pyramid.

- Complete the shading and rendering of a cylinder and a cone.

- Complete the shading and rendering of a sphere.

- Complete the shading and rendering of the above shapes in color pencil.

- Complete the shading and rendering of the above shapes in color marker.

Sign, date, and place this work in your portfolio.

D E S I G N   B R I E F

Design and develop a chart to display a color analysis for each of the members of your team of classmates (four to five people).

- Use sheets of colored paper and fabrics of various colors and tones.

- Focusing on one person at a time, place the swatches of color close to his/her face.

- Determine the degree that the color harmonizes or contrasts with each person's skin tone and hair and eye colors.

- Conclude which kind of colors are the most flattering for each person.

- Develop a color analysis chart for each person for his/her portfolio.

- With this information, discuss what colors would be most appropriate to use for a team T-shirt or emblem.

REVIEW AND ASSESSMENT

- Identify the elements of visual design and an application of each.

- Identify the principles of visual design and an application of each.

- Describe the four ergonomic factors to be considered in the design of objects to be used by people.

- Describe the steps you would take in order to prepare a Gantt chart for setting up the concession stand at a school sporting event.

- Identify the different types of pictorial drawings and the advantages and disadvantages of each.

- Describe how you develop a presentation portfolio using the IDEATE model.

CHAPTER 3

Models and Model-Building

Figure 3.1 ■
Computer-generated models add realism to objects under development. (© 1992 Charly Franklin/FPG International)

INTRODUCTION—TRANSLATING IDEAS INTO REAL OBJECTS

Transforming your ideas into a **model** is an important way of communicating with others, and is also useful for translating your ideas into real objects. Modeling your ideas can help you see how those ideas might be improved.

55

KEY TERMS

adhesives
appearance model
architectural model
computer model
concept
conceptual map
conceptual model
development
electronic model
electronic prototyping
   board
fabrication
forming
functional model
master
mechanical model
model
physical model
production model
prototype
rapid deposition modeling
   (RDM)
representation
scale model
shaping
stereolithography
structural model
subsystem
systems
systems model
theoretical model
theory
tooling up
transferring

Models take many different forms, but all are simplified **representations** of something from the real world. Three types of models are used extensively throughout this book. These include **conceptual models, theoretical models**, and **physical models**. They are all useful in design and technology work for representing ideas. You will see and use models as you progress through your design work and this book. We will turn considerable attention to physical models shortly. First, we need to touch briefly on conceptual and theoretical models.

TYPES OF MODELS

Conceptual Models

Each of us uses conceptual models to make sense of the real world. For example, the idea of a "tree" serves as a model that helps us deal with thousands of instances of trees found in nature. We recognize many different large plants as trees when we see them. We know that trees have parts—roots, a trunk, branches, and leaves of some form. We also know that these parts go together in identifiable ways—trunks connect to roots that reach into the ground, branches grow out from the trunk and are tipped with leaves or needles. The parts of a tree work together in specific ways, with each part performing its function. Although trees may differ somewhat, all of them have similar parts.

The **concept** of a "tree," however, is not limited to trees that are found in nature. We use our experience with trees to represent the idea of "trees," such as family trees and decision trees. We also use "tree-related" language to describe other concepts. For example, we talk about the branches of a stream or river, the roots of a problem, the leaves of a book. Figure 3.2 illustrates the concept of a tree in relation to a conceptual model. If someone told you about the "trunk lines" of a telephone system or railroad system, would you have a general idea of what they were discussing? These are instances of **transferring** an idea from what we know to explain and explore something new and unfamiliar.

Currently, you use hundreds of conceptual models to deal with the unfamiliar. They help you understand and cope with the complexity of the world in which you live. Through this book, you will gain new conceptual models of technology and its major components. These models include the *process, resources, systems*, and *impacts* of technology. You will also gain new models of each of these

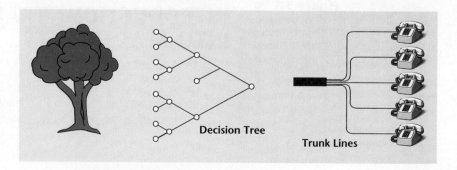

Figure 3.2 ■
Example of a tree and its parts used in conceptual models.

Decision Tree

Trunk Lines

components that include *tools, materials, energy, information, structures,* and *control.* Your collection of conceptual models will help you understand and communicate about the technological aspects of the world within which you live.

One use of conceptual models that will be of help to you in your design and technology work is the **conceptual map**. All maps are themselves models that show the connection and positions of objects in relationship to each other. The most common maps are those we use in our travels or in finding information about other countries. Conceptual maps can take a variety of forms. Two specific examples, the question web and the planning web, are shown in Figures 3.3A and B.

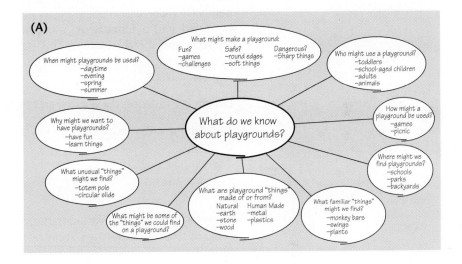

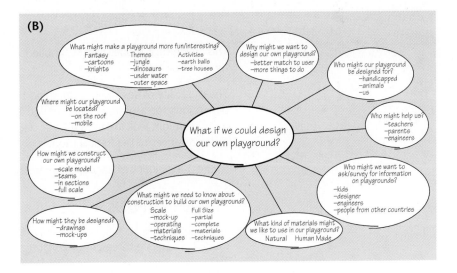

Figure 3.3A, B ■
Examples of conceptual maps:
(A) question webs and
(B) planning webs. *(Courtesy of Project UPDATE Team)*

Conceptual models and maps provide you a means of getting ideas recorded so you can think about them more clearly or so that you can discuss the ideas with others. The two conceptual maps above were developed by a group of teachers and students as they looked at the problem and opportunity for designing and developing a new play area. The first map, the question web, was used to identify what activities might be included in a play area. The second map, or planning web, was used to generate questions that would guide the teachers and students as they investigated and developed possible solutions.

Theoretical Models

Theoretical models are similar to conceptual models but go a few steps further. Theoretical models spell out in some detail how the parts of the model work together and thus describe how the theory works. As a **theory** becomes more and more clear and has been tested over time, it can be expressed by using numbers. Theoretical models are usually expressed as statements that can be translated into a formula. An example of a theoretical model relates time, distance, and speed (velocity). This statement indicates that "distance equals velocity multiplied by time." It is expressed in the following mathematical formula:

(Where D = distance, v = velocity, and t = time) D = vt

This statement can be arranged to indicate that velocity equals distance divided by time:

$$V = \frac{D}{t}$$

If you had ridden your bike a distance of 30 miles in a time of 80 minutes, you could calculate your average velocity by substituting these quantities in the above formula.

$$V = \frac{D}{t} \quad \text{or} \quad V = \frac{30}{80 \text{ mins.}} \quad \text{or} \quad V = \frac{30}{1.3 \text{ hours}} \quad \text{or} \quad V = 23.8 \text{ mph}$$

This theoretical model can be applied to trains, planes, runners, or any other moving object.

The capability of describing something by using a formula, such as the ratios and dimensions for a garment, allows you to use a computer to model what that garment might look like for a specific size. In later chapters, you will see these and other instances of theoretical models.

Figure 3.4 includes examples of physical, conceptual, and theoretical models.

Physical Models

In the remaining sections of this chapter, we will turn our attention to physical models. These are three-dimensional representations of something that is real.

Physical models tend to fall into three general categories—**functional models**, **appearance models**, and **production models**.

Functional models. Functional models are used to see if an idea will work. Functional models help you investigate an idea or concept quickly, with little concern for the appearance of the model. You can make these models from

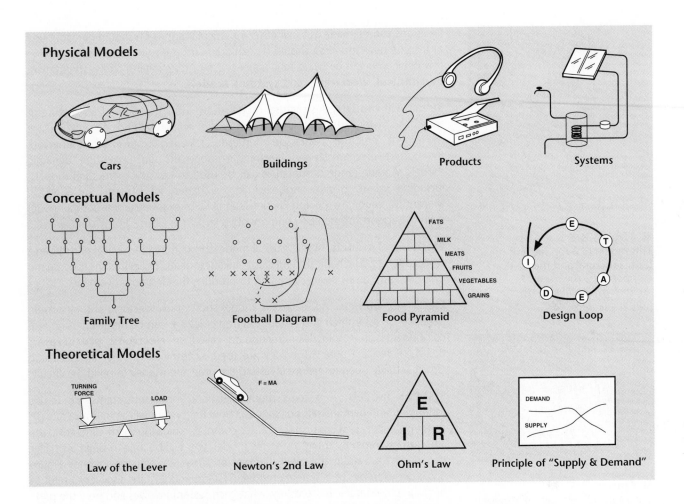

Physical Models

Cars   Buildings   Products   Systems

Conceptual Models

Family Tree   Football Diagram   Food Pyramid   Design Loop

FATS
MILK
MEATS
FRUITS
VEGETABLES
GRAINS

Theoretical Models

TURNING FORCE   LOAD $F = MA$

E
I   R

DEMAND
SUPPLY

Law of the Lever   Newton's 2nd Law   Ohm's Law   Principle of "Supply & Demand"

Figure 3.4 ■
Types of models.

cheap or scrap materials. Models help you visualize an idea and reach a solution to a problem. You may make several such models. You may use them to check not only how the device functions, but also sizes, layouts, and proportions as well.

Functional models are an effective means for you to think in three dimensions. For example, you may want to design and build something for your room; perhaps a storage area, a workbench, or a lighted mirror. You will find that it is faster and less costly to work out your ideas in a simple, functional model before you try to build the object. Developing a functional model is similar to sketching and drawing, except your ideas take on a more realistic three-dimensional form.

Any material can be used in developing a functional model. A good guideline is to use materials that will allow you to develop the model as quickly and as inexpensively as possible. You may find it useful to show your functional models to others so you can discuss what you are trying to accomplish. These models can be very helpful if you are working as part of a team. The models can help reduce misunderstandings between team members, and each person can more easily contribute ideas to improve the design.

Functional models can also be used to demonstrate the principles of a device or machine. Models of this type can help determine if the basic principles are sound.

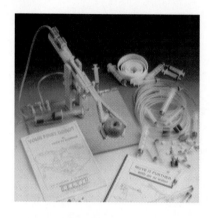

Figure 3.5
This model of a robotic arm uses both structural and mechanical components. *(Courtesy of Technology Teaching Systems, Cambridge, MA)*

Figure 3.6
Electronic model built on a special board for prototyping. *(© Michael Dzaman Photography)*

Figure 3.7
An example of a systems model. *(Courtesy of Plastruct, City of Industry, CA)*

Decisions can be made as to whether a particular approach should be continued or if some new approach should be explored. Through the activities included in this book, you will be using four major types of functional models: **structural**, **mechanical**, **electronic**, and **systems models**.

Structural models. A structural model shows how the parts of a structure fit together. Many structural models are static in nature and do not involve movement. Structural models are often used, however, to provide the framework to support mechanical and electrical components when creating dynamic models. A wide range of materials can be used to construct structural models, including wood, paper, cardboard, and a variety of kits such as K'NEX®, Mecanno®, Lego®, Lasy®, and Fischer Technik®. An example of a structural model used to support mechanical components is shown in Figure 3.5.

Mechanical models. A mechanical model shows how mechanical parts fit and work together. Mechanical models use linkages, gears, pulleys, cams, pneumatic cylinders, and other devices required to produce the desired movements.

Electronic models. An electronic model shows how electrical and electronic parts will be assembled and tested. Electronic devices can be modeled using real components on a solderless breadboard called an **electronic prototyping** board. (Refer to Figure 3.6.) They are used to test out whether an electronic circuit actually operates before investing the time and money to build the circuit.

Systems models. A systems model illustrates how subsystems are to be assembled and demonstrates or suggests how the system will actually operate. In many instances, **systems** can be rather complex, involving many different parts and **subsystems**. By using a systems model, you will be able to build a smaller part of a large system and see if that part or subsystem actually operates. Figure 3.7 is an example of a systems model. Systems models are also very useful in helping to communicate to others what each system involves and how the parts of the system relate and work together.

DESIGN ACTIVITIES

Identify a toy that operates mechanically. Without taking it apart, try to figure out how it works. Make several sketches of how the parts might move and work together. Construct a model to represent how the toy operates.

Appearance models. Appearance models are used to show what an object looks like when it is produced. Appearance models are used to show the final form of products, buildings, and environments. Product models represent what a commercial product will look like when produced. Interior models often include models of furniture and are intended to show how a space might be used in a home, business, or industry. Interior models are often used in planning for the renovation of space. This might be for a new kitchen or addition to your home,

(A)

(B)

■ **Figure 3.8A, B**
Several appearance models: (A) an architectural presentation and (B) an anti-pollution cycle mask. *(A. Courtesy of Hewlett Packard Company, B. © Philip Forrest Smith)*

■ **Figure 3.9**
This production model uses the computer to determine the most efficient layout for the pattern on the cloth before cutting. *(© 1994 William Taufic/The Stock Market)*

a commercial show of electronic products, or an improved science and technology exhibit.

Architectural models are commonly used to show the buyers or financiers of a project what a building or development will look like once it is built. Good model-making can be teamed up with the creative use of photography and video cameras to provide a simulated trip through a planned building. Another use of architectural models is to study what changes the new building might make in air movement and patterns of sunlight in surrounding areas. This type of modeling helps the designers look at larger environments rather than a single building.

The appearance models illustrated in Figures 3.8A and B represent things that have not yet been produced. Models can range from handcrafted to computer-generated objects. These models help provide insight into what these objects might look like when and if they are produced. As we become more concerned about the production of the object, particularly if it is to be produced in quantity, the use of a production model becomes more critical.

Production models.   When a product is produced in quantity, it is important that all copies of a part are as similar as possible. In many cases, parts will be made in molds or dies that provide cavities to be filled with a metal or plastic. In this way, the finished shape is produced. Molds and dies tend to wear, and after a certain amount of use, they no longer produce parts that are exact enough. At this point, it becomes necessary to return to the production model, whether a real device or computer file, to produce another die or mold. The production model, in this sense, is a **master** that can be used for the **tooling up** necessary for manufacturing the product in quantity. (See Figure 3.9.)

Prototypes

Prototyping is an advanced form of model-making. Developing a prototype represents an idea for a product, system, or environment as close to the finished form as possible. Prototypes are the final models used to test out a product before it is actually put into use or into production. An example of a prototype is shown in Figure 3.10. If a product is to be one of a kind, the prototype may be adequate as its final form. In this case, the prototype can fulfill its function, such as a one-time display. If you have developed a new piece of furniture or a piece of clothing for yourself, the prototype may be quite adequate for your use. As much as possible, a prototype uses the actual materials and dimensions and is the same as the finished product.

In some cases, the prototype brings an idea a step closer to a mass-produced product. Products that are to be mass-produced require prototypes for working out the "bugs." Examples show how the product operates and how to set up the equipment and system to produce the product. The prototype provides insights as to what the product should look like and the possible processes for mass producing it. In this sense, the prototype is useful in designing the production process. It can be helpful in selecting the tools and equipment and other resources that will be needed to make many identical copies of the product.

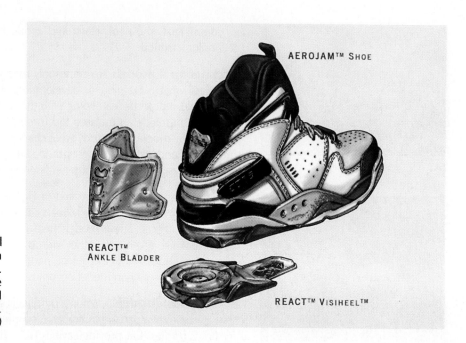

AEROJAM™ SHOE

REACT™ ANKLE BLADDER

REACT™ VISIHEEL™

Figure 3.10 ■
Building prototypes is an important step in mass production. In this case, the parts of the product will be manufactured separately and then assembled. *(Courtesy of Converse, Inc.)*

MODEL-MAKING

Materials

As you become more involved in developing models, you will become adept in collecting materials that will be helpful. For the model-maker, materials that other people call scrap take on a different value. Materials that are often thrown away become useful in developing models. The plastic containers from fast-food restaurants, empty food containers from your home, and other unwanted packaging materials are particularly useful. Cardboard boxes can be taken apart and used quite nicely for making the shapes and forms that you want to develop. Cheap fabrics, old sheets, and other scraps can be cut up for use.

General Materials for Models

The materials you use in developing models need not be complex or expensive. In the illustrations that follow, you will see a range of common materials that can be very useful in your work. They save you time and effort and also reduce the cost of your model-building activities.

These materials can take many forms. They will include materials that can be molded and sculpted such as plasticine, clay, papier-mâché, or flour polymer (bread and glue mixture). You will also use materials that can be cut and shaped through the use of cutting and forming tools. These materials include foam plastic, including rigid and flexible, hard and soft, and sheet and blocks.

In model-building, it is important to have several different means of assembling the parts of the model. You will find that some **adhesives** work very well with

Figure 3.11 ■
An effective modeling center includes a range of equipment and materials for translating ideas into a tangible form.

some materials but rather poorly with others. Some adhesives will dissolve selected materials if you try to use them together. Some adhesive materials you will want to have available include rubber cement, contact cement, polyvinyl acetate (PVA), polystyrene cement, acrylic cement, and hot glue and a hot glue gun. Other methods of joining may include tapes, pins, staples, needles and thread, and cord.

Safety Note:
It is important that you be careful with adhesives. Some adhesives may irritate your skin or may be toxic. Some should only be used in a ventilated area while wearing a mask.

Special Materials for Appearance Models

There are a variety of materials that will be helpful as you develop appearance models. Appearance models are a little like special effects in a movie. They intend to trick the viewer into imagining something different than what they actually see. The trick of appearance models is to look as real as possible without the work and time required to produce the real object. This places a lot of importance on finding easy-to-use materials that can represent something else. These materials include sheets of plastic that are relatively easy to cut and form, plastic wood and fillers that can be shaped and formed, and sheets of textured materials, such as fabrics, sandpaper, cork, or floor tile, that can represent different surfaces.

Other materials that you will find helpful in your model-building include different types of lettering, such as dry transfer lettering, vinyl cutout lettering, molded plastic letters, and embossed lettering (made on a hand-operated machine). Such lettering will allow you to add details to your models in a professional manner.

Rolls of tape of different widths and colors are also helpful. Drawing tapes and car stripping tapes are particularly useful. Many products have lines that are used to enhance or highlight specific features. Many of these lines can be represented in your model by using pieces of tape. Tape can also be used for creating a nice edge on models of industrial machines, displays, and furniture. A complete collection of tapes for model-building might be a little expensive. However, because only a small amount of material is used on any one model, the tapes will last a long time.

SPECIAL MATERIALS FOR APPEARANCE MODELS

- sheets of foamcore, polystyrene, Plexiglas™, Lucite™, and Perspex™

- fillers and plastic wood (ready-mixed and powder)

- sheets of textured materials such as fabrics, sandpaper, cork, and floor tile

- dry transfer lettering, vinyl cutout lettering, molded plastic letters, and decals

- embossed lettering (made on a hand-operated machine)

- drawing tapes of different widths and colors, and car stripping tapes

- spray paints, car touch-up paints, acrylic paints, food coloring, and textile paints

The finishes you use on your models will be important as you attempt to create the image of the product or object. These finishes include paints of many kinds, but the most useful will be spray paints and car touch-up paints. These paints are relatively easy to use and come in a variety of colors. When applied in an even coating to produce a smooth final surface, paint contributes a professional look to your model.

Scale model parts can be particularly helpful in the development of models that require detail work. For example, a wide range of parts are used in architectural models. These parts include details of the building as well as models of trees, shrubs, vehicles, and humans. The careful use of scale model parts can add some very realistic finishing touches to an architectural model.

■ **Figure 3.12**
Scale model parts can be helpful in adding realism to different models. (*Courtesy of Plastruct, City of Industry, CA*)

Scale model parts are also useful in developing models of furniture or interior designs of a home, store, or exhibit. The detail that can be added, similar to architectural models, can be very effective in adding realism to your work.

Finally, scale parts can make the development of systems models easier. Systems models tend to include considerable detail. Large systems models may represent an industrial layout of part of a factory or a whole plant. Smaller systems models show several related subsystems as they work together. For example, a solar heating system might include a subsystem for energy collection, another subsystem for converting that energy into useful forms, and a third subsystem representing the distribution of that energy as heating or cooling throughout the building.

Special Materials for Functional Models

Functional models are intended to show how something works. The materials you collect and use should help you create the movements or actions (functions) that you want to show. These materials will range from simple and inexpensive to complex and costly. You need to decide which are appropriate for your model and what you are trying to accomplish.

DESIGN ACTIVITIES

Return to the toy for which you made the mechanical model. Determine changes that you would make in its appearance. (This should include the use of the elements and principles of visual design discussed in the preceding chapter.) Construct an appearance model that incorporates these changes.

Refer to the illustration in Figure 2.41 of the personal communicator. Develop an appearance model for that product.

Tools and Equipment

In most cases, model-making requires building something that is smaller than the real thing. Building models in smaller scale is made easier by using tools and equipment that are designed for that purpose. In a modeling center, a variety of

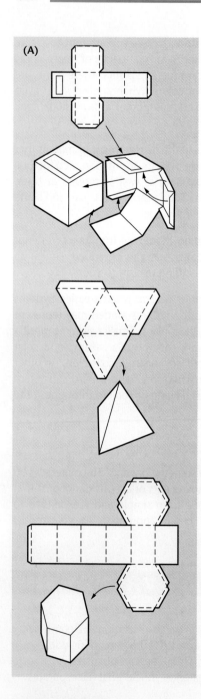

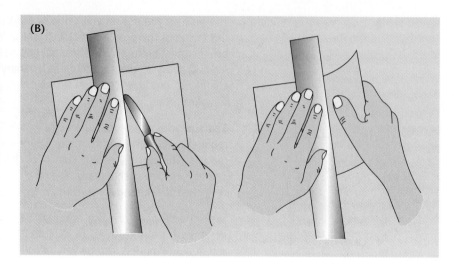

model-making tools for cutting materials are needed. These include saws, knives, scissors, scoring tools, drills, and punches. Model-making equipment can include electric saws, sanders, drills, foam cutters, and sewing machines.

A variety of tools and machines exist for shaping the materials for your models. These include forming tools, files, and sandpaper. Some of the machines that are used for shaping materials include vacuum formers, strip heaters, rulers, steam irons, and straight edges. These are only a few of the tools and machines that might be useful.

Another important process in model-building is joining together the parts. This is most often done by adhesion. A variety of adhesives are currently available. The liquid used in adhesives can also affect your materials. Some adhesives dissolve certain plastics. Some adhesives wrinkle paper and cardboard. It is important that you experiment with each new adhesive you use. Record your findings in your journal or log as reference information for future work.

Techniques

Constructing models from two-dimensional materials.
Many models are made from two-dimensional materials. The word **development** is often used in design and technology to describe how a three-dimensional form is made from two-dimensional (flat) materials. Models can be made through this process in a quick, easy, cheap, and effective manner.

The use of developments for model-making places considerable importance on drawing skills. Particularly important are the orthographic projection skills introduced in the preceding chapter. In Figures 3.13A and B, a step-by-step process is shown for developing several three-dimensional forms.

The use of development is also very helpful in creating models of packages. In many instances, packages are special boxes designed so that they can be cut out of a single piece of card material. First, an image is printed on the material and then the shape is cut out. A package can be formed by folding and then gluing

■ **Figure 3.13A, B**
(A) Making three-dimensional models from two-dimensional materials using different developments. (B) Using a blunt knife to score and then fold card stock.

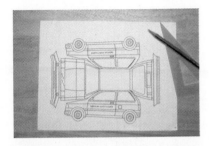

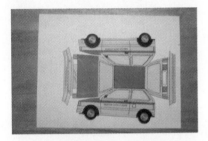

■ **Figure 3.14**
Development and final version of a three-dimensional electric car model. *(Courtesy of Wendy Snyder, photos by Michael Dzaman)*

the box together at the appropriate places—usually the tabs. Windows can be included in the design as open holes or covered in transparent plastic. (See Figure 3.14.)

Construction kits. Construction kits provide a helpful source for parts in model-building. Each of the different systems is designed so that parts are compatible. This allows you to build rather complex devices easily and quickly. (See Figure 3.15.) If you modify existing kits, you need to make only special parts for those not included in the sets. You may encounter some difficulty if you try to combine parts from several different sets. The parts of the different sets often are not compatible. You must determine whether you can modify pieces to connect the parts from one set with the parts of another.

Scale models are particularly helpful when you try to create a sense of realism in your models. Scale models are used extensively in architectural and industrial models. The cost of the parts are well worth the time and labor saved and the overall effects that can be achieved.

■ **Figure 3.15**
Scale models made from different construction kits. (*Courtesy of Creative Learning Systems, Inc., San Diego, CA, and K'NEX by Connector Set Toy Company, Hatfield, PA)*

Constructing models from basic materials. **Fabrication** involves making models from the materials that were identified earlier in this chapter. These materials need to be processed to create the desired mode; that is, the materials must be cut, shaped, formed, assembled, and then finished in some appropriate manner. Tools and related processes for changing materials will be introduced in Unit 2. Some of the general fabrication processes related to model-making are presented here, however.

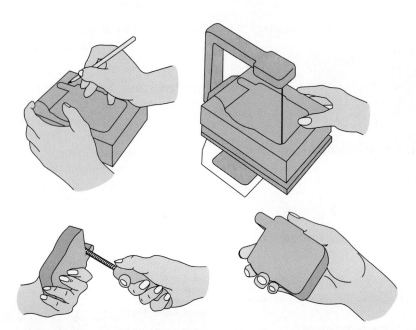

Figure 3.16 ■
Cutting and shaping high-density foam to make a model of the personal communicator shown in Figure 2.41.

Cutting and **shaping** materials can be done through a variety of processes. High-density polystyrene foam is an excellent material for creating solid models. The foam can be cut easily with a hot-wire cutter. The foam can also be cut and shaped using standard hand tools used for cutting other materials. These include such tools as saws, knives, scissors, and shears. (Refer to Figure 3.16.)

Forming refers to the processes that are used to change the form of materials without necessarily removing any of the material. A commonly used forming process in modeling is vacuum forming. In this process, plastic is warmed until it becomes very pliable and then is formed over or into a mold. The forming is accomplished by vacuum and air pressure. Forming can also be done on wood, metals, and other materials, but, because of their unique internal structures, different forming processes must be used.

Safety Note:
Be sure to use hot-wire cutters in well-ventilated areas.

DESIGN ACTIVITIES

Using one of the available construction kits, develop a functional model of a device that includes movement. This might include a model of an amusement park ride, an elevator or escalator, a conveyor system for moving feed to livestock or storage, or a system for handling and cleaning dirty dishes in the school cafeteria.

■ Figure 3.17A
A computer model for testing car designs, to improve aerodynamic features and increase fuel efficiency. *(Courtesy of Ford Motor Company)*

Computer Models

The term **computer model** is often used to describe a computer-generated, perspective drawing that is rendered electronically to add realism to the object. Although these illustrations are technically not physical models, their detail allows them to play a similar role. Computer models provide a dynamic function that allows the object to be manipulated and modified as needed. As shown in Figure 3.17A, modeling through computers allows objects and phenomena to be represented in a manner not possible through other approaches.

Physical modeling is possible through computer-controlled processes called **stereolithography** and **rapid deposition modeling (RDM)**. In stereolithography, a part is fabricated using computer-driven lasers that traced out the part in a bath of liquid plastic (polymer). As the plastic hardens, layer by layer, a 3-D model emerges from the bath. In RDM, the layers are sprayed on one by one to achieve a similar result, as shown in Figures 3.17B and C. Developing models in the future using virtual reality and the next generations of computers promises to be very exciting.

Figure 3.17B, C ■
Developing three-dimensional models of products using the computer-controlled process called rapid deposition modeling. *(Courtesy of Iven E. May, CNC Software, Gig Harbor, WA, and Stratasys Inc., Eden Prairie, MN)*

SUMMARY
CHAPTER 3

Modeling provides a means of visually representing your ideas and plans. Models can help you translate your ideas and designs into progressively more detailed and specific objects and plans. Some models can be presented in two-dimensional form and are useful for recording abstract ideas. Such models may be sketches, mathematical formulas, or computer programs. Three-dimensional models are useful in clarifying and sharing ideas about products, systems, and environments. Such models can range from simple presentations in paper and cardboard to sophisticated working models and computer models. Modeling is a useful adjunct to your drawing skills. For some purposes, modeling is a preferred means of presenting ideas.

Models can be helpful at several places in the design process. You may build them as you design and develop a new piece of clothing, a new tool or machine, a proposed building or bridge, a projected road, or a communication system. In the early stages, the final form of every new product, system, or environment is unclear and uncertain. Model-building is an essential skill and process for helping to translate possible ideas and dreams into specific forms for discussion, planning, development, testing, and improvement. Modeling can save you a great deal of time and money as you search for the best solutions to the problems you face.

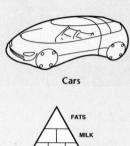

Cars

Food Pyramid

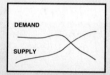

Principle of "Supply & Demand"

■ **Figure 3.18**
Examples of physical, conceptual, and theoretical models.

ENRICHMENT ACTIVITIES

Model-building is a valuable set of skills that you should find useful in your design and technology work. Refer to the section on modeling in the *Portfolio and Activities Resource* and complete the following activities:

- Develop an activity web for projects such as a miniature golf course, a picnic area, a concession stand, or a planning and resource area for your D&T facility.

- Develop a question web for the project you chose in the above activity.

- Design and develop a display that illustrates one of the theoretical models introduced in Figure 3.4.

- Develop a list of materials you can collect to help in your model-building.

- Develop a chart to show which tools can be used with which model-making materials.

- Develop a three-dimensional model of a building or vehicle from a thin piece of cardboard.

- Develop a functional model from one of the available construction kits.

- Fabricate an appearance model of a handheld product, such as a glue gun, electric knife, or hair dryer, using high-density foam and appropriate tools.

REVIEW & ASSESSMENT

- Identify the three different types of models and how they can be used.

- Identify the four categories of physical models and contrast them in terms of their use.

- Give two reasons why someone might develop appearance models.

- What is the difference between a model and a prototype?

- Compare some of the advantages and disadvantages of computer models versus physical models.

Resources of Technology

Introduction to Unit II

By now, you have some understanding of the process by which technology is developed (the design and problem-solving approach). In this unit, our examination turns to some of the content of the technology. Over time, we have developed an understanding of some of the elements that go into solving problems and producing things that humans want and need. These ideas can be used to increase the effectiveness of any design that is made.

In order for any technological event to occur, resources are needed. This unit identifies these resources and helps you understand the basic aspects of each. Chapter 4 describes various tools, devices, and machines. All tools and machines are alike in that they have systems for guidance and control, for support and structure, and for energy transmission. Tools are used to extend human capabilities and do the work that we want. They can be used to convert materials, information, and/or energy. Several aspects of machines are examined in this chapter so that you can make better decisions about tools as you implement your designs.

Materials come from natural sources or have been converted to be more usable. The structure, characteristics, and functions of materials are treated in Chapter 5. Materials can be designed and made for specific uses. Often, they are combined into material systems. Choosing an appropriate material for your design requires that you understand some of the characteristics of the material and the effects of combining different materials to achieve the results you want.

Chapter 6 looks at the forms of energy found in the energy spectrum and the sources of energy. Energy is converted into usable forms in order to accomplish the work that we want. It can also be sampled to produce information used to control a technological event. Renewable and nonrenewable materials are used to produce energy.

Each design and problem-solving activity requires information. Information is needed for decision-making and the guidance and control of tools and machines. Chapter 7 examines the use of signals and symbols and symbol systems. Signals are required in order for humans to communicate with machines and for machines to communicate with each other. Many symbols and symbolic systems have been developed over time by humans as we try to communicate with each other.

Humans guide technological events and often supply the necessary information and energy. Several aspects of human dimensions, development, and motivation are discussed in Chapter 8. Understanding more about the physical, emotional, and social development of humans can be helpful in designing technological products and services. The results are more likely to be what you intend and to be user-friendly.

The resources of time, space, money, and labor are reviewed in Chapter 9. Management of all the resources used in developing a product or service requires trade-offs. Decisions often involve questions about how much time, money, and space are available for producing the design. When the product is made in quantity, decisions about acquiring resources, using them efficiently, and maintaining an adequate supply are part of the management process. Aspects to consider when working in a production group are explored in this chapter as well.

Once you have studied both the process of technology and the resources required, you will know more about the technological environment. Also, we hope that you will be better informed for making choices between resources and using them in your own designs, as well as in the choices you make as a consumer. Later units will expand on this awareness as we look at the systems of technology, and the consequences and future of technological development.

Tools, Mechanisms, and Machines

Figure 4.1 ■
Tools, mechanisms, and machines are used in every aspect of technology. (© 1992 Dale O'Dell/The Stock Market)

INTRODUCTION—DEVICES TO EXTEND OUR ABILITIES

• •

This chapter can be very important to you. Whether you carry out a design or otherwise engage in a technology activity, tools, mechanisms, or machines are involved. These key resources of technology are used in every technological activity.

77

KEY TERMS

accumulator
air motor
bell crank lever
belts
bevel gears
cam
cam shaft
chain
compressor
control system
crank
cylinder
driven gear
driving gear
energy conversion
flow control regulator
fluidics
force
fulcrum
gear
gear reduction
gear train
generator
hydraulics
information conversion
lever
linear motor
linkage
load
lobe
magnetic poles
materials conversion
mechanical advantage
mechanisms
mitre gears
moment
pneumatics
pulley
rack and pinion gears
reservoir
solenoids
sprocket
spur gear
structure and cover system
subsystems

From the beginning, people have used tools to make work easier and to improve their lives. Archaeologists who study prehistoric remains look for evidence of tools to know if humans once lived there.

Some scientists believe that it was the use of tools that helped early humans develop more advanced thinking and complex skills. Although some animals use sticks or stones as simple tools, humans are the designers and developers of a vast array of tools, mechanisms, and machines. You are gaining new knowledge and skills in designing and developing technological products, systems, and environments. Learning to use tools, mechanisms, and machines will be an important step in that process.

TOOLS EXTEND OUR ABILITIES

Why do people use tools? Tools, including mechanisms and machines, help people extend their natural abilities. A simple tool such as a rake or a flyswatter can increase our ability to reach. Eyeglasses, electron scanning microscopes, telescopes, and television sets extend our ability to see. Optical fibers can be inserted into the human body through natural openings or small incisions. These sensors can allow a physician to see inside the body or make a diagnosis of the blood and other body fluids. Surgery may be accomplished by sending a laser beam down the optical fiber.

Photos taken from a satellite can be used to sense troop movements in a battle or patterns of deforestation from logging operations. Hearing aids can be placed in your ear. "Bugging" devices used by spies and radio telescopes turned toward outer space are other tools used to increase our hearing capability.

Tools are also used to extend our abilities to smell and taste. Smoke detectors and heat-sensing devices are special electronic tools that give us early warning of overheating or fires. Some tools extend our ability to touch. Thickness gauges, sensors, and other measuring devices help us determine size, weight, and volume much more accurately than we can with our unaided senses. There are tools to detect movement and tools to protect our bodies from many kinds of discomfort and danger. We use tools to live more enjoyable lives.

The computer has had an enormous effect on people and how they live and work. Some say it has had more impact on people than did the revolutionary development of the wheel and the steam engine. Computers and electronic calculators are machines designed to augment our thinking and our ability to make logical decisions. Continued developments in microprocessors have led to smaller and more powerful electronic devices. Tiny bionic parts for humans are now made and attached to the body. Attaching these parts requires intricate operations, using other specialized tools. Currently, extremely small machines are being developed that may someday be released inside the human body to do selective maintenance and repair on our internal organs.

The term "tools" is used here as a broad term or category that includes not only hand tools, but all types of mechanisms (mechanical, fluidic, optical, and electrical). The term tools also includes machines. Machines usually are comprised

**toothed belt and pulley
   system
torque
valve
velocity ratio
windlass
worm gears**

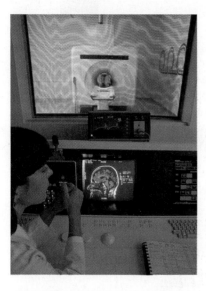

■ **Figure 4.2**
The scanner, as a tool, extends our
ability to see inside objects, like
our bodies. Scanners, operating in
this manner, use different energy
sources, such as X-ray, sound, and
nuclear energy. *(© David Joel/Tony
Stone Images)*

of several different mechanisms and are, consequently, more complex than
simple tools. In this book, all of these words will be used to indicate the same
thing—tools which extend our capabilities. Figure 4.2 demonstrates a scanner
being used as a tool.

How Tools Work

People often use tools without knowing how they work. Today, many technical
problems are solved through the use of tools and machines, but if you ask the
operators how and why a machine works, they may not know. They only know
that it works! This has been true for a long time.

For many centuries, tools were often developed before the underlying principles
were understood. Usually, it was not necessary to understand why the tool
worked, only that it did. Later, when people wanted to improve a tool, it became
more important to know why it worked as it did. Only recently have the scientific
principles been discovered first and then tools designed to apply those principles.
Early humans, for example, used stones to crush nuts, to stun or kill animals or
enemies, and to build shelters. This was long before they understood the
principle of kinetic energy, trajectories, or gravity.

People began to understand the "whys" of tools and their operation through trial
and error and later, through scientific experimentation. Such knowledge allows
us to be much more effective as we develop new tools and other products of
technology. The laser is a modern example of this. The laser (which stands for
light **a**mplification through the **s**timulation and **e**mission of **r**adiation) was
developed by scientists studying light. From that study, the scientists gained an
understanding of the laser and how it could operate. Then they developed a
functioning laser. After this time, some practical uses of the laser were identified
and laser tools were developed.

TOOLS AND MECHANISMS
• •

Tools and mechanisms are designed to increase our ability to do work. Tools may
be very simple or very complex. Even complicated machines can be broken down
into simple parts. All mechanical tools function using only a few basic principles.
The first of these is the principle of the **lever**. The concept applies to the wheel,
the wedge, and the other simple machines. (Refer to Figure 4.3.)

Levers

The principle of the lever is a key element to understanding tools, mechanisms,
and machines. More than two thousand years ago, the ancient Greeks identified
the importance of the lever. They showed how it could be used to make other
basic devices. It is interesting to note, as the Greeks did, that all mechanical tools
are built on the concept of the lever. That suggests that all machines spring from
the same ancestor. If so, then all machines are related, and understanding a few
machines could help you understand all machines.

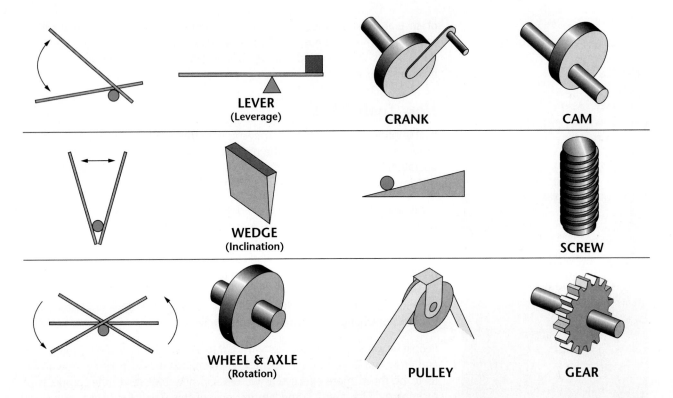

LEVER
(Leverage)

CRANK

CAM

WEDGE
(Inclination)

SCREW

WHEEL & AXLE
(Rotation)

PULLEY

GEAR

Figure 4.3 ■
By moving a lever around a fulcrum in different ways, you can transform it into all of the other simple machines.

The lever, an amazingly simple device, involves four related ideas: **load**, **fulcrum**, **force**, and **moment**.

(a) The load is the force to be moved by the lever.

(b) The fulcrum is the pivot or the point around which the lever turns.

(c) The force is the effort that is applied to the lever.

(d) The moment is the magnitude of the turning force.

The three components of the lever can be arranged in three different orders, as shown in Figure 4.4.

The lever is a very common mechanism that provides a mechanical advantage, such as in the "claw" at the back of a carpenter's hammer. Have you ever tried to pull a nail from a board with your hands before you were able to find a hammer? Were you surprised at how much easier it was with the hammer? The pop-top on beverage cans is a lever. Have you ever tried to open a can if the "lever" is broken? The hammer and the pop-top opener are devices that incorporate a lever. Both provide you with a **mechanical advantage** that makes the task easier to perform.

In using a lever, you decide on the placement of the components. This means that you will choose where to put the fulcrum (pivot), where the load will be placed, and where you will apply the force (effort). The "class of lever" designation of 1st, 2nd, or 3rd is a shorthand way of referring to the arrangement of the components. The different classes of levers allow you to choose a different mechanical advantage to accomplish the job to be done.

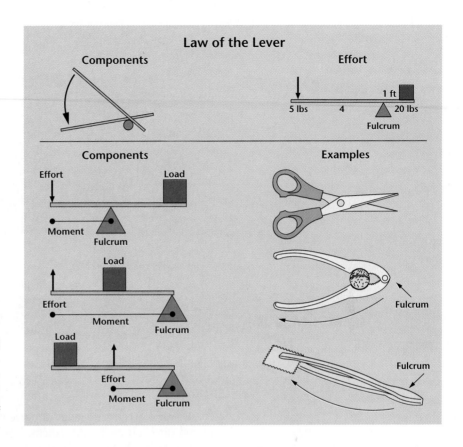

Figure 4.4 ■
The Law of the Lever in practice. The components of the lever (load, force, fulcrum, and moment), when placed in different order, produce the different classes of levers.

You may, for example, want a large mechanical advantage to make it easier to move a heavy load over a short distance, as in lifting a heavy stone into place for building a stone wall. This problem would be quite different than using a fishing pole to cast a lure out into a lake, representing a smaller mechanical advantage. (A similar lever could be devised to throw a stone, attached to a long wire, across a stream or gorge to set up a telephone line.) In the sections that follow, you will see many uses of tools, mechanisms, and machines. Your *Portfolio and Activities Resource* includes several opportunities to develop devices of your own design.

If you have ever tried to remove a flat tire and found the lug nuts were stuck, you may have used the principle of the lever. You cannot remove the nuts with just your fingers, so you use a lug wrench. If the nuts won't budge when you apply all the force you can to the end of the wrench, maybe you tried standing on it (to apply more force). A longer wrench would help increase the leverage.

Bell Crank Lever

The *bell crank* lever is a special example that is worth mentioning. The bell crank is a right-angle, Class 1 lever that allows force and motion to be transmitted through a right-angle. The name bell crank came from the device illustrated in Figure 4.5A that was used to ring a bell to call servants. The effort of pulling the handle down moves the load (the clapper of the bell) to make the bell ring. Bell cranks are still used when you want to use force applied in one direction to cause force and movement in another direction. Another more familiar example can be found on the brakes of a bicycle. (See Figure 4.5B.)

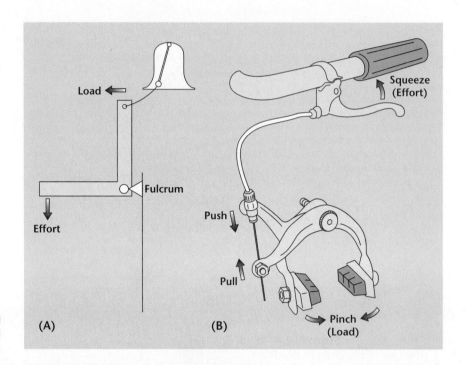

Figure 4.5A, B ■
(A) The bell crank lever and
(B) how it is used in bicycle brakes.

Linkages

Often, we want a complex series of movements, some of which may involve a change in direction of the movement. Linkages are special cases of levers that allow us to create force and motion at one point and transmit that force and motion to another point. The purpose of a **linkage** is to transmit forces and linear motion of appropriate and manageable size from a force applied as an input.

If you open the door on a toaster oven and the tray slides out, there is probably a linkage involved. Playing a piano involves applying a force to keys (hopefully, the right ones). The keys are linked to hammers that strike wires of different lengths to produce sound as musical notes. Linkages are also involved when the windshield wipers are turned on in an automobile. Another more visible example of linkages is found in a baby stroller that can be folded up to make it easier to store or transport.

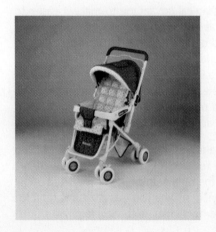

■ **Figure 4.6A**
This baby stroller makes interesting use of linkages, allowing the device to be folded for easy transport and storage. *(Courtesy of Kolcraft Enterprises, Inc.)*

D E S I G N   B R I E F

Using the device and components illustrated in Figure 4.6B-E as a starting point, design and assemble linkages that will provide:

• an output force that is greater than the input force,

• an output motion that is greater than the input motion,

• an output force that is equal to and in the same direction as the input force, and

• a rotary output motion from a linear input motion.

(B)

Input Motion

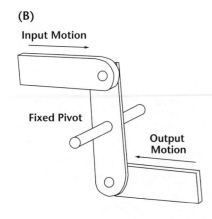

Fixed Pivot

Output
Motion

MATH/SCIENCE/TECHNOLOGY LINK

Using the devices that you designed in the above activity, show how you would modify them to achieve:

- an output force that is twice as great as the input force,
- an output motion that is three times greater than the input motion, and
- an output force that moves in the same direction, but is one quarter the input distance.

(C)

Input Motion   Output Motion

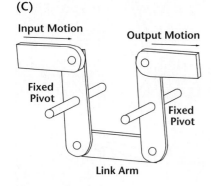

Fixed
Pivot

Fixed
Pivot

Link Arm

(D)

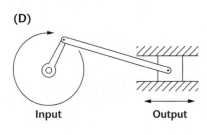

Input   Output

(E)   Effort

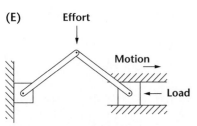

Motion

Load

■ **Figure 4.6B, C, D, E**
Some examples of linkages: (B) reverse motion linkage, (C) push-pull linkage, (D) treadle linkage (crank and slider), and (E) toggle linkage.

Wheels and Axles

Imagine a lever on its fulcrum or pivot going around in a complete circle. Now you have a wheel, and the axle represents the fulcrum or pivot. The wheel and axle can take many forms used to create and transmit circular motion. Variations of the wheel and axle include belts and pulleys, sprockets and chains, and gears, as illustrated in Figure 4.7.

In some mechanisms, the wheel turns on the axle. In these cases, the axle provides the support for the wheel, which can then turn freely on the axle. If energy is to be transmitted through the axle to the wheel or through the wheel to the axle, they must be joined. An example of this is the **crank** or **windlass** found in the steering wheel of a car. The steering wheel is fixed to a rod (the axle) that extends down through the steering column. This allows the force exerted by the driver to the steering wheel to be transmitted through the steering linkage. This force, consequently, controls the front wheels.

The windlass is used for raising water from some wells. Energy is exerted on the handle to cause the axle to rotate. The rotating axle, in turn, winds up the rope and lifts the bucket of water (the load). The mechanical advantage of the windlass is determined in part by the diameter of the wheel in relation to the diameter of the axle. The leverage provided by a crank is determined by the length of the arm of the crank and the diameter of the axle around which the rope is wrapped. A long arm (a large moment) provides a larger leverage; a short arm (a small moment) provides a smaller leverage.

To determine mechanical advantage (MA), it is necessary to consider the loss caused by friction. Calculating the MA of different devices involves comparing the effort it takes to move a load, or (LOAD/EFFORT). (See Figures 6.18A and B.)

Examples of windlasses that you may be familiar with include grinders, such as those used for pepper, coffee, or meat. This is not the electric kind, but one that has a handle that is turned round and round to move the grinding blades. A familiar and simpler example of a device that uses a crank or windlass is the hand-operated pencil sharpener. A pencil sharpener is designed to give you a mechanical advantage when you turn the crank.

Figure 4.7 ■
This example shows an historic example of a large lifting crane that uses the concept of the wheel and axle. *(Courtesy of George Bauer, from* De re Metallica.*)*

A—Reservoir. B—Race. C, D—Levers. E, F—Troughs under the water gates. G, H—Double rows of buckets. I—Axle. K—Larger drum. L—Drawing-chain. M—Bag. N—Hanging cage. O—Man who directs the machine. P, Q—Men emptying bags.

Belts and Pulleys

Belts and **pulleys** are slight variations of the wheel and axle and, as you will see later, are also very similar to gears. Essentially, the belts can link two or more pulleys that are some distance apart, as shown in Figure 4.8.

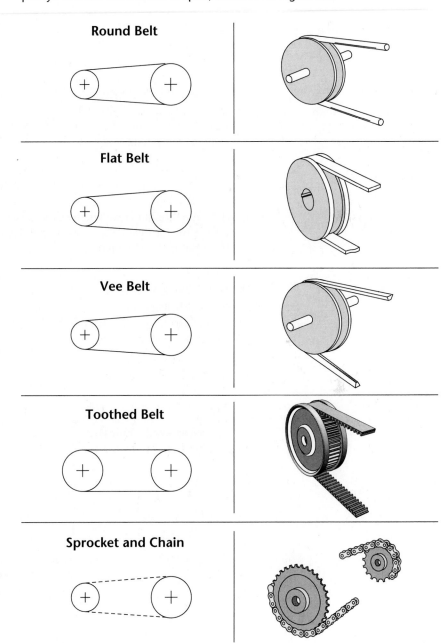

Round Belt

Flat Belt

Vee Belt

Toothed Belt

Sprocket and Chain

Figure 4.8 ■
There are a variety of devices that can be used for transmitting rotary movement from one point to another.

Safety Note:
Be careful when using mechanisms with pulleys, gears, and other moving parts.

Similar to wheels and axles, the main use of belts and pulleys is to create and transmit rotary motion from a driving shaft to a driven shaft. Belts and pulleys are generally cheaper to manufacture than gear mechanisms, and are used when relatively small forces are transmitted and when a small amount of "slip" between the driver pulley and the driven pulley does not affect the operation of the machine. For example, on devices such as vacuum cleaners and sewing machines, round belts that fit in grooved pulleys can be used.

The force that is needed to create a rotary motion is often called **torque**. In order to move an object in a circular direction, you must apply enough energy to overcome the resistance of the object to move. Applications that require higher torque, such as fan belts in a car or drive belts in a drill press, use vee belts and pulleys. This ensures there is little slippage and little loss of the energy applied to start the movement.

Early machines used flat belts to transmit power to the driven pulley. In some museums, you can see old factories and shops where flat belts were used to drive many different machines. Flat belts were also used on old farming equipment, such as a threshing machine. The belts looked like giant ribbons. The power from a large tractor (first steam driven, then later gasoline driven) was transmitted to the input pulley for the thresher.

On modern equipment, the **toothed belt and pulley** system is often used. The toothed belt is a variation and improvement of the flat pulley. Because the toothed belts do not slip as easily as flat belts, they can be used when higher speeds and torques are necessary. This system is used in riding mowers, washing machines, and some hand power equipment. Some automobile manufacturers are using toothed belts and pulleys in the timing systems of their engines and the power transmission systems of their cars.

Sprockets and Chains

Sprockets and **chains** are similar to belts and pulleys. They evolved as an improvement to belts and pulleys for transmitting larger forces while eliminating slippage. The pulley is replaced with a toothed wheel called a sprocket and the belt is replaced with a chain. At one time, chains and sprockets were used in trucks, but now their most common use is in bicycles and motorcycles.

Gears and Gear Trains

Gears have evolved to become the most effective way of transmitting rotational motion from one shaft to another. The many different types and sizes of gears provide considerable flexibility in designing **gear trains**. A "train" is formed by combining pairs of gears so that they mesh with each other. By using gears with a different number of teeth, the speed that the input shaft turns can be different from the speed of the output shaft. Depending on the requirements of the machine or mechanism being developed, gear trains can be used to provide high output speeds (with low torque) or low speeds (with high torque).

Spur Gears

Have you ever opened a can with a hand-operated opener? If so, you may have used a **spur gear**. The teeth of spur gears are cut on the outside edge of the gear and look something like the spurs that cowboys wear on their boots. Spur gears are set up so the teeth of the gears mesh. In this manner, the rotary motion of one shaft is transmitted to a parallel shaft, as shown in Figure 4.9. Spur gears are quite common and may be clustered or stacked to save space. In some instances, an idler gear may be used so that the **driving gear** and the **driven gear** can turn in the same direction.

The term **gear reduction** is used when you pair gears together in a gear train in order to slow down the motion. A gear reduction allows you to reduce the rotary speed of the output gear as compared to the rotary speed of the input gear. The relation of the two rotary speeds is called the **velocity ratio**. For example, when a bicycle is in a low gear, you will be pedaling very fast, but the back wheel will be turning rather slowly. In this case, there is a high velocity ratio, perhaps 5 to 1. If you are riding your bike in high gear, the pedal crank and the back wheel may turn at the same **velocity**. This would be described as a velocity ratio of 1 to 1. This can be important when a motor operates at a fast speed, but you want the work to be done slowly.

Bevel and Mitre Gears

Bevel gears and **mitre gears** are used when it is necessary to transmit motion through a right angle. The teeth of both bevel and mitre gears are usually cut at a 45° angle so when fitted together they make a 90° turn. Mitre gears are always the same size and when used make no change in the rotary speed. Bevel gears always have different sizes for the driving and driven gears. Bevel gears are used when the rotary speeds of the input and output shafts are to be different.

A hand-operated egg beater gives you a ready example of the use of bevel gears. Perhaps you have also seen an antique butter churn that uses the same types of gears. These simple machines take the force from one direction and apply it in another direction. In both examples, the beaters rotate faster than the speed at which the handle is turned.

Worm Gears

Worm gears are particularly useful when low speeds and high torques are needed. As shown in Figure 4.9, the input must come from the worm gear that is used to drive the spur gear (sometimes called the worm wheel). The input shaft turns the worm gear so that it continuously threads itself into the teeth of the driven gear. The simplest worm gear has one tooth that is similar to the thread on a bolt so that for each revolution of the worm gear, the worm wheel advances the distance of one tooth.

Some windows close by turning a crank. The window moves slowly on a worm gear mechanism, pulling the window closed so you can hook the latch (another lever). Most electric beaters use a worm gear to slow the high rotary speeds of the motor to the speed required for the beaters.

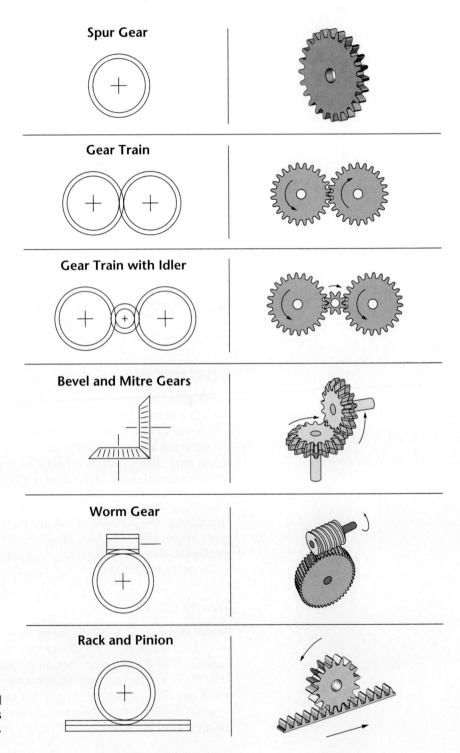

Spur Gear

Gear Train

Gear Train with Idler

Bevel and Mitre Gears

Worm Gear

Rack and Pinion

Figure 4.9 ■
Examples of several different gears
are shown here.

Rack and Pinion

Rack and pinion gears provide the capability of converting rotary motion to linear motion or from linear to rotary motion. The rotary motion of the pinion can cause the rack to move or the linear motion of the rack can cause the pinion to move. A rack and pinion gear is shown in Figure 4.9.

Cams

Cams are important mechanisms because they are able to change their motion into another type of motion. Cams are uneven in shape. The illustrations in Figures 4.10A and B show some different forms of cams and the followers they move. The shape of a cam is important. The raised portion on a cam is called a **lobe**. The lobe acts like a wedge and determines when the cam follower will move. The cam follower rides in contact with the cam and follows the shape of the cam as it moves. As the lobe moves, it causes the follower to move.

Many cams operate using a rotary motion. It may be helpful to look at these cams as a special form of wheel and axle. The lobe on the cam makes its follower move each time the cam turns. In an automobile engine, cams are used to control the movement of the intake and exhaust valves. The cams are machined from a shaft of metal that is connected to the engine by gears or pulleys and belts. This **cam shaft** controls the timing of when the valves open and close.

In the chapter on Information, you will see how the lobe on a cam actually contains information. Later yet, in the chapter on Control, you will see how cams can be used to control other mechanisms.

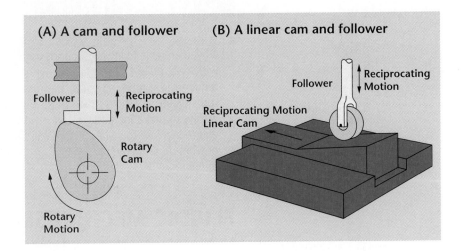

■ Figure 4.10A, B
Cams and followers can be used to produce linear motion from (A) rotary motion and (B) linear motion.

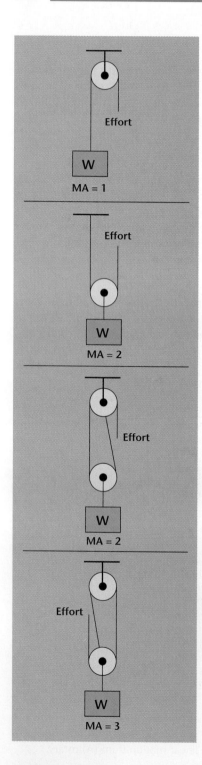

■ **Figure 4.11**
Pulleys can be used in different combinations to achieve different mechanical advantages (MA). (These examples ignore the loss of mechanical advantage due to friction.)

Pulleys

Pulleys, a special form of the wheel and axle, are used to lift heavy loads. Pulleys can also be used to change the speed of rotation. The major advantage of pulleys is their ability to magnify the force that is used to drive the system.

In Figure 4.11, several simple pulley systems are illustrated. As more pulley wheels are used, it is possible to lift a heavier load without increasing the input force. However, there is another problem that is introduced as more pulleys are used. The more pulley wheels you use, the more friction you encounter.

D E S I G N   B R I E F

Using pulleys, chains and sprockets, and/or gears, design and model devices that

- increase the output rotary speed of a fan relative to the input by a factor of two,

- increase the output rotary speed of a fan by a factor of ten,

- increase the M.A. of an elevator by a factor of ten, and

- increase the M.A. of an elevator by a factor of 200.

MATH/SCIENCE/TECHNOLOGY LINK

If you are able to lift 100 pounds using one pulley, you can lift 200 pounds with two pulleys, 400 pounds with four pulleys, and so on. This does require more time, however, because more pulleys require more loops of rope. Determine how much more rope you will need to pull through the pulleys to move the same distance with two and four pulleys. Will this allow you to increase the M.A.? Why? Will this allow you to increase the amount of work done? Why?

FLUIDIC MECHANISMS

A source of power that is used in machines involves force from compressed gases and liquids, or **fluidics**. Fluidics includes both pneumatics and hydraulics. **Pneumatics** deal with producing movement through the use of compressed air. When high forces are required, liquids such as oil, rather than air, are used. Using liquids under pressure is called **hydraulics**. Hydraulics are widely used in industry in machines that are used to manufacture products. Also, many of the products use fluidics to do work. Pneumatics and hydraulics are found in the brakes and

Plunger Sleeve

Pump
Cylinder Gasket
Plunger
Plunger U-Cup Seal
Check Valve

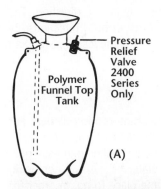

Pressure
Relief
Valve
2400
Series
Only

Polymer
Funnel Top
Tank

(A)

(B)

■ Figure 4.12A, B
The hand-operated pump and storage shown in (A) is a simple form of the electric air compressor shown in (B). *(A. Courtesy of RL FLOWMASTER, Lowell, MI 49331. B. © Michael Dzaman Photography)*

suspensions of cars and other vehicles we travel in, the automatic doors and elevators in the buildings we use, and many of the machines we use in our work and play.

This section deals with the different pneumatic mechanisms and how they operate. In later chapters, you will learn how to build and control pneumatic systems as you create systems of your own design. As a point of interest, the word "pneumatic" is derived from the Greek word for "wind" or "air."

Compressors

Compressors provide the source of compressed air required to operate pneumatic devices and systems. Essentially, compressors are pumps that push air into a storage tank. The air in the tank is available as needed. The force of the compressed air can be used to cause different pneumatic devices to move. We will return to the reasons that gas can be compressed and the principles involved in the chapters on Materials and Energy.

> **Safety Note:**
> Be sure to use pneumatic devices at low pressure settings.

Illustrated in Figure 4.12A is a simple compressor that, through a simple hand-operated pump and storage tank, can provide compressed air at low pressures. These are pressures of less than 30 psi (pounds per square inch) or 207,000 Nm^2 (newtons per square meter), the normal amount used in automobile tires. Electric-driven compressors and tanks similar to the one shown in Figure 4.12B can provide compressed air at high pressures. This pressure can exceed 100 psi (690,000 Nm^2). More detailed attention will be given to forces and pressures in the chapter on Energy.

> **Safety Note:**
> Every bike rider who has had a punctured tire knows that compressed air can produce a great deal of force. Filling a "doughnut" spare tire to the same capacity as a regular car tire can create a very dangerous explosion. You should wear safety goggles when connecting compressed air lines. Be careful not to point the air lines or direct a blast of air toward yourself or anyone else.

Air Motors

Air motors are essentially mechanisms that produce rotary motion from the force of compressed air. The air motor is actually simpler than the common electric motor that we find almost everywhere. Both types of motors rotate at relatively high rpms (revolutions per minute). An air motor works in a manner similar to a pinwheel. The harder you blow on the vanes of a pinwheel, the faster it turns. With an air motor, the harder the air from a compressor hits the vanes, the faster the motor will turn. Rotary air tools, like those used to remove or replace the lug nuts on car wheels, use air motors.

Cylinders

Cylinders are the most important of the pneumatic components for producing movement, especially for the products you will be developing in your design and technology activities. Cylinders are used to produce linear movement and force to accomplish the desired work. A bicycle pump is a simple pneumatic device that is very similar to a single-action cylinder.

Pneumatic cylinders are either single-action or double-action. The single-action cylinder is moved in one direction by air pressure; the return action is caused by a spring. The single-action cylinder can create force only in one direction. The double-action cylinder can be made to move with force in both directions. By using valves, air can be directed first against one side of the moveable piston and then against the other side.

Valves

Valves are devices used to control the operation of air motors and cylinders. For example, compressed air can be allowed to enter a cylinder by turning a valve to an open position. You may have turned on the gas for a propane barbecue grill. A standard on/off valve similar to those used for water or gas will not work, however, because the compressed air must have some place to go. In order for the air cylinder to operate, you will need to use a valve similar to the ones shown in Figures 4.13A and B. Each of these allows the trapped air to escape from the cylinders. More accurately, the valves allow the air to exhaust to the atmosphere.

A single-action cylinder can be controlled by a push-button, three-port valve. In the first position, pushing the button directs air from the compressor tank to the piston of the cylinder. This moves the piston to the extended position. The pressure will hold it there as long as the push button is held down. When the button is released, the air supply is cut off and the air in the cylinder is allowed to escape through the exhaust and the spring in the cylinder forces the piston to return to its original position.

Flow Control Regulators

Flow control regulators are used to adjust the amount of air entering or escaping a cylinder or other air-operated mechanism. Two kinds of control regulators are considered here: flow restrictors and flow regulators. The flow restrictor is a relatively simple device that squeezes down the opening through which air flows. This action is similar to pinching the neck of a balloon to slow down the escape of air. As shown in Figure 4.13B, the squeezing down (or opening up) action is accomplished by turning a threaded needle to adjust the amount of flowing air.

The flow regulator, also shown in Figure 4.13B, works as an unidirectional flow control valve. This means the flow of air is controlled in one direction while the flow of air moves freely in the opposite direction as though the control device wasn't there.

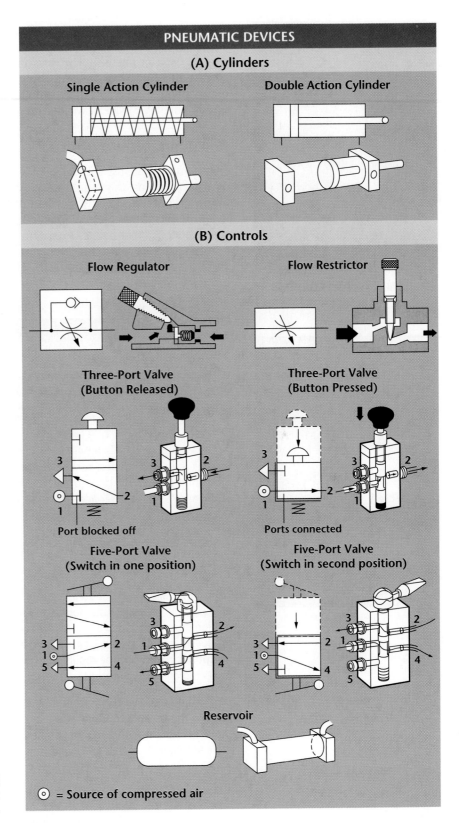

Figure 4.13A, B ■
Some of the (A) basic pneumatic
cylinders and (B) control devices.

(A)

(B)

■ **Figure 4.14A, B**
An animated robot that uses many pneumatic devices. *(Courtesy of Gene Poor, LifeFormations, Bowling Green, OH)*

This works because the internal, unidirectional valve allows a free flow of air in one direction only; in this case, to the left. If the air flows in the opposite direction (to the right), the internal valve closes and the air must pass through the opening created by the adjustable needle. This provides a controlled flow of air. The rate of flow through the regulator depends on the position of the needle. The rate of air flow entering a cylinder will determine how fast the cylinder will move.

Reservoir (Accumulator)

In pneumatic systems, it is sometimes necessary to create a time delay. The devices used for this purpose are called **reservoirs**, sometimes called **accumulators**. When air is directed through a pneumatic system, there is a slight delay as the air pressure fills the lines and builds to a point that will cause the cylinder or other devices to operate. The time delay can be lengthened by increasing the volume to be filled—that is where the reservoir comes into play. The reservoir in the circuit must be filled before the air pressure can build up and cause the cylinder to operate.

DESIGN BRIEF

You have been asked to help design a display that will include an animated figure that will move slowly back and forth in a slow and controlled fashion. Identify the devices from Figures 4.13A and B that you would use.

Design and sketch a pneumatic circuit using the above devices to create the back and forth motion needed for the display. Can you modify the circuit to introduce a time delay before the figure returns to its starting point?

ELECTROMECHANICAL MECHANISMS

Electromechanical devices use electricity and electromagnetism to create motion, and in some instances, use motion and magetism to create electricity. Creating motion from electricity is accomplished by harnessing the magnetic fields that are produced around a wire that is carrying electrical current. The attraction of unlike **magnetic poles** and the repulsion of similar poles provides a push and pull that will cause an electric motor to rotate.

Electrical current can be generated by moving a coil of wire through a magnetic field. As the wires cut through the magnetic field, electrical current is caused to move in the wire. This magnetic energy can be converted into electricity by moving magnetic metals and conducting wire through the magnetic field.

Safety Note:
Be sure to use electrical devices at appropriate voltages.

■ **Figure 4.15**
These small motors are useful in design and technology work. *(Photo by Michael Dzaman)*

■ **Figure 4.16**
This scanner uses a stepper motor to rotate the cylinder in small, controllable increments. *(Courtesy of Frank Cappelle, photo by Michael Dzaman)*

Electric Motors

Electric motors are normally used to create rotary motion. Some of the motors that are useful in design and technology project work are shown in the illustration in Figure 4.15.

Motors come in a wide range of sizes, power requirements, and power outputs, so it will be helpful to do some research before selecting a specific motor for a task you have in mind.

There are two basic types of motors. One type uses direct current (DC); the other type uses alternating current (AC). Most small motors are DC, while larger motors that produce more power are usually AC. The difference in the design of the two types of motors is the way that electricity is fed into the moveable coil to produce the rotary movement. The brushes, the parts that conduct the electrical current to the coil, remain in contact with the slip rings. The fluctuations in the alternating current cause a change in the magnetic field which in turn causes the coils to rotate.

Another type of motor that is useful for project work is the stepper motor. They are called stepper motors because they rotate a small amount (a step) each time a signal is sent to the motor. A stepper motor is made with many small coils of wire that can be energized one after the other. Each step is actually caused by energizing a coil so that the motor will rotate a small angle, usually .6 or 1.2 degrees. This means that one full turn of a stepper motor will require either 600 or 300 stepped movements. Stepper motors are fairly expensive, but are very useful in robotic applications and other devices that you might like to control by computer or other driver system. (Refer to Figure 4.16.)

One other type of motor is of importance for design and technology project work. This type of electric motor moves in a straight line and is described as a **linear motor**. Small linear motors are used to open and close window drapery. Although these motors are fairly expensive, they do provide a movement that is unique for an electric motor.

Solenoids

Solenoids operate by harnessing magnetic energy. When an electrical current is passed through a coil of wire, a strong magnetic field can be produced. A solenoid is the name given to a coil made of many loops of wire. Solenoids are made stronger by including a metal core. This core is made of iron that loses its magnetism quickly, called "soft iron." A piece of steel could be used, but usually is not, because the steel tends to become permanently magnetized and does not move as freely in the coil. The diagram in Figure 4.17 illustrates how the moveable soft iron core will move inside a solenoid.

Solenoids are used to create a short, linear movement electrically. A common use of a solenoid is in the door chimes used in many homes. The "ding-dong" that signals someone has pressed the button is caused by a solenoid that is first moved in one direction to hit a chime and then moved in the other direction to hit a second chime, usually by a spring.

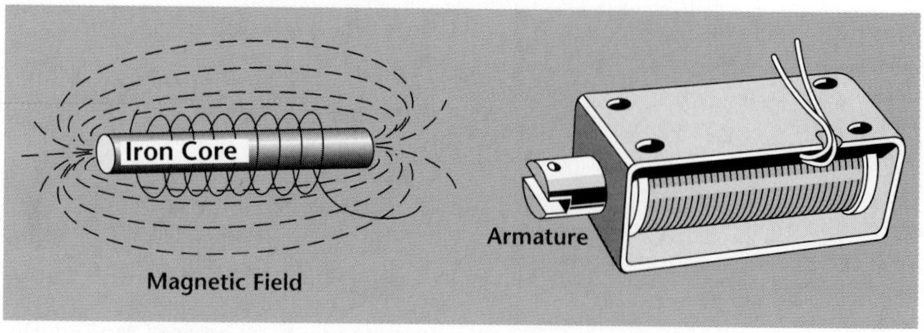

Figure 4.17 ■
Solenoids use magnetic force to create linear movement.

Iron Core

Magnetic Field

Armature

Another common use of the solenoid is in the starter system of automobiles. The solenoid is engaged by the flow of electricity caused when the key is turned to the appropriate position. The solenoid then creates the contact needed to complete the circuit between the battery and the starter motor. The starter motor must create high torque in order to crank and start the engine. The solenoid circuit allows a small amount of current to turn on the heavy-duty switch that can conduct the high-current circuit of the starter. Electric cars use solenoid controls because of the large current flows involved.

Electric Generators

A final important type of electric device to be considered in this section is the electric generator. Electric generators are similar in construction to electric motors in that each has a moveable coil. As stated earlier, an electric motor rotates because of the magnetic action caused by the flow of electricity through the wire loops in the motor. In an electric generator, the opposite effect is accomplished. (See Figure 4.18.) The moveable coil is turned in the generator by an outside energy source (muscle power, water power, wind power, and the like). The wires are turned within a magnetic field of the generator. This causes an electrical current to flow in a process commonly called "generating electricity." Try it with a small DC motor designed for toys or models.

■ **Figure 4.18**
This generator, mounted on a bicycle, is a simple version of larger machines. (© Michael Dzaman Photography)

MACHINES
. . . .

Early tools were designed to increase human capability, often to enhance human muscle power. The following statement is compatible with the definition of technology:

> Technology is the use of our knowledge, tools, and skills to solve practical problems and extend human capabilities.

Throughout history, humans served as the major source of energy. Later, people learned to attach animals to tools and machines so they could be controlled to do work. The water wheel and windmill were developed a long time ago as important sources of energy. However, the use of animals continued for centuries and is still a major source of energy in many countries worldwide today.

Tools and mechanisms were improved to become more efficient. Machines were developed by combining different mechanisms. The first machines were mechanical in nature. Progressively, they were made more complex to harness water and wind power and later, steam power.

Continuing improvements in production and use of energy were very important to humans. With the introduction of wind and water power and later, steam engines, electric motors, and other energy devices, people no longer had to depend solely on themselves or on animals to provide power. These work-saving devices were available to reduce human toil, if the user could afford to pay for them and the required fuel. Designs that use various approaches to increase mechanical advantage have continually decreased the amount of human energy required in order to do the desired work.

The Subsystems of a Machine

Although machines may look very different from each other, all are similar. All machines, from electric can openers to jet aircraft, have similar arrangements of parts that we will call **subsystems**. Within the following pages, these subsystems will be introduced. As you learn about these similarities, you will understand better how machines operate, how to use them, and even how to design and develop new ones.

To help us understand how machines work, let us look at a machine that should be familiar: the bicycle. Some parts of a bicycle support the rider and the bike. Other parts are used to cover the chain, sprocket, and tires. These parts make up the **structure and cover system** of the bike. The cover parts of the system provide protection for the rider and the parts of the machine.

Any machine, including a bike, must also have a system to make it operate. These parts are called the **energy transmission system.** The energy transmission system transmits energy from one place to another so the machine can do its work. This system usually provides the power, but on a bicycle, the rider does that.

The bicycle's energy transmission system uses the energy generated by the legs of the rider. The energy is transmitted from the pedals to the rear wheel and on to the ground, so the bike goes forward. Note that the rear wheel provides the support and transmits the power. Often, one part does several jobs, as you will see in other examples.

If a bike was designed with only the support and cover and the energy transmission systems, it would probably hold you up, and you could make it go forward by pedaling. But you would have some serious problems with no way to steer or to stop it. Brakes and steering devices are examples of a **control system.** In the bicycle, some parts provide guidance and control as well as support, cover, and energy transmission.

In summary, all machines must have a means to hold themselves up, to protect their parts, to move energy to where it is to be used, and to provide for their direction and control. In other words, all machines must have systems for structure and cover, energy transmission, and control. (Refer to Figures 4.19A-D.)

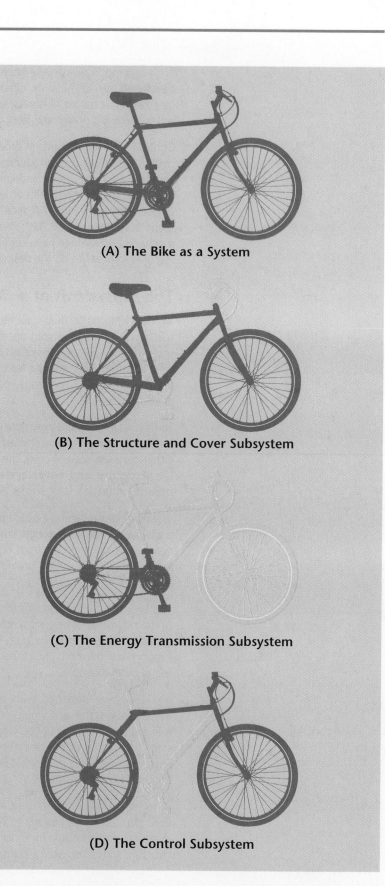

Figure 4.19A ■
This bike, like all machines, must have three subsystems. All machines must (A) hold themselves up and protect their parts, (B) move energy from a source to where it is to be used, and (C) provide for direction and control.

(A) The Bike as a System

Figure 4.19B ■
The *structure and cover* system of the bike must hold up the weight of the rider and the bike. It must provide protection for important parts of the machine and the rider.

(B) The Structure and Cover Subsystem

Figure 4.19C ■
The *energy transmission* system of the bike must move the energy from its source to the point where it will be used.

(C) The Energy Transmission Subsystem

Figure 4.19D ■
The *control* system of the bike must provide a means of guiding or directing the bike. It must control the bike's speed, as well as start and stop it.

(D) The Control Subsystem

The bicycle has gone through countless changes. The first bike had no pedals or brakes. What do you think it was like to ride one of these? The guidance and control system of the early bike was very primitive. Shown in Figure 4.20 is part of a very unusual, modern bike. You can be certain that this bike requires a very efficient energy transmission system, a very light support and cover system, and a very dependable guidance and control system. Why? Because it is a bike that flies!

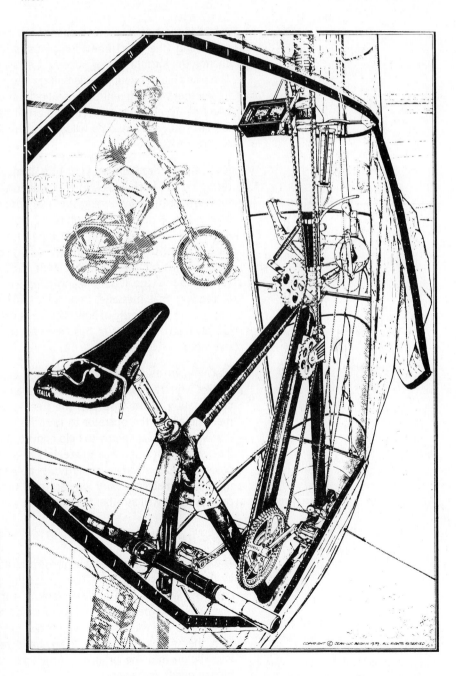

Figure 4.20 ■
Bicycles, like other tools and machines, go through many changes and improvements. This unique bike is part of the cockpit of the Gossamer Albatross. It converts energy in order to fly. *(© 1979 Jean-Luc Beghin. All Rights Reserved)*

Function of Machines

Another similarity of all machines is that they convert something from one form to a different form. The "something" that machines convert to different forms is energy, information, and materials. That is all that machines can do, nothing more and nothing less. If you wanted, you could adapt your bike to perform all three functions of converting energy, information, and materials into different forms.

A bike is most often used to change the rider's energy into energy for moving something (the rider) from one place to another. Because the bike is used to change the form of energy, it could be called an **energy conversion** machine. The bike changes your muscle energy into motion. You could also use your bike for another very different purpose. Let us imagine that you are on a long-distance bike ride. You have an accident and break a leg, making it impossible to ride any farther. You are alone on a wilderness bike trail, but there is a small town some distance away.

If you have a headlight on your bike, it may be operated by a small **generator** driven by the back wheel. In this situation, your bike could be used to send a message for help. By turning the bike upside down and cranking the pedals by hand, you can generate enough energy to operate your headlight. You can cover and uncover the light with your hand in a pattern of dot-dot-dot, dash-dash-dash, dot-dot-dot. The message (... --- ...) is an international signal for help. The signal stands for SOS in Morse code. SOS is said to mean "Save Our Ship." Perhaps in this case, it is more appropriate to have SOS stand for "Save Our Skin." In sending that message, your bike has been used to change your muscles' energy into the energy of light. You have used the light to send information (the call for help). The bike in this case was used as an **information conversion** machine.

Machines are also used to change materials. An ingenious person might combine a bike with an old-fashioned ice cream freezer to make ice cream. The combination of the bike and the freezer then becomes a new machine. This machine allows the operator to turn the freezer more easily than by hand. The bike uses muscular energy to help convert the cream mixture into ice cream. The bike is now operating as a **materials conversion** machine. In times past, or in places where electrical energy is not readily available, human-powered machines—machines very similar to a bicycle—are often used for materials processing as well as other kinds of work.

Looking Ahead

The ideas and skills that have been introduced in these chapters will be expanded in the next three chapters on materials, energy, and information. In those chapters, we will take a closer look at the different tools, mechanisms, and machines that perform the processes of materials, energy, and information conversion. We will also see how tools, mechanisms, and machines provide us with the means to convert these basic resources of materials, energy, and information into the products, structures, and environments that help satisfy human needs and wants.

SUMMARY
CHAPTER 4

All of the ideas covered in this chapter can be brought together in a diagram. Figure 4.21 is a diagram which describes what tools and machines are and what they do. You can see that humans, materials, energy, and information are all necessary for a tool, mechanism, or machine to function and reach a desired goal or output. These devices and machines go through a process that changes (converts) the energy, materials, and information into (1) some different form of material (such as a product, like a candy bar or a milk container), (2) a different form of energy (such as heat or light), or (3) a different form of information (such as a tape recording or a signal to another machine). The human element is involved in designing, operating, and making use of the machine.

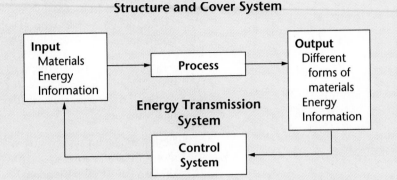

Structure and Cover System

Input
Materials
Energy
Information

Process

Energy Transmission System

Output
Different forms of materials
Energy
Information

Control System

■ **Figure 4.21**
This model includes the essential functions and systems of all machines.

All machines are similar. They are designed to help us reach goals using materials, energy, and information. Machines change these inputs through many different processes to meet the goals. The output of machines are new products, materials, energy, and information. For any machine to do its work, it must contain all three systems: (1) a support and cover system to hold it up and protect the machine itself, (2) an energy transmission system to get the energy to where the work is to be done, and (3) a control system that uses information to make the machine operate. Every part of a machine, no matter how complex, fits into this simple scheme.

ENRICHMENT ACTIVITIES

- You have been asked to build a seesaw that will allow young children to use it with larger children. Design a seesaw that can change the distance the seats are from the fulcrum to compensate for the different load exerted when a child who weighs 40 kilograms is on one end of the board and a child who weighs 30 kilograms is on the other end.

- A bicycle racing team is trying to design a new gear system for one of its bikes. One of their ideas is to replace the sprocket and chain with a toothed belt and pulley. Identify some of the advantages and disadvantages of each system.

- Design a gearing system that will lift a one-kilogram load using a small motor from a Lego™, Mecanno™, Lasy™, K'NEX™, or Fisher Technik™ kit. Can you lift two kilograms?

- Design a hydraulic system for a gripper hand on a robotic arm. The device must be able to grasp and hold a one-kilogram load while it is being raised to the height of one meter.

- Design a device that will allow you to slide a bolt on your bedroom door to unlock it without getting out of bed.

MATH/SCIENCE/TECHNOLOGY LINK

- In the seesaw design problem described previously, you determine through a survey that the average weight of the young children who will use the ride is 25 kilograms. The average weight of the larger children is 50 kilograms. You also determine that for the safety of the small children, their side of the board should not exceed three meters. What will be the required distance from the fulcrum where the larger children will sit?

50 kg times X = 25 kg times 3 meters, or

(where X is the distance in meters where the larger child must sit)

50 kg x X = 25 x 3

50X = 75

X = 1.5 meters

In the seesaw design problem, you decide to make the board six meters long and provide three places for the fulcrum. For the average weights of the children (25 and 50 kilograms), where must one of the pivot points be located?

- We use many tools and machines almost every day. In many cases, we don't understand how they operate. Conduct a survey at home and school to see what tools are available for you to use. Make sketches of the tools on small cards (3" x 5"). Write the name or function of the tool at the bottom of the card. Determine the value (or replacement cost) for each tool.

  Analyze each tool and determine in which of the six categories (lever, wheel and axle, wedge, pulley, windlass, screw) it belongs. Describe how each of the tools operates.

  Identify which of the tools you have used. Add them to the list in your *Portfolio and Activities Resource*.

REVIEW AND ASSESSMENT

- Identify the contribution that tools, mechanisms, and machines make for people.

- Identify how the different classes of levers are created using the three ideas of load, fulcrum, and force.

- Identify at least three uses of each of the classes of levers.

- Identify and sketch some of the practical uses of linkages.

- Identify the different rotary devices presented in this chapter and describe some of the advantages and disadvantages of each.

- When would hydraulic devices be used rather than pneumatic devices?

- Compare the operation of an electric motor and a solenoid. How are they different? How are they similar?

- What are the subsystems of a machine and what role do they perform?

- What are the different functions of a machine? Give at least five examples of machines that perform each of the functions.

CHAPTER 5

Materials and Materials Processing

Figure 5.1 ■
Materials, whether simple or complex, are the basic ingredients of technology. (© Gill G. Kenny/ The Image Bank)

INTRODUCTION—THE BASIC INGREDIENTS OF TECHNOLOGY

Materials are important for our survival and for realizing our dreams. People use materials to make things that are needed for safety and comfort. People use materials to feed themselves, build shelters, raise cities, and journey across oceans and into space.

105

KEY TERMS

acoustical processes
action
addition process
adhesion
alloying
alloys
amorphous materials
atoms
attraction
basic materials
casting
ceramics
chains of molecules
changes
characteristics
chemical processes
coating
cohesion
combustion
composite compounds
conversion
crystals
dynamic materials
electrical processes
elements
energy conversion
expandable materials
extruding
families of materials
fastening
ferrous metals
firing
forming
functions
functional materials
gas
green
information conversion
interlacing
joining
kiln
knitting
layering
liquid
magnetic processes
material layers

To understand technology, we must also understand materials. Materials are used to build our machines. Materials are then changed by some of these machines to make products. Materials are used to supply energy that we need for production, transportation, and communication. Materials are the basic ingredients of technology. Sometimes materials are fairly simple and natural, such as soybeans, which provide oil and food. Materials may also be complex and manufactured, such as a custom-designed ceramic for shielding a space vehicle.

Early humans had only natural materials to work with. Later, manufactured or **synthetic** materials were developed. Often, synthetics are combined with natural materials. T-shirts, for example, are sometimes made with a blend of polyester (synthetic) and cotton (natural).

Many of the materials we use today are manufactured or processed. To make these synthetic materials, energy and raw materials are necessary. The wide variety of materials from which to choose increases the complexity of decision-making when designing a product. We spend a great deal of effort trying to get materials into the form we want, and maintaining that state over time.

There are many properties, forms, uses, and varieties of materials. It helps us to group the almost endless variations of materials. Grouping helps us talk about materials and understand them more easily.

One of the purposes of this chapter is to help you develop a way of looking at categories of materials. This will help you understand and use materials wisely. We will also show how some materials take on different properties when two or more materials are combined. Finally, we will explore what happens to materials when they are used. The four key concepts in this chapter, therefore, are (1) the **characteristics** of materials, (2) their **structure**, (3) their **function**, and (4) the **conversion** of materials to change the characteristics, structure, and function.

CHARACTERISTICS OF MATERIALS

There are thousands of materials available now that were unknown only 50 years ago. New ones are being developed almost every day. If a new job calls for a quality that existing materials do not have, researchers try to make a new material to fit the need. Astronauts' food, for example, looks very different from the ordinary food you might find in your refrigerator.

The product's function and special requirements determine what material should be used to make the product. Land vehicles such as automobiles must be strong and relatively light. The strength protects the passengers, while the lightness is needed to make it easier to move the vehicle. A construction crane, on the other hand, must be very strong and it must be heavy enough to balance the weight of the load being lifted. An airplane must be both strong and light.

To make women's stockings, hosiery manufacturers may choose a fiber that can be heat-set to retain a given shape. To make athletic socks, the manufacturer may choose a different fiber, one that absorbs and cushions. Once, cotton was

KEY TERMS

material mixtures
material systems
materials conversion
mechanical fastening
mechanical processes
metal
metamorphic materials
mixing
molding
molecules
monolithic materials
natural materials
optical processes
penetration
platelets
pliability
polymer
pressing
process
properties
repulsion
results
sawing
separation processes
sequence
shearing
solid
structure
synthetic
textile
thermal energy
thermal processes
vitrification
weaving
wedge
wood
zero force

considered the best fabric to choose if you wanted to stay cool. Textile systems developed for athletic clothing today include fibers that stretch and move with the person without being bulky or constraining. In addition, fibers are woven to wick the excess heat and moisture away from the body.

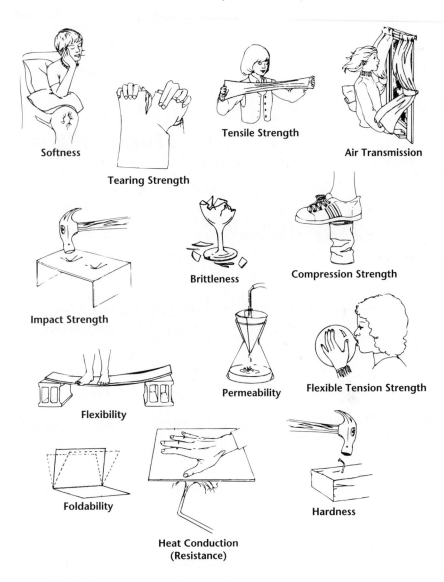

■ **Figure 5.2**
The choice of an appropriate material must consider the characteristics of the material.

Often when you are talking about someone you have met, you identify that person as being dark-haired, heavy, friendly, strong, noisy, or any of several other descriptive terms. Candidates for a space exploration program must have characteristics such as stamina, intelligence, technical knowledge, and an ability to adapt to new conditions. When we talk of people, we often speak in terms of their **characteristics**.

Materials, too, have characteristics. Within any group of materials, some can be identified as soft, rough, heavy, or brittle. Figure 5.2 shows various materials and the characteristics which permit them to be used for making different products. Engineers call these characteristics of materials **properties.** Peanut brittle, saltwater taffy, and fudge all have some very different properties, yet each is sweet. Many of the types of woods that we use are relatively soft and easy to cut and finish. Metals are generally hard and will polish to a high gloss. Leather, however, is both soft and tough. Softness and hardness are properties of materials. Other examples of properties are brittleness, density, flavor, color, odor, and ability to conduct heat.

THE STRUCTURE OF MATERIALS

Materials often look very different from each other, yet in some ways all materials are quite similar. To understand what makes materials similar, you must use your imagination.

MATH/SCIENCE/TECHNOLOGY LINK

Imagine you are stepping into an incredible shrinking machine. Each minute you are in the machine, your size will be reduced to one-half of what it was the minute before. After you are in the machine for a minute, those of you who are about one and three-quarter meters tall (5' 8") will be seven-eighths of a meter (2' 10"). We plan to keep you in this machine until you are small enough to walk around inside a sheet of aluminum. You will become so small that you will be able to stand between the atoms. If you continue to shrink at this rate, it would take you nearly 33 minutes to be small enough to fit between the atoms of aluminum. If you had been growing at the same rate, you would be more than two million miles tall and would bang your shins by tripping over the moon. Inside the aluminum, it would probably look much like the sky at night because of the great expanse of space in the materials.

Much of our information about materials at the atomic and subatomic level has come from scientific investigation and theory, as well as technological developments in tools and machines. The scanning tunneling microscope lets researchers create an image of the electron clouds that surround atoms. The atoms are held apart and in position by forces of **attraction** and **repulsion**, somewhat as the planets and stars are held in space. These forces can be compared to the way two magnets will attract each other or repel each other, depending on how you hold them.

To help you understand the nature of materials, we will give you a brief tour of the **atoms** and **molecules** of a few materials. Molecules are combinations of atoms that are linked together. We will consider how some atoms link together.

We will also consider what shape molecules might take as they combine to form a material.

The atomic or molecular structure of a material determines the properties or characteristics of the material. If you were to buy a knife for carving wood, you would not need one with the same molecular properties as a knife used by a surgeon for an operation. The wood you select to carve or the wood you place in the ground as a fence post would be different in molecular structure from the wood used for constructing a fine violin.

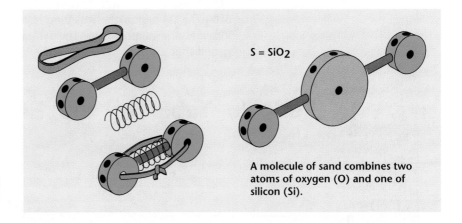

S = SiO_2

A molecule of sand combines two atoms of oxygen (O) and one of silicon (Si).

Figure 5.3A ■
These models represent the bonds between atoms in the molecules.

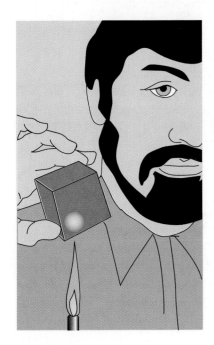

■ **Figure 5.3B**
Insulation tiles made of silica (SiO_4) shed heat quickly. A hot tile can be picked up at the edges with a bare hand, just after the tile comes from an oven at 2000°F.

A simple model of the atom and atomic structure might help us. (See Figure 5.3A.) Here the atoms are represented by small balls. The rod represents the distance between the atoms. The rods would be invisible in a real set of atoms. The rubber band represents the attraction between atoms, while the spring represents the forces of repulsion between atoms.

Our model gets very complicated if all the lines of attraction and repulsion are shown. We will use the simplified model, knowing that each rod represents attraction and repulsion as well as the distance between atoms. Even though our model is rigid, remember that the atom is pulsating. The push and pull keep the particles in constant motion.

The rods are important parts of our model. They represent the bonding energy that must be overcome to separate molecules of materials. When we cut metal or fabric, or when we burn the materials, we are breaking the bonds between molecules.

Most people use materials without knowing about the atomic structure. As people try to solve more difficult problems and synthesize more materials, they must study materials in greater depth.

Compounds

Some materials, which we classify as chemical **elements**, are made of only one type of atom. Copper is made only of copper atoms, for example. There are many chemical elements. A few other familiar elements are gold, silver, iron, carbon, oxygen, helium, sodium, neon, and uranium.

Most materials are combinations of elements. These are called **compounds** and are made up of many molecules. Each molecule, however, is made of atoms of specific elements. All molecules in that compound are alike.

Sand, for example, is a compound made of silicon and oxygen atoms. One silicon atom collects two oxygen atoms to form silicon dioxide, a molecule. Many molecules of silicon dioxide are in each grain of sand.

It is quite clumsy to spell out the names of the elements and compounds, so chemists have developed a shorthand system of symbols. Silicon is always designated as **Si.** Oxygen becomes **O**. The shorthand name of silicon dioxide is **SiO_2**. The **O_2** designates two atoms of oxygen. "Di" is the standard prefix which indicates the number two. The symbols for the elements are shown in the periodic table below.

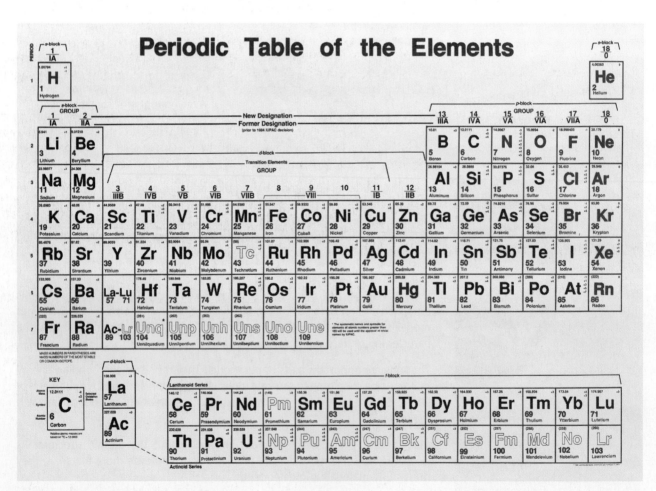

Courtesy of Sargent-Welch Scientific Company

Common table salt is a compound made up of sodium **(Na)** and chlorine **(Cl).** Each molecule of salt contains the atoms **Na** and **Cl** combined in the same way. It is interesting to note that both Na and Cl are, by themselves, poisonous if eaten. When the NaCl molecule is formed, it becomes a material vital to human survival. If someone calls you "the salt of the earth," they are complimenting you. Salt is so valuable it was used in earlier times as a medium of exchange (money).

Over the years as more and more elements were discovered, scientists compiled a table of the names and properties of the elements. All materials, natural and synthetic, are formed by combining these basic elements. Some of the combinations are rather simple in nature, others are more complex. Refer to the Periodic Table of the Elements on the previous page.

For example, copper atoms **(Cu)** on a copper surface often combine with the oxygen **(O)** in the air. This combination makes molecules that appear as a dark film on the copper object. Some people spend a lot of time polishing their copper goods to remove this film of oxidation. Copper is sometimes used for roofing. It is not cleaned, but allowed to build up a coating of oxidation. The coating will eventually turn green and become harder than the copper itself. The Statue of Liberty was cleaned a few years ago. Once the layer of pollutants was removed, the green copper oxide was more apparent.

Atoms and molecules (material units) combine in relatively sensible ways. Some material units have almost no attraction for each other. Some have tremendous forces of attraction. Other material units fall somewhere between these extremes. A simplified set of connections is shown in Figure 5.4.

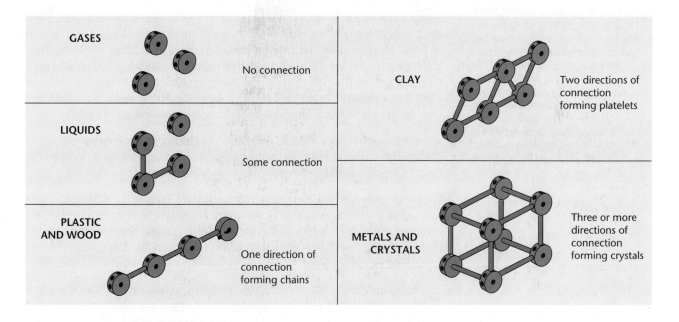

Figure 5.4 ■
These models show the structure of materials. The amount and direction of connections determine the structure.

In a **gas** such as air, each material unit (molecule) has almost no attraction for other units in the gas. Because of the space within the gas molecule, gases can be easily compressed. There is more attraction between the material units in **liquids**. The attraction is greater in liquids that have greater density, such as mercury. Have you ever noticed how hard it is to separate a blob of mercury into smaller blobs? Generally, the molecules of **solid** materials have the greatest attraction for one another.

Material units may combine in one, two, three, or more directions. These connections and the strength of these bonds determine the properties of every material.

Chains of Molecules

The basic building units of materials can be combined in different ways. Some materials are made of units connected in long chains. These chains are held together by the molecular forces of attraction.

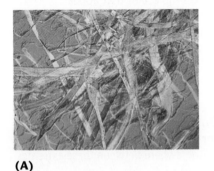

(A)

(B)

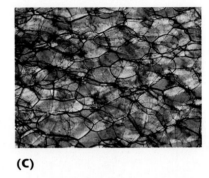

(C)

Figure 5.5A, B, C ■
The structure of the bonds formed within materials becomes evident at the microscopic level. Shown here are (A) paper made from wood, (B) metal, and (C) polyurethane packing material. *(A. Courtesy of Astrid & Hanns-Frieder Michler/Science Photo Library/Photo Researchers Inc. B. © Manfred Kage/Peter Arnold, Inc. C. © Alfred Pasieka/Peter Arnold, Inc.)*

Wood is one example of this. It is made largely of long cellulose molecules. The basic units of the giant cellulose molecule are glucose molecules. Several hundred of these glucose molecules may be joined to make a cellulose molecule. New molecules are formed each year as a tree grows, producing wood as a form of natural polymer. The cells of wood and the long cellulose molecules are similar to a bundle of drinking straws held together by paste. The straws can be bent, but they do not break easily. They can be separated lengthwise quite easily.

A second example of a material that has long **chains of molecules** is polymer plastic. The long chains of the synthetic material are very similar in form to the long molecules of cellulose in the natural **polymer**, wood.

The long chains of polymer plastic molecules are created by linking together many simple molecules over and over again. Polyvinyl chloride, for example, is a common plastic used in tires and cassette tapes. It is made of many vinyl chloride molecules. Our toy model may help again to explain the basic vinyl chloride units. (See Figure 5.6.)

The basic building block of this plastic is the material unit $CH_2 CHCl$ which is sometimes called a "mer" or **monomer**. When many monomers are connected to make a long chain, the term "poly," which means "many," is added. The result is the more commonly used term "polymer."

Plastics vary in the way that the chains are formed and in the strength of the bonds that hold them together. Plastics may be tough, fragile, fireproof, easily burned, light, or dense. These qualities depend on the way the molecules are linked together.

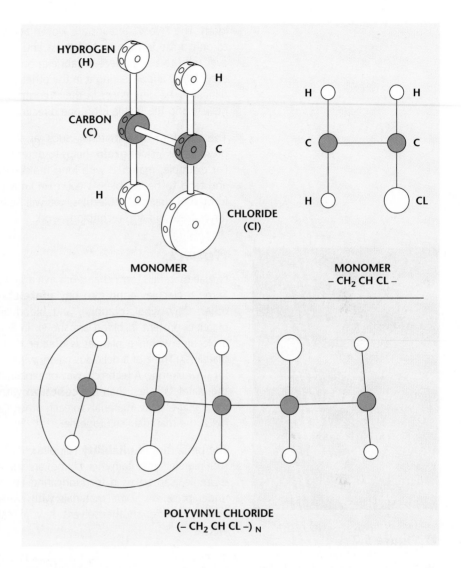

Figure 5.6 ■
A model of a "monomer" and the combination of "monomers" to make a polymer.

These properties become important when choosing a material. You wouldn't use Styrofoam (often used in cups) if you wanted to make a table or a chair. Styrofoam is used for "one-time" crash helmets for cyclists because of its capability of absorbing energy of a crash or impact.

Because polymers are long molecules, they tend to be strong and to have some "give" when pulled lengthwise. These plastics are not nearly as strong if they are pulled in the other direction, so that the long chains separate from one another. You can prove this by taking a plastic bag and trying to tear it into pieces. Try it first in one direction and then try to tear it at 90° to the first tear. This weakness is similar to the weakness of wood when you split it along the grain, between the cellulose molecules. As another example, woven textiles use both strong and weak fibers on a weaving loom. The "warp" threads must be strong because they pass through the loom and must withstand the pulling and stress caused by the loom. The "woof" threads are woven across the warp threads and are subjected to only a modest amount of stress and need not be nearly as strong. Have you ever tried to tear a piece of fabric in one direction (with the grain) and found it easy? What about tearing it in the other direction (across the grain)? Was it more difficult? The differences in the strength of the warp and woof fibers determine how strong the textile is on one direction versus the other.

The structure of a material, such as the presence of long molecules or the presence or lack of **grain**, helps to determine the characteristics of that material. For example, materials with long molecules tend to have good tensile strength and tend to be flexible. As you gain knowledge and experience using materials formed of chains of molecules, you will be better able to select the best materials for your design and technology work.

Platelets

Not all materials form chains or have a grain direction when they bond together. Some molecules bond into tiny **platelets** that slip and slide easily over each other. Clay, wool, graphite, and blood are examples of such materials. The platelets present in blood are shown in Figure 5.7. The attraction between the molecules within a platelet is greater than the attraction between two whole platelets. (Think of a delicious pastry with layers and layers that can be separated for slow eating.) A technical way of explaining this is that the **adhesion** between the platelets is less than the **cohesion** within the platelet itself. Adhesion is the force that attracts materials to each other. Cohesion is the force of attraction that holds the material itself together.

The properties of **pliability** (slipping of platelets over each other) and strength (the platelets maintain their shape) are valuable for some purposes. Graphite, for example, can be used for lubricating and can continue to slip and slide even under pressure. Some materials with platelets may be made more pliable by adding water or another solvent. A wool scarf stretches and changes shape when wet.

■ Figure 5.7
In this photomicrograph of blood, the platelets look like soggy cereal. *(© Manfred Kage/Peter Arnold, Inc.)*

In most of these materials, the platelets can be changed by heating them. The wool scarf becomes very stiff and is no longer so good for keeping you warm if you iron it at too high a temperature. Pounding on wet wool can make the material so compact it becomes quite wind- and water-resistant. These same processes for producing change can be used for other advantages as well. Let us consider ceramic materials.

Ceramics, in the form of pottery and glass, are some materials that have been used for centuries. Museum displays of ceramics and glass indicate that very beautiful and functional products were made and traded by ancient civilizations.

Ceramics can be considered in two forms, the heated and the not yet heated. When ceramics are heated to a predetermined, high temperature, we say they are **fired.** Ceramic materials that have not been fired are called **green.** If green ware is mixed with water, it will become pliable and soft again. In this state, the ceramic molecules are formed by a weak attraction and take the shape of platelets. Platelets are very small and thin and look very much like microscopic, wet cornflakes of clay. These platelets slide over each other very easily. This accounts for the ease with which wet clay can be worked by hand or by machines to make many different products.

During firing of ceramics, however, changes occur in the materials. First, the water is driven off and the clay begins to dry. As the temperature rises in the **kiln** (an oven-like device capable of reaching temperatures of 2000°F and more), the platelets melt and fuse together. Crystals form and the material begins to look like stone. This point is called **vitrification**; the platelets are no longer evident. The clay has undergone a change that cannot be reversed. Ceramics that have been well fired will have a new molecular structure. This structure is more like the structure of metals than that of the slippery original material.

Ceramics can be cured in other ways as well. Heat, light, and drying can be used to form the molecular structure that is needed to create the desired characteristics. In some cases, this curing creates crystals.

Crystals

One of the most useful kinds of materials with **crystals** is metal. When you think of a metal, you probably think of a material with a smooth, hard surface. If you could see the metal through a powerful microscope, you would find an uneven surface made of many crystals. These crystals bond closely together to form the dense materials we know as metals.

When metal is obtained from the raw materials found in the earth, it is often heated to refine it. Chemicals are added to give the metal new properties. The properties of the metal depend upon the way the atoms of the metal link together to form new molecules as they cool. The crystals may be modified to the desired form by processing while cooling. (This is what you do when you stir fudge while it cools, to make the crystals small and the texture smooth.)

In metals, atoms and molecules pack together in a simple manner. You can build a model of a metal material by using spheres as the particles. Some metals seem to have particles that stack together in the shape of cubes. (See Figure 5.8A.) Other materials pack together in a six-sided, hexagonal shape.

In all metals, the material units are packed close together. They form a relatively dense material. Engineers cr technicians can decide how to cut or shape the metal if they understand the structure of the units. The structure of the metal also influences what type of product can be made from it.

The drawing of the body-centered cube is not what would appear under a microscope. It is a model that shows how the marbles would fit together. In the actual material, the basic units are actually some distance from each other, as represented in Figure 5.8B. In a metal, close packing makes the molecules very stable because of their internal attraction and repulsion. The molecules will not separate from each other unless a very strong force is applied. Such metals are very durable and resist bending or being deformed.

The properties of the material depend upon the way the atoms combine to form molecules. The material's properties also depend upon the way individual molecules join to form the entire piece of material. Of all materials, only metals

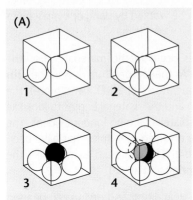

Building a Body-Centered Cube of Marbles

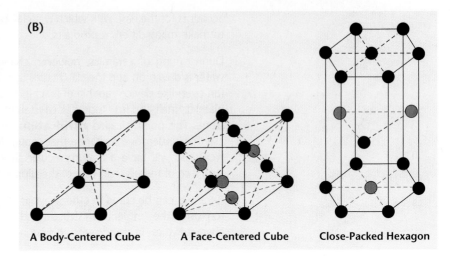

A Body-Centered Cube   A Face-Centered Cube   Close-Packed Hexagon

Figure 5.8A, B, C ■
Metals form regular patterns of atoms and molecules because they pack close together. (A) The close packing of marbles illustrates a body-centered cube, one form of close packing. (B) These drawings of a body-centered cube, a face-centered cube, and a close-packed hexagon represent common patterns in metals. (C) Because close-packed molecules slide over each other more easily, the materials can be shaped or molded more easily than loose-packed molecules.

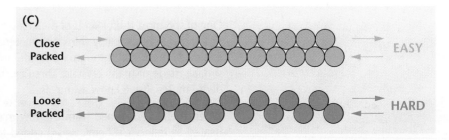

have both tightly packed atoms in each molecule and tightly packed molecules in the material as a whole. The best use of the metal, how it can be cut, and how easily it can be shaped all are determined by how closely the molecules pack together.

In some metals, the molecules are packed together in tight layers. In other metals, there is more space between the molecules within the layers. (See Figure 5.8C.) In close-packed metals, the molecules move fairly easily over each other. If the molecules are packed together loosely, however, it becomes more difficult to move the layers past one another. These loose-packed materials can not be cut, molded, or shaped as easily as close-packed materials.

Some materials other than metals are more easily identified as crystals. Diamonds, amethysts, other jewels, sugar, mica, and salt are a few. In recent research on the making of artificial diamonds, both graphite crystals and diamond crystals are formed. If you want to make a pencil or a lubricant, you would want the graphite. However, if you want to make a grinder or a piece of jewelry, the diamond is what you want.

Not all materials form chains, crystals, or platelets. New materials may be developed as materials are given new structures. The properties will change as the structure changes.

D E S I G N   B R I E F

Materials have many characteristics. A characteristic that will be important in your design and technology work is the strength of the materials. Strength of materials relates to how well a material holds up under tension, compression, shear, and torsion. (These are called the tensile, compressive, shear, and torsional strength of materials.) You will remember that the idea of torque was introduced in Chapter 4 as the force that is needed to create a circular motion. This is called torsion. Shear is the force needed to create a cutting action.

The drawings below show the concept of testing a material for tensile and compressive strength.

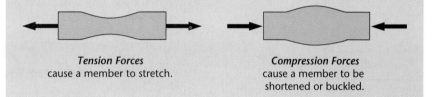

Tension Forces
cause a member to stretch.

Compression Forces
cause a member to be
shortened or buckled.

Using the specific material sample assigned by your teacher, determine if it should be tested for tensile or compressive strength or both. Design a device that can be used to determine its strength.

• If appropriate, determine the tensile strength of the material by using weights that will pull the sample apart.

• If appropriate, determine the compressive strength of the material by using weights that will crush the material sample.

• Discuss with your classmates how the other types of strength of the material could be tested.

• List some ways in which each material might be put to other uses.

Design and build a structure, using only three sheets of thin card stock, 36″ of tape, and a pair of scissors, that will support a common house brick as high off the ground as possible. After you have built your structure, test it and then:

• Use a spring balance to determine the weight of the brick in pounds and in newtons.

• Determine which parts fail first and why. Do the parts tear or crumple?

• Determine which parts of the structure require added compressive strength.

• Determine which parts of the structure require added tensile strength.

• Determine how the structure can be improved.

MATH/SCIENCE/TECHNOLOGY LINK

Tensile Testing of Materials

- Determine the weight* required to pull the sample piece of thread apart.

- Measure and calculate the cross-section size of the material sample.

- Determine the force required to break (dislocate) the sample. Express this in pounds per square inch.

- Determine the force required to dislocate the sample in newtons per square centimeter.

Compression Testing of Materials

- Determine the force required to break the toothpick sample by squeezing (compressing) it.

- Measure and calculate the cross-section size of the toothpick sample.

- Determine the force required to break the material sample in pounds per square inch.

- Determine the load required to dislocate the sample in newtons per square centimeter.

*Weight is a force we measure in pounds or newtons.

FUNCTIONS OF MATERIALS

The next aspect of materials to consider in choosing which to use in a design is their function. Materials can be grouped into four categories that relate to how they function over time.

Monolithic materials are made of only one ingredient, element, or compound. They do not change during use. There are two kinds of monolithic materials: natural and basic. **Natural materials** are used as they are found in nature. **Basic materials** are natural materials that have been changed to make them more useful for certain products. These materials are seen as basic to the making of certain products. Basic in this sense means "important to the whole product."

A raw material such as iron ore is not used in its original form to make products. First, it is processed to extract the pure iron. The iron is then used with other materials to make many kinds of steel that go into bridges, automobiles, machinery, and many other products. The raw material, iron ore, is called a natural material. Pure iron, because it has been processed from its natural state, is a basic material.

CHART OF MONOLITHIC MATERIALS

| **Natural** | **Basic** |
Produced or existing in nature	Important to a whole product
Wood tree	Lumber
Limestone	Lime
Iron ore	Pig iron
Cotton fiber	Cotton thread
Wool fiber	Wool thread
Wheat	Flour

A natural or basic material is called monolithic because it is used in one (mono) state, as found in nature. It is not combined with other materials at this point. When it is used, you can expect it to stay about the same size and shape. You can also expect the material to stay relatively inert (inactive), not reacting to the materials around it.

Nearly all the products we use are made from basic rather than natural materials. Flour used in cakes, for example, is a basic material which is made by grinding (processing) the natural material, wheat. A diamond crystal may be admired in its natural state, but it is usually "cut" before being used in jewelry or in a bit for drilling or cutting.

Amorphous materials lack any specific structure or organization. As their name suggests, their structure and form are not organized as in a crystal or in a string-like molecule. Amorphous materials are rather nondescript and disorganized in their form.

Amorphous materials may take a variety of forms. For example, there are amorphous forms of carbon that are significantly different than the crystalline carbon we know as diamonds or the platelet form we know as graphite. Three amorphous forms of carbon are activated carbon, carbon black, and carbon fibers. Activated carbon is used regularly in filters, from cigarettes to gas masks and water purification systems. Carbon black is used in paints, inks, and rubber tires. Carbon fibers are used for making composite parts that have tremendous strength but are very lightweight. The Voyager, the airplane that flew nonstop around the world, made extensive use of carbon fibers to reduce the weight of the aircraft.

Material systems have two or more materials combined into a system. The ingredients are combined for a specific function or to have a desired characteristic. Material systems are not really new. Clay brick reinforced with straw was used thousands of years ago to form a material system for constructing shelters. So, too, were mud, manure, straw, and branches used in making walls of wattle (a material system still in use in some parts of the world). Today we use many material systems. Some examples are plywood, plasterboard, fiberglass, concrete, reinforced carbon fiber, and fabrics backed with a thermal lining.

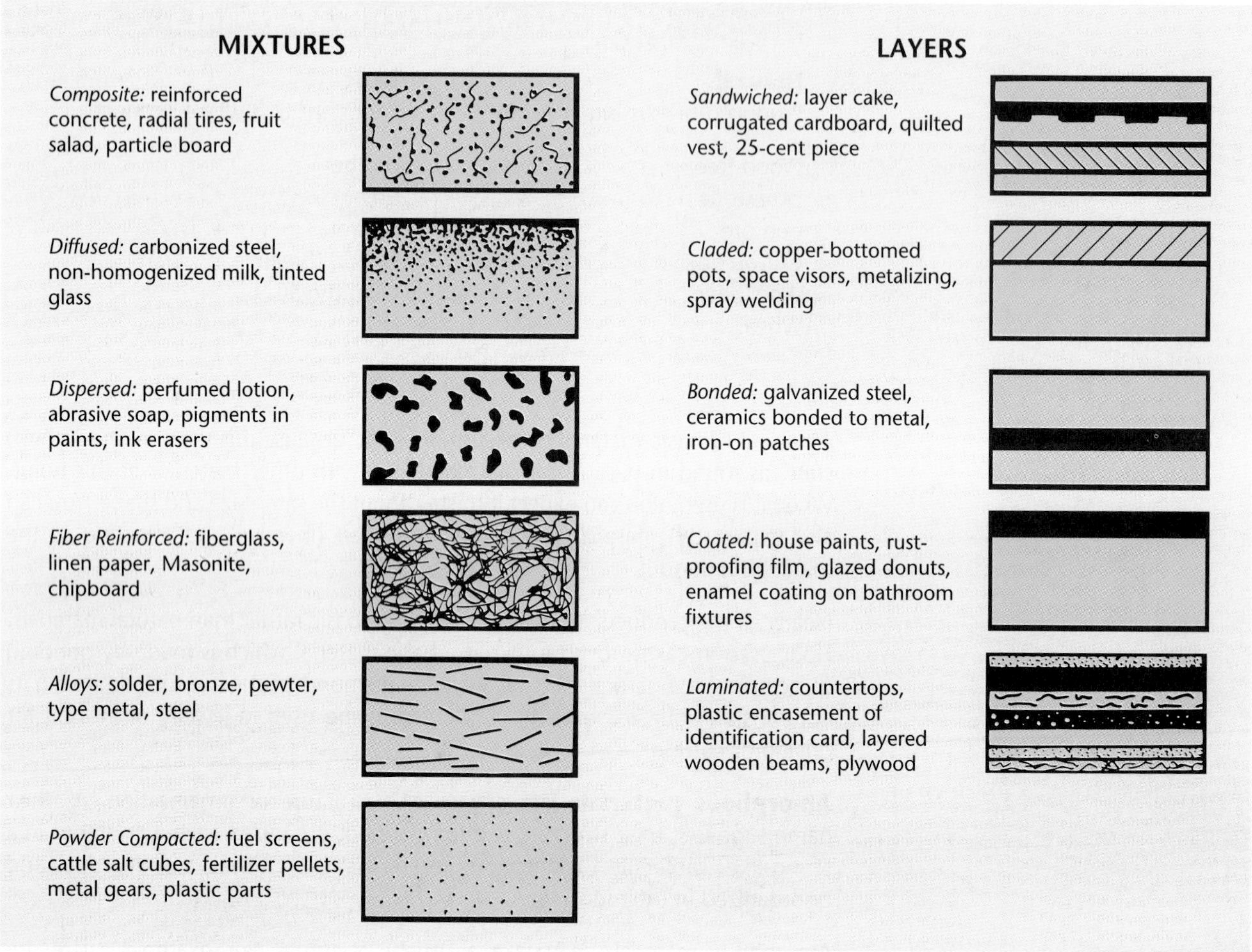

MIXTURES

Composite: reinforced concrete, radial tires, fruit salad, particle board

Diffused: carbonized steel, non-homogenized milk, tinted glass

Dispersed: perfumed lotion, abrasive soap, pigments in paints, ink erasers

Fiber Reinforced: fiberglass, linen paper, Masonite, chipboard

Alloys: solder, bronze, pewter, type metal, steel

Powder Compacted: fuel screens, cattle salt cubes, fertilizer pellets, metal gears, plastic parts

LAYERS

Sandwiched: layer cake, corrugated cardboard, quilted vest, 25-cent piece

Claded: copper-bottomed pots, space visors, metalizing, spray welding

Bonded: galvanized steel, ceramics bonded to metal, iron-on patches

Coated: house paints, rust-proofing film, glazed donuts, enamel coating on bathroom fixtures

Laminated: countertops, plastic encasement of identification card, layered wooden beams, plywood

Figure 5.9 ■
Material systems—those that are mixtures and those that are in layers.

Material systems form the most-used types of materials. We define material systems as two or more materials that differ in form, that work together to become a new, more desirable material when they are combined. The resulting material system may be, for example, stronger, lighter, or longer-lasting than any one of the single materials used alone.

All material systems fall within two categories: those that are **material mixtures** and those that are **material layers.** These two terms reflect the general ways that material systems are made. (Refer to Figure 5.9.)

A material mixture is a **composite** in which the ingredients of the system are mixed. They may be mixed mechanically, chemically, or physically. The ingredients may form a random or an oriented pattern. Heat, pressure, or both may be required to form a mixture. Many food products are material mixtures. A recipe for mixing can be used to get consistently similar products.

A material layer is composed of materials that are different from each other that are put together without actually blending the materials. Each material lends to the final product the special characteristics it possesses. Your home is made of many material systems. If it is painted, the paint layer is bonded to the wood or other material, but most of the paint could be removed without destroying the base materials. In some cases, the materials are combined with the expectation that the layers will blend over time, forming a mixture as well as layers.

A material layer you may be familiar with is "bubble wrap." If you have enjoyed popping the bubbles, you know that air is trapped between layers of plastic film. The air can be compressed to absorb the shock of dropping or bumping products during transport.

Dynamic materials change during use. You can choose the material you want if you know that these changes will occur. The three types of dynamic materials—**metamorphic materials**, **expandable materials**, and **functional materials**—are shown in Figures 5.10A-C.

Metamorphic materials change slowly from their original characteristics to something different as they are used. The metal in the blade of a bulldozer is often made of a metamorphic metal. The blade gradually becomes harder as it is used to move dirt and rock. Cedar house siding, when left unpainted, gradually turns a beautiful gray color as it reacts to weather. We call these metamorphic changes.

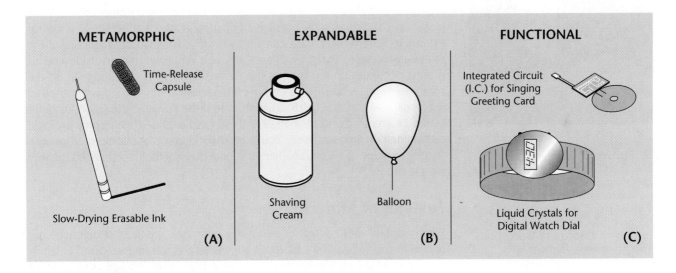

Figure 5.10A, B, C ■
Dynamic materials go through a process of change. (A) Dynamic materials that change slowly over time are metamorphic. (B) Others are expandable and change by growing and/or contracting. (C) Others that change to perform different tasks are functional materials.

Expandable materials are of two types: those that flex and those that foam. An example of a flexible material is a rubber balloon. Shaving cream from an aerosol can is an example of a foamable material. Both types of materials change more rapidly than in materials where composition or structural changes occur. Materials like the rubber balloon will change and repeat the process. Flexible materials can expand and contract. Flexible materials are used in tents, clothing, and as expandable containers for gases and liquids.

Most foaming materials, like the shaving cream, expand only once. Some plastic and rubber materials use specially chosen ingredients that bubble up like a runaway dishwasher when mixed together. Depending upon the ingredients, the new expanded material may become solid and hard or spongy and soft.

■ **Figure 5.11A**
Functional materials change when exposed to energy and return to their original state when the energy is gone. These glasses darken according to the level of sunlight. *(Courtesy of Corning, Incorporated)*

■ **Figure 5.11B**
This solid-state refrigeration device gets cold on one side and warm on the other. When the electrical current stops, the temperature on both sides of the device becomes the same. *(Courtesy of Melcor Thermoelectrics, Trenton, NJ)*

These foamable materials have many different applications. They can be used to provide a lightweight protective packing material or "peanuts" that serve the same function as bubble wrap. Foamable materials have reduced the time and cost of insulating refrigerators. Because the bubbles retain air, expandable foams have been used to insulate older homes. Expandable materials can reduce the cost of heating in winter and cooling in summer. Food additives in artificial whipped cream make and keep the dairy product fluffy.

Baking powder is another expandable material. When mixed with water, baking powder changes chemically and releases a gas. This gas expands. If it is trapped in a cake flour mixture, it causes the cake to rise. This makes a light and fluffy cake.

Functional materials change back and forth from one state to another as an outside force is applied and then removed. Photosensitive eyeglasses are made of a functional material. The action of sunlight causes the materials in the glasses in Figure 5.11A to change and gradually become less transparent, so that less light can get through. Once the sunlight is removed (when you go indoors), the glasses become more transparent and allow more light to reach your eyes.

A solid-state refrigeration unit, as shown in Figure 5.11B, is another example of a dynamic functional material. When electrical energy is fed to the unit, one side of the unit absorbs heat from the other side. The second side becomes cooler. When placed in contact with the storage compartment of a refrigerator, the device extracts heat from the storage compartment. Heat is drawn from one side of the solid-state device to the other. There the heat is ventilated to the outside air. This process continues until it reaches the desired temperature (determined on the control/sensing switch). These changes take place with no moving parts. Another functional material that holds great promise is the "organic transistor" made of plastics, currently being developed. If these functional devices become available as commercial products, they could be used in some applications to replace transistors made of silicon, another functional material that has changed the way we live and work.

Families of Materials

When we talk about a family, usually we refer to a group of people who are related to each other. Some materials are related and can be considered as family members. Many of the materials we have been studying are really part of **families of materials**. For example, the term "plastics" is a name for a family of related materials. This family includes many types of plastics such as vinyl, styrene, and acrylic.

We use such terms as **metal**, **ceramic**, **textile**, and **wood** to indicate families of materials. We could name many individual members of these family groups. We may talk about the food family, for example. To help people understand the nutrients their bodies require, the food pyramid has been developed. An example of the food pyramid is shown in Figure 3.4 of Chapter 3, page 59. This model groups together the many different foods that belong to each category, along with the daily requirements for an average person. Figure 5.12 shows the different categories of food listed on the food pyramid.

■ **Figure 5.12**
The examples above represent the basic food groups from the food pyramid. *(Courtesy of USDA)*

One important material family is the **ferrous metals**. These are irons and steels. Iron is made from processing coal, iron ore, and limestone, which are natural materials. Iron is seldom used in its basic state. Instead, it is mixed with small quantities of materials such as carbon, silicon, manganese, nickel, chromium, molybdenum, sulfur, and phosphorus to form **alloys**. Within each group there may be dozens of individual steels.

We mix various ingredients in various amounts to make different types of cakes. This is much like making different types of steels. Most cakes have flour, sweeteners, and milk. All steels have pure iron, carbon, and silicon. We can vary the amounts of the basic materials and add other materials to change the texture and lightness of a cake. The same is true for making steel.

Just as cookbooks contain hundreds of recipes for different types of cakes, so do the steel companies have formulas to make many different types of steel. These steels are designed to do different jobs and make different products.

This is true for other metals. There are more than 200 commercially available alloys of aluminum, for example. The plastics family is large and continues to grow. The ceramics family is expanding as well. Chemical engineers formulate new plastics, ceramics, and other materials with the needed characteristics for specific jobs.

D E S I G N   B R I E F

Understanding the different functions of materials is important if you and your classmates are to use materials effectively in the design and development of products and structures. In order to share what you learn about materials and their functions, collect samples of different materials. Design and develop a display of materials that presents and describes each of the following:

- natural and basic materials—monolithic and amorphous

- material systems—mixtures and layers

- dynamic materials—metamorphic, expandable, and functional

Analyze the products provided by your teacher and determine the different materials that are used in the manufacture of their parts. Next:

- categorize each of the parts according to the materials used,

- identify which parts might perform better if changed to a different material, and

- write a short defense of why the new material would be better for the part.

MATH/SCIENCE/TECHNOLOGY LINK
Functions of Materials

- Devise and conduct simple tests to determine the structure of the samples of materials you collected (crystals, platelets, polymers, or amorphous).

- Using a microscope, see if you can determine whether the material samples you have collected are comprised of crystals (grains), platelets, or polymers, or if the material is amorphous in form.

- Develop drawings to represent the structure of your material samples.

CONVERSION OF MATERIALS

Just as new materials may be synthesized, existing materials can be converted into other forms. Materials conversion processes can be a little difficult to understand because they are not always visible. If you watch a pizza baking in an oven, you do not actually see the chemical action (the baking) taking place. What you do see is the sequence of changes that the pizza goes through during baking. After the process of baking, you see the results of the changes that occurred. You can taste and smell the differences as well. (See Figure 5.13.)

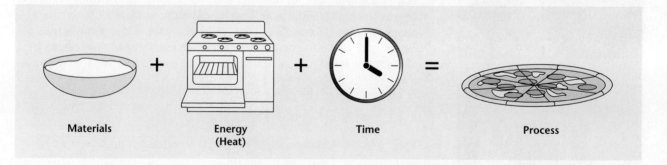

Materials   Energy (Heat)   Time   Process

Figure 5.13 ■
An example of a materials conversion process—baking.

The term **process** as used here simply refers to a sequence of actions that lead to some result; in this instance, the changing of materials (baking the pizza). Materials can also be converted not only to form new materials, but also to generate energy and to create information. All three types of processes (conversion of materials, energy, and information) are similar and involve **sequence**, **action**, **changes**, and **results**. These are the four essential parts of any process.

The sequence of actions involves placing the right materials together in the proper places and treating them in a certain way over a given time. Usually, the recipe of actions has been developed over several sequences of trial and error. A comparison of the changes and outcomes is made to determine the process that seems to work best. The change in materials may be of several kinds. The result is the desired outcome—a conversion to new materials, information, or energy.

Coal can be burned. This is an example of an energy conversion process. When burning takes place, a chemical reaction releases energy from carbon-containing molecules. The initial heat allows those molecules to break some molecular bonds and combine rapidly with oxygen, an action we call **combustion.** By changing coal through burning, we can produce **thermal energy** (heat). This energy is much greater than the small amount of heat used to start the process. The best methods of burning coal have been developed over time. Carefully controlled amounts of initial heat, the proper oven, and measured amounts of added oxygen can be used for the most efficient way to get energy from coal. Compare this with picking up a piece of coal and trying to get it to burn with a match or lighter. Using the process that has been developed by others will probably decrease your frustration.

In a flashlight battery, just as in burning coal, the materials are combined and changed to release the energy which exists in the materials. In one type of battery, zinc and carbon provide the energy. In coal, the carbon burns to provide energy. The process used is different, but the outcome of conversion of materials to energy is the same.

Other processes are used to change materials into new materials and into products. If we have an adequate supply of thermal energy (such as that from burning coal) we can take a sample of tin and a sample of brass and melt them together by applying heat (thermal energy). A new and different material (bronze) will be produced. This process is called **alloying**.

You can take wood pulp and mix it with other ingredients to make paper. Wood pulp can also be processed in a different way and added to bread as a source of fiber. Other **materials conversion** processes take energy and new materials and change them into useful products. These processes always require some form of tool or machine. There are many processes that can be used to create an almost endless number of products.

Popcorn offers a simple example of a useful materials conversion process. Place the popcorn and oil (the materials) in a pot on the stove, or place the paper bag in the microwave oven. Now we are ready to start the process.

On a gas stove, **energy conversion** takes place as burning gas (materials) generates heat (thermal energy). The heat warms the pot and corn and converts the water in the popcorn. It becomes steam. (In the microwave oven, the microwaves excite the molecules of water more directly.) In either way, the steam expands and pushes out from within each popcorn kernel. When enough pressure builds up inside the kernel, it explodes with a pop. The oil helps to keep the corn from burning and helps to transfer the heat more efficiently from the pot to the corn. After popping, we can finish the process by adding salt and have a useful product. The popcorn, now an expanded material, has been put through a process or series of processes to change it into its desired form.

You may know that not all corn will pop efficiently. Some is more suited for eating as a vegetable, some for making hominy grits and cereal. Corns for different purposes have been developed over the years.

(A) Mixing

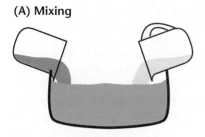

(B) Joining/Fastening (adhesion)

(C) Joining/Fastening (cohesion)

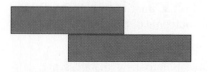

(D) Coating

(E) Weaving/Knitting/Interlacing

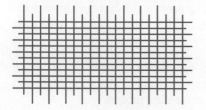

■ **Figure 5.14A, B, C, D, E**
Examples of the four addition processes.

A third type of process, **information conversion**, has been developed to convert materials. Films, paper, and other chemicals can be used to make photos or videos that record or copy information. A stage set can be built using paints and other materials to create the mood and impression for a play or speech. Materials for light systems and sound systems are vital to a concert. Printing creates signals and symbols with materials like inks, paints, paper, and the like. Textiles can be dyed or woven with patterns that remind you of a certain cultural group or time period. In a different example, keys are made when the material is impressed with information that will fit a specific lock.

Materials Conversion Processes

The rest of this chapter will focus on processes that can be used to change materials. Luckily for us, the thousands of different processes used to change materials fall into four categories. Materials can be changed by (1) adding materials to each other, (2) separating materials, (3) making contour changes, or (4) making internal changes.

A "black box" idea of processes can be applied to a specific example of materials conversion. The raw materials (along with energy, in some cases) are the input. The changed material is output in the form of a finished product. The process is what happens in between. We can divide each of the four materials conversion processes into smaller categories. Perhaps this will help you identify similarities and differences among the various processes. Knowledge about these processes can make you a wiser consumer of products. It can also allow you to choose the outcome you want when you construct or repair products yourself.

Have you ever wondered why bread must be kneaded before it is baked? Or why muffins should be mixed only until the ingredients are moistened and no longer? Why are cake mixes beaten at high speed for several minutes while pastry is to be handled as little as possible? All this has to do with the sequence of actions in the mixing process and the desired outcome. Recipes have been developed to identify the sequence of actions as well as the duration and conditions under which the actions are to be applied. Because of this, the outcome (food) is more likely to be consistent and desirable.

Addition Processes

Materials are added to each other in many different ways. Previously, we discussed material systems. Material systems are the product of the processes identified here. Adding materials together can form a new material or product that combines the characteristics of each of the ingredients.

It is possible to group the addition of materials into four categories. These include (1) **mixing**, (2) **joining** or **fastening**, (3) **coating**, and (4) **weaving** or **interlacing**, as shown in Figures 5.14A-E.

Mixing is one of the most obvious ways of combining materials. You have done it many times. Water is mixed with water-based paints to make them thin enough to use. Sugar, flour, and other ingredients are mixed to make cookies. Sand, gravel, and cement are mixed together to make concrete.

These are all examples of combining materials through mixing. In some cases, materials that are mixed interact with each other so that the original materials cannot be retrieved. In other cases, the original materials may be recovered for later use.

Joining or fastening materials are added together through adhesion, cohesion, and mechanical processes. Joining and fastening materials may be done for structural or decorative purposes. (See Figure 5.15.)

In an adhesion process, the connection formed between materials is a surface bond only. The atoms and molecules of the two surfaces remain separate. The adhesive material is "sandwiched" in between. Adhesive processes include gluing, soldering, brazing, printing, painting, and putting together a peanut butter sandwich. Many new, specialized adhesives have been developed in recent years.

Figure 5.15 ■
Joining materials often involves using an adhesive material.

Joining and fastening can also take place at a microscopic level. Cloning is a process applied to living material in which segments of DNA are joined together to form a replica of the original. Gene-splicing also requires using a fastener to attach two portions of a gene together. The adhesive in this case is also a segment of living material.

Cohesion is a joining process activated by heat, pressure, or both. A bond is formed by the attracting forces between molecules at the surface of each material. The process makes the molecules of the materials themselves stick together. Cohesion processes include oxyacetylene welding, thermal welding of plastics, and pressing together a hamburger patty. The welding process is demonstrated in Figure 5.16.

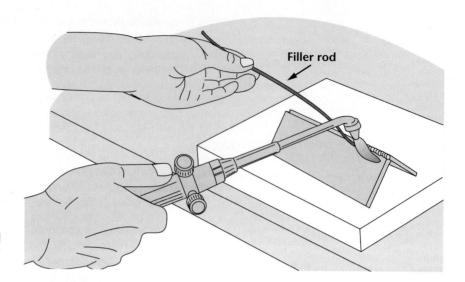

Figure 5.16 ■
Materials can be joined together by a welding process.

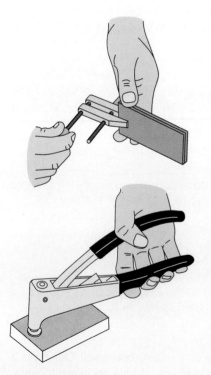

In **mechanical fastening** processes, materials are joined with fasteners such as screws, bolts, zippers, rivets, staples, clamps, tapes, velcro, thread, or pins. (See Figure 5.17.) With the development of flexible plastic materials, hinges and closures can be repeatedly opened and closed without breaking or wearing out. New developments in fasteners will increase the ease of assembly and disassembly to aid in recycling materials.

Coating is an addition process used to cover a base material for protection or decoration. We coat materials by adding a layer of a second material. (See Figure 5.18.) Painting is one example. The paint has little penetration and sticks to the surface by adhesion. Glazing doughnuts, putting polish on your toenails, and waxing a car are other examples of adding a coating. Rubber gloves, plastic films, and other forms of shielding may be used to keep materials from combining as they are used together. New processes for creating small diamond crystals can provide an economical material for a coating that is highly resistant to wear.

■ **Figure 5.17**
Fasteners come in many types and are used to mechanically join materials.

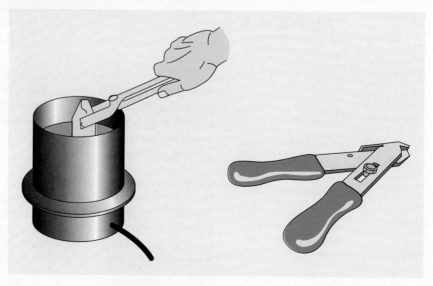

Figure 5.18 ■
One material is applied to coat another in order to achieve and maintain the results you want.

There are coating processes that alter the surface of a material. In these instances, the second material penetrates by physical or chemical means, and the separation of the coating and the base material is less distinct. Examples of this type of coating are staining of wood or dying your hair (by physical absorption), chemical etching or bleaching, and anodizing of aluminum (by chemical reaction).

Coating may be used for protection and decoration. Can you identify any other processes that involve coating by **layering**? How about some others that are coated by **penetration**?

Weaving, **knitting**, and **interlacing** materials are brought together so they are mechanically locked together. Obvious examples are threads combined to make cloth and wires combined to make screen. Weaving and interlacing are processes used by people for centuries. Ropes, cords, and strings are examples; so are fire hoses and baby booties. (See Figure 5.19.) More recent applications of the weaving process involve making shielded electrical cable and weaving the body of an experimental car made of composite materials.

Addition processes that have an intended and identifiable pattern are called weaving or knitting. Interlacing refers to materials added in random or non-patterned ways. We can take glass fibers and weave them together to make fiberglass cloth, for example. We can combine the glass fibers in a random pattern to produce fiberglass insulation. Netting materials can be developed with a great deal of stretch, flexibility, or rigidity, depending on the way in which the materials are locked together.

Processes for Separating Materials

Shearing and **sawing** are two very common separation processes. Both of these processes separate materials by use of a **wedge**. They are different because the saw separates the material by using a blade with cutting teeth. This produces waste in the form of chips. (See Figure 5.20.) Shearing, however, uses two smooth blades (a pair of scissors, for example) that work together to separate the materials without producing a chip.

Shearing can be done two ways. Two cutting edges may cross one another, as scissors cut hair. A single cutting edge may go through the material, as a knife slices meat. Other examples of shearing tools include paper cutters, cheese graters, tin snips, bench shears, squaring shears, solid punches, hollow punches, wood chisels, and cold chisels.

Sawing is done using a tool that has pointed, wedge-shaped teeth evenly spaced along the edge of a blade. Examples include hacksaws, crosscut saws, rip saws, band saws, jig saws, and jewelers' saws. Other processes similar to sawing include grinding, filing, shaping, drilling, milling, turning, sanding, and polishing.

Shearing and sawing require some form of mechanical force. We can separate materials through nonmechanical processes as well. Chemical or thermal processes are examples. These are called **zero force** processes because they do not use mechanical force. Bleaching, etching, dry cleaning, oxyacetylene torch cutting, carbon-arc cutting, chemical milling, laser cutting, and ultrasonic cleaning are examples of zero force processes.

■ **Figure 5.19**
This fish trap is woven together with wire. Workers construct the trap while standing inside. When the top gets narrow, they lift themselves out with a rope attached to the ceiling above them. *(Courtesy of Aramco World)*

■ **Figure 5.20**
This process separates the material with a wedge (the wire) but does not generate a chip. *(Courtesy of Beth Deene, photo by Patricia Hutchison)*

A great deal of human effort goes into separating materials we want from those we do not want and keeping them separate. Think of all the time you spend on sorting and cleaning. Think of the amount of materials and energy used for keeping bugs, germs, dirt, and other materials we call weeds or pollutants where we want them. This can be as simple as sweeping the floor or winnowing the chaff from the grain using the wind. The effort can also be as complex as adding a microbe to an oil spill in the ocean so that it will "digest" the oil into less harmful materials.

D E S I G N   B R I E F

Materials Conversion—Addition and Separation of Materials

Addition—In the design and development work you will be doing, it is important that you know which materials conversion processes are appropriate to use in making models, prototypes, and products. To provide you with a better background for this work, design and develop reference pages for your portfolio. Through drawings, narratives, and examples taken from old magazines, develop a short presentation for each of the following addition of materials processes—(1) **mixing**, (2) **joining** or **fastening**, (3) **coating**, and (4) **weaving** or **interlacing.** Illustrate each example included. Write an explanation of how the materials are added to each other and why that process was selected.

Separation—Design and develop a reference page for the following separation of materials processes—(1) **separation with a chip**, (2) **separation without a chip**, and (3) **zero force separation.**

MATH/SCIENCE/TECHNOLOGY LINK

- For the processes you selected above, determine the relative cost of the materials required for each process.

- Using sample materials and sample adhesives provided by your teacher, adhere two pieces of the materials with each adhesive. Use the test device provided by your teacher and determine how much force is required to break the adhesive bond.

- Develop a chart to show the relative strength of the different adhesives when applied to different materials.

- Compare your results of these tests with those of your classmates and discuss how and why the results are similar and different.

- If you were to repeat this testing, what factors (variables) would you want to control to ensure that this is a fair test of the strength of the adhesive bond only?

- Using a cutting tool and sample pieces of wood, plastic, and fabric provided by your teacher, determine the force needed to cut (shear) the samples.

- Develop a chart to show the relative strength of the different materials (with the grain, across the grain, and diagonally across the grain—on the bias).

- Identify if there are any factors that interfere with your making accurate measurements on these tests. How would you take these into account?

Contour Change Processes

Contour change processes alter the outside or surface of materials without removing or adding any materials. We can pour warm gelatin into a mold that determines the shape or contour of the mixture as it cools. This change stays after the gelatin is removed from the pan. This is the process of **casting**, even though the pan is often called a gelatin mold.

There are five general types of contour change processes: casting, **extruding**, **forming**, **pressing**, and **molding**. All of these processes can be used to change the shape of a material.

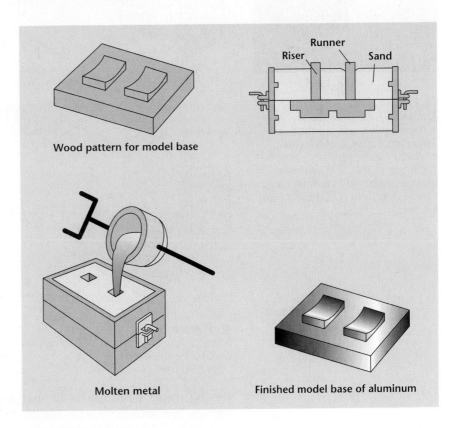

Figure 5.21 ■
Using the casting process to produce the shape you want.

Wood pattern for model base

Riser   Runner   Sand

Molten metal

Finished model base of aluminum

Casting is the process of forming a desired shape by liquefying a material and pouring it into a mold. When the material cools, or sets, it retains the shape of the mold. (See Figure 5.21.) You may have used the casting process to make candies or candles. Many plastic toys are produced by casting.

Extruding involves forcing a material through an opening (a die). This controls the cross-sectional area and the shape. (See Figure 5.22.) Toothpaste is extruded from a tube. Cookie dough can be extruded through a press. Plastic or metal is often extruded, as when making parts of storm windows.

Forming is a process which shapes materials by impact or pressure. Forging is an example of impact forming. Forging is used to change the shape of a piece of metal while increasing the metal's structural strength. (See Figure 5.23.) Horseshoes and very strong metal tool parts are forged. Bending, another example of forming, uses uniform pressure around a form. We use forming to shape a pie crust inside a pan.

■ **Figure 5.22**
This snack food is extruded. Shown is the machine without the automatic cutter in place. The operator can periodically remove the blade so that any cornmeal that has collected in the die tubes can be removed. *(Courtesy of Frito-Lay, Inc.)*

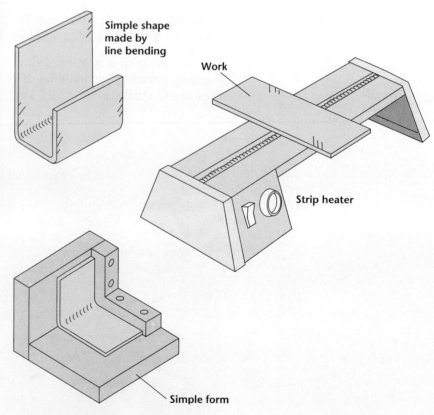

■ **Figure 5.23**
Many products are made by forming the materials into the shape that is needed for the design.

Pressing shapes sheet material by squeezing it between two dies or textured surfaces. The surfaces force the material to take on a three-dimensional shape. (See Figure 5.24.) Waffles are pressed in a waffle iron, metal bottle caps are pressed, and hamburger patties can be pressed with dies.

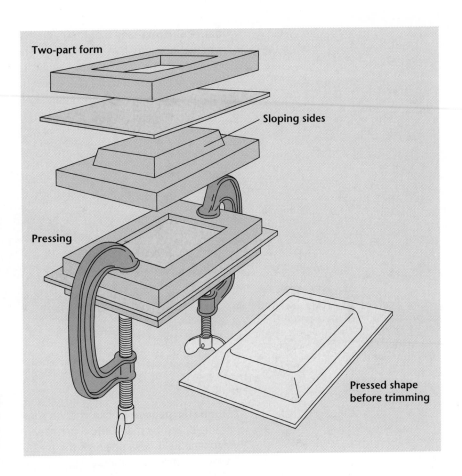

Figure 5.24 ■
The pressing process requires at least two dies, used to give the material a three-dimensional shape.

Molding is a combination of extrusion and casting. The material is forced into a mold (casting) under pressure (extrusion). Plastic cups and wastebaskets may be made through the molding process.

Internal Change Processes

Some processes are used only to make changes inside materials. These processes are not always visible to the eye. The outcome of the process is often evident, however.

Thermal processes such as baking, boiling, and freezing produce internal changes. These processes are called thermal processes because they depend upon a change in temperature. A decrease in temperature is needed to freeze ice cream. Ice cream making, therefore, is a thermal process. The energy for the thermal process can come from a number of sources.

Mechanical processes can make internal changes. Hammering, peening, and explosive-forming are mechanical processes that use shock to cause the change in materials. The making of felt for clothing is done by a special process of hammering. Stretching a material can increase its strength as it attempts to return to its original shape.

Chemical processes that make internal changes include electrolysis, tattooing, fermentation, and film developing. The images on ordinary camera film are developed in the darkroom by chemical processes. Medicines and drugs are used to make or prevent internal changes in the body. Hormones are sometimes added to the food of cattle and other animals to make them grow faster.

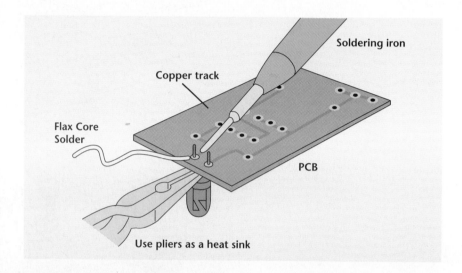

Figure 5.25 ■
Chemical processes were used to produce this printed circuit (PC) board. The thermal process of soldering is used to mount the components mechanically to the circuit board.

Magnetic processes can change some materials internally. A screwdriver, for example, can be made into a permanent magnet by passing it through a strong magnetic field. This polarizes the metal atoms. A needle can be magnetized to make a compass. Computers use electromagnetic processes for sorting, storing, and retrieving information.

Acoustical processes also create internal changes in materials. Sound, in the range we can hear, causes changes in the molecular bond of specific materials. The most common acoustical processes use sound outside our range of hearing (ultrasonics). Ultrasonics can be used to increase the magnetic properties of selected materials. This is useful in materials testing. Recently, ultrasonics in health care and medicine has been used to create images that can supplement and sometimes replace X-ray photographs.

Optical processes refer to the change of materials that is caused by light. Some materials, such as plastics, change if left in sunlight for long periods of time. The addition of a material that decays with sunlight is a process used for increasing the biodegradability of some plastics.

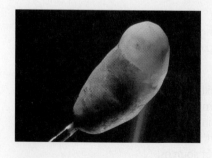

■ **Figure 5.26**
Another way to peel potatoes. This process uses a laser beam to vaporize the skin. The potato itself is not cooked. *(Courtesy of Battelle Memorial Institute)*

Photography is an example of using the effects of light on light-sensitive materials. Light also causes changes in materials that are important for electronics. These include the photoelectric, photoemissive, and photoconductive changes.

A laser, used in a compact disc player, quickly sorts and selects the track of music that is indicated. You may be familiar with using a lens to focus light rays in order to start a fire.

Electrical processes cause changes in materials by passing a stream of electrons through the materials. A flow of electrons can cause materials to change internally and vary their properties. Materials can be made to change their magnetic, conductive, resistive, light transmission, and light emission properties.

If you have ever handled an electrical appliance carelessly, you know that the process can create a shock to the body that grounds the circuit. With living organisms, electrical processes can stun, shock, and kill. Yet the same processes, used with understanding and control, can produce desired outcomes in non-living materials.

D E S I G N   B R I E F

Materials Conversion—Contour and Internal Change of Materials

Contour Change—It is important that you know which materials conversion processes are appropriate when you wish to modify the shape (contour change) and structure (internal change) of materials. Design and develop reference pages for your portfolio that include drawings, narratives, and examples to illustrate each of the contour changes—**casting, extruding, forming, pressing,** and **molding.**

Internal Change—Design and develop a reference page for each of the following internal change processes—**thermal, mechanical, magnetic, chemical, acoustical, optical,** and **electrical.**

MATH/SCIENCE/TECHNOLOGY LINK

Materials Conversion

Using the sheets of material provided by your teacher, modify the contour of one piece without removing any material. Determine how much weight an unchanged piece of material can support. Determine how much weight the new form can support. Compare the two sets of data and determine how the strength of the material has changed.

The tools (provided by your teacher) have been produced using contour and internal changes in the materials used for the handles. Conduct the three tests below on the reactions of and use of the tools by your classmates.

Determine the reactions of your classmates on how they like or dislike the look of the standard and formed tools.

Standard tool   ☐ Like   ☐ Dislike

Formed tool   ☐ Like   ☐ Dislike

(Refer to page 209 and you will note that these data are nominal in nature.)

Determine how your classmates rank the "feel" of the tools on a five-point scale ranging from very comfortable to very uncomfortable.

Standard tool   Very Uncomfortable ☐ ☐ ☐ ☐ ☐ Very Comfortable

Formed tool   Very Uncomfortable ☐ ☐ ☐ ☐ ☐ Very Comfortable

(Refer to page 210 and you will note that these data are ordinal in nature.)

Determine how the different tools perform in increasing the torque generated by the user.

Standard tool   _____ Foot/pounds

Formed tool   _____ Foot/pounds

(Refer to page 210 and you will note that these data are interval in nature.)

What difference is there in the data of the three tests and what they tell you about the reactions to and use of the standard and formed tools?

SIMILARITIES IN MATERIALS CONVERSION PROCESSES

The materials conversion processes presented in this chapter may appear to be different from each other. Yet all of them are doing something quite similar. The processes change the way the materials are bonded together. We use these processes to create new bonds, break bonds, or modify existing bonds in some way.

Perhaps this is easiest to understand if you again consider materials at the molecular level. The molecules are shown as very small balls with invisible connections (lines of force) that hold them together, yet keep them apart. When we separate materials, we are attempting to break these lines of force.

We can use mechanical energy to separate bonds when we cut or etch to remove some of the material. We can combine thermal and chemical processes to break bonds between molecules so they can be sorted or removed.

As we change materials through addition, we are trying to create new bonds. Paint forms new bonds between itself and the base materials during the drying process. If we paint a dirty surface, we may get a poor bond because the paint bonds with the dirt rather than the surface. In a different example, it is possible to melt some metals together. When they cool, new bonds are formed in the resulting alloy.

During contour changes of materials, we usually rearrange bonds. We may pack some of the molecules more tightly, and we may shorten and strengthen the bonds. For this reason, some machine parts are forged rather than cast.

We may also stretch materials apart, sometimes weakening them. A plastic trash can, for example, may tear at the corners or edges where the plastic was stretched during the forming process.

We often use contouring processes to take advantage of these changes. A good example of contour change of materials is the compacting of powdered metal in a press. The same process is used to put eye shadow or face powder into a makeup compact, or to make aspirin tablets.

Modification of materials by contour change usually means no loss of material. We can pack finely powdered metal together to form a usable shape. The pressure developed in the press causes the granules of metal powder to hold together until they are heated and fused (called "sintering"). Or the powder can be loosely packed and then fused to make a very fine metal screen. A screen of this type can be used for separating dirt from gasoline in an automobile fuel line. We can pack the metal even more tightly and fuse it (an internal change) to form very strong bonds. This material can be used to make gears for automobiles.

Internal changes can also be caused by other forms of pressure and impact. Such internal changes often cause a material to take on different characteristics. For example, metal that needs to be tough and strong, as in a crankshaft, is strengthened by impact in a process called drop forging. Now you know why some hand tools are stamped "drop forged." The manufacturer wants you to know the tool is a strong one.

Materials can also be changed internally by creating new bonds or modifying old bonds. A chemical action is used to make yeast-raised bread. As the yeast reacts with sugar in the bread mixture, a gas is released and trapped in the dough. Kneading develops the protein, making strong walls which can contain the fine gas bubbles. The walls are stabilized during the baking, which produces several other internal changes.

Any process you select will make some kind of change in the materials you are using. It will be important to know what those changes are so you can take advantage of them. If you want to make an object that requires considerable strength, you need to know what processes should be used. When you buy a product, you want to avoid one made by processes that could weaken the material.

SUMMARY
CHAPTER 5

Materials come in many natural and synthetic forms. Manufactured materials far outnumber the natural ones. Materials have quite predictable structures, characteristics, and functions.

Materials have identifiable internal structures that form regular patterns. Materials have characteristics that can often be changed to fit specific needs. Synthetic materials, especially, can be formulated and produced to take on desired characteristics.

Tools, machines, and materials are related in some very obvious ways. Tools or machines could not be built without suitable materials. The sophistication of the tools that we can make is directly related to the materials that are discovered and the improvements that have been made. We often name the phases of historic development after materials (the Stone Age, the Bronze Age, the Iron Age). This is an indication of the importance these materials have played in the development of humankind.

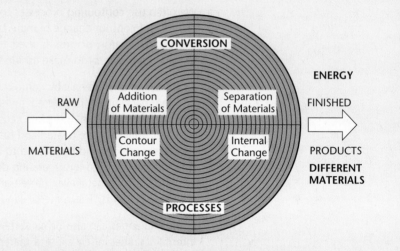

■ **Figure 5.27**
A conceptual model, showing the various ways for processing materials in order to produce energy, other materials, or information.

There is another important relationship here. Materials are converted into useful objects through the processes that the tools and machines perform on those materials. Tools and materials interact through one or more different processes. Processes are the sequence of actions that lead to some result.

Processes can be used to reach a desired goal or output, such as new energy, information, or materials. Examined in this chapter are the processes that change materials by adding or separating materials, or by making contour or internal changes. These materials conversion processes change the structure and thus the characteristics of materials. Within limits, you can select and use specific material processes to create the characteristics required in your designs.

ENRICHMENT ACTIVITIES

Develop a chart of the different characteristics of materials to include:

- strength—compression, tension, shear, and torsion
- stiffness, hardness, brittleness, flexibility, toughness, and density
- thermal properties of conductivity and expansion
- electrical properties of resistance and conductivity
- magnetic properties of natural and induced magnetism
- optical properties of transparency, translucency, and opacity
- properties of reflection, radiation, and absorption of heat and light

Obtain from your teacher samples that represent the different characteristics and functions of materials. Identify which of the samples fit best in the categories of properties in your chart.

DESIGN BRIEF

Using the information and findings recorded in your chart, choose and complete one of the following design briefs:

- Design and develop a model of a piece of clothing that will conserve body heat by reducing wind penetration.
- Disassemble and analyze a simple product, such as a toy. Determine the materials used in its manufacture and the appropriateness of each. Choose one of the products and illustrate how you would improve its design through the materials and processes used for its production.
- Revisit the design project on the seesaw from the preceding chapter. If the seesaw was to be made of wood, what properties would be most appropriate? What specific woods are most suitable? Remember to take into account the availability and cost of the different woods.
- You have been asked to determine which material would be best for the seesaw—wood, metal, or plastics—based on the criteria of strength, weight, cost, comfort of the user, and durability. How would you determine which material is most appropriate? What specific material would you choose?
- Using your new knowledge of materials, look at a range of products—toys, furniture, clothing—and predict what part of the products will fail first.
- What design changes would you propose in the products to improve their performance?

MATH/SCIENCE/TECHNOLOGY LINK

Obtain from your teacher samples that represent the different characteristics and functions of materials.

- Test each of the samples and rank order them in terms of their strength—compression, tension, shear, and torsion.

- Rank order the samples in terms of their properties of stiffness, hardness, brittleness, ductility, toughness, and density.

- Test and rank order the samples in terms of the thermal properties of conductivity and expansion; electrical properties of resistance and conductivity; magnetic property of magnetism; and optical properties of transparency, translucency, and opacity.

- Finally, identify which samples exhibit the best properties of reflection, radiation, and absorption of heat and light.

REVIEW AND ASSESSMENT

- Describe the four concepts of (1) the characteristics of materials, (2) their structure, (3) their function, and (4) the conversion of materials.

- Identify examples of each of the following characteristics of materials—thermal, mechanical, magnetic, chemical, acoustical, optical, and electrical.

- From the periodic table, identify the major groupings of elements.

- What are some of the different types of polymers and their characteristics?

- What is vitrification of ceramics, and how does it relate to green and fired ceramics?

- Describe and illustrate the close packing of crystals.

- What are some of the key differences of materials that are identified as monolithic materials, amorphous materials, material systems, dynamic materials, and families of materials?

- Give an example of sequence, action, changes, and result for each of the types of processes (conversion of materials, energy, and information), and indicate how they are similar.

- Identify and describe how materials can be changed by (1) adding materials to each other, (2) separating materials, (3) making contour changes, or (4) making internal changes.

- Using the conceptual model of materials shown in Figure 5.27, describe how the different materials conversion processes are similar.

Energy and Energy Processing

Figure 6.1 ■
Humans have found many ways of converting the prime energy sources found in nature to energy forms more suitable for their use. *(© Eric Meola/ The Image Bank)*

INTRODUCTION—THE DRIVING FORCE OF TECHNOLOGY

Energy is so much a part of our everyday lives that it is easy to ignore. Yet without energy directed to the purposes we want, even life itself would be impossible. Certainly, we could not stay comfortable, grow, work, be entertained,

141

KEY TERMS

atomic energy
biological energy
bond energy
chemical energy
compression
cosmic energy
efficiency
electromagnetic energy
energy conversion
energy differential
energy spectrum
entropy
fission
fossil fuels
frequency
fusion
geothermal energy
glucose
gravitational energy
Hertz
infrared
kinetic energy
magma
mechanical energy
molecular level
nucleus
oxidation
photosynthesis
potential difference
potential energy
primary energy sources
protein
radiant energy
ratio
secondary energy source
slow oxidation
solar energy
stress
tension
thermal energy
tidal energy
ultrasonics
waterhead
x rays

travel, or communicate without energy. Energy is needed to make things happen. Understanding energy becomes very important within your design and technology work because energy is the driving force for changing the materials you use. Any tools, devices, or machines you make or use will require energy.

Energy is available to us from several sources and can be converted into many different forms for storage and use. These three concepts—(1) sources, (2) forms, and (3) conversion—represent the major sections in this chapter. The general intent of this chapter is to help you understand energy, where it comes from, and how to convert it into more useful forms. This chapter lays the groundwork for understanding energy, so that you can put your knowledge about energy to practical uses.

Energy can be difficult to understand because its source is sometimes hidden. Also, energy is often converted through several different forms. One example, found in Figure 6.1, shows energy from coal being released through a chemical process (burning) to produce heat (thermal) energy. This energy can then be used to heat water to form steam. The pressure from steam can then be used to drive a generator. The generator, in turn, transforms mechanical energy into movement. This movement uses magnetic energy to produce electrical energy. How many examples of the use of electricity can you find in your neighborhood?

SOURCES OF ENERGY

If we trace the energy we use back to its source, whether it is the electricity we use, the food we eat, or the fuel we burn, we almost always end up at the sun. The sun is our biggest primary energy source. The sun works like a giant energy machine that releases energy from the reactions that occur deep within it. These reactions free the energy in the atoms and molecules that make up the sun.

Figure 6.2A is a simplified model showing the energy in the bonds within and between atoms and molecules. Much of the energy we use on Earth is also given off when bonds are broken between atoms or molecules of materials, and the bonds are reformed to make new materials. Most of the time, these bonds are broken by burning (rapid oxidation). Often our energy sources are in carbon, such as coal, gas, oil, and food, and are called fuels. (There are fuels other than those that are carbon-rich, such as uranium. However, the term usually means a source that has energy within its many carbon bonds.)

Energy is released from a fuel in a fairly straightforward way. Fuels have material units with high-energy bonds. With most fuels, these bonds are broken during burning and new, low-energy bonds are formed. The difference in the old and new energy bonds is released as (converted to) heat. This process is called rapid oxidation. (Refer to Figure 6.2B.) The process is similar when food is the fuel, but the energy is released through a slower chemical process called slow oxidation.

As bonds of a carbon fuel are broken, new bonds between carbon and oxygen are formed. This process results in gases of carbon dioxide and carbon monoxide, as well as other by-products. Neither of these gases requires as much energy to exist as did the original fuel. The extra energy is released in the form of heat.

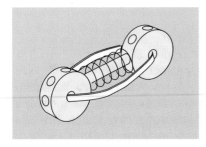

■ Figure 6.2A
This simplified model shows the basic source of energy in the bonds within and between atoms and molecules.

Energy goes through many changes of form, but the basic source of all energy is the bond between material units. Large usable reservoirs of energy are called **primary energy sources** from which other **secondary energy sources** derive their energy. The sun serves as the major primary source. Solar collectors that capture the heat from the sun or plants that use its light and heat to grow are secondary sources. Our discussion starts with primary sources. Much of our attention then turns to identifying or developing secondary sources for energy conversion and use.

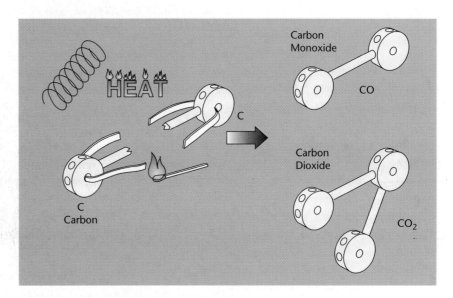

Figure 6.2B ■
The model also shows how a fuel releases energy. Through burning, carbon breaks down into new compounds of carbon dioxide and carbon monoxide. Since these require less energy in their molecular bonds, the excess energy is converted to heat.

Energy from the Sun

Solar energy is radiated from the sun. Solar energy includes visible light, **infrared** waves, microwaves, ultraviolet, and several other energy wave forms, as well as atomic particles. Solar energy heats the oceans and land surfaces and helps create air movement (wind). Solar energy also supplies plants with the necessary light energy for **photosynthesis**. Finally, solar energy keeps our planet warm enough to maintain life.

Solar energy provides all our energy except what comes from nuclear and geothermal sources. A solar energy power station that collects and concentrates energy from the sun is shown in Figure 6.3. Wind power, most water power, **fossil fuels**, and fuel from plants or animals are sources of energy derived from solar energy.

■ Figure 6.3
This device collects and concentrates energy from the sun. *(Novosti Press Agency/Science Photo Library/Photo Researchers, Inc.)*

The sun operates like a gigantic hydrogen furnace. Deep inside the sun, millions of tons of hydrogen are converted to nearly the same amount of helium. A large portion of the hydrogen is converted to energy which eventually emerges on the sun's surface. Much of that energy is in the form of heat and light.

There is no other energy source available for human use that comes even close to the reservoir represented by the sun. The amount of solar energy delivered to the Earth annually is five thousand times the energy available from all other usable sources.

Energy from within the Earth

Geothermal energy is a vast reservoir of heat that comes from molten rock, called **magma.** Magma makes up almost all of the volume of the Earth. The Earth's surface is a hard crust of rock only a few miles thick. This crust floats on a gigantic ball of molten rock. You have probably seen pictures of small amounts of that rock when it has surfaced as lava coming from a volcano. We use only a very small amount of the heat from this geothermal energy source. The heat comes from large domes of the molten rock that have been pushed up close to the Earth's surface—within six miles of the surface. Water from the surface of the Earth trickles down through cracks in the rocky crust. The water is heated to very high temperatures, as high as five hundred degrees Fahrenheit. Because the water is under tremendous pressure, it remains a liquid instead of turning into a gas. As it is driven to the Earth's surface, however, the pressure decreases. The water then boils and is converted to large amounts of steam.

A huge amount of energy is stored in magma. There are relatively few places on Earth that provide easy access to geothermal domes from which energy can be used. New technological developments are making geothermal energy more usable. Home heating systems, for example, can take advantage of the temperature difference between the ground and the outside air to achieve efficient heating and cooling. Someday we may develop techniques, such as a proposed nuclear "mole," that could bore through the rock of the Earth's mantle to get close to the magma. Such a machine would help us extract usable thermal energy even in places where the magma is more than six miles under the Earth's surface.

Energy from Movement of Air

Wind is one of our primary energy sources. The movement of the Earth itself and of the air, as it heats and cools, creates wind. The Earth tilts on its axis and changes the angle at which the sun's rays hit the Earth. More energy is received from direct rays from the sun, so some areas of the Earth absorb more energy than others at a given time. (You recognize the effects of this difference in seasonal changes.)

In addition, some materials absorb and lose energy more easily than others. Air, being a gas, becomes warm and cool faster than water does. Land masses also heat up and cool down faster than do bodies of water. This unequal heating of the land and water creates **energy differentials** in the surrounding air. This differential is usually noted as high- or low-pressure zones. Air expands when it is heated; therefore, a zone of hot air weighs less than an equal volume of air in a cold zone.

The larger the energy difference between the high and low zones, the greater is the speed or flow of the winds. Violent storms can be created in which the winds are a destructive force. Hurricanes and tornadoes result from extreme differences in the energy potential of two bodies of air.

The wind has been used for centuries as a source of energy. For instance, wind energy was the primary source of power for sailing vessels in the past. (Refer to Figure 6.4A.) Wind has been used in rural America for pumping water and

■ **Figure 6.4A**
The sail on this ship catches and uses wind energy.

■ Figure 6.4B
Elaborate "wind farms" are used to capture energy from the movement of air. *(U.S. Department of Energy)*

generating electricity. Today, wind farms are being developed and wind is being considered again as a usable source of energy to turn large windmills (for generating electricity). (Refer to Figure 6.4B.) Wind is also being harnessed as a source of additional energy to drive large ships.

Energy from Movement of Water

Energy from moving water can be traced to two influences of the sun on the Earth—gravitation and evaporation. **Gravitational energy** between the sun and Earth creates moving water in the form of tides. Energy, as heat from the sun, causes water to evaporate. This water eventually falls back to Earth as rain that can be stored in natural and artificial reservoirs.

Tidal energy is unique. Although tidal energy is linked with the sun, it is one of the few forms of energy that does not depend on radiated solar energy. The tides draw their energy from the gravitational forces of the sun and moon. The gravity of the sun and the moon serve to make oceans pulse in long, steady counts.

The intensity of the pulses depends on the position of the sun, moon, and Earth. If they all fall in a generally straight line, the high and low tides become more exaggerated. At these times the tides are more powerful. If the moon and sun form an angle with the Earth, the water is pulled in more than one direction. The tides are then less extreme and less powerful.

Tidal power takes two general forms. The most common is in the form of waves. These can be seen and heard breaking on coastlines around the world.

Tidal power is also obtained by the gradual filling and emptying of a bay or estuary. A dam can be built across the opening to hold water at high tide. At low tide, the water can be released. The dropping of the water creates energy which can be used to drive machines.

Water power is also available as a result of the Earth's gravitational force, which causes water to fall to a lower level when unhindered. The water energy reservoir is replenished by the natural weather cycle. Surface water evaporates because of the sun's heat and is redeposited at higher elevations in the form of rain, snow, or dew. These simple mechanics renew the water energy source and make this a continuous source of energy.

■ Figure 6.5
The potential energy in this waterhead (height of the water in the dam) can be used to drive a turbine at high speeds. *(U.S. Department of the Interior/Bureau of Reclamation)*

Water stored in a mountain lake or a mill pond represents **potential energy**. This energy has potential for doing work, but is basically at rest. As the water flows from the lake or mill pond, it falls to a lower level. Falling water has energy from movement similar to the energy of a falling brick. The moving energy of the falling water is called **kinetic energy**. A falling brick has energy to break open the shell of a walnut. The falling water has the energy to turn a water wheel or a turbine, which can drive an electric generator. In this way, the kinetic energy can be converted to mechanical energy.

The greater the fall of the water, the greater is the force that can be developed. This height is called the **waterhead.** Tall dams allow the waterhead to get very high. The fall of the water from this height can turn a turbine at great speeds and with great power, as shown in Figure 6.5. Most moving rivers have little waterhead, but the river's energy can be harnessed by water wheels. A low waterhead means the water has only a short way to fall and less potential energy to convert.

Energy from within the Atom

Nuclear energy is one of our newest forms of harnessed energy, available only since the 1940s. It is energy from the **nucleus** of the atom. Nuclear energy results from tearing the nucleus of the atom apart (**fission**) or from combining the nucleus of one atom with the nucleus of another (**fusion**). The forces that hold the nucleus of an atom together are very complex and extremely powerful.

Initially, nuclear energy was used for destructive purposes, in atomic bombs. Today, nuclear energy can be controlled and used as a heat source to make steam for generating electricity and for many other uses. A diagram of the "Tokamak," the giant donut-shaped nuclear fusion reactor designed to generate electrical power, is shown in Figure 6.6. There are two sources of nuclear energy.

1. In nuclear fission, the nucleus of large atoms is split apart. Tremendous amounts of energy are released.

2. In nuclear fusion, the nuclei of small atoms, such as hydrogen atoms, unite or fuse together. This produces a new nucleus, other particles, and a fantastic amount of energy.

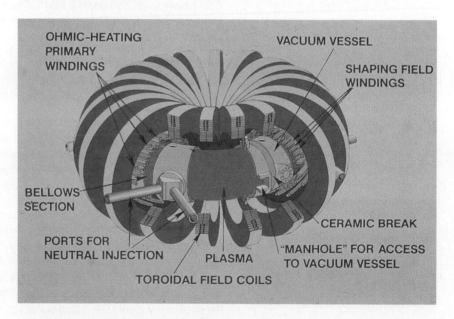

Figure 6.6 ■
Recent breakthroughs with the "Tokamak" nuclear fusion reactor hold great promise for generating electrical power. *(Courtesy of Princeton University: Plasma Physics Laboratory)*

The fusion reaction, not yet available on a large scale, requires extremely high temperatures and can only be done with a special nuclear fuel.

It is difficult to control the vast amounts of energy required for nuclear fusion reactions. The waste products from fission remain dangerous for many years—even centuries. These concerns have created many arguments about the use of nuclear power as a major source of energy.

Energy from Fossil Fuels

Fossil fuel is vegetable and animal matter that has been buried for millions of years under a certain set of conditions. Usually, vegetable and animal materials decay in the presence of oxygen—one of the gases in air. The carbon in the decaying material combines with oxygen in the form of carbon dioxide gas. This is called **slow oxidation**.

Fossils fuels, however, are formed when organic materials decay but do not have enough oxygen to decompose entirely. The incomplete oxidation and incomplete decay result in high-carbon deposits. These may take the form of peat, coal, oil, shale, petroleum, or natural gas.

The sun begins the sequence of events which create fossil fuel. Green cells in plants capture small amounts of the solar energy that reaches the Earth. The plants carry out photosynthesis to convert light energy to chemical energy which is stored by the plants. Over millions of years, only a tiny portion of the chemical energy in plants and animals remained in unused forms. Dead plants and animals were gradually covered over by dirt or buried in volcanic ash or water. Over time, the pressure of this covering built up and any decay that continued occurred without oxygen (anaerobic). This buried chemical energy became fossil fuel.

The process of fossilization is still going on but at a very slow rate. Peat is an example of a fossil fuel that is still forming. It would take a million years to produce less than one percent of the fossil fuels that have already been formed and that are being extracted now. Because it is so easy to burn these fossil fuels, they have been our major source of energy for the last two hundred years.

Based on estimates of our current use, the fossil fuels could be depleted in only a few more centuries. Oil and natural gas will be used up first. Coal reserves may last a century or two longer. In other words, limited amounts of fossil fuel now exist on Earth. Because little new fossil fuel is being generated through natural means, it is inevitable that we will someday exhaust the supply of this convenient type of fuel, and will need to develop ways to use other forms of fuels and energy.

Energy from Life Forms

Another source of energy is from plants and animals that are still alive. **Biological energy** comes from living things. Living plants and animals create food, fuel, and muscle power. To do this, they use minerals, organic compounds, and in the case of plants, sunlight. (Refer to Figure 6.7.)

Biological energy is based on solar energy and **chemical energy**. Green plants are the primary food producers. They are the only organisms that can convert light energy to chemical energy. Through **photosynthesis**, plants produce **glucose**. They use glucose to live and grow.

Glucose is the basic energy ingredient of the foods we eat. The energy is in fats, carbohydrates, and proteins. When animals consume more glucose than they use, it is converted and stored in the body, primarily as fat. Fats or oils in plants are usually stored in the seeds. Fat and oil are a concentrated reservoir of energy. In other words, they have high energy density. Glucose from carbohydrates and

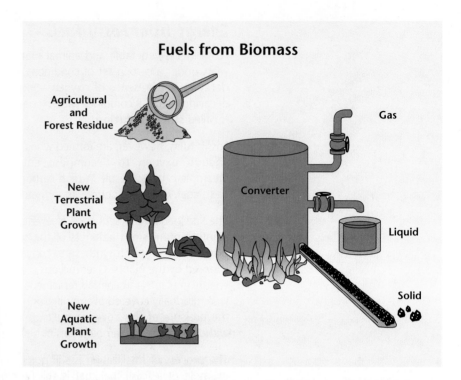

Fuels from Biomass

Agricultural and Forest Residue

New Terrestrial Plant Growth

New Aquatic Plant Growth

Converter

Gas

Liquid

Solid

Figure 6.7 ■
Organic materials may be converted into usable fuels of solids, liquids, and gases through a biomass process.

proteins produce four calories per gram when burned as muscle energy. Fat, however, produces nine calories per gram.

There is great variation in the **efficiency** of plants and animals that use and provide biological energy. Of the solar energy that falls on plants, only about 1 percent is converted into chemical energy. About 10 percent of this chemical energy in green plants can be digested and used by animals, including humans. Of that 10 percent, only about one-quarter is used to power the work done by our muscles. Most of the remaining energy is dissipated as heat (normal for warm-blooded animals).

D E S I G N   B R I E F

There are many sources of energy you can use to operate the devices you design. To help you better understand the different sources of energy and their uses, the following class activity is proposed. You (or you and your team members), as part of the class, are to design and develop a device or system that will measure energy from devices that produce movement, electricity, and heat. Three suggested devices are shown on the next two pages.

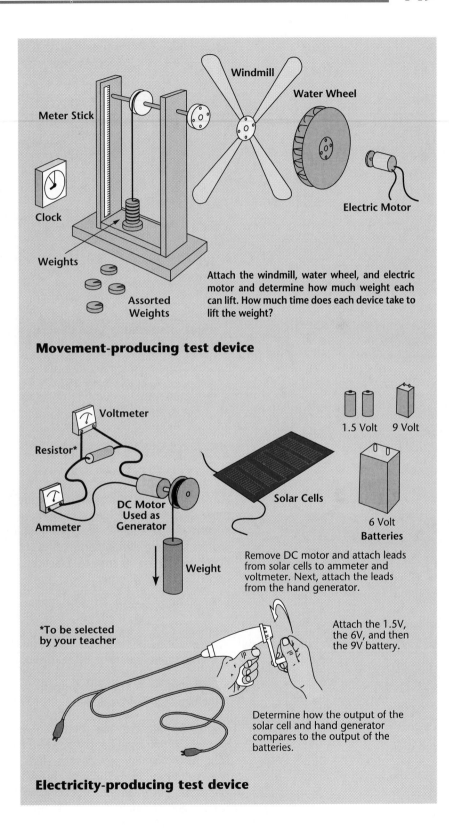

Attach the windmill, water wheel, and electric motor and determine how much weight each can lift. How much time does each device take to lift the weight?

Movement-producing test device

Remove DC motor and attach leads from solar cells to ammeter and voltmeter. Next, attach the leads from the hand generator.

*To be selected by your teacher

Attach the 1.5V, the 6V, and then the 9V battery.

Determine how the output of the solar cell and hand generator compares to the output of the batteries.

Electricity-producing test device

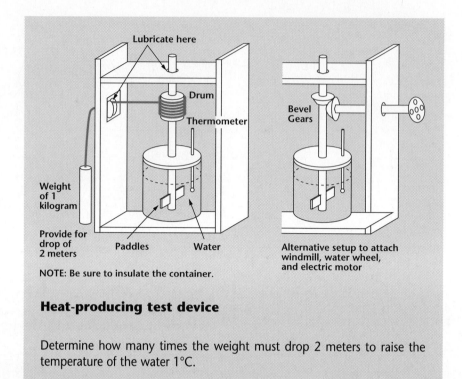

Heat-producing test device

Determine how many times the weight must drop 2 meters to raise the temperature of the water 1°C.

FORMS OF ENERGY

As you have seen, energy sources are reservoirs from which we draw power to move ourselves, our tools, and our machines. The energy sources must be converted into usable forms. There is little practical value in merely knowing the sources of energy. An understanding of the forms of energy will be another step closer to putting energy to practical uses. Four related ideas will help you understand energy forms.

• All energy is similar.

• In theory, any energy form can be changed into any other form.

• Energy cannot be made or destroyed. Energy can only be changed.

• In the process of converting energy to work, not all the energy is converted to useful work. The remainder is usually dissipated as heat.

The portion of energy that cannot be used for work is described as **entropy**. For example, as you use a battery-powered flashlight, the bulb creates heat as well as light. As the battery provides the electrical energy through a chemical process, heat is also produced. This heat energy is no longer available, in the sense that it cannot be put to use in producing light. Entropy is an important concept of science and technology and has been expanded to include information and signal loss. This concept of entropy will be revisited in future chapters.

MATH/SCIENCE/TECHNOLOGY LINK

- If your design dealt with lifting, identify how much weight it can move. Express this weight (force) in terms of newtons (N). (Here on Earth, the mass of one kilogram exerts a force of 10 N.) Determine how high your device can lift this weight, as shown in the previous illustration, "Movement-producing test device." Calculate the work done when:

work done = force × distance moved (in the direction of the force)

$$w = f \times d \qquad \text{Work is expressed in units called joules (J)}$$

The relationship represented in the formula indicates that a force of 1 newton applied through 1 meter will equal 1 joule. For example: 2 N through 3 m = 6 joules. If your device can lift 300 grams a distance of 75 centimeters, you would then calculate:

work done = 300 grams × 75 cm   300 grams = 0.3 kilogram (because 1 kg exerts a force of 10 N, 0.3 kg = 3 N)

= 3 N × 0.75 m   100 g = 0.1 kg = 1 N

= 2.25 J

- If you want to express this in terms of power, you must consider the time it takes to do the work. The relationship is 1 joule per second = 1 watt. For example: 3 N through 2 m in 4 seconds = 6/4 watts, or 1.5 watts.

If your device took five seconds to move 300 grams 75 cm, its power will be:

power = rate of doing work = $\dfrac{\text{[work done]}}{\text{time required (seconds)}}$ (expressed in joules per second)

$$= \dfrac{2.25 \text{ J}}{5 \text{ s}} \qquad = .45 \text{ J/s} \qquad = .45 \text{ watt}$$

- It will be helpful to remember that the unit of one joule per second (J/s) is also expressed as a watt, named after James Watt, designer of early steam engines. Knowing that 1 watt = 1 J/s will allow you to compare the power produced as electrical energy with the power produced as mechanical energy. Power related to electricity is expressed as:

power = voltage × current, or $P = V \times I$
where
power is measured in watts (W), voltage in volts (V), and
current in amps (A or I)

Current (amperes) is already expressed in terms of time (the amount of electrons that flow past a point in a second). Determining how many watts are developed by your device allows you to translate that amount directly into joules per second J/s. Conduct a test that uses an ammeter and a voltmeter, as shown in the previous illustration, "Electricity-producing test device," for determining the electrical energy that can be generated from the power output of your device or system.

152 Unit 2 The Resources of Technology

• If you plan to measure the output of energy of your device as heat, you will need to determine how many calories were generated. A calorie (cal) is designated as the heat required to raise a gram (1 g) of water one degree Celsius (1°C). The calorie is equivalent to 4.187 J. Use the test device shown in the previous illustration, "Heat-producing test device." The device uses the stirring of water to see how much heat your device can generate in calories.

Energy Spectrum

One way to study the range and relationships of energy is to arrange energy forms into a theoretical spectrum. The **energy spectrum** provides a "clothesline" on which we can hang the major categories of energy forms. The spectrum ranges from energy as extremely small units of matter (atomic particles), to energy of incredibly fast vibrations (cosmic rays).

Near the center of the spectrum is the energy activity that we can experience through our senses. The energy particles are large and travel slow enough to be detected. Energy forms in the middle range (heat, light, and sound) have been used for a long time. Use of energy in the outer ranges is less common because fewer tools and machines have been developed that can harness that energy form.

The spectrum has six major bands, including **atomic energy**, **chemical energy**, **mechanical energy**, **electromagnetic energy**, **radiant energy**, and **cosmic energy**. (See Figure 6.8.) Each band is unique, but those forms of energy at the boundaries of the bands may have characteristics of both energy bands.

Figure 6.8 ■
This matter-to-energy spectrum ranges from the very small atomic particles on the left to the very fast cosmic rays on the right. All forms of usable energy fit between these two extremes.

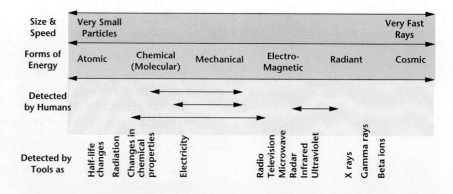

CAUTION

All forms of energy are potentially dangerous to humans if the levels of exposure become too high. Mechanical and electrical forms of energy, such as moving devices and machines or high-voltage devices, can cause direct and immediate bodily injury. Many of the other forms of energy have effects that are evident only in the long term.

BE CAREFUL WHEN YOU ARE USING ANY FORM OF ENERGY!

Atomic Energy

The chapter on Materials and Materials Conversion stresses the importance of bonds, which are what holds the insides of atoms together and what holds atoms to one another in molecules. A bond is a balance between a repulsive force and an attractive force. An earlier example used an elastic band to demonstrate a bond: When you stretch an elastic band between your hands, there are two forces—the pulling force you generate and the band attempting to return to its original shape. Energy is stored in the bonds of the band. If you cut it, the bonding energy is suddenly released, and the band stings you.

Chemical energy is released, as when gasoline burns, when the bonds between atoms and molecules are broken. Inside each atom, which is mostly empty space, an immensely hard nucleus is held together with immensely high forces (high energies). We are able to release this energy slowly in nuclear reactors or very quickly in nuclear weapons. (The old-fashioned and wrong name was atomic energy or weapons; the true atomic energy is chemical energy.) The energy in the nuclear bonds in the nucleus can be released by splitting the nucleus (fission) or by combining nuclei (fusion).

MATH/SCIENCE/TECHNOLOGY LINK

In nuclear fission, when the nuclear bonds are broken, the resulting total nuclei actually weigh a little bit less than what the nuclei did at the start of the process. The energy released by this change of mass (m) is very large and is determined by Albert Einstein's law of relativity, $E = mc^2$, where (c) is the speed of light or 3 times 10^8 meters per second. Let us consider what happens when a very small mass of nuclei (say one millionth of a gram or roughly the weight of a period at the end of a sentence) is converted into energy. What would be the total amount of energy released? (For more details, refer to the Energy chapter in the *Portfolio and Activities Resource*.)

The sun, a typical star, produces energy by the nuclear fusion of hydrogen that eventually produces helium. (Helium is named after the Greek god of the sun, Helios, because helium was detected on the sun before it was discovered on Earth.)

Chemical Energy

At the **molecular level,** the bonds between molecules involve the energy of attraction and repulsion, or **bond energy**. Chemical processes such as burning, photosynthesis, and digestion release the energy that is stored in high-energy bonds within the molecule. As new molecules that contain lower-level energy bonds are formed, the excess energy is released. As indicated before, these changes can often be detected as heat.

A common use of chemical energy is in the battery. In each cell of the battery, chemical reactions occur when the chemicals are allowed to combine. This combination of materials results in a flow of electrons. In this process, an excess of electrons occurs in one of the materials that becomes negatively charged. The electrons are then attracted to another material that is positively charged and is deficient in electrons. Batteries operate because the electrons cannot move from one plate to another within the battery, but they can move relatively easily through circuits that are connected between the positive and negative poles of the battery. (Refer to Figure 6.9A.)

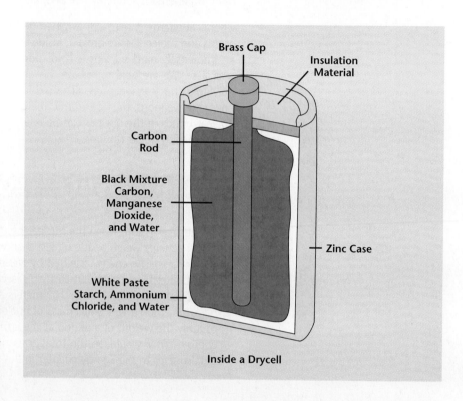

Brass Cap

Insulation Material

Carbon Rod

Black Mixture Carbon, Manganese Dioxide, and Water

Zinc Case

White Paste Starch, Ammonium Chloride, and Water

Inside a Drycell

Figure 6.9A ■
An ordinary battery produces energy (electricity) through a process of chemical changes among materials. *(From Herman, Delmar's Standard Textbook of Electricity, ©1993 Delmar Publishers, Inc.)*

Some chemicals combine with a slow release of energy, others with an explosion. As with other forms of energy conversion, it is important to control the rate at which energy is released so that it can do the intended work. Heat is often the by-product of chemical processes.

CAUTION

Although most chemical processes are safe, some can be quite dangerous.

DO NOT ATTEMPT TO CONDUCT ANY UNSUPERVISED EXPERIMENTS!

■ Figure 6.9B
Food is a form of fuel that can be converted readily through chemical processes (digestion and metabolism).

Thermal Energy

As indicated earlier, **thermal energy** is found in all the energy bands. Thermal energy relates to the level of kinetic energy of the molecules of a material. Heat energy is produced by almost all enegy conversion processes, either as a by-product or as the desired form. Heat is transferred by three means—conduction, convection, and radiation. Heat is one of the most used forms of energy in common everyday uses such as cooking, as well as in more exotic technological applications such as sintering.

To understand some other interesting applications of thermal energy, it is necessary to understand what happens when a solid is turned into a liquid and then to a gas. When this happens, energy is needed to overcome the attraction between the bonds. This is not just a matter of cooling the material to condense it or heating it to expand the space between molecules. Two concepts come into play called the "latent heat of evaporation" and the "latent heat of fusion." For example, in the boiling of water, the temperature stays the same (even though heat is still being added) until all of the water has turned to gas (water vapor). The temperature remains constant while the energy that is added will equal the latent heat of evaporation. The same phenomenon occurs as you take the heat away from the water vapor to convert it back to water. The process is similar as water is cooled sufficiently to freeze and turn to ice. The energy that is extracted from the fluid must equal the latent heat of fusion. The concepts of latent heat of evaporation and fusion are put to use in liquids called "coolants." These coolants readily absorb heat and in doing so, turn to gases. For example, thermal energy can be extracted from the contents of your refrigerator or your air-conditioned room. When allowed to return to their natural gaseous state, the coolants release that energy (the latent heat of fusion) and are ready to start the cycle over again to extract and release more heat.

Mechanical Energy

To understand mechanical forms of energy, you need to understand the forces called **compression** and **tension**. Although the ideas of compression and tension were introduced in some detail in the chapter on Materials, an example is appropriate here. When you bounce a ball, the energy from your arm throws the ball to the floor. The material of the ball absorbs the impact with the floor and the molecules are compressed closer together. The molecules of the rubber in the ball resist the change and attempt to push back to their original state. This causes **stress**. The stress is released in the bounce of the ball. If you catch the ball before all the energy in the stress is released, you feel the impact on your hand. If you bounce the ball for a longer period of time, you will notice that it gets warm as part of the energy of compression and tension is changed to heat energy.

MATH/SCIENCE/TECHNOLOGY LINK

Sir Isaac Newton developed three basic concepts (now known as Newton's Laws of Motion) that are useful in explaining movement.

Newton's 1st Law—In the absence of force, an object at rest will stay at rest.

Unless you throw the ball or pull on the rubber band, they will just sit there.

Newton's 2nd Law—Objects in motion tend to stay in motion.

Granted, the force of gravity and friction will slow down your rocket and cause it to stop flying if you are on Earth. In outer space, with very little matter to rub against and little gravity, your rocket would just keep going.

Newton's 3rd Law—For every action, there is an opposite and equal reaction.

This explains the force you must use to keep your hand in place when you catch the ball. It also is evident in the heat that is formed in your bare hand when you catch the fast-moving ball and the ball quickly loses its forward motion.

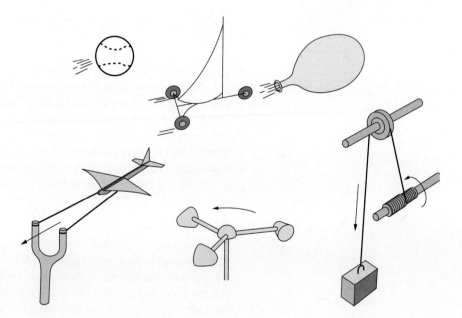

Figure 6.10 ■
Some simple examples of Newton's Laws of Motion.

Suppose we first compress and then stretch a material. The change in the internal structure produces a rhythmic vibration. Some high-quality watches keep time by using the rhythmic vibrations generated by materials such as quartz. In industry, devices such as vibrators and oscillators can make similar rhythmic changes. Some materials have high tensile strength and will come back to original form when stretched. The longer the stretch, however, the greater the force when it pops back. (Opposite and equal reactions, again.) If the force becomes too great, the materials will break into pieces.

If we stretch and compress something a little more rapidly, internal strain becomes detectable movement. We actually create a motion which is constantly changing with a regular tempo. A tree swaying back and forth in the wind has a tempo, as does a vibrating tuning fork.

Have you ever strummed a guitar? You set the string in motion with your pick or fingers. The string doesn't move in only one direction, however. It moves back and forth. The distances traveled each way are nearly equal. The movement continues to decrease until the guitar string is at rest. This back and forth movement is called vibration. The number of vibrations in one second is called **frequency**, which is measured in cycles per second, now known as **Hertz (Hz)**.

As we increase the frequency of the vibration, we reach a point on the spectrum in which the energy can be heard as sound. This is a name given to vibrations in materials with frequencies in the range of 20 to 20,000 Hz or cycles per second. Continued increases in frequency send the pitch of the sound higher and higher. Eventually, the frequencies are out of the audible range and into the area of ultrasonics.

Electromagnetic Energy

Even if a material can stand the strain of the increasing frequency of vibration, there is a point at which a given material just can't move fast enough. There is a physical limit to mechanical vibration. Beyond this point is the electromagnetic band of the spectrum.

Most frequencies up to 7,000 Hz are generated by means of electrical or electromechanical techniques (generators). The most common generator consists of a coil rotating in a magnetic field. This coil or wire cuts magnetic lines of force and creates an electrical current. Each time the moving wire changes direction, it creates an electrical current. The faster the coil is rotated, the greater the frequency.

Eventually, there comes a point at which a generator coil cannot rotate any faster. This point is the top frequency limit of the electromechanical band of the spectrum. Frequencies above 7,000 Hz normally must be produced by other devices. For example, ultrasonic transducers are used to generate frequencies greater than 20,000 Hz. The most common instance of **ultrasonics** is sonar for underwater detection of submarines and fish or through scanners used to map internal body parts. (See Figure 6.11.) The higher frequencies of ultrasonics overlap the lower radio frequencies.

The frequency of energy waves increases through the spectrum of radio waves, television waves, and microwaves. (Refer to the diagram of a microwave oven in Figure 6.12.) When food is cooked in a microwave oven, for example, some materials become quite hot. The oven itself and, hopefully, the dish will stay cool. Solid masses, such as body organs or tumors, respond to waves such as those emitted by a CAT (Computer Axial Tomograph) scan or an MRI (Magnetic Resonance Imaging) in a different way than body fluids and bones respond. These differences can be captured on a computer image or film.

Devices that generate higher frequencies are needed to produce **infrared** energy. This is the beginning of the spectrum of visible light. The spectrum changes in color as we continue to increase frequency. The light spectrum goes through the reds, yellows, blues, and into the ultraviolet.

■ Figure 6.11
In the future, physicians might use virtual reality technology to provide an improved view of an unborn child, using ultrasonic imaging.

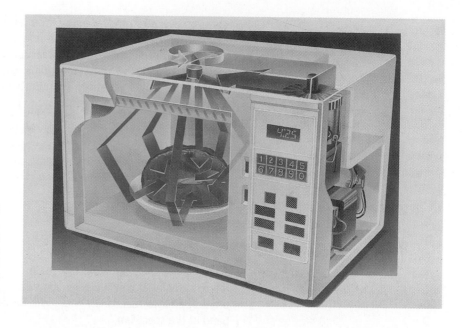

Figure 6.12 ■
In the microwave oven,
electromagnetic energy is
converted into microwaves. The
waves excite water molecules and
the rapid movement produces heat
that cooks the food.
(Ian Worpole/Discover Magazine)

Lasers are highly focused light energy. The use of the laser for precise cutting has made some major changes in surgery on humans as well as in cutting nonliving materials with extreme accuracy and little heat.

CAUTION

High energy forms can be very dangerous to humans. Light from lasers, ultraviolet rays from the sun and some lamps, and X rays can cause injury or disease.

DO NOT SUBJECT YOURSELF TO EXCESSIVE AMOUNTS OF HIGH-ENERGY RADIATION!

USE PROTECTIVE MATERIALS, SHIELDS, AND GLASSES TO BLOCK THE RADIATION.

Radiant and Cosmic Energies

Now we reach the energy level of the radiant band. **X rays** and gamma rays are in the radiant band. We have only a few tools that use or generate energy at this extreme level of frequency. Energy at the extreme end of the known spectrum is called cosmic energy. Just as we know very little about energy at the extremely small level of matter (subatomic), we also know very little about generating or using energy at the extremely fast level of rays (cosmic).

Because we know so little about the conversion and control of energy in this portion of the energy spectrum, there is concern for potential side effects for humans and other life forms. Exposure to X rays and ultraviolet light are known to produce unwanted changes in the body that increase the likelihood of developing cancerous cells. Regardless of the form of the energy, it is important to control the focus and quantity of the energy that is emitted.

D E S I G N   B R I E F

There are many different forms of energy included in the energy spectrum. The spectrum was described as a "clothesline" on which to hang the major categories of energy. Of those forms of energy, six are the most common and available for people to use. These include chemical, thermal, mechanical, sound, electrical, and light. All of the energy forms can have undesired effects on humans, if the amount of exposure exceeds the levels of human tolerance.

• Refer to the illustration of the Environmental Tolerance Zones (ETZ) shown in Figure 2.18 and provided in your *Portfolio and Activities Resource.*

• Match the appropriate factors of the ETZ that fall within the categories on the energy spectrum.

• Design and develop a display that illustrates the range from comfort to tolerance limit for one of the ETZ factors.

MATH/SCIENCE/TECHNOLOGY LINK

After you have completed your study of the different forms of energy, determine:

• which forms of energy are potentially most dangerous to humans,

• which of the forms are linked to the energy from the sun, and

• which of the forms are linked to sources that are not directly "sun-related."

Through your study of energy, its sources, forms, and conversion, determine:

• what is meant by the conservation of energy,

• what is meant by energy differential, and

• what the concept of entropy means as it relates to energy.

CONVERSION OF ENERGY

This section on the conversion of energy brings you to an important stage of your design and technology work. As you design and plan the objects you wish to build, you need to determine the energy requirements. Our attention will be turned to energy systems you might use most often in your design and development work. These include mechanical, fluid, and electrical systems. Other systems may be useful, depending on the outcome you design.

We use the idea of **energy conversion** to indicate that energy is being changed from potential, or stored, energy to kinetic energy or from kinetic back to potential energy. In the chapter on Materials and Materials Conversion, we saw how potential or stored energy is released through the conversion of materials where fuels are burned.

Converting Energy for Storage

The storage of energy may appear to be a simple matter, but generally that is not the case. For example, it is not practically convenient to store large amounts of electrical energy. It has to be stored in other forms of energy and converted back to electrical energy when we need to use it. This is also the case with some other forms of energy. Descriptions of some of the key energy storage systems are presented below.

Batteries. Batteries are an important method of storing relatively small amounts of electrical energy. Batteries use chemical processes to convert materials into the flow of electrons as direct current (DC). Similarly, the flow of electrical current can cause materials in some rechargeable batteries to convert back to their original state. This conversion of the materials within a battery is commonly referred to as charging the battery.

Hydrogen gas. Hydrogen gas can be produced by the flow of electricity. This process, called electrolysis of water, separates the hydrogen and oxygen gases so they can be stored. The hydrogen can then be converted back into electricity when burned as a fuel. The only by-product of using hydrogen as a fuel is water, making hydrogen a very clean burning fuel.

Hydrogen is also used experimentally in fuel cells to produce electrical energy. Although support of research and development has been modest, there have been a number of important developments in the use of fuel cells over the past few years, but much remains to be done before it will represent a major source of energy.

Water storage. Another method of providing energy when needed is water storage. This has been used for many years when dams have been built on rivers to create reservoirs. The water from the reservoir can then be used to turn electrical generators to produce power as needed. Recently, there has been the increased use of electricity for pumping water back into the storage reservoirs. During periods of low demand for electrical power, pumps are driven to move water back through the system for use during high-demand periods. This practical approach is limited, however, to sites that have suitable reservoirs and systems for generating electricity by hydropower.

Flywheels. Flywheels are a practical means of harnessing and storing mechanical power. A common use of a flywheel is found in the internal combustion engine. The flywheel stores energy between the thrust of each piston and smoothes out the energy transmitted to the drive wheels. In stationary machines, the spinning mass of a heavy flywheel can then be used to drive a generator and produce electrical energy. Flywheels can also be used to store mechanical energy to propel a vehicle. Capturing the energy as the vehicle is slowed down could help reduce the amount of energy lost as heat when brakes are applied. As the energy in the spinning flywheel is used to get the vehicle moving again, less energy would be required from the power system of the vehicle.

Heat storage. A commonly used means of holding energy for future use is heat storage. In standard and solar buildings, selected materials in the buildings are used to absorb heat during the heat of the day. This heat is then released into the interior of the building as the outside temperature drops at night. For example, selected salt compounds are used as storage media because of the high levels of thermal energy they can absorb, compared to water or stone. Another instance of heat storage is found in the use of heat pumps that draw energy from underground. In winter, there is a usable energy differential between the cold temperatures above ground and the relatively warmer temperatures below ground. Heat pumps are designed to convert that thermal difference to usable energy. Some research is being conducted on warming water by solar energy so that it can be pumped underground to create a reservoir of heat for use during the winter months.

Converting Energy for Use

Energy is generally converted from one form to another through the use of two types of devices—converters and transducers. Most of our attention in the remainder of this chapter will be on converters. We will then introduce transducers, which will be treated more in-depth in the chapter on Information.

A converter uses one energy form to produce a new form or to change the nature of the original energy form. We have already indicated that living things convert energy from the sun and from the consumption of food. Converters include people, plants, generators, engines, motors, and transformers. Any machine, animal, or plant can be considered an energy converter since they all use energy to make things happen. In converters, one form of energy is the input while the output is another form of energy.

Input \ Output	Chemical	Heat	Mechanical	Sound	Electrical	Light
Chemical	foods, plants	fire, food, steam boiler, furnace	rocket, gas engine, animal muscle	explosion, smoke detector	battery, fuel cell	candle, oil lamp, phosphorescence
Heat	gasification, vaporization, distillation	heat pump, heat exchanger	turbine	explosion, flame tube	thermopile, thermocouple	fire, arc lamp, lightbulb
Mechanical	gunpowder	friction, brake	flywheel, pendulum, water wheel	voice, musical instrument	generator, alternator, steam power plant	flint, spark
Sound	hearing	sound asbsorber	ultrasonic cleaner	megaphone	telephone, microphone, hearing	light amplifier, color organ
Electrical	batteries, electrolysis, electroplating	toaster, heat lamp, spark plug	electric motor relay	horn, thunder, loudspeaker	diode, rectifier, transformer	television, fluorescent light, light-emitting diode
Light	plant photosynthesis, camera film	laser, heat lamp	photoelectric door opener, electroscope	movie soundtrack, video disk	solar cell, photoelectric cell	laser, light-wave repeater

Figure 6.13 ■
This Energy Conversion Chart shows examples of how energy can be changed. Energy is usually changed into different forms, such as electricity to heat. Energy may be changed from one form of itself to another, such as when electricity is transformed from a high voltage to a lower voltage. Energy may also be changed to allow us to measure the amount of an energy form, such as determining the temperature of a room.

Mechanical Energy

There are several reasons for using a tool, device, or machine. Often the intent is to expend as little energy as possible from your muscles and make a machine do the work. You also want to use as little energy as possible in as efficient a manner as possible. There are some principles that will allow you to multiply the effects of the energy by your design. You will be achieving a mechanical advantage (MA).

Over a period of time in using machines, formulas have been developed to allow calculations related to the use of devices to obtain a desired MA. For example, if you want to determine the speeds required and the amount of force needed, there are formulas you can use. The calculations may seem somewhat confusing at first, but they are straightforward and based on a simple model.

MATH/SCIENCE/TECHNOLOGY LINK

You have already been introduced to some of the major ideas and skills related to mechanical systems in the chapter on Tools, Mechanisms, and Machines. Now it is appropriate to revisit some of these ideas and apply them in the conversion of energy. As indicated earlier, the principle of the lever is a key element to understanding tools, mechanisms, and machines. The lever involves four related ideas: load, fulcrum, force, and movement.

(a) The load is the force to be moved by the lever.

(b) The fulcrum is the pivot or the point around which the lever turns.

(c) The effort is the force that is applied to the lever.

(d) The movement is the turning effort acting around the fulcrum.

Load is the amount of mass of the object to be moved. If a load is at rest, it will stay in place as long as the downward force (gravity) is the same as the upward force (support). The support may be provided by a structure or by the Earth itself. If the forces are balanced, the object is in equilibrium and will remain at rest. If you are designing a structure, you will be pleased if it is stable or in equilibrium. If you are designing a playground ride, a vehicle, or other moving structure, you will want to overcome that equilibrium in order to create the desired movement. This means your design will have to overcome the inertia of the object at rest so that you can move the object.

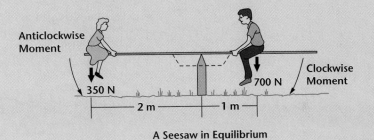

A Seesaw in Equilibrium

■ Figure 6.14
This seesaw is an example of a lever system in equilibrium.

The concept of equilibrium will be important in your design and technology work. It can be represented in a rather simple form. As you use a lever, you can think of the fulcrum (pivot) as the equal sign (=) in an equation. If the object is in equilibrium, whatever is on one side of the fulcrum (or =) is equal to what is on the other side. Specifically, that means that the force multiplied by the distance from the fulcrum equals (=) the load multiplied by its distance from the fulcrum. In the example illustrated in Figure 6.14, the young man can be considered as a load of 700 N that is 1 meter from the fulcrum. The load he represents is in equilibrium with the force of 350 N at 2 meters from the fulcrum, generated by the young woman.

Turning Forces

Suppose you took that lever and kept moving it around the fulcrum, or pivot point. Soon you would have an arc or a full circle. You can see that as you move an object on a lever, you are moving it through part of a circular arc. As the knowledge of the lever and mechanics evolved, the principle of the lever became the Law of the Lever. This law actually applies to movement in an arc or circle (wheel). Movements in a circular direction are caused by turning forces. It is probably evident that wrenches, cranks, screwdrivers, and doorknobs all use turning force (called torque in such instances). Turning forces are an important part of the Law of the Lever, as shown on the next page.

THE LAW OF THE LEVER

The **Law of the Lever** states that:

a clockwise turning	=	**a counterclockwise turning**
force of effort		**force of the load**

This means the input force to move the load will equal the force created by the load.

This law can also be stated as:

clockwise moment of force = **anticlockwise force of load**

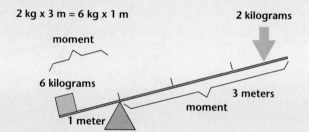

Turning force = force x distance from the turning point (moment)

■ **Figure 6.15**
Turning forces and moments.

If you look closely, it becomes apparent that seesaws and wheelbarrows also use turning forces. (They move in a small portion of the total circle.) The turning effect of the force is called the moment of force and sometimes the moment arm. The moment of turning force around a point is described in the equation in Figure 6.15.

We use the lever to provide us with a mechanical advantage (MA) to make it easier to move a load. In the instance shown in Figure 6.15, the mechanical advantage is approximately **3**, or sometimes stated as 3 to 1. The energy output, however, is never exactly the same as the energy input, the work. Can you identify why the mechanical advantage is always less than 3 to 1?

MATH/SCIENCE/TECHNOLOGY LINK

Turning Forces

turning force = force times distance (at which the force is applied) from the turning point

The units of these quantities are:

turning force	distance	newtons (N)
force	newton meters (Nm)	meters (m)

turning force = newton meters (Nm)	force = newtons (N)
distance = meters (m)	

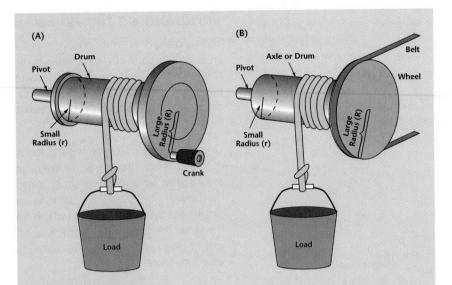

■ Figure 6.16A, B
These examples show (A) a windlass and crank, and (B) a wheel and axle. Both have approximately the same moment arms.

Turning Forces and Moments

The turning forces in a windlass or a wheel and axle are shown by:

load × r = effort × R

where small r is the radius of the axle or drum and large R is the radius of the wheel or crank. The size of the moments is determined by the size of small r and large R.

The formula below shows a shorthand way of describing the mechanical advantage (MA) of a wheel and axle. The same formula can be applied to pulleys, sprockets, and gears. Remember that belts, chains, and gear teeth serve to transmit motion from one wheel to another. Complicated setups allow for a large mechanical advantage to be achieved.

$$MA = \frac{load}{effort}$$

Because devices are not 100 percent efficient and some useable energy is converted to heat, you can describe mechanical advantage using the above statement.

You will see in the next Math/Science/Technology Link how the relationship of

$$\frac{R}{r} \quad or \quad \frac{radius\ of\ wheel}{radius\ of\ axle}$$

allows you to compare the velocity of the wheel and the velocity of the axle. This is called the velocity ratio, or VR.

Variations on the Wheel

The windlass is a common device, but we usually see it in the form of a crank (a lever, moving in a circle around a pivot). The fulcrum of a lever, the pivot for a crank, and the axle or drum in a wheel all serve the same function. In some machines, the axle is a wheel itself. This is the case in examples using gears. In all cases, the intent is to provide a mechanical advantage for turning something in a rotary motion.

One key point in determining the mechanical advantage of a crank is the length of the arm of the crank. A long arm (a large moment) provides a larger mechanical advantage; a short arm (a small moment) provides a smaller mechanical advantage. The same is true for other variations of the wheel. The smaller the wheel where the force is applied, in relation to the wheel that does the work, the greater the mechanical advantage. If you apply the force to a small axle or drum and that force is transmitted to a large wheel that moves the load, your input force is multiplied.

MATH/SCIENCE/TECHNOLOGY LINK

Velocity Ratio (Relative Speed)

Let us now turn attention again to the idea of relative speed or velocity of the wheel (or crank) and the axle (or drum). A formula similar to the one above can be used to determine the **ratio** of velocity of the wheel and axle.

The velocity ratio (VR) is determined by using:

$$VR = \frac{\text{circumference of wheel}}{\text{circumference of axle}} = \frac{2\pi R}{2\pi r} = \frac{2\pi R}{2\pi r} = \frac{R}{r}$$

This allows you to determine the velocity ratio simply by comparing the two radii or diameters.

This principle and formula applies to pulleys, sprockets, gears, and the networks of these if they are linked together. Chains and gears have an advantage over belts in that no slippage (therefore, wasted energy) occurs. The concepts are the same when applied to these variations on the wheel. Velocity ratios are determined by the radii or diameters of the pulleys.

When using the formula with gears, it is not necessary to use a measurement to calculate the radius or circumference of the wheels. The teeth will be of even size—a tooth on one wheel will engage a tooth on the other wheel. Therefore, it is convenient to compare the ratio of the number of teeth on one gear with the number of teeth on another. The formula then becomes:

$$VR = \frac{\text{number of teeth on the gear that moves the load}}{\text{number of teeth on the gear where the force is applied}}$$

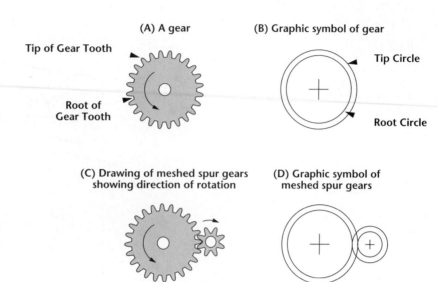

(A) A gear

Tip of Gear Tooth

Root of
Gear Tooth

(B) Graphic symbol of gear

Tip Circle

Root Circle

(C) Drawing of meshed spur gears
showing direction of rotation

(D) Graphic symbol of
meshed spur gears

Figure 6.17 ■
Gears can be used to change the
direction of movement and to gain
a mechanical advantage. Less
energy is needed in order to
accomplish the work.

It does not matter if the gear is a sprocket or if it is a bevel, miter, spur, or worm
gear. Remember, the worm gear is a special case (most have only one tooth that
looks like a long thread or worm), and the calculation is comparing one tooth on
the worm gear to the number of teeth on the worm wheel. Because of the big
ratio difference (e.g., one tooth to 40 teeth), a worm gear provides a significant
gear reduction requiring little space.

Rack and pinion gears provide the capability of converting rotary motion to linear
motion or from linear to rotary motion. The basic function of a rack and pinion is
to convert a back-and-forth movement to a circular one, or convert a circular
motion to a back-and-forth one. The movement of the rack is determined as:

$$\text{movement of rack} = \frac{\frac{\text{number of teeth}}{\text{on pinion}} \times \frac{\text{number of times}}{\text{pinion turns}}}{\text{number of teeth per meter on rack}}$$

D E S I G N   B R I E F

You have been asked to work in a design group that has the responsibility
of providing a means for moving physically challenged and elderly people
from one floor to another in a six-story building. The specific team you have
joined has been asked to develop a working model of an elevator as a
possible solution to the problem. You have been asked to create the
movement of the elevator so it moves from one floor to another in the
same scale of time as the finished, full-size elevator would. (If a full-size
elevator takes ten seconds to move from floor to floor, your model elevator
should take ten seconds to move from one floor to another.)

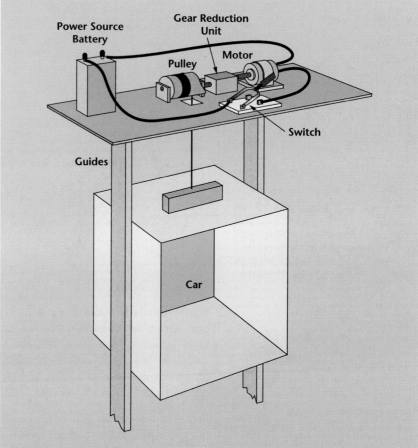

■ **Figure 6.18**
Conceptual design of an elevator.

- Using the modeling kits provided, develop the working model so that it travels at the required speed.

- Also using the modeling kits, develop a working model that shows how the elevator doors would open at an appropriate rate of speed.

- Identify some of the problems you can anticipate if the modeled solutions were to be translated into a full-size prototype.

Pulleys

A pulley can be considered as a lever in the form of a wheel with the arms of the pulley the same length (equal to the radius of the pulley). In this sense, pulleys are a form of Class 1 levers. A single pulley does not normally provide a mechanical advantage when used as a lifting device. It serves to change the direction of the effort and the load. If the force on the effort and the force on the load are the same, the pulley system is said to be in equilibrium and the pulley will remain at rest. If additional effort is applied, the load will start to move.

MATH/SCIENCE/TECHNOLOGY LINK

For the model elevator and doors you are developing, determine the following:

- What will be the speed that the elevator will move from floor to floor?

- If you use a small LEGO motor for your model, what type and size gears will you use in the gear train?

- What will be the VR between the motor and the pulley lifting the car?

- If the motor to lift the elevator is not big enough, what size will you need as a substitute?

- How would incorporating a counterbalance help to improve your design?

- Suppose you decide to use a rack and pinion to operate the elevator door. What gear ratio will you need to achieve an appropriate speed for the door?

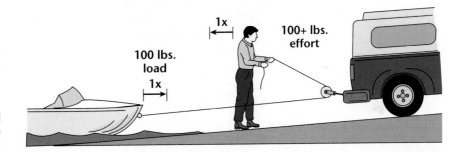

Figure 6.19A ■
A single pulley on a boat is used to change the direction of motion.

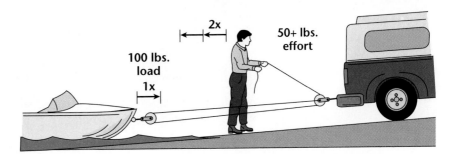

Figure 6.19B ■
Two pulleys can be placed to achieve a mechanical advantage of approximately 2:1.

If you use two pulleys, you will increase your mechanical advantage. Two pulleys allow you to lift nearly twice the load with the same amount of effort and results in a mechanical advantage (MA) of almost 2. Some of the effort will be lost in overcoming friction, and the MA is always less than the velocity ratio (VR).

MA is always = $\dfrac{\text{load}}{\text{effort}}$

It is possible, however, by passing a rope around several pulleys to gain a higher MA. By this line of reasoning, three pulleys will provide a VR of 3 and an MA of close to 3, and so on. As a specific example of this, Figure 6.19C shows four pulleys that have a VR of 4 and an MA of something less than 4. The MA of such pulley arrangements is dependent on the number of ropes involved in moving the load, and the amount of effort lost because of friction.

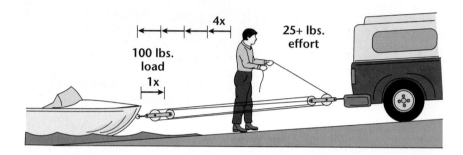

Figure 6.19C ■
Four pulleys in a system can achieve a mechanical advantage of 4:1 (ignoring the loss due to friction).

Fluid Mechanisms

Compressors. When you use levers and wheels, the force is applied to one place. When force acts over an area, it is called pressure. We discussed the high- and low-pressure zones of air produced by uneven heating of the Earth's surface. The air in the atmosphere has gravity acting on it. Normally, you cannot feel this pressure unless you change from a higher or lower level of pressure. You can notice this change as you ride up or down an elevator in a tall building or as an airplane you are in moves higher or lower. Most of the time your body is in equilibrium—the pressure inside of you is approximately equal to the weight of the air around you.

MATH/SCIENCE/TECHNOLOGY LINK

In the seventeenth and eighteenth centuries, several scientists investigated how the pressure, volume, and temperature of a gas were related. One of the earliest relationships (identified by Englishman Robert Boyle and Frenchman Edme Mariotte) linked pressure and volume and states:

If the temperature is constant, the volume of a fixed mass of gas is inversely proportional to the pressure.

In English-speaking countries, this is known as Boyle's Law.

One instance of this law in operation is seen in automobile tires. Inflated at the same pressure, tires will look underinflated on a cold day, but will appear more fully inflated on a hot day or when the tires, and the air inside them, have warmed up after a long trip.

A second instance is seen in the results of increasing the air pressure supplied to an air cylinder. If the pressure is doubled, the force on the piston of the cylinder is also doubled.

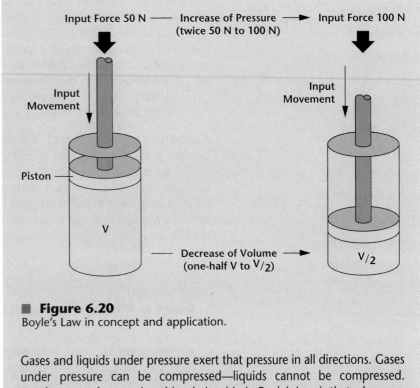

Input Force 50 N ⟶ Increase of Pressure ⟶ Input Force 100 N
(twice 50 N to 100 N)

Input Movement

Input Movement

Piston

V

Decrease of Volume ⟶ V/2
(one-half V to V/2)

■ **Figure 6.20**
Boyle's Law in concept and application.

Gases and liquids under pressure exert that pressure in all directions. Gases under pressure can be compressed—liquids cannot be compressed. Another way of expressing this relationship in Boyle's Law is that when you double the pressure on a gas, the volume becomes one-half. Liquids, however, will simply transmit the additional force in all directions and will not become smaller.

The pressure exerted depends also on the size of the area on which the force is focused. Walking on a hardwood floor in high heels with a tiny tip can press a dent in the floor. The same person walking on snow in snowshoes makes very shallow footprints in the snow.

You may have used compressed air in the form of a bicycle pump and tire. The pump compresses the air and stores it in the tire. This compressed air absorbs the impact created when you ride over bumps or rocky terrain by compressing even further. If you fill your tires from a tire pump at a service station, the air you use comes from a compressor and tank that is usually located inside the building.

CAUTION

High pressures of air or liquid can be dangerous.

**BE CAREFUL WHEN USING FLUID POWER!
DO NOT POINT AIR HOSES OR CYLINDERS AT ANYONE!**

The flow of air can be used to produce movement when compressed air at high pressures is released to a zone of low pressure. This is basically what happens when you let the air out of a balloon while stretching the rubber opening. You can create sound when the air escapes, or you can regulate the flow to produce movement of a pinwheel or other device. Air pressure is measured in Newtons per square meter. An electric-driven compressor and tank that provides compressed air at high pressures can exceed 690,000 N/m^2 (100 psi). This pressure can do work or damage, depending on how it is controlled.

Calculating piston force.   You may decide to use a pneumatic or hydraulic cylinder to apply force in one of your designs. In this case, you need to decide what force is to be transmitted by a piston. The chart in Figure 6.21 shows some example uses of pneumatics and hydraulics and their general pressure requirements. Some of the forces, such as the brake and clutch requirements, will exceed any project you will design. (As you refer to the devices in Figure 6.21, it may be helpful to remember that a Newton is approximately one-quarter of a pound.)

Figure 6.21 ■
General pressure requirements for different tools.

Device or Machine	Function	Piston Force (approximate)
pneumatic vise	holding a workpiece	500 N–1,000 N
pneumatic stamping machine	forcing a stamp into a material (copper)	200 N–500 N
conveyor belt	parts/package transfer	10 N–1,500 N
heavy vehicles	brakes or clutches	100,000 N
shops/businesses	door-closing system	100 N–1,000 N

MATH/SCIENCE/TECHNOLOGY LINK

As you prepare to calculate the force generated by a piston, you will need to know the pressure of the air supply and the piston area. This will allow you to determine the following:

$$pressure = \frac{force}{area} \quad or \quad P = \frac{F}{A}$$

where   pressure is measured in N/mm^2
force is measured in newtons (N)
area is measured in square millimeters (mm^2)

Based on the above, it follows that

$F = P \times A$

The area (A) of a round piston $= \pi r^2$

or $= \dfrac{\pi D^2}{4}$

where   r = radius of piston in millimeters
        d = diameter of piston in millimeters

If you study the above formula, it shows that the greater the pressure, the higher the force on the piston. Similarly, if the area of the piston is increased, the force on the piston will be increased.

In your work, the cylinders and their pistons are going to be very small. It will be easier, therefore, to measure the pistons in square millimeters rather than square meters. One square meter is equal to 1,000,000 square millimeters, or:

$1,000,000 \ N/m^2 = 1 \ N/mm^2$

What will be the force on the piston of a small air cylinder that is 10 mm in diameter and is supplied air from a hand-operated compressor at 10 psi ($69,000 \ N/mm^2$)?

D E S I G N   B R I E F

Developing a pneumatic device for moving materials.

You have been asked to help develop a pneumatic device that will move packages from a conveyor line to a loading operation where the products will be placed on a pallet for transport by truck to the users. The packages weigh 25 kilograms each.

An air pressure of 200,000 N/m^2 is currently available to service the pneumatic device. The air cylinder will need a stroke of 60 mm to move the packages the required distance. What is the minimum diameter of the cylinder to move the packages, one at a time, onto the loading station? (Ignore the frictional losses in the piston/cylinder and between the package and the surface on which it rests.) Refer to Figure 6.22 to see a drawing that shows how the operation would work to include the placement of the pneumatic cylinder. Make a sketch to show how you would set up the cylinder and controls. After an analysis of the overall process, what suggestions would you make to improve the packaging operation? How might you solve the problem if you have a fairly powerful cylinder that has a stroke of only 20 mm?

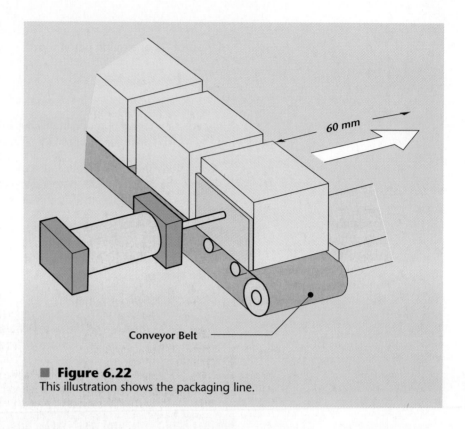

Figure 6.22
This illustration shows the packaging line.

Electromechanical Mechanisms

You may recall that the forces of magnetism can be converted to electrical energy, which in turn can be converted to mechanical energy for movement. This movement can also take the form of heat, light, or sound as well as physical movement of a load. As you develop designs that include the use of electrical components, you will need to understand and use appropriate measurements and formulas.

Electrical measurements. It is possible to make direct measurements of two properties of electricity. These include the flow of electricity, or the current, and the force, or voltage, that pushes the electricity through a conductor. (Force in electricity refers to electromotive force, or EMF.) EMF as electrical potential difference is measured as voltage or volts (represented by the letter "V").

By using an ammeter, you can measure current in amperes, or amps (represented by the letter "A"). When you take measurements of current, ammeters must be connected in series with the components. When you take a voltage measurement, the voltmeter is connected in parallel with the components. (See Figure 6.23.)

You can place an ammeter anywhere in a series circuit and the flow of current will be the same. In a parallel circuit, the amount of current flow will depend upon the number of branches and the resistance in each branch. If you add up the current flowing in each of the parallel branches, it will equal the current flowing in the series part of the circuit. (See Figure 6.24.)

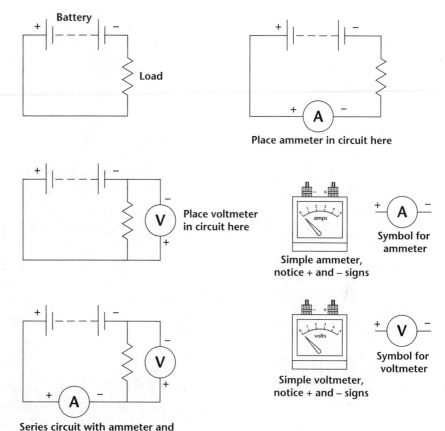

Figure 6.23 ■
These diagrams show ammeters and voltmeters in a series circuit.

Series circuit with ammeter and voltmeter in proper places

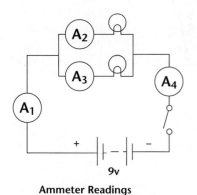

Ammeter Readings

Ammeter	Reading (A)
A_1	1
A_2	0.5
A_3	0.5
A_4	1

■ **Figure 6.24**
Ammeter in a parallel circuit.

As you connect a voltmeter across different components in a circuit, the voltage is different. This difference depends on the resistance of the components and where the voltmeter is placed in the circuit. (See Figure 6.25.) The voltage reading across any component is described as the **potential difference (pd)**. The potential difference measured across a component represents the amount of electrical energy that is changed within that component from one form of energy to another. For example, if 6 volts of pd is measured across the lights in the circuit shown in Figure 6.25(B), it indicates that 6 volts of electrical energy is being converted to light (and heat).

By using a voltmeter, you can also find the voltage of a battery. The total voltage across the battery is described in terms of the EMF and can be imagined as the pressure available to push the electricity through the circuit. In this example, the voltage of the battery is 9 volts.

The resistance of a material indicates how much it restricts the flow of electricity. Electrical components have different resistances, depending on how well they conduct electricity. Good conductors have low resistance. Poor conductors have high resistance.

Electrical power.   Electrical power refers to the rate at which electrical energy is converted into another form of energy. Electrical potential energy is measured in joules and electrical power is measured in watts. Watts of power can be

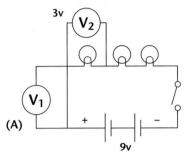

(A)

Potential Difference (voltage drop)
(pd) = 3v

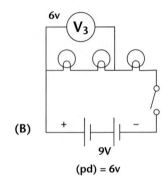

(B)

(pd) = 6v

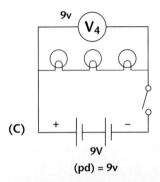

(C)

(pd) = 9v

Voltmeter Readings	
Voltmeter	Reading (V)
V₁	0
V₂	3
V₃	6
V₄	9

Figure 6.25 ■
Voltmeters across different
components of a circuit.

MATH/SCIENCE/TECHNOLOGY LINK

Resistance can be determined by measuring the voltage across a component and the current flowing through it. By using **Ohm's Law**, you can determine the relationship of voltage, current, and resistance. Resistance was originally indicated by the Greek letter omega (Ω), but since computers have come into common use, the letter "R" is used. Ohm's Law, introduced as a theoretical model in Chapter 3, indicates that:

voltage = current times resistance, or $V = I \times R$

You can rearrange this formula so that:

resistance = $\dfrac{\text{voltage}}{\text{current}}$ or $R = \dfrac{V}{I}$

There are many instances when it is important to reduce current flow through the use of a resistor. The example that follows shows a resistor used to reduce the voltage to a light-emitting diode (LED). LEDs are small devices that normally operated at about 2 volts and require 20 mA to function. (One mA is equal to 1/1,000 of an ampere.) To ensure that the voltage is correct for the LED, it is placed in series with a resistor. The circuit shown in Figure 6.25A is powered by a 9-volt battery, and 7 volts must be dropped over the resistor so that 2 volts are available to operate the LED. Use Ohm's Law to calculate the size resistor that is required.

Light Emitting Diode

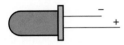

Symbol for LED

(B) Standard LED

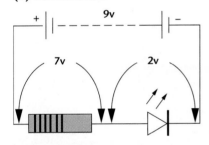

(C) Flashing LED

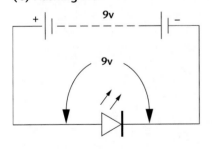

■ **Figure 6.26B, C**
(B) Circuit with a resistor and LED.
(C) Circuit with a flashing LED that does not require an external resistor.

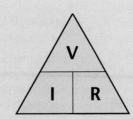

V = voltage
I = amperage
R = resistance

$V = I \times R$

$I = \dfrac{V}{R}$

$R = \dfrac{V}{I}$

■ **Figure 6.26A**
Ohm's Law shows the relationship between volts, amperage, and resistance.

When two or more components are placed in a circuit as shown above, they form a potential divider. They divide the voltage from the battery between them. If the LED requires 2 V, then the resistor must create a drop of 7 V. As indicated above, the LED requires about 20 milliamps (mA) to operate. Because the flow of current through a series circuit remains the same, the value of the resistor can be calculated as:

$$R = \frac{V}{I} = \frac{7}{.020} = 350 \text{ ohms}$$

Resistors come in standard values and you can select a 350-ohm resistor. You can refer to Figure 6.28 to see what the color code will be for this resistor.

compared to the more general unit of energy called the joule. One watt of electrical power is equivalent to one joule per second. For example, if an electric light uses 60 watts of power, then 60 joules of electrical energy will be changed into light (and heat) every second. The heat is often seen as an undesired side product of the light and is considered unusable and sometimes unsafe.

MATH/SCIENCE/TECHNOLOGY LINK

In the example of the 60-watt lightbulb, the relationship of watts and joules is:

60 watts = 60 joules/second, or 1 watt = 1 joule/second

You can calculate electrical power by using the power rule that shows the relationship of voltage, current, and power to be:

power = voltage times current, or $P = V \times I$

when   voltage (V) is measured in volts (V)
        current (I) is measured in amps (A)
        power (P) is measured in watts (W)

Resistors in circuits. Resistors, by design, are able to withstand the heat they generate when electricity flows through them. All resistors have a power rating within which they will operate. If that rating is exceeded, the component will burn out or fail.

Resistors are manufactured to provide different power ratings. The cheapest resistors are made of a carbon composition and are rated for about 0.25 watts. Resistors made of carbon film can withstand about 2 watts. Resistors made of metal oxide are rated at only .5 watts, but are more stable in operation than carbon resistors. Resistors that must perform at high power ratings are usually made of wire. These are described as wire-wound resistors and withstand up to 50 watts.

MATH/SCIENCE/TECHNOLOGY LINK

Combining Resistors

Resistors can be combined in series, as shown in Figure 6.27A. The total resistance in this use is determined by adding the resistance of all the parts.

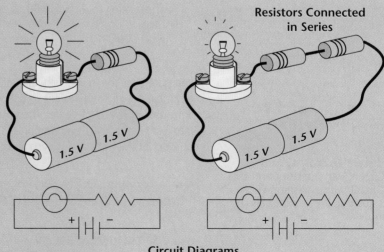

Resistors Connected
in Series

Circuit Diagrams

■ **Figure 6.27A**
Resistors in a series.

The total resistance is determined as:

$R_T = R_1 + R_2 + R_3$, etc.

Resistors can also be combined in parallel, as shown in Figure 6.27B.

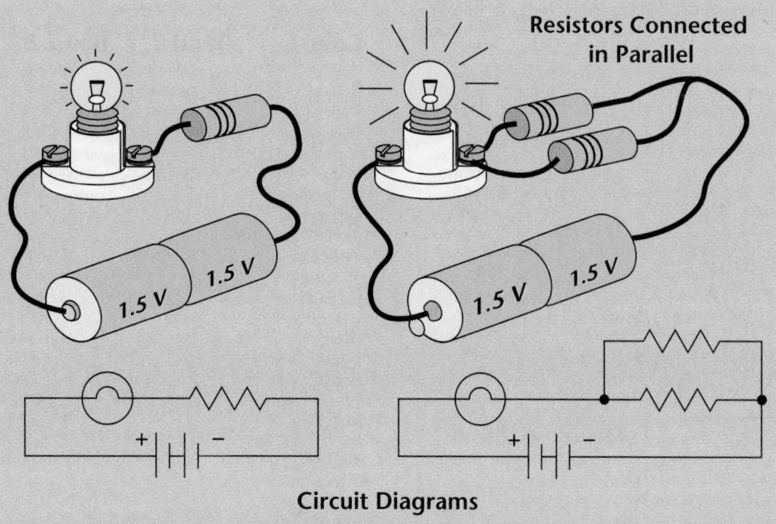

Resistors Connected in Parallel

Circuit Diagrams

■ **Figure 6.27B**
Resistors in parallel.

The total resistance is determined as:

$$\frac{1}{R_T} = \frac{1}{R_1} + \frac{1}{R_2} + \frac{1}{R_3}, \text{etc.}$$

Resistance value and tolerance.   Resistors come in many different values of resistance (electrical sizes). This means that some are manufactured to have a few ohms resistance while others have many, many ohms resistance. The chart presented in Figure 6.28 shows the codes for determining the resistance of a resistor. The first three color bands indicate the value of the resistor.

To determine the actual value of a resistor, the numbers represented by the first two color bands must be multiplied by the number represented by the third color band. The resistor needed in Figure 6.26B could be 350 ohms. If you use a 350-ohm resistor, it will have the color code of "orange, green, brown." The color orange designates a 3, green designates a 5, and brown designates a multiplier of 10. This results in 35 times 10, or 350 ohms.

When resistors are manufactured, they cannot be made exactly alike. For example, even resistors made from the same batch of raw materials vary. The differences are acceptable if it stays within 10 percent of the value. (A resistor of 350 ohms could range from about 315 ohms to about 385 ohms.) Those that vary too much are discarded. The acceptable difference of resistance in a resistor is referred to as tolerance. The tolerance is identified by the fourth color band. If a resistor has a fourth band, it will be one of three colors. If it is red, its tolerance will be 2 percent. If it is gold, the tolerance will be 5 percent, and if it is silver, it will be 10 percent.

Color	Band 1	Band 2	Band 3	Band 4
Black	0	0	times 1	Red 2%
Brown	1	1	times 10	Gold 5%
Red	2	2	times 100	Silver 10%
Orange	3	3	times 1,000	
Yellow	4	4	times 10,000	
Green	5	5	times 100,000	
Blue	6	6	times 1,000,000	
Violet	7	7	times 10,000,000	
Grey	8	8	times 100,000,000	
White	9	9	times 1,000,000,000	

Figure 6.28 ■
Resistor color codes and tolerances.

D E S I G N   B R I E F

Developing Your Energy Conversion Chart

The Energy Conversion Chart, introduced earlier in this chapter, will be most useful in your design and technology work if you add instances and examples that mean something to you. Using the blank format of the Energy Conversion Chart provided in your *Portfolio and Activities Resource*, use names and sketches of the devices that you have used, as well as the processes and devices that you have learned about in the class. Enter other devices and processes as you proceed in your individual and class activities.

SUMMARY
CHAPTER 6

This chapter focuses on the primary sources of energy: solar, geothermal, wind, water, atomic, fossil, and biological. These primary sources are related to a general energy spectrum with bands of energy which include atomic, molecular, chemical, mechanical, electromagnetic, radiant, and cosmic. Two ideas are essential to understand the energy spectrum: bonds and frequency. Energy is captured in the bonds between and within elements and compounds. These are released by nuclear, molecular, and chemical processes. Energy is also found in the frequency of movement from very slow mechanical forms of energy to the very rapid cosmic forms. At extremely high frequencies, the energy is referred to as rays. Examples are X rays and cosmic rays.

The usefulness of energy depends in part on how available it is for our use and how efficiently it can

Input \ Output	Chemical	Heat	Mechanical	Sound	Electrical	Light
Chemical						
Heat						
Mechanical						
Sound						
Electrical						
Light						

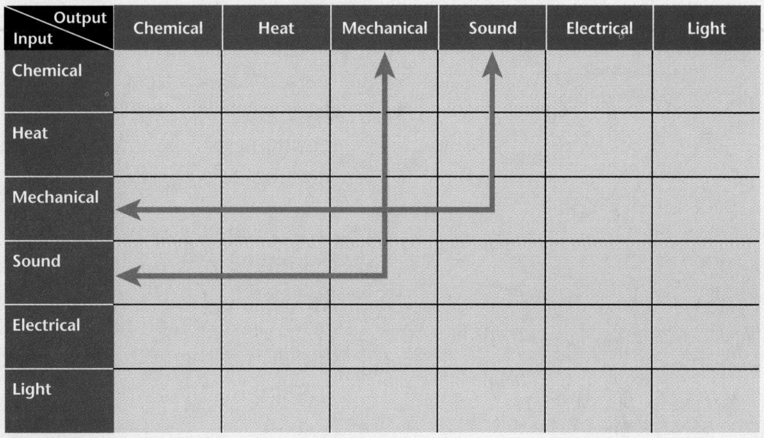

■ **Figure 6.29**
Energy conversion chart.

be converted from one form to another. Currently, we are limited in our knowledge of atomic and cosmic energy. We know very little about how to convert them to other energy forms. Most of the energy conversion processes we use are applied to the energy band within the middle of the energy spectrum. Converters are the

devices that are used to change energy from one form to another. As you will see in the next chapter, transducers are unique converters used to change energy into information in the form of signals. Signals are necessary for the guidance and control of machines.

ENRICHMENT ACTIVITIES

Design Briefs

- Identify five different tools you use at home or in school that operate on electricity. Determine the power ratings for each. Rank them according to their ratings from high to low. Compare two devices such as a hair dryer and an electric drill. Determine which has the higher power requirements. Design a sticker that will inform users about the power requirements of different electrical devices.

- Identify what you would eat in the morning as your favorite breakfast. Determine where each of the products for that breakfast was grown, processed, and packaged. Discuss what forms of energy were used to grow, prepare, and deliver the food to your table. Design and develop a display that shares this information with others. What changes would you propose to improve the efficiency of this system?

- Design a safety card that identifies the key safety features of a tool or machine. The card should help people understand the precautions that should be taken or protection that should be used to shield the user from the energy used by the machine. Consider the protection required for electrical energy, thermal energy (extreme heat or cold), microwave energy, or mechanical energy in the form of moving gears, belts, pulleys, motors, and cutting tools.

- Using electricity and gas bills, determine how much energy is being used in your home each day and each month. Can you determine the difference of the energy used between the seasons, particularly winter and summer?

- Although the Earth receives more energy from the sun each month than the energy stored in the known fossil fuel reserves, solar energy remains underutilized in this country. Design and construct models of different devices and systems that will put energy from the sun to practical use.

- Identify and record the kinds and amounts of foods you consume in a 24-hour period. Determine the calorie content of the different foods you have eaten. Identify the activities you do during the same day and determine the calories used for that day.

MATH/SCIENCE/TECHNOLOGY LINK

As indicated earlier, something interesting happens when a single pulley is connected so that it is moveable. With the resources provided by your teacher, set up such an arrangement so that the pulley is supported by the rope similar to the illustration below.

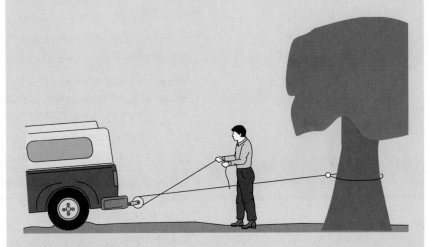

■ **Figure 6.30**
Moving a load with a movable pulley.

• Measure the weight of the load to be moved or lifted.

• Place the load on the pulley and lift it, measuring the force that is necessary to move the load.

• Lift the load and measure the distance it moved.

• Measure how much rope you had to pull through the pulley to move the load through that distance.

• From this data, determine the mechanical advantage (MA) and velocity ratio (VR) provided by a single moveable pulley.

• From what you have learned, how would the MA of a pulley system relate to the number of pulley wheels that were used? How would MA relate to the number of times the rope loops back and forth between the pulleys?

REVIEW AND ASSESSMENT

- For each of the following prime sources, identify at least one example of application: solar energy, geothermal energy, movement of air, movement of water, atomic energy, fossil fuels, and energy from living things.

- Describe the differences between slow and rapid oxidation and provide an example of each.

- Compare potential energy and kinetic energy as stored energy and energy in motion.

- Provide practical examples of devices or systems that use energy from the categories of the energy spectrum—atomic, chemical, mechanical, electromagnetic, radiant, and cosmic.

- Describe how the latent heat of evaporation and the latent heat of fusion can represent energy that is wasted or energy that is stored.

- Different forms of energy can be detected by humans with their unaided senses. Identify these forms and describe how humans detect and use these energy forms.

- Identify the additional examples you would add to the Energy Conversion Chart.

- There are many important terms related to energy and its use. Describe the following and provide an example that shows the application of entropy, friction, efficiency, and equilibrium.

- Describe and illustrate how mechanical advantage (in the use of levers) involves loads, moments, and moment arms.

- Describe and illustrate how mechanical advantage and velocity ratio are related.

- Compare the differences and similarities of using wheels and axles, belts and pulleys, and chains and sprockets.

- Describe and illustrate how mechanical advantage operates in pneumatic systems.

- Describe and illustrate the relationship between force and pressure.

- Describe how the resistance in series circuits differs from resistance in parallel circuits.

- There are a number of important concepts related to electromagnetism such as amps, ohms (resistance), potential difference, voltage, and electromotive force (EMF). How do these different concepts relate to each other?

- Consider energy that is stored as potential energy. Identify specific examples and devices that can be used to store energy in a mechanical, electrical, or fluid (pneumatic, hydraulic) form.

- What is meant by the idea of converting energy into information? What do transducers have to do with this process?

CHAPTER 7
Information and Information Processing

Figure 7.1 ■
Symbols and signals of information provide codes by which humans and machines can interact.
(© Rob Atkins/The Image Bank)

INTRODUCTION—BUILDING BLOCKS FOR COMMUNICATION AND CONTROL

R ight now, you are probably being bombarded by thousands of bits of sensory information. Colors, sounds, temperature, textures, and smells come from all directions. These pieces of information are a buzzing sea of confusion that we

187

KEY TERMS

binary
black box
Boolean Logic
cam
clutch
codes
communication
control device
convert
data
decode
dwell
electromagnetic
information
input sensor
integrate
interval
levels of measurement
logic gates
logo
message
Morse code
nominal
ordinal
output device
pictogram
processor
program
ratio
relay
ROM
sampling
semiconducting
sensors
signals
symbols
telegraph
template
thermocouple
transducers
truth table
variable resistor

almost automatically attempt to organize. Our brain and sensory mechanisms screen out and organize experiences to protect us and to make sense out of what is happening. This chapter examines some of the ways in which we organize data from our own experiences and use it to serve our wants and needs. We will also note how humans in the past have processed **information** to make it easier to handle, remember, and communicate. The tools and materials that are used in these technological endeavors will be explored.

Humans and machines require information to do work. As with many other aspects of technology, information has been used since the first living organism sent a **message** to one another. Information was processed and used long before we understood what we were doing or why it worked. The accumulation of knowledge itself is a product of information processing and use.

Humans are not the only species to use information for communicating. However, we seem to have developed the most complex systems and devices for communication and control. Also, it seems we are the most driven to know and organize and push the limits of understandings. The development of human self-awareness, a sense of meaning and order, and communication with others have been called human needs—necessary for survival and well-being.

PUTTING INFORMATION TO USE

For what purposes is information used? All of the information humans can use comes through our sensory mechanisms. We see, smell, taste, hear, and feel. The information that comes from each of our sense organs must be processed and coordinated. (In the chapter on Humans, we will develop these ideas further.) Over time, people have developed language and other forms of symbols in order to coordinate and categorize their own experiences. As humans perceive, store, and retrieve information, they do so with **symbols**. These symbols are also used as we attempt to share our experiences and their meaning with other humans. In addition, we use **signals** to communicate with machines and to connect machines with each other.

As we indicated in the chapter on Tools, all machines have a subsystem for guidance and control. Machines, from the most simple to recently developed "artificial intelligence" devices, do not process information in the same way that humans do. Signals are needed. Signals and symbols are the basic building blocks of information. Through them, you are able to communicate ideas and control actions. Let us first take a look at signals and then at symbols. Following this introduction, we will return to each for more in-depth discussion and applications.

Signals, whether they are signs or impulses, are intended to cause action. When we flip a switch, we send a signal to something—a light, motor, pump, or television set. Usually, the word "signal" is used when information is sent from a person to a machine or from one machine to another. Signals, as used in this book, mean the signs or impulses to communicate or control machines by humans or by other machines. As we talk about collecting units or bits of information, the term **data** may be used. Data refers to units of any type or quantity.

Symbols, on the other hand, are signs that infer meaning. Symbols represent something more than the symbol itself. For example, some older light switches have two buttons, one white and one black. The white button is a symbol representing a lighted room; the black button represents a darkened room. Computer equipment often has buttons that are white (open/on) and black (closed/off). Symbols, therefore, are the signs used by humans and machines to communicate with humans. Symbols can be a part of a complicated system that allows us to represent a large quantity of information. The symbols provide us the bits and pieces through which we can communicate.

Figure 7.2A ■
These analog information devices use symbols that humans can understand.

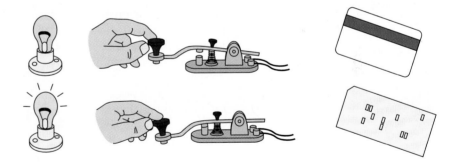

Figure 7.2B ■
Digital information devices send signals to machines.

The simple switches illustrated in Figure 7.3 are good examples of how a signal can be used. When a common light switch is turned to on, the electrical circuit is complete. Electricity flows and "tells" the bulb to emit light. A movement to the off position breaks the circuit. The switch sends the information for the bulb to become dark. On and off, light and dark, and open and closed—this is all the information that a switch can provide. The codes for two conditions, off and on, are similar to the conditions of other devices. Each of these examples really represents a "yes" or "no" condition that can be shown by the **binary** symbols of 0 and 1. The "0" symbol indicates that something is missing (off), while the symbol "1" indicates a condition where something is present (on). A little later in the chapter, we will revisit these ideas as they relate to the idea of machine logic.

This may seem to be a simple system of signals. Yet, the binary system is the basic language of the computer, our most sophisticated information processing machine. Millions of signals can be organized, stored, and retrieved. Signals can be turned into symbols that are user-friendly, if humans are to use the information. If the information is sent to another machine, the signals do not need conversion to symbols.

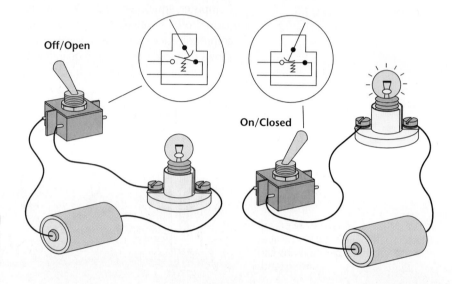

Figure 7.3 ∎
Switches are used to complete the electric circuit and turn on the light.

Off/Open

On/Closed

Let us consider another simple example of using information. Suppose your grandfather needs to take medication in the form of several pills each day. However, a side effect of some of the medication is forgetfulness. Also, some of the pills are to be taken only every other day. There are many ways to solve this problem. You might make seven boxes and label them according to the day of the week. Once a week you fill the boxes with the prescribed medicines for each day. Then you might just phone your grandfather, remind him of the day of the week, and ask if the box is full or empty. This may seem like a variation of the on/off signals of the light switch. However, you can probably detect many instances in this problem in which symbols are used.

SIGNALS: INFORMATION FOR MACHINES

Signals do not need to be complicated in order to cause machines to take some action. Suppose you manage a car wash and want to save money and water by designing a **control device** for the washing section of the machine to turn on only when a car is passing under the spray nozzles. The setup of this operation is fairly simple. A light source is directed toward a photoelectric cell. The photoelectric cell is attached to a sensitive relay or a digital device, both of which act as an electrical switch. When the light is turned on and shines on the photocell, the water turns off. (Refer to Figure 7.4.)

∎ **Figure 7.4**
Automatic machines can use simple switches as the information devices necessary for controlling processes such as washing cars.
(© Michael Dzaman Photography)

Why will this work for us? If the light can reach the photocell, there is no car to block the beam of light. When a car does move down the line, we want the photocell to sense the change (from the light on to the light off). If the light is blocked, the photocell stops producing its small amount of electricity. This causes a transistor, an electronic device that acts as a switch, to turn on and trigger a sensitive **relay**. The relay controls the water pump, and the water is pumped into the spray nozzles as long as the car blocks the beam of light. Once the car has moved past the light, the photocell again sends electrical current to the relay. The relay (an electronic switch) turns the pump off. The water stops until the next car comes through the car wash line. We have saved water by using a simple control system comprised of an **input sensor** (the photocell), a **process device** or **processor** (the transistor), and an **output device** (the water pump).

In general terms, this and all other control and communication systems can be represented as shown in Figure 7.5. As indicated several times before, all systems must include inputs, processes, and outputs. These three categories are used to show the different components used in control and communication systems.

Input Signals from Sensors

Just as humans receive information from the outside world through their sensory mechanisms, **sensors** have been developed for tools and machines. These sensors detect a portion of the energy spectrum and **convert** that energy into new usable forms—in this case, the signals (the input) can be modified and changed (by the processor) to create some form of action (the output). In the chapter on Energy, we explored how energy can be changed from one form to another. We can build upon that experience to show how energy conversion is very similar to information conversion.

In information conversion, **transducers** are used to take a small sample of energy and change that energy into information (signals) for communicating with or controlling a machine. Transducers are devices that transmit energy from one system to another system. Sensors are specific examples of transducers that sample energy in one system (thermostat sampling temperature in a room) and provide a signal that can be used to control another system (controlling the amount of energy supplied to an oven). For example, we use a **thermocouple** as the sensor to provide input to control an oven. As the oven gets hot, the thermocouple takes a small amount of that heat and converts it into electricity. When the heat in the oven becomes high enough to produce enough electricity, a processor triggers a switch controlling the heater, which is then turned off. When the heat level drops, the contact is broken and the switch turns the oven on again. This cycle of on and off continues until a timer or a person turns off the oven's main switch.

Many machines are controlled through the use of sensors. Even the most sophisticated machines use sensors. They may use several sensors and **integrate** the signals through a computer or microprocessor.

The sensor is the major link between energy and information. A signal is generated by **sampling** some form of energy, whether heat, light, sound, or movement. Sampling and generating that signal are done by a sensor. This is especially important in machines that are designed to operate automatically.

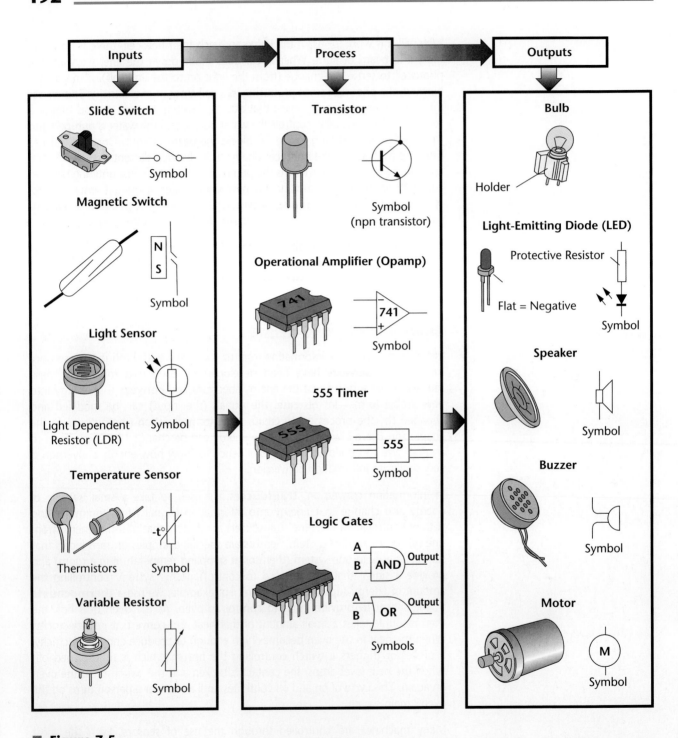

■ **Figure 7.5**
Sensors and switches provide **input** signals that are used by devices that provide a **process** to generate control signals to operate **output** devices.

Machines such as furnaces cannot be put on automatic control without sensors. Thermocouples, pressure switches, and the like provide information in the form of signals. The devices then send that signal to the guidance and control subsystem of the machine. Such a signal can be used to control a machine (turn it off, speed it up, or make it hotter). The machine then becomes somewhat self-directed or automatic.

Another example of using a sensor can be seen in some security devices. Suppose a store is closed for the night and someone enters. A sensor can detect the movement or maybe the heat from a body. A signal can be sent to turn on the lights in the room or an alarm in the police station.

The beat of your heart and the movements of internal organs can be detected through sensors and recorded as blips on a paper or screen. These signals can be compared with those from other people or from your own body at previous checkups. This monitoring is useful to see whether everything is working as expected.

Some signals that are used in machines are very simple and direct. Others require a sequencing to tell the machine to go through a series of more complicated moves. These signals are put together into a code that the machine is programmed to understand. **Codes** allow people to make sense out of signals. But the codes must be prearranged and cannot vary from the original form. For example, a toaster makes our toast light, medium, or dark, depending on how we set the selector switch. The setting on the selector switch is the code that we can **program** into the machine. Selecting the code for "medium" mechanically transmits a signal to the heating element in the toaster to stay on a little longer than it would for "light". If, however, you use a piece of bread that has already been toasted, your toast may not look medium when done. Because the code for the toaster controls only the time during which the heating element stays on, your toast may be burned.

The program that is established when you set the toaster is quite similar to a program used with a computer. The difference is in the tremendous complexity of the computer program.

Sensing Mechanical Energy—Movement

Some of the earliest devices that required switches for controlling the energy they used can be found in mechanical machinery driven by wind and water power. In order to use the energy from a water wheel, it was necessary to tap into a portion of the energy generated by the water wheel system and make the necessary connections without stopping the process. Some ways of switching the mechanical energy on and off were large and cumbersome when compared to what we use now.

One of the simplest of mechanical devices that is used for information and control is the common **cam**. In the chapter on Tools, the cam, a lopsided wheel, was described. The cam has a specific shape that contains certain information. In Figure 7.6, the sharp point of the cam lobe indicates that the action on the control rod will be rather short in duration. If the cam had a larger lobe, it would cause the control rod to stay open longer. This is called a longer **dwell**. The information contained in the shape of the cam relates to the length of the signal sent to the receiving device. In this case, this device is a timer that can be used to release a given amount of oil, grain, or other material.

Mechanical switches can be used to change the rate of movement or to activate different parts of a machine. A common device for switching mechanical energy on and off is a **clutch**. An illustration of a simple friction clutch is shown in Figure 7.7. The spinning disk on the power side of the mechanism and the disk of the

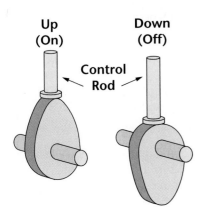

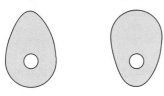

■ **Figure 7.6**
As a cam moves, its shape can turn a switch on and off for different periods of time, depending on the shape of the cam and its speed of rotation.

machine to be powered must be moved together to form a friction drive system. The clutch can be released at any time to turn off the machine. This system works fairly well as long as the force does not exceed the friction of the clutch. If the clutch begins to slip, the power to the machine is lost.

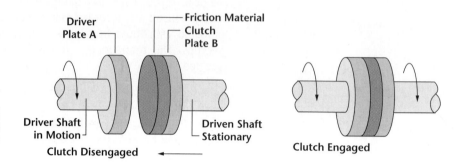

Figure 7.7 ■
Mechanical clutches provide on/off information and operate like switches.

Sensing Electrical Energy

The most used of the energy forms for information and control is **electromagnetic**. Electrical devices are now most commonly used for switching devices for signals and controls. Electricity provides the most flexible of the energy control systems. The trend has been to use these systems wherever possible. This energy can be tapped by plugging into the electrical systems available in homes and other buildings. Electrical energy from batteries is also highly accessible.

Electrical energy, because of its flexibility and ease of distribution, is an ideal form for generating and transmitting signals for control. Some of the switches that can be used to generate electrical signals are illustrated in Figure 7.5.

Sensing Magnetic Energy

There are a number of switches that operate by sensing magnetism as an energy form. For example, the reed switch operates when a magnetic field is strong enough to attract one of the reeds in the switch to come into contact with the other. A simple bar magnet can be used to make the switch operate.

A simple reed switch is closed (the two reeds touch each other) when triggered by a standard magnet, while a relay is operated by an electric coil or magnet. The relay is turned on and off by the magnetism generated by the flow of electrical energy through the coil of the relay.

Signals generated by the magnetic pattern on a credit or debit card can be relayed directly to a machine at a bank or credit card company. This information activates your account and can record the information about the amount of your purchases. It can also transfer that amount from your account to that of the store, or activate the machine to dispense some cash.

Magnetic patterns can activate a door to open or a parking gate to raise. The electromagnetic impulses of a person who is responding to a series of questions activate the machine that is recording a lie detector test.

Sensing Light Energy

Other sensors can convert light energy into usable forms for controlling electrical energy. For example, photo devices contain materials that are sensitive to light energy. In the photovoltaic cell, the light energy falls on **semiconducting** materials and causes a flow of electrical current. In the case of the photoresistor, the light energy causes the materials to become more or less resistant to the flow of electrical current. The device operates as a **variable resistor** that changes, depending on the level of light energy that hits it. The changes that occur both in the photovoltaic cell and the photoresistor can be used as inputs to a control device or system. For example, a photocell, using a color filter, can be used to activate control devices to sort green apples from red ones.

Sensors can be used to detect expected patterns and variations from them. Your tests may be scored by an optical scanner. Your answers are compared with the key of right answers. Printed pages and photos can be scanned for entry into a computer for duplication and for comparisons. The image may be a fingerprint that can be compared with millions of prints on file.

Fingerprints can be surgically altered and voices can be recorded. Therefore, a recent development in security systems has been to scan and use the pattern on your retina (on the back of your eyeball) as a means of identification. The retina is unique and few people would be able to alter it to mask their identity.

Sensing Thermal Energy

Other devices are able to sense a change in the temperature of selected materials. The thermocouple is one type of temperature sensing device that actually generates an electrical current with an increase in temperature. Thermistors, the other type of temperature device, change their resistance to electrical current as the temperature changes. (See Figure 7.5.)

Thermal sensors respond to a change in temperature and can be used to turn on an air-conditioning system or a sprinkler. A strip of chemicals that change heat to color can be used to measure body temperature.

Sensing Sonic Energy—Sound

One of the simplest sensors that use sonic energy is the microphone. Voice-activated recording machines can save energy by only working when someone is talking or singing. Tiny bits of sound can be amplified and transmitted by telephone and radio to someone in the house next door or to someone traveling in space.

Voice patterns can be compared and synthesized. A sample of the waves made by your voice speaking a given message can be obtained. A machine can then be patterned to reproduce your voice and message. Similarly, computers now have voice recognition capabilities that allow you to give programming commands by voice. In a more simple form, many cars now have a microprocessor that can generate a voice to gently remind you, "The fuel is low" or "Your seat belt is not fastened."

■ **Figure 7.8A**
A computer-based synthesizer can make a variety of musical "voices."
(© Teri Bloom. All Rights Reserved.)

■ **Figure 7.8B**
This speech trainer helps a child with a hearing impairment to improve his ability to speak.
(Courtesy of International Business Machines Corporation)

D E S I G N   B R I E F

You have become part of a small team that will be responsible for developing systems that can be used for the remote control of machines. As a part of that work, you are expected to know the different types of inputs and outputs, how they can be used, and how they operate.

From the components provided by your teacher, identify which are the input components. (You may want to refer to Figure 7.5 again.) Work as a team to determine what changes in resistance take place in each sensor as you use it. Record any changes you can detect on the worksheet provided in the *Portfolio and Activities Resource.*

Input	Process	Output
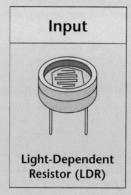	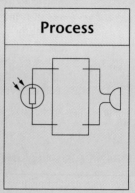	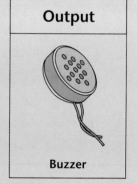
Light-Dependent Resistor (LDR)		**Buzzer**

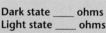

Dark state ____ ohms
Light state ____ ohms

Silent ____ v
Low level ____ v
High level ____ v

■ **Figure 7.9**
Block diagrams showing the devices used for the input and output of a system can help clarify what function the processor part of that system must perform.

Again using the set of components provided by your teacher, identify which ones are output components. Using the low-voltage power source provided, test to see if you can make each device operate.

Develop a block diagram, similar to the above example, that shows an input device and an output device of your choice. Describe the input signal that will be provided by your sensor. Describe the output signal needed by your output device. Keep this data for use in design activities that will follow.

MATH/SCIENCE/TECHNOLOGY LINK

Devices, like the light-dependent resistor (LDR) illustrated in Figure 7.9, operate because the material used in their design and manufacture can change. The cadmium sulphide used in the LDR is affected by light energy. For the other input and output devices, determine the following:

* What is the energy form that caused the device to change?

* What is the minimum energy level needed to make the device change?

* What is the nature of the change that happens in each device?

Processors, Codes, and Logic

Processors take the signals provided by sensors and converts them into electrical signals that can cause output devices to function. Up to this point, we have considered the processor part of the information system as a **black box** where the contents are unknown. The idea of a black box is used when you will not be certain of how a process operates. Many people use machines and devices they do not fully understand. For example, we may use a television set, but for most of us the internal workings are unknown. It will be important that you develop insights into how simple processor circuits operate, particularly if you are going to design and develop such circuits.

There are some standard approaches to developing electronic circuits to serve as processor units. Illustrated in Figure 7.10 are two such standard circuits that could serve as processors. Both circuits operate on the basis of a change in resistance. The first circuit, in Figure 7.10A, indicates how a switching function can be accomplished as the resistance of the sensor increases. As the sensor samples the appropriate energy form, its resistance increases and the input voltage to the transistor also increases. When the input voltage rises above a specific point (0.6 V) the transistor will operate the output device.

Figure 7.10A, B ■
Many sensors used in control circuits operate on the principle of a "change of resistance." Circuit A operates as the sensor resistance increases (when a photo resistor is in the shade). Circuit B operates as the sensor resistance decreases (when a photo resistor is in the light).

(A)

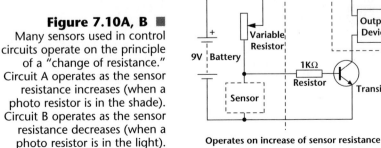

Operates on increase of sensor resistance

(B)

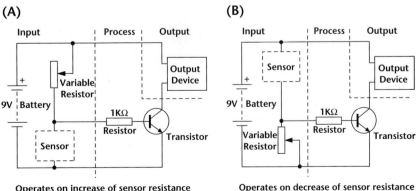

Operates on decrease of sensor resistance

The second circuit, in Figure 7.10B, operates in an opposite manner and accomplishes a switching function as the resistance of the sensor decreases. In this sensor, the sampling of the energy form causes its resistance to decrease, but the input voltage to the transistor to increase. Again, when the input voltage rises above 0.6 V, the transistor operates the output device. Experiments and problems included in the *Portfolio and Activities Resource* will help you design and develop circuits that can be used to perform communication and control functions.

Signals provided by single-purpose processor circuits have interesting but somewhat limited applications. Combining signals from several sensors and processors to increase the sophistication of what machines do, presents a different type of problem. In order for machines to use this information, it must be presented in a "logical" way. The term "logic," as used here, has a special meaning that applies to the patterns and sequences of signals transmitted to machines. The signals must be combined in appropriate codes and sequences.

Every time we use a machine, we send it information as signals in some form of prearranged code. The old-time player piano, shown in Figure 7.11A, uses a roll of paper with rows of holes punched in it to store information (musical notes). Each row on the paper is for a specific key on the piano. A small vacuum develops behind the roll of paper. This vacuum comes from the bellows pumped by the operator's feet. When a hole for a given note passes in front of the vacuum, the outside air pressure trips the hammer for that note. The hammer then causes the note to be played. (Refer to Figures 7.11B and C.)The player piano is an early example of a **read only memory** (**ROM**) device now quite common in computers. The piano reads only what is already coded on the punched roll of paper. The Jacquard weaving loom, one of the first machines to use punched cards, is another example of a ROM device. You may recognize already that the computer is one of our most sophisticated information processing machines. Computers can also be used in conjunction with other media devices. For example, interactive video involves a video machine connected to a computer that allows the user to select from a wide range of programs quickly and easily.

■ **Figure 7.11A**
The player piano was a model for later, more sophisticated tape-controlled machines. The information for the notes was recorded as holes in the paper roll.
(© Michael Dzaman Photography)

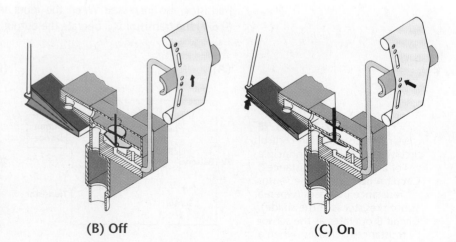

Figure 7.11B, C ■
As a hole in the paper roll passes over the sensing hole, (B) the vacuum causes the bellows to collapse. The rod then moves to hit the right note (C).

(B) Off   **(C) On**

Machine Logic for Decision-Making

Machines can be given a decision-making capability. This can be accomplished by a variety of devices, including mechanical and fluidic components, but the most commonly used is electronic. Machine logic is based on **Boolean Logic**. Machine or Boolean Logic refers to the combinations of 0's and 1's produced by the different **logic gates**. The term *Boolean* comes from the name George Boole (1815–1864), who first studied this digital logic. Boolean Logic provides the theory that describes how computers and control systems function. Electronic devices provide logic gates in which the output of a circuit is dependent on the input of the circuit. For example, in the simple circuit shown in Figure 7.12A, the circuit will only work if both switches "A" and "B" in Figure 7.12B are on. In decision-making or "logic" terms, this is known as an AND gate.

The word "gate" is used because the switches are described as being open or closed. In logic terms, the input of the circuit refers to whether the switches are turned on (are closed) or are turned off (are open). We can use a shorthand way of talking about the state of the switches, as shown in Figure 7.12C. When a switch is on, we describe it with a number 1. If a switch is off, we label it with a number 0.

(A)

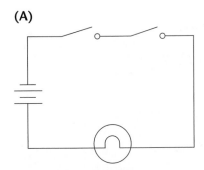

■ **Figure 7.12A, B, C**
This simple AND circuit is shown as (A) a symbolic drawing, (B) a practical application for helping Aunt Sarah check the size of cans, and (C) a truth table for an AND logic circuit.

(B)

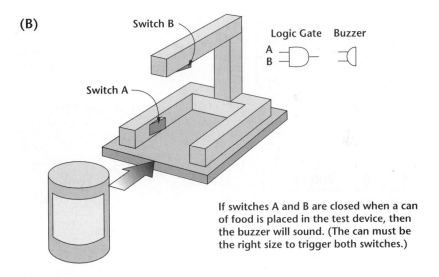

If switches A and B are closed when a can of food is placed in the test device, then the buzzer will sound. (The can must be the right size to trigger both switches.)

(C)

Switch A	Switch B	Bulb	A	B	Output
off	off	off	0	0	0
off	on	off	0	1	0
on	off	off	1	0	0
on	on	on	1	1	1

One of the ideas that April and Enrique tried early in their design project (from Chapter 2) was to use an AND circuit to help Aunt Sarah determine the size of the cans in her cabinets. It is possible to use this simple AND circuit for decision-making in a simple system. For example, an AND circuit could be used to help Aunt Sarah determine the size of a specific can, as shown in Figure 7.12B. A short can could close switch A to on, but would not be adequate because the circuit remains incomplete with switch B still off (0). If a can of appropriate height is used, as shown in the second figure, switch B is turned on (1) and, with both switches on, the buzzer would sound.

The logic behind the circuit is obviously rather simple. The two switches can either be on or off, which will result in the buzzer being on or off. A standard way of describing the logic, actually the logical functions of a circuit, is by using a **truth table**. The truth table for an AND circuit, shown in Figure 7.12C, indicates the state of the output based on the state of the input. In this case, it refers to whether a lightbulb is on or off based on whether the switches are on or off.

The truth table for this AND circuit includes more information than necessary because the information on the right-hand side of the table means exactly the same thing as the left-hand side. In the truth table for the OR gate shown in Figure 7.13C, only the standard logic notations of 1 and 0 are used.

(A)

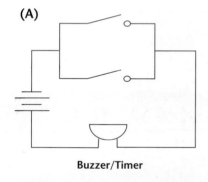

Buzzer/Timer

(B)

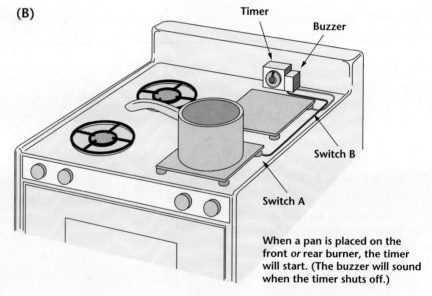

When a pan is placed on the front *or* rear burner, the timer will start. (The buzzer will sound when the timer shuts off.)

(C)

A	B	Output
0	0	0
0	1	1
1	0	1
1	1	1

■ **Figure 7.13A, B, C**
This simple OR circuit is shown as (A) a symbolic drawing, (B) a practical application for helping Aunt Sarah keep track of the time a pan is on a burner, and (C) a truth table for an OR logic circuit.

In the following design brief, we will use simple logic gates to provide an alternative means for dealing with the problem of identifying the content of the cans in Aunt Sarah's kitchen. In the chapter on Control, we will return to more sophisticated applications of logic gates and how they can be used for machine decision-making.

DESIGN BRIEF

Based on your new knowledge of logic gates and logic systems, you have been asked by April and Enrique to help improve their design for identifying cans in Aunt Sarah's kitchen. They believe that a relatively simple device could be designed and developed that would help Aunt Sarah check the contents of a can by its size.

After some experimentation, the three of you come up with an idea to add two additional switches to the overhead arm so the device could be used to measure cans of three different heights, as shown in the sketches below. The design of the logic system should signal Aunt Sarah by buzzer for the shortest cans, by bell for the medium-size cans, and by both buzzer and bell for the tallest cans.

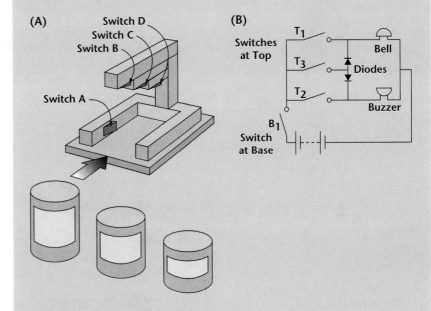

■ Figure 7.14A, B
This device uses several AND gates to help Aunt Sarah check the size of three different size cans. (NOTE: The diodes serve as one-way electrical valves so that the bell and buzzer sound appropriately.)

MATH/SCIENCE/TECHNOLOGY LINK

- Investigate and report on how Boolean algebra relates to machine logic.

- Determine and describe how an invertor gate works electronically and mathematically.

- Identify how you can use an invertor to change the function of the different logic gates introduced earlier.

- Develop truth tables for combinations of different logic gates.

- Determine how you can simplify the process of developing truth tables that involves the use of several combinations of gates.

SYMBOLS: INFORMATION FOR HUMANS

As indicated earlier, symbols have inferred meaning; they stand for something. For example, the numbers in an elevator are symbols representing different floors. The number "3," obviously, is not the third floor itself. It only represents the third floor. It is a symbol. We interpret what the "3" means when we see it on a display of numbers in an elevator. The elevator only notes the position of the button that is pushed; the number means nothing to the elevator.

If you saw a "3" in the address of a building, you would assume that the "3" indicated something other than the third floor. A "3" as a symbol means something quite different on a street sign, a birthday cake, or a three dollar bill. Symbols have meaning only when people bring that meaning from their previous learning. (Refer to Figure 7.15.)

Figure 7.15 ■
Different symbols can have very different meanings of the concept of "3," or more accurately the characteristic of "3-ness."

Symbols only have meaning and communicate information if the person who receives the symbols can interpret them. There is some evidence that the human infant may be "programmed" to process information in certain ways. Newborns attend to patterns that are similar to faces. They show a startle response and other similar simple reactions to the environment. And if you have ever been around a baby, you know that they do not have to be taught to respond when they get internal signals of discomfort—too hot or cold, low on fuel, wet, etc. A sensitive

■ **Figure 7.16**
From early history to the present, many different symbol systems have been developed to support trade. The future suggests that additional new systems will be developed. *(© Michael Dzaman Photography)*

caretaker soon learns to distinguish between cries of the growing infant. Some communicate a need for care. Others express a need for stimulation and company. The caretaker learns to interpret the infant's signals at the time the infant is learning the symbols used by the caretaker.

Recording Information

Humans are not satisfied with processing only their own experiences. We modify our sense of self by communicating with other humans. We seek to understand our place and our time. We want to transcend our own time and know more about the past and future. We want to know about other places, whether real or imaginary.

Centuries ago, people attempted to talk with others in their group and developed oral language systems. Words came to stand for the actual object. Expressions, gestures, and other movements and sounds came to symbolize emotions and actions. Have you traveled outside your neighborhood or attempted to communicate with someone much older or younger than you? If so, you know the importance of having a common meaning for words. **Communication** of information depends on shared agreement on meanings. Direct communication requires the processing of information from the context, the words, and the nonverbal expressions. With expansions in travel and trade, symbols become more complex and ways for recording and exchanging information are needed.

How many? Trade also required some sort of system to keep track of transactions. Using a stone to symbolize a pot of oil, for example, meant that the ancient trader could account for the pot. Numerical systems were developed to keep track of quantities. Marks on papyrus and other materials were some of the first records of trade. The invention of the zero to stand for "nothing," or as a placeholder in a decimal system, was a major breakthrough for counting, measuring, and communicating quantity. The zero in the binary system used in signals to machines indicates that there is no signal present.

What is it worth? In addition to keeping track of the number of items traded, it is necessary to keep track of their relative value. If the value is about the same, people can trade one product evenly for another. This is called "barter." Usually, symbols that represented a measure of worth were needed. Various monetary systems were developed over time. These symbols allowed for a sale to occur.

Stones, coins, letters of agreement, and pieces of plastic all supply information about a sale. A critical point is whether other people will honor a given symbol of exchange. Do you think you could use a peso, pound, or ruble to buy a soda? Forms of exchange now include electronic transfer of funds and credit accounts. No tangible evidence of the exchange is required other than a computer message on a screen, or directly to another computer.

What time is it? Can you imagine what your life would be like if no one thought about time? Babies and toddlers have little sense of time except that imposed by changes in their own bodies. "It's time to eat" is a welcome message if the child is hungry. The numbers on a clock mean little. You, however, may eat because of the time, whether you are hungry or not. If you don't, other scheduled activities may interfere with eating when you are hungry.

■ **Figure 7.17**
Stonehenge is one of the oldest observatories used to keep track of time. It works like a celestial calendar.

Early concepts of time came from observations of natural phenomena. Ancients measured time and regulated their lives by patterns they observed in their environment, especially patterns of change in the sun, stars, and seasons. People were active when it was daylight and sought shelter when it was dark. People planted seeds at times that had worked best in previous years. Examples of early attempts to record and use natural patterns range from simple calendars and records of planting and harvest, to such ancient observatories as Stonehenge, shown in Figure 7.17. People adjusted their lives to take advantage of these rhythms.

Measurement of time allows people to find their location on Earth. Later in this book, we will discuss some of the tools and inventions that influenced travel. Putting together the measurements of motion with those of time gives us a notion of how fast we travel. This information was coupled with an understanding of the relationship of the Earth to the sun and stars and used to determine the location of explorers and navigators as they traveled long distances across the continents and oceans. New advances in timekeeping allow measurement of time accurate within seconds over a million years.

Sharing information. Pictures drawn on early dwellings were another attempt to record information. These forms became stylized and over time conveyed information to others even when details of the drawing were minimal.

Letters in alphabet systems, pictographs, numbers, and other symbolic systems allowed flexibility in freezing information in a form that could be saved, transported, and exchanged. (Refer to Figure 7.18.) People can learn about the experiences of other groups and avoid some of their mistakes. The generalizations and biases that you may have learned can be modified by the new information. Pictures expand this potential awareness. Songs, costumes, and objects that people use in everyday life allow others to imagine what life might be like for people from that culture, place, and time.

■ **Figure 7.18**
This merchant is using the Arabic alphabet and numbers. These symbolic systems could be written in a form that was easily transported. *(Courtesy of Aramco World Magazine)*

An important point to emphasize again is that the receiver of the information must have the background to interpret the symbol. If not, it may as well be gibberish. Try talking to someone who does not speak your language. Read a Mayan pictograph. Interpret an abstract painting. Dance like a bee, revealing the location of a good source of nectar. If you cannot break the symbolic code, you cannot interpret or repeat the message.

Only a few centuries ago, the major way of sending messages to someone else was to talk or yell. Communicating over long distances was very difficult until the invention of the **telegraph** by Samuel Morse. This was an important development during the westward movement in the United States and the settling of the western states and territories. A telegraph operator had to both send and receive messages in **Morse code**. This code was a language system used to send and receive messages through a wire. However, someone had to be at the receiving end, waiting, or the message would be missed. Morse developed the simple recording machine, illustrated in Figure 7.19, that helped solve the problem. It was now unnecessary for the receiving operator to **decode** the message while it was coming in. The recording machine made a mark on the paper tape as it was drawn past a marking pen. Every time a signal was sent, an electromagnet drew the arm forward. The pen made a mark each time a signal

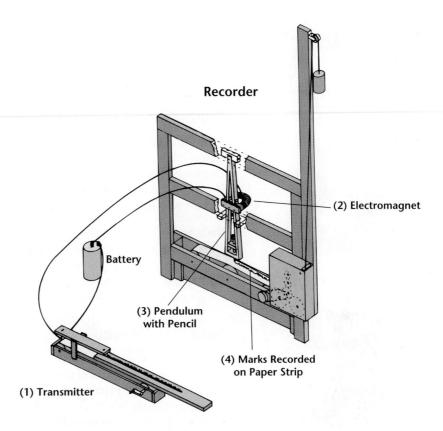

Recorder

(2) Electromagnet

Battery

(3) Pendulum
with Pencil

(4) Marks Recorded
on Paper Strip

(1) Transmitter

Figure 7.19 ■
Morse, in addition to developing
the telegraph and Morse code,
invented this recording instrument.
Electrical signals sent by wire cause
an electromagnet to move a
pendulum with a pencil attached
to make marks on a strip of
moving paper. The marks served as
a record of the telegraph message.
(Can you determine what is wrong
with this drawing?)

was transmitted. The record of the message was a long tape of these marks. The
telegraph operator could decode the tape during the transmission of the message
or at some later time.

Signals developed as the common language between two machines are not
usually detectable or understandable by humans. When a machine is to be
controlled by information, the human must try to "think" like a machine; that is,
each step of the process must be considered objectively and logically. Information
must be broken down from the symbols a person might understand into tiny bits
of data.

Consider, for example, how a voice-activated machine responds to the message,
"Open the door." The machine does not interpret this message. In order to get
the machine to respond, a person must have analyzed that message into each of
the sequences of required action the machine must make. A separate signal for
each of the steps in the process must be programmed into the machine. In other
words, a **template** within the machine sorts out the signal that will be received
and used. The signal must match the predetermined template. An even more
complicated code is needed if the machine is to respond only to the voice
patterns of one or two people.

Machines can also be programmed to provide signals for people—signals they
can convert to symbols. Have you ever been in a hotel when the fire alarm went
off? A loud, offensive sound has been chosen that makes continued sleep almost
impossible. Yet, it takes a little time to interpret the signal and take the

■ **Figure 7.20**
How many of these logos and pictograms do you recognize? *(Courtesy of Technology Student Association, Denver Broncos Football Club, Apple Computer, Inc., and Nike, Inc.)*

appropriate action of leaving as soon as possible. Some alarm systems also have a recorded voice that warns against taking an elevator. This whole warning system has been designed to make it likely that people will be convinced to leave the hotel as quickly as possible.

Symbols, Codes, and Messages

If you are to communicate through symbols, it is not enough to only understand each symbol you use. Symbols are put together in different ways to provide meaningful messages. You need to know the code for those symbol sequences. You need to know how to use the code and symbols to create messages you want to share with others.

The way you select your haircut, style of clothing, and manner of walking and speaking are another form of message system. Line, form, color, patterns, and textures can be carefully combined to indicate that you belong to one clique or social group, and not others. Uniforms, insignias, and brand names are chosen and displayed because of the message they convey.

Individual Symbols—Logos and Pictograms

Some symbols are relatively easy to understand and use because they have a limited meaning. **Logos** and **pictograms** usually have a single meaning. Logos are developed as a graphic representation of a company or organization. Some organizations have logos that include acronyms—the initials of the organization. Other companies and organizations combine a picture or representation as part of their logo. This is a pictogram. Examples of logos and pictograms are shown in Figure 7.20.

DESIGN ACTIVITIES

Logos, with or without pictograms, are used to give an organization a readily recognizable visual image. The image is intended to convey a desired message concerning the organization.

• Identify at least five products or companies that use a recognizable logo or pictogram.

• Determine what ideas the designers of the logos were trying to get across to people who see the logos.

• Determine which logos or pictograms appear to do the best job in representing visually what the company or product is about.

• Design and develop a logo or pictogram for something you think is a marketable product.

■ **Figure 7.21**
Universal symbols are helpful as
people from different language
systems and cultures travel, trade,
and communicate. Pictograms
such as these are very useful.
(© Michael Dzaman Photography)

Collections of Symbols

In some cases, a collection of pictograms are used to convey simple messages to the users. Each symbol has a limited meaning, but collectively they can provide an impressive amount of information. For example, the traffic signs we see while we are driving are a collection of such symbols. These symbols are such an important part of driving that drivers must pass an examination that tests whether they can interpret these symbols.

The shape and color of the highway signs take on important meaning. Square shapes mean something different than round shapes. Triangular shapes suggest something different than octagonal ones. In the signs, red is a more powerful color than white, blue, or yellow. Shape and color become an important part of how the sign collection gets its message across to drivers.

Over the years, other collections of signs and symbols have been designed and developed to serve a wide range of uses. Examples of these are included in Figure 7.21.

Each of these collections was designed to play a unique role in a specific kind of setting. The success of the design depends on how well the signs are able to convey the intended meaning to the people who use the signs. Which of the signs included in Figure 7.21 do you think are particularly good? Which do you think present a problem for users?

Symbol Systems

The next level of sophistication in using symbols involves the use of a symbol system. A system of symbols indicates that the individual images can be combined in a prescribed manner to construct messages. One of the most familiar to all of us is the alphabet of letters that we use in our writing and reading. To most of us, the alphabet has become a natural part of our communication skills. The simplicity of the alphabet belies the long period of development that it has gone through. The development of our number system went through a similar evolution. The importance of the development of the idea of zero, mentioned earlier on page 203, is an example of that evolution.

Many symbol systems have been developed to help those involved in design and development activities communicate ideas easily and effectively. You have already been introduced to some of these. The symbols introduced in the preceding chapters, and the rules presented on drawing in Chapter 2, can help you develop different types of drawings and plans. Symbol systems, such as those used for electronics, pneumatics, and other devices, can help you construct an image to record information for your own or to share with others.

D E S I G N   B R I E F

- Identify and use a current symbol system for communicating ideas about your design and technology work. What suggestions would you make to improve the symbol system?

- Develop a drawing and symbolic representation of an electronic system you or your team has developed.

- Design a symbol that will help draw attention to the safe use of a tool or machine.

- Design a symbol that will help people understand the basic principle or concept behind a tool or machine.

- Design a symbol system for a major sporting event.

- Design a symbol system to help store and retrieve resources.

USING SYMBOLS IN DESIGN AND TECHNOLOGY

In Unit I of this book, the design process was presented through illustration of the IDEATE model. If you are solving a problem for yourself or for one other person, you may not need to be concerned with symbols and symbol systems other than those involved in direct conversation. Suppose you want to show how your problem relates to a larger social context or you want to repeat your solution for several people. In this case, careful collection, analysis, and presentation of data are needed.

Sometimes you are really not sure if a problem exists or just how widespread it may be. For example, you may notice that one leaf on your favorite apple tree has developed black spots on it. Is that a problem worth addressing? As you examine your tree further, you may count the number of leaves that are affected. You find some leaves that have brown spots on them. As you examine the trees in the neighborhood with the same condition, you will get a better idea of the problem. Is this different from last year's condition? Are any other people worried?

If you decide to just take action as an individual, you can use the IDEATE model in a very informal way with only the informal data you have. If, however, you think the problem requires action on the part of the whole neighborhood, or the region, much more precise information will be needed. More formal methods of organizing and presenting your findings will be required if you are to convince others that there is a problem and they should contribute time or money to carry out your design for action. If you plan to repeat the process in other parts of the state or next year, you will need formal information for evaluation of results.

The "informal" or "formal" data refer to just how precise the procedures are for collecting information. What kinds of measures are made? Over the years, numerical systems have become important, even essential, for communicating meaning about quantity. Everyone who uses the units of measure needs to know what they stand for. As you gather, organize, analyze, and prepare data for presentation to others you may use these numbering systems in various ways, depending on your purposes.

Levels of Measurement

Sometimes when we collect information, it is enough to just group things and give them a name—"My apple tree has green leaves, green ones with black spots, and green ones with brown spots." Share this information with an agronomist, and she may be able to diagnose what is wrong. She may indicate it is a natural occurrence with apple trees. This level of measurement of data is called **nominal** and means that things are only given names. You might draw the trees and use colors closely matched to the leaves to increase the amount of information you present.

Your Tree   **Neighbor's Tree**

Nominal Level of Measurement

Leaves	Leaves
Yellow w/black spots	Green

Ordinal Level of Measurement

Leaves	Leaves
Most are green	All are green
Some are yellow w/black spots	
Few are brown w/holes	

Interval and Ratio* Levels of Measurement

Leaves (sample of three limbs)		Leaves	
Green	207	Green	261
Yellow w/black spots	32		
Brown w/holes	8		

*The ratio level of measurement requires an "actual zero." In this case, zero would mean no leaves.

Figure 7.22 ■
Data can be gathered by nominal (naming), ordinal (ordering), and interval (counting) measurements.

The next level of measurement is when you place things in order from most to least. This is called the **ordinal** level of measurement. For example, you could determine which of the trees had the most spotted leaves, which had the least, and which were in between. This might be important for deciding whether you should spray with an insecticide, when you might spray, and which trees to spray first.

As another example, suppose you decide to design a better menu plan for the school cafeteria. You may give potential customers a survey in which they tell you what they like best and least. You may be able to use this information to decide that you will serve liver and onions the least number of times, and tacos less often than pizza. Your data may be organized to present the foods in order of preference. In presenting your data, you have the number of people and how they responded. You can say that five times as many prefer tacos than prefer liver and onions.

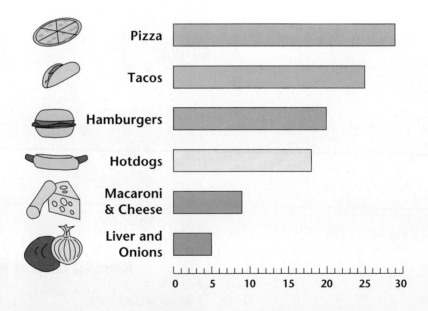

Figure 7.23 ■
A bar graph can be used to organize and present data from a survey to show preferences and choices.

The third level of measurement is called **interval** and requires measurements that use equal units. You can set out to count how many of each type of leaves your tree has. You might count them all or count a sample of the leaves.

There is a fourth level of measurement that is even more precise. This level, the **ratio** level, requires an actual zero. At the interval level, you may have decided to use a zero, but if a quantity can fall below that point, the zero is considered artificial. For example, if someone asks you how much money you have, you may say none (zero), but you could owe money to your sister and therefore have less than no (zero) money. Temperature may be measured on a scale with a zero, yet it can drop even colder than the zero point. One example of ratio measurement where an actual zero can be established, has to do with height and weight. Suppose you want to decide whether your little sister is growing as she should. You can measure her height and weight with standard measurement devices. You could compare these results with those of the average girl her age. At this level of measurement, you can tell which is bigger or smaller, and you can tell by how many units the measurement varies. Your sister is two inches taller than one of her friends. This two inches represents the same amount of difference when you find out that your sister is two inches shorter than another friend.

Interval and ratio levels of measurement allow us to add, subtract, and calculate averages and proportions. The difference between 25 pounds and 30 pounds of flour is the same quantity as the difference between 140 pounds and 145 pounds of stones.

MATH/SCIENCE/TECHNOLOGY LINK

Levels of Measurement Put to Use

Enrique and April had to decide just how precise they needed to be in their design project for Aunt Sarah that we considered in Chapter 1. As you recall, they wanted to ensure that Aunt Sarah could find and identify a can of food from her cabinet before she opened it. They wanted to design a solution so that when she prepares a meal, she is accurate in her choice of cans, even though she cannot see well enough to read labels. One way to deal with the problem could be to designate different cabinet spaces for different foods. When someone delivers her groceries, they can simply put soups in one spot, vegetables in another, fruits in another, and meats in still another part of the kitchen cabinet. When Aunt Sarah goes to the cabinet closest to the door, she will know that cans of meat are there. This represents a **nominal level of measurement**.

■ **Figure 7.24**
A storage system for storing cans by name only is a nominal system.

Suppose Aunt Sarah wants to tell the difference between a can of peas and a can of tomatoes, however. Another level of precision will be needed. They will both be in the vegetable area. She could decide that in the future, peas will be bought in small cans and tomatoes will be purchased in middle-sized cans only. Vegetable juices can be bought in large cans, since she likes to drink juice by the glassful. Now when she goes to the storage area for a can of vegetables, she can distinguish between the vegetables by feeling for the size of the can. This represents an **ordinal level of measurement**.

■ **Figure 7.25**
Cans that are stored according to size illustrate an ordinal system. Cans are grouped in order, from small to large.

But since she is dependent on other people to help her with shopping, Aunt Sarah would like to keep track of how many cans she uses of each kind of food. When someone calls her and asks if there is something they can pick up for her at the grocery store, she would like to know. Enrique and April designed a labeling system in which raised, 3-D labels were placed on each can to indicate its contents. As Aunt Sarah uses a can of food, she can save the labels for reuse at a later time. Counting the number of labels from what she has used gives her a means of keeping track of her inventory of canned food. This data is at an **interval level of measurement**. This also satisfies the requirement for **ratio level of measurement**. She can say whether she used twice as many cans of pineapple as she did tuna. Her responses about what she needs to replenish her stock of foods can be much more precise.

■ **Figure 7.26**
The labels from cans are removed as the can is opened. Labels are stored according to the type of food. To replace the inventory of canned food, the number of each type of label is counted. This is an interval level of data.

We will return to these ideas when we consider the evaluation and re-design of the solutions from your design and problem-solving efforts.

Units of Measure

Early attempts at measurement used units that appeared in nature, such as a handful or the length of your finger or foot. You can see the problem when you buy a handful of rice from two different people. You would certainly get more in exchange if the person's hands were bigger. A foot was supposedly designated as the size of the foot of the king. A yard was the length from the tip of the nose to the tip of the middle finger. This may work out if you only need a guess or an approximation. However, if you want to be consistent as you repeat the process over and over, an interval level of measurement is needed.

A great deal of work went into developing a common system of measurements that will help us share information more readily. For example, by using units of feet and inches or meters and centimeters we are able to share information about length. A few selected units of measurement, including distance, time, force, weight, mass, energy, and power, are listed on the next page.

Units of Measurement		
Quantity	**Name of Unit**	**Symbol**
distance	meters	m
time	seconds	s
force	newton	N
weight	newton	N
mass	kilogram	kg
energy	joules	j
power	watts	w

Throughout all parts of your designing process, you will need to select how much precision is appropriate. If the design is to address the needs or wants of only one person for only one time or if the required resources are not very expensive, little precision may be necessary. When designs address the needs of many people or are to be replicated in several places or many times, several issues should be considered. A pinch of salt may be okay in making biscuits. You want a pharmaceutical company to be much more precise when making the pills that will become your medicine.

Gathering and Presenting Information for Design Work

Information is needed at several points when you are conducting your design and problem solving activities. Your research in developing an understanding of the problem and in deciding on the specifications for any design solution produces information of many types and from several sources. At this point, it could be said that you are using information in order to diagnose the problem and its dimensions.

When you gather information to predict the effectiveness of different strategies for solving the problem, you may talk about averages and variations from one person to another and from one time to another. A given solution may not work all the time, but just how does it perform usually? How much error is acceptable?

When the design solution is evaluated, you will want to try it several times. Gathering the data about the performance is needed if you are to decide how to improve the design solution. Communicating your results can be used to justify your use of resources, especially if they were contributed by someone else or are expensive.

Charts, Graphs, and Maps

Several forms for organizing and presenting information have been developed over time. After you have collected the data relative to the ideas you want to communicate to others, you can select the most appropriate form or forms for sharing that information. Three general means of sharing information with others that may be helpful are charts, graphs, and maps. (Refer to Figures 7.27, 7.28, and 7.29.)

MAJOR USES OF LAND, 1982
TOTAL UNITED STATES
(MILLION ACRES)

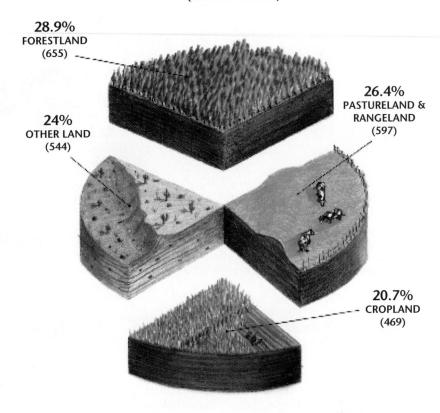

28.9%
FORESTLAND
(655)

24%
OTHER LAND
(544)

26.4%
PASTURELAND &
RANGELAND
(597)

20.7%
CROPLAND
(469)

Figure 7.27 ■
This pie chart uses images and numbers to represent how land was used in 1982. *(Courtesy of Wendy Snyder)*

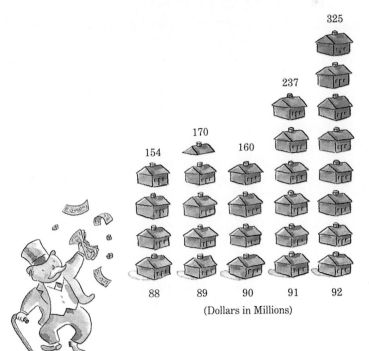

325

237

170   160

154

88   89   90   91   92

(Dollars in Millions)

Figure 7.28 ■
This graph shows the relationship of net revenues and operating profit for a corporation. *(MONOPOLY®, the distinctive design of the game board, as well as each of the distinctive elements of the board and playing pieces are trademarks of Tonka Corporation, for its real estate trading game and game equipment. ©1935, 1995 Parker Brothers, a division of Tonka Corporation, Beverly, MA 01915. All rights reserved. Used with permission.)*

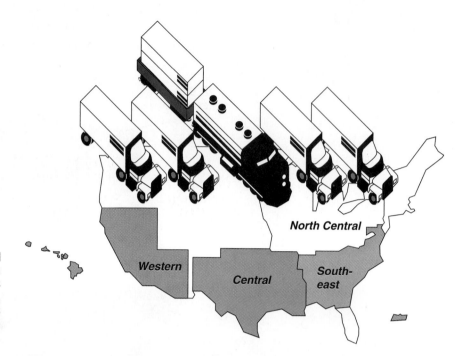

Figure 7.29 ■
Maps can be used to show a variety of information such as weather patterns from a satellite, crop distribution, or in this example, the service regions of a transportation company.

North Central

Western

Central

South-east

SUMMARY
CHAPTER 7

Information exists as patterns of material and energy in the forms of signals and symbols. In the form of signals, information is used to cause some intended action. Information in the form of symbols is used to infer meaning.

Information is the basic building block from which communications and control are constructed. Both communications and control require that some sharing of meaning takes place as people and/or machines interact. Such meaning is possible only when an understood code specifies what the units of information are to be.

The resources of technology are closely interrelated with information. Energy can be turned into information when small samples of

energy are used in generating signals. The sampling of energy is accomplished by small devices called transducers. Sensors can be used to sample mechanical, electrical, magnetic, light, thermal, and sonic energy to make signals for a variety of uses. Information as signals is sent to the guidance and control systems of machines.

Materials can be converted to information in the form of symbols and signals. Materials and information interact through on-off devices, such as solid-state switches. The switches are made of dynamic materials that are changed by a signal.

Information provides a major link between humans and machines. Symbols provide a form of infor-

mation that humans use with each other. Signals must be converted to symbols to allow machines to communicate with humans. Symbols must be converted to signals for humans to communicate with machines and for machines to communicate with each other.

Complicated systems of symbols in the form of alphabets, numbers, money, codes, and records of time and place have evolved over the years. Various forms of graphs, maps, and charts can be used to present data in a form that will most easily communicate to other people. The level of measurement needed for design and problem solving depends on the audience and how the information will be used.

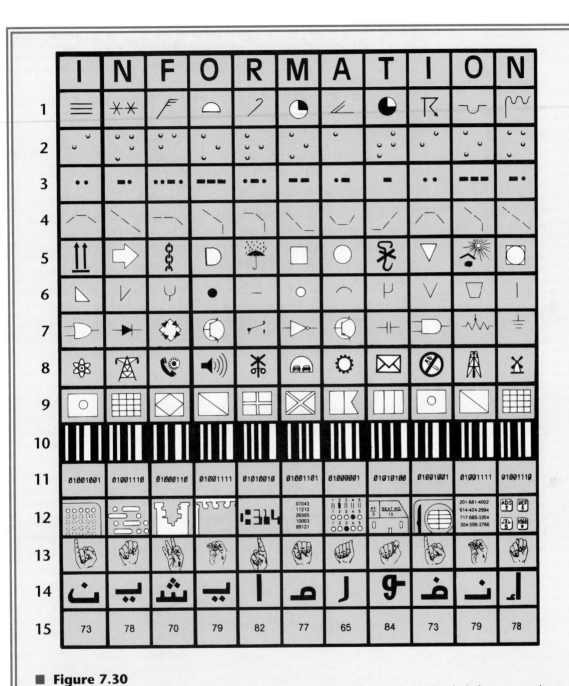

■ Figure 7.30

A number of codes and symbols are shown in the illustration above. Line 1: Symbols from a weather map. Line 2: Braille code for the blind. Line 3: Morse code used first on the telegraph. Line 4: French semaphore code used for sending messages from hilltop to hilltop over long distances prior to electronic communication. Line 5: Symbols used in material handling to indicate how a part should be transported and handled. Line 6: Symbols used to indicate the different welding processes that are to be used during fabrication. Line 7: Symbols used in electronics and computers. Line 8: International traffic symbols. Line 9: The International Maritime Code used with watercraft. Line 10: Optical bar code. Line 11: The binary code used in computers. Line 12: Different devices that are used to store information for our use. Line 13: The American Sign Language for the deaf. Line 14: The Arabic alphabet. Line 15: The hexadecimal code used in computers.

ENRICHMENT ACTIVITIES

Design Briefs

- From your experience in design and technology, design a portfolio format that represents symbolically one activity you wish to pursue or one product you wish to develop.

- Analyze the problems you have in keeping your log and journal up to date. Design new page formats and methods to improve your record-keeping.

- Mass-transit systems, particularly bus and subway routes, tend to be difficult for new users. Design and develop a map and supportive brochure that would help a visitor from another country use the system in your community easily and safely.

- Design and develop a survey questionnaire to collect information about a product you plan to develop. In the design of the survey and questionnaire, consider the following:

    What is the purpose of the survey?

    What information do you need?

    What people will you include in your survey?

    What specific questions do you plan to ask?

    How can you make the survey easy to analyze?

    How will the survey be conducted?

- Design a graph or chart that conveys information gained from the survey questionnaire described above.

- Design and develop an electronic map to be used in a shopping mall that will help people find the location of products they want to purchase. All stores that sell a product should be identified when the user requests the information.

- Design and develop a map to help people travel in your area on unmarked back roads, on streets within the largest city in the area, or on highways that have many entrance and exit points over a relatively short distance.

MATH/SCIENCE/TECHNOLOGY LINK

- Identify common articles that can be used to help you and other students become familiar with the metric system of measurement. Develop a display of the objects and the related measure they represent.

- Identify the units of measure that are used for the sensors that have been introduced in this chapter.

- Develop a display of the standards of measurement that have been adopted internationally.

- Identify the levels of measurement—nominal, ordinal, interval, and ratio—as they apply to your classmates, the products you buy, or the project you are currently planning or implementing.

- The most common temperature scales we use are centigrade (C) and Fahrenheit (F). The zero point on these two scales is different and temperatures can fall below zero. The Kelvin (K) scale is based on absolute zero. Theoretically, temperatures below zero on this scale are impossible. Identify which levels of measurement (nominal, ordinal, interval, or ratio) are possible for each scale, and describe why.

REVIEW AND ASSESSMENT

- Describe how signals represent essential information for machines.

- What are the major components for communication and control and how do they work together?

- What are the sensors and input signals for sensing:

    mechanical energy
    electrical energy
    magnetic energy
    light energy
    thermal energy
    sonic energy

- Identify the most common devices used in machine logic and how they can be combined to provide more sophisticated logic decisions.

- Identify how truth tables work and how they relate to Boolean algebra.

- Identify some of the symbols humans use and describe how they use them.

- Describe some of the problems you are likely to encounter in recording information and sharing information.

- Describe some of the advantages and disadvantages of using pictograms and logos as a means of representing something.

- Identify the four levels of measurement and two examples of each.

- Describe some of the most common means of gathering and presenting information in design and technology work.

- What are the advantages and disadvantages in using charts, graphs, and maps for sharing information with others?

CHAPTER

8

Humans as Designers and Consumers

Figure 8.1 ■
We can learn more about technology, and also use technology to learn more about ourselves—our strengths and limitations. (© Larry Keenan/ The Image Bank)

INTRODUCTION—THE CREATORS AND USERS OF TECHNOLOGY

Throughout this book, we have indicated that technology is a uniquely human endeavor. Using the best information, materials, tools, and energy available at the time, people have changed the natural world enormously. Human goals

221

KEY TERMS

achievement
additives
auditory
autonomy
belonging
cognitive templates
coordination
cultural templates
delete
distort
enlightened self-interest
food additives
generalize
generativity
human control subsystem
human energy transmission
    system
human motivation
human structure and cover
    system
individual template
insulation
integrity
intimacy
kinesthetic modality
microorganisms
nutrients
nutritional supplements
patterns of growth and
    development
processed food
psycho-social stages
sensory templates
shelters
symbiotic relationship
synthetic skin
tolerance limit
trust
virtual reality
visual modality
vulnerable

and needs are reflected in the applications of technology and its outcomes. People themselves have been changed by the products and the processes of technology.

Attempts at understanding and controlling the forces around us often involve using people's past experiences for comparison. Tools and machines have been developed as extensions or imitations of the human. Information processing has changed as we develop processes and tools that mimic the human brain. We are the only species that has developed such complex systems for technological events. The human species also has the ability to use the outcomes of technology for both constructive and destructive outcomes.

WHAT IS A PERSON?

It is probably safe to assume, since you are reading this page, that you are a human. If not, and you are a visiting life form from another planet, you may want to know more about these things called humans. The designer/engineer/inventor Buckminster Fuller provided this somewhat tongue-in-cheek description of humans:

> A self-balancing, 28 jointed adapter-based biped; an electro-chemical reduction-plant, integral with segregated stowages of special energy extracts in storage batteries, for subsequent actuation of thousands of hydraulic and pneumatic pumps, with motors attached; 62,200 miles of capillaries; millions of warning signals, railroad and conveyor systems; crushers and cranes (of which the arms are magnificent 23 jointed affairs with self-surfacing and lubricating systems), and a universally distributed telephone system needing no service for 70 years if well managed; the whole, extraordinarily complex mechanism guided with exquisite precision from a turret in which are located telescopic and microscopic self-registering and recording range finders, a spectroscope, et cetera, the turret control being closely allied with an air conditioning intake-and-exhaust, and a main fuel intake.

Whew! Did you recognize yourself? Although there is much more to being human than that, let us start with the basic physical equipment. Any time people try to design replacements for body parts or tools and materials for human use, taking an objective view can help.

DESIGNING FOR THE HUMAN BODY

In the chapter on Tools, you learned that there are three basic systems to all machines: structure and cover, energy transmission, and control. These systems can also be found in the human body. The **human structure and cover system** includes our skeleton and skin.

■ Figure 8.2
This robot was designed to mimic human capabilities in order to test safety clothing. It can breathe, sweat, walk, and maintain a human-like skin temperature. It is even programmed to speak in order to announce problems! *(Courtesy of Pacific Northwest Laboratory)*

The skeleton forms a framework of bones which supports your body. Some parts of the skeleton, such as the skull and ribs, play both a support and cover function. The bones in your arms and legs act as levers that are moved by pulling muscles to move you from place to place. Your skeleton is made up of more than 200 bones and provides you with a frame capable of being loaded with well over a ton of weight. Yet, the bones in your skeleton probably weigh no more than 20 pounds! Did you realize that your skeleton is such a superbly designed structure?

The major part of the human's cover system is the skin. It has an area of approximately 20 square feet. The skin provides a protective covering to keep harmful germs and dirt out of the body. It keeps itself moist and stretches to move as you move. It repairs itself when damaged. Skin works to keep the temperature of your body within safe limits. Skin is even waterproof.

The **human energy transmission system** includes our muscles, tendons, and cartilage. Since the human is self-propelled, the digestive system provides an inboard power plant to change food to energy and replacement materials.

The muscles are attached to the bones by tendons. The nerves that control the muscles transmit signals to make the muscles contract. When the muscle contracts or shortens, it pulls the bone. The muscle cannot push the bone back in place, so another muscle paired with it contracts to pull the bone back. We use muscle energy to move ourselves and our tools. Some muscles are much more powerful than others. For example, the large muscles in the legs provide much more power than do the muscles in the face and head.

The joints are a unique part of your skeletal system. Some joints move only back and forth as a hinge. Other joints, such as in your shoulders and hips, work as a ball and socket hinge and move in several directions. These living hinges always provide their own lubrication. A built-in sac encloses the joint to keep the oily liquid in and abrasive materials out.

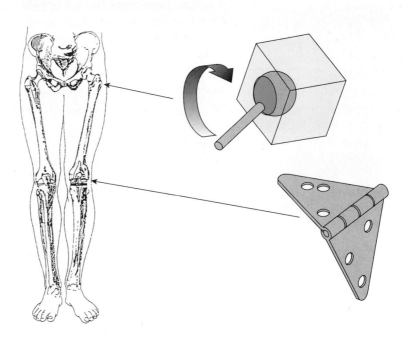

Figure 8.3 ■
Replacement parts for humans are made by comparing how machines and body parts are alike and different.

The **human control subsystem** includes the nervous system and all of the senses—sight, sound, smell, taste, and touch. Some parts of the body work automatically. For example, we do not need to think to breathe or blink our eyes. Other parts of our body require conscious control. We must decide to get up and walk. We must also decide when to stop.

Once we have decided, our brain sends signals to make our muscles move. The process of moving is very complicated. Babies must learn gradually to control parts of their bodies. They learn this control from head to toe and from inside systems to outside. The first movements involve the large muscles, and later infants learn to control fine movement. If you have ever watched a three-year-old, you will understand just how rapidly this control is learned.

Some movements require fine **coordination**. Our control system usually takes data from several senses at once and coordinates it in a split second. Thought processes occur much faster than muscle processes. Think of all the times you have forgotten what you were going to say before you could get the words out. Scientists are just beginning to understand how the brain works. The more knowledge we obtain, the more respect we have for the extremely intricate processes of the human brain.

DESIGNING FOR HUMAN USERS

Knowledge of the body is important when industrial designers fit a product to a user. Tools and machines must have an appropriate size and shape to be useful to humans. The size and shape of many of the things we use came about without much planning. Most household scissors are made the way they are chiefly because the shape is easy to manufacture. Only recently have designers developed new shapes that will better fit the average user, whether left- or right-handed. Many products have never really been designed, but just sort of happened. The size and shape of a pencil, a toothbrush, cups, and dishes probably just evolved without much planning.

■ **Figure 8.4**
This fabric system is lightweight, but was able to stop a chain saw blade before it slashed into the leg of the operator. *(Photo courtesy of DuPont)*

D E S I G N   B R I E F

A Real Design and Technology Project

A school class decided to design a garment that would protect landscapers at work. The garment needed to protect from exposure to toxic chemicals, dirt, and other dangers. The first thing the class did was observe current workers. An all-white garment was made and worn by workers on the job. Spots from the chemicals and the dirt were evident. The garment was taken apart and the amount and location of stains from the various chemicals were measured. Also considered were the stresses and strains on the garment, how to ventilate the heat of the body, and which parts of the body needed to be shielded from the machines and flying objects. Fabrics and fabric systems were designed. Any garment had to be washable, durable, and inexpensive, as well as look good. Several were constructed and compared. The best customized garment for the worker was made and tested. Imagine the delight of the class when a large lawn care company bought the design for their employees.

In the chapter on Design, we discussed how important it is to gather and use information about the user and the setting in which the product will be used. We explored ergonomics and the value of this data in designing for humans.

The size of the human body is only one type of information. Other physical characteristics include movement, reach, and strength. To understand movement of the body and its parts, it will be helpful to compare the body to machine parts. We can make the following comparisons:

1. **bones**—support
2. **skin**—cover
3. **joints**—bearing surfaces
4. **joint linings**—lubricant
5. **muscles**—motors, shock absorbers, locks
6. **tendons**—transmission cables
7. **nerves**—control and feedback circuits

As you may recall from the chapter on Designing, your strength varies, depending on the muscles you use and the movements you make. You can lift more weight with your biceps than you can push with your extensor muscles (triceps). There is also considerable difference in the strength of the wrist as compared to the hand. The hand's grip is usually stronger than most of the wrist's movements. Our legs are stronger in pushing and stretching than in bending. For a short period of time, the leg can generate forces exceeding 450 pounds. The design of any tool or machine for human use should consider which muscle group will exert force in which direction.

In spite of our immense complexity, humans are extremely vulnerable. Perhaps this has motivated our attempts to control our environment through technology. Human beings can survive in some places on our planet with few adaptations. We have even developed technical means to survive in space and on other planets, as well as under water.

Human operation is affected by the physical variables of the environment. The human body has **tolerance limits**. Beyond these limits, great discomfort, physical harm, and even death take place. Time, motion, and stress studies have been used to arrange workplaces, improve sports equipment, and protect human workers from toxic substances. Some methods of protecting ourselves from harmful exposure include monitoring for temperature control, shielding from viruses and bacteria, filtering the sun's rays, cushioning to prevent excess noise and vibrations, and screening for privacy.

■ **Figure 8.5**
The strength and flexibility of the hand are considered in the design of a hand-operated tool. *(Courtesy of NASA)*

GROWTH AND DEVELOPMENT

Patterns of growth and development also influence a person's vulnerability and strength. During prenatal development and in the first two years of an infant's life, the brain and nerves are growing rapidly. People continue to learn throughout their lives, but the early years are important in developing the potential to learn and think intelligently. You could say that our control system develops early in life. Toys that stimulate the infant's learning of patterns, as shown in Figure 8.6, have been developed to enhance the early learning of the child.

The ages from about three to middle childhood are important times for the growth of internal organs, such as the heart and digestive organs. Children this age tend to grow steadily and need to store reserve nutrients for later growth. Just before the onset of puberty, the genital and sexual organs grow and change rapidly. These changes may not be evident until external physical changes occur. The timing on these changes varies widely and is different for males and females.

Next comes the rapid growth of the skeletal structure as the body begins to assume adult proportions. Girls tend to reach their adult physical size earlier than boys do. Most people have reached their adult height by their late teens or early twenties. Adults continue to grow and change, but the patterns may not be as rapid or noticeable as in children. With advanced age, some loss in height and some rounding of body contours occurs because of decreased activity, disease, and poor nutrition, as well as aging.

Products and designs for people at different stages of growth need to consider the expected changes that will occur. Ergonomic calculations are needed for several stages if the product is to be used by people of all ages. It also makes little sense to make a very expensive product that will be outgrown quickly; unless you intend for the product to be recycled and reused by other people. The developmental stage of individuals also influences their interests and vulnerabilities.

■ **Figure 8.6**
Since the growth and development process occurs early in life, toys that stimulate and encourage sensory development can improve the infant's ability to learn.

D E S I G N   B R I E F S

- With a partner, measure your height, length from top of head to shoulders, length from shoulders to waist, length from waist to knee, and length from knee to floor. Measure from the top of your shoulder to your elbow, from your elbow to your wrist, and from your wrist to your fingertips. Then, place a piece of stiff cardboard so that it protrudes out in front of you on either side of your head at your eye level. Measure the width of your head at this point. Repeat the procedure to measure the width of your neck, your shoulders, your chest, waist, hips, knees (while standing comfortably), and your ankles. Compile and record these data in the chart provided in your *Portfolio and Activities Resource*. Add your numbers to those of your entire class. Form three groups and each make an ergonomic figure. One figure will represent the 95th percentile of class size, another, the 50th percentile and the third, the 5th percentile.

- Repeat the process and make ergonomic figures to represent these percentiles for a kindergarten class or a sixth-grade class. Compare the figures for height, width, and proportions.

- As a class, become a school safety consulting firm. Form groups to survey and assess safety aspects of areas of your school such as buildings, grounds, equipment, furniture, and cafeteria. Identify potential hazards and unsafe practices. Categorize and list your findings according to the source: the tools, materials, energy, and other humans. Also, categorize and list who is most likely to be affected: your classmates and other students, children who might visit, elders who may use the facility for meetings, other people, and/or environmental conditions. Make signs that use signals and symbols to communicate the potential hazards to the groups most likely to be affected. Take action to reduce the dangers, wherever possible. Talk with teachers, administrators, suppliers, and others who may need to change their products, delivery, schedules, and the like. Document your work and share it with other classes, the administrators, and other appropriate groups.

HUMANS AS MATERIAL USERS

In the chapter on Materials, we examined many ways in which humans convert materials to products and resources such as other materials, tools, energy, and information. Some of this use is to protect basic physical survival. One aspect of survival is breathing. We must inhale enough oxygen and exhale waste gases. We may need to be shielded from some materials, such as toxic gases and particles. Scrubbers on the smoke stacks of some factories have decreased the pollution that once coated buildings downwind when the particles in the air settled. Filters on cigarettes may decrease the toxic gases inhaled by the smoker. (Others in the environment are not shielded, however.)

Although we walk around as distinct units, we constantly interact with our environment. Our skin is usually considered the boundary line, where the self stops and the rest of the world begins. Yet, there is a continual exchange of materials between our bodies and the environment around us. Just as we depend on some plants and animals for food and fiber, we also need some **microorganisms** in order to survive. You might be surprised to see the little plants and animals that are living on you and in you right now. Did you know that you are the host to many types of "flora and fauna"?

When one living being needs a living host for its existence and that host gains benefits from its guest, the relationship is called a **symbiotic relationship**. The host and guest need each other for the well-being of each. But have you ever had a guest become a pest? Did they overstay their welcome or use resources that you and your family need? Did they leave their messes around and make your own life uncomfortable? The same thing can happen with unwanted microorganisms, especially those we call germs and viruses.

Some organisms that come to live on and in us compete with our own cells for the nutrients we need. Their wastes and by-products are irritating or toxic to us. In other words, although we need some microorganisms to survive, others threaten us with illness and infection. Understanding the needs and life cycles of these microscopic beings has led to new and better medicines to kill them or to resist their moving in. In addition, sanitation systems and other preventive measures have been developed to shield us from many germs and toxins. Using pesticides and poisons against cockroaches, flies, rats, and mosquitoes does not mean we hate them! We just do not like the diseases they sometimes carry.

The human body's natural immune system protects us from some disease-causing organisms. As we understand more about the immune system, we can learn more about how to stimulate a more effective protection against invasion. Research on AIDS, cancer, and several other diseases study what can go wrong with natural immune processes. Currently, the most effective protection from most diseases is the practice of everyday health habits.

Nutrients

A delicate balance must be maintained within our cells. Required nutrients must be supplied and wastes and poisons must be removed. Perhaps the most important material for human use is water. Many of the chemicals necessary for life are water-soluble. Your body may be able to go without food for weeks, but a lack of water would threaten your survival in a few days. We wash and cook our food with water. Plants and animals must have water as a nutrient and for cooling and washing. Think of all the uses you make of water.

Clean water for drinking is in short supply in some areas of the world. Processes to extract drinking water from sea water are used in some of these areas. Some people must walk great distances to obtain drinking water for their family. Water for bathing and washing is also needed in order to protect the skin and decrease the spread of diseases.

Assuring the safety of water and air is a continual concern. Effective systems for treatment of human and animal waste were not developed until we understood more about germs and disease. Since the Industrial Revolution, the wastes that are released to pollute the water and air are often materials not found in nature. Some of them are not naturally broken down in the environment. Yet, even if the material is one that naturally decomposes over time, the volume of wastes released into the environment is often greater than the environment can absorb. An example of such a contaminant is lead. Deposits are found in old houses that were painted with lead-based paint and in the soil where there is a great deal of auto and truck traffic. Children who live in these areas can be slowly poisoned as they accumulate lead in their bodies. Changes in gasoline for automobiles have reduced the amount of lead in exhaust emissions. The recipe for house paint has been changed, and people have been warned to protect themselves when they remove paint as they restore old houses.

When we think of survival and its relation to materials, we often think of food. Food is a form of material that we consume, burn, and store. Unused food is excreted as waste, along with the by-products of the food that was used. Methods for handling and processing food to ensure a secure, year-round supply

■ **Figure 8.7**
Obtaining water that is clean and safe is essential for survival.
(© Jacques Jangoux/Tony Stone Images)

■ **Figure 8.8**
Over the centuries, humans have discovered and developed food-plants. Pictured here are some of the many sources for complex carbohydrates and vegetable protein that are staples for the diets of people around the world. Other foods include wheat, rice, potatoes, and casaba.

were some of the first technological endeavors. Over the years, several methods of food preservation and preparation have been developed to keep harmful microorganisms from growing. Currently, safety in food handling is an important issue, as fewer people grow and process their own food. We rely on others to transport and prepare the food.

Humans can satisfy their requirements for **nutrients** in many ways. However, in order to survive and stay healthy, certain chemicals are required. These chemicals are found in proteins, carbohydrates, fats, minerals, and vitamins. You can get these nutrients from slugs and grasshoppers as well as from chicken or shrimp. Plant sources can supply the nutrients as well as foods from animal sources, if you pay attention to appropriate combinations of foods. Different cultures have developed distinctive eating patterns. The ways people eat and the utensils and mannerisms for eating may be different from one culture to another, but the basic material requirements for food are the same.

In the United States, the accessibility of fast food and the increased dependency on **processed food** has changed many family eating patterns in the last few decades. Ask a grandparent if he/she remembers when McDonald's had sold only a few thousand hamburgers. Now you can find fast-food stores in countries all over the world, and the sales are in the millions.

In general, most people in the United States can receive the nutrients they need if they eat regularly from a wide variety of nutritious foods. Artificial vitamins, **food additives**, supplemental diets, synthetic flavors, and sweeteners have been developed in recent years. Some of these are needed in special cases, such as when an imbalance has been caused by illness. Some changes with aging may require special medicines or **nutritional supplements**.

The following are calculations based on a five-year span, from 1987 to 1992, showing the changes that have occurred:

World Population	up	444 million people
World Top Soil Supply	down	120 billion tons
World Grain Production per capita	down	7.6%*
World Meat Production	up	9.3%
Gross World Product	up	11.2%
Gross World Product per capita	up	2.0%
U. N. Peace-Keeping Expenditures	up	500%
World's Nuclear Warhead Arsenal	down	17.2%*
World Oil Production	up	6.3%
Wind Power Generating Capacity	up	86.9%

*Net change from 1986 to 1991

Source: W. W. Norton and World Watch Institute <u>Vital Signs</u>, 1993, and Zero Population Growth Council calculations

A major activity of many people in developing nations is the production, storage, and preparation of food. In the United States, however, only a small percentage of the population works at growing food. Instead, many workers are involved in importing, exporting, packaging, distributing, and processing food. The development of hybrid seeds and improved agricultural processes have increased the available food supply for the world. However, the size of the world population has increased even more rapidly. Distribution is an even bigger problem, since some people have more food than they need (and worry about losing weight), while others starve.

Materials for Protection

Materials to maintain temperature are used in clothing and other means of insulating and ventilating the body. People in warm climates wear considerably less clothing than people do in colder climates. You probably wear much lighter and briefer clothing in summer than in winter. Clothing expresses many other aspects of the person and the social group. The major purpose of clothing, however, is to protect the body and skin from temperature extremes and potential injury. Clothing may be considered a **synthetic skin**.

A tough pair of pants can help protect your legs if you hike through thistles and other weeds. Foot coverings have evolved according to terrain and available materials. Hats and caps can be extremely important in hot or cold weather because your head absorbs and loses heat readily.

Bulky materials were used by early humans for protection from the cold. Animal skins or hair, wool, or fur trapped air and provided an **insulation** that kept in body heat. You may know the importance of layering clothing in the winter time. Modern textiles are much lighter. Some materials used in the space program are designed to keep the wearer warm or cool, depending on which side of the material is next to the skin.

Humans have built **shelters** out of many kinds of materials. The purpose of human shelters is to protect us from outside forces, as well as to keep us warm or cool. Early shelters included caves, mud huts, and many types of portable structures. They were built with available materials and according to the type of protection needed.

Humans need materials for nutrition, protection, and maintenance of body temperature. Certainly, we use materials for many other reasons. We may be able to survive without communication and entertainment, but most of us would not choose to. We use materials for adornment and self-expression.

You also recognize that a great deal of material is wasted and used to destroy. Human history is full of stories in which materials are used for human and environmental destruction. Competition for resources and fear of others led to the development of guns, knives, bombs, and other weapons that can kill, maim, and destroy. Often, innocent people starve, while valuable croplands are unused because a war is being waged. Some people waste resources, while others in their neighborhood lack basic needs for health and safety. People in the United States spend more money on cosmetics and legal drugs (such as cigarettes and alcohol) than they do on education.

■ **Figure 8.9**
The materials systems in this garment allow the worker to function, even though the temperature is extremely high in the processing of this molten metal. *(Courtesy of American Iron and Steel Institute)*

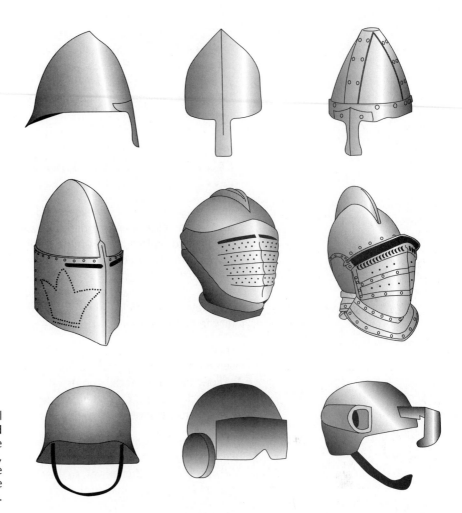

Figure 8.10 ■
For centuries, humans have waged war with each other. These helmets have some similar features, even though they reflect the materials that were available at the time of their design.

DESIGN ACTIVITIES

- Create menus for one week for the following groups of people:

    local nursery school

    local nursing home—(be sure to include selections that suit different potential medical problems)

    sports team

    teenager on a weight-reduction plan

Be sure to include sufficient basic nutrient needs for each group. Calculate the nutritional components for each diet, using either a table showing the nutritional values of each food or a computer program designed for this analysis.

Calculate the costs of each menu. Divide into groups and survey two supermarkets for costs—one that is known to be expensive and one that is thought to be moderate in prices.

• There is an environmental disaster coming your way. You can take with you only one portable suitcase of personal items. List what you will include and why. Compare your lists with those of classmates. What do your items indicate about what you value most?

HUMANS AND ENERGY USE

In the chapter on Energy, we identified some of the tools and machines that allow humans to utilize energy from several sources. Early technological activities were designed to replace human muscle power with other sources of energy. When early humans used tools, the person was an essential part of the "human/machine system." People supplied the control part of the system. When animals were harnessed to do work, humans had access to more power. Wind and water power helped humans do greater amounts of work. Atomic energy may free us even more, but there are many potential dangers involved.

Energy affects our lives in many other ways. We produce sound and use it to communicate thoughts and feelings. Sound is a form of energy. As we control the airflow over our vocal chords, they stretch or relax to produce different frequencies (pitch). Control of muscles in the larynx, mouth, and lips form words and other utterances.

Our hearing provides feedback so that our sound production can be corrected. People vary in the degree to which they can accurately reproduce sounds. Deaf people can't hear themselves, so their speech is often flat and off-pitch. They lack the feedback of information that is needed for control.

Simulation of speech is a complicated process attempted by designers of computers. People who have lost their speech capabilities through disease can sometimes be taught to use devices that simulate the original voice box.

Sound is also used by humans to communicate over great distances. Telephones, television, and radio have extended our knowledge of people in other parts of the world and allowed us to communicate with them. As a consequence, some people say that the world is "shrinking."

Light energy is important in vision. The ability to visually sense direction, movement, color, and shape is important to human behavior. Imagine what your world would be like without colors. Images are quite different in black and white than in color. Can you tell when an old movie has been colorized? Light energy affects us in other ways too. You may turn down the lights when you want to relax. Strobe lights and laser light shows can be used to create excitement.

■ Figure 8.11
Entertainment groups use lights, sound, and other effects to create excitement and change the mood of the audience. (© Frank Ricelotta/Retna Ltd.)

People convert materials to energy in order to cook food and to keep warm. Some developing countries currently face serious energy problems as the supply of firewood becomes scarce. Early in history, it was a warm fire that led groups of people to gather together. This need to keep warm may have led to other forms of social learning. Ceramics, glass-making, metalworking, and language may have begun with a cooking fire. If you have ever been camping, you may have gained a sense of belonging to the group by learning camp songs and games. Maybe you even invented some new words!

HUMANS AS INFORMATION PROCESSORS

Information Processing

Our use of language and our ability to learn set us apart from other animals. Several other animals have communications systems and many animals are capable of learning. Some chimpanzees can even be taught to use language similar to that of two-year-old humans. However, the infinite complexity of the human's ability to receive, process, store, and retrieve information is unmatched by any other animal in the world. (Or maybe we just lack the technology to know how complex some of them are!) The processes we use to make sense out of our experiences are the same processes that we often need to correct. Many technological activities have been designed to "correct" for human tendencies in processing information.

Making Sense Out of Our Experiences

Part of what goes on both outside and inside our body is experienced through our senses. In order to adapt to or change the environment, we must gather information about it. Our senses gather information and our brain interprets it. In order to detect something, enough stimulation must be received. Otherwise, we do not pick up the information and it is not passed on to the brain. You usually cannot hear a mosquito in another room. When you are tired, or when you are watching the same movie for the third time, you may miss a message.

On the other hand, when too much stimulation occurs, the body acts to protect itself. Sometimes, our senses become overloaded and turn off (fainting, for example). The information must be just right—not too much or too little. We will either fail to notice the information or defend against the stimulation. In other words, in order to get us to process information, you have to get our attention, but don't overwhelm us.

We have already considered this idea when examining how to get your ideas across to others in your design work. Suppose there is a great deal of competing stimuli in the environment, or your receiver is otherwise occupied. Then you need to choose an approach that is in sharp contrast to the background noise. This may be similar to yelling louder, or it may be similar to being the one, quiet voice in an argument. In some cases, when the stimulation is not wanted, you shield against it. Perhaps you turn your radio up to keep yourself from listening to your neighbor's loud dog. "Elevator" music in the background enables you to turn your focus away from other noises.

The human brain seems to be "programmed" to process information in given ways. Have you seen a sewing machine that uses a die or template for making buttonholes? Or remember the templates used to draw circles and ellipses in the activities in Chapter 2 on Designing, Documenting, and Drawing? The brain seems to have **cognitive templates** that are used to screen and organize information. Let us use this model to explain how we think the brain works.

Remember, if we are to receive the information, you must first get our attention. Secondly, the information that is coming into the brain has to match our templates. If it does not, we will pay attention just long enough to make some adjustments. This means we must either change the template or change the data. Normally, we find it easier to change the information we receive than to change our template. We **delete** and **distort** information—to fit what the brain expects to receive. If the experience will not fit the old pattern, then we have to change the brain pattern. In other words, we learn. Or to be more accurate, we must reconstruct our previous learning.

When there is not enough difference between the information coming in and what the brain expects, we simply assume this is the "same old stuff." In other words, we pay attention just long enough to note that this is the same as past experiences. We **generalize** that there is nothing new. These processes are called "assimilation" (make the new experience fit our old ones). If the information is very different, we must make an "accommodation" (change our ideas, etc., as a result of the new experience). In either case, some energy must be expended in order to process the information. Once things match close enough, we go on to something else.

What does this mean in practical terms? Humans are good at conserving mental energy—we try to think as little as possible. We change as we interact with our environment, especially with new experiences. We must make sense of the people and events around us in order to survive. If the new experience is very inconsistent with what we expect, we have to work our brain, hard! Mental work does make you tired. We are self-correcting and constantly learning. Learning, however, is a personal thing. It is the individual's brain that works on the stimuli he/she experiences. No one else can learn for you. Also, we are all prejudiced. We approach a new situation with a prejudgment, a preconceived notion of what we will find.

Your sense of the world around you is uniquely yours. It must be—it is even different from that of your twin (if you have one). The two of you are not at the same place and time. You bring slightly different cognitive templates to any new experience.

Information processing machines can be developed to correct some of the prejudgments and expectations we bring. We still must interpret the symbols produced by the machines, however. Even with a video of a fight, bystanders and participants can disagree on what happened.

Over time, people have developed several models to attempt to understand and communicate about how we think and learn. If you go back to the quote from Buckminster Fuller, you can see that he compared the brain to a telephone switchboard. More current models use examples from computers. Attempts to develop artificial intelligence try to mimic the human brain. Computers that continue to process information all day and night are used. They don't need a rest and they don't get bored. Through years of research and development, the machines have been improved. They now are able to process enormous amounts of data and use some strategies of logic and feedback. They still cannot reproduce the creative thought processes of even a very young child, however.

Figure 8.12 ■
The can has fallen off the counter and nearly hit Aunt Sarah's foot. Even though the event is the same, people perceive it quite differently.

We organize data for perception, storage, and retrieval of information. In doing so, we must delete, distort, and generalize. Technological devices and events have been developed to "fill in these blanks" and correct some of the distortions.

How Do You Know What Is Real?

Let us now turn to the kinds of cognitive templates and how we develop them. The first level for processing information comes from the templates we have as human beings. The second level of filtering and organizing information is learned from our social groups and culture. The third is determined by our individual traits and experiences. By organizing information through each of these types of templates, we are able to organize information that we experience at a given time.

Before we can put together an idea of what is happening, information from the real world is filtered several times. First, we interact with the world outside of us through our senses. The world that you experience is somewhat different from that of a dolphin or a dog. Different species have different sensory receptors. Have you ever thought of what you would see if your eyes were placed where your ears are? The first screening comes through what we can see, hear, smell, taste, and touch. To do so, you use your human **sensory templates**.

Figure 8.13 ■
This model is used to explain how humans construct their ideas and feelings about what is happening and what has happened to them.

■ **Figure 8.14**
We are each unique, even down to the level of our fingerprints and DNA. (© 1992 Ed Kashi)

The second kind of filtering and organization is shaped by our experience with other people. These are our cultural expectations. We all live in human groups. In those groups, we have learned language, a sense of place and belonging, and have developed a sense of what to expect from others. These are our **cultural templates**. If you lived in some parts of the world a few centuries ago, people would be shocked and outraged to know that you take a bath—with no clothes on. Even now, in some parts of the world, people would be shocked if you sat in a tub of water to bathe. In still other parts of the world, a family and any other person who wants to join them will bathe in a common pool. Rules and language to label and describe events can be quite different, even for something as ordinary as getting off the dirt and grime from everyday living.

The next level of filtering comes through what you bring to the situation as an individual. This **individual template** includes your own individual traits, past experience, and how you interpreted that experience. Each person is unique. The genetic code each of us carries distinguishes us from other living beings. An analysis of a single hair can indicate your gender, race, and some of your family connections.

We have already indicated several ways in which this uniqueness is used as a means of identification. Not only will your DNA identify you, but so will your fingerprints, voice pattern, and retina. This uniqueness is also found in the style and patterns you use to process data. In addition, your experiences are unique. No one else has been in your shoes each moment of your life. The meaning you have made of your experiences influences the way you approach a new event. Even though there are billions of people in the world, the chances of finding another you is nonexistent. Therefore, you have your very own individual cognitive template.

What does this mean in practical terms? Even two humans from the same family, only a few years different in age, will disagree on what is real. Have you ever been very sure of the words to a song, only to have your sister insist that you are wrong? It is a wonder that people ever come to an agreement and share meanings. We each have our own ideas, attitudes, and experiences that influence how we process a new situation and learn from it. Often we need to discuss something over and over again. Referring to technological products such as tapes, videos, and photos can help. There is no guarantee that people who are in the same situation will agree on reality, however.

Figure 8.15 ■
CALVIN AND HOBBES copyright 1993 Watterson. Dist. by UNIVERSAL PRESS SYNDICATE. Reprinted with permission. All rights reserved.

DESIGNING AND HUMAN COGNITION

We have already considered some of the ways in which technology has been used to extend human abilities. Products have also been designed to correct or guard against human tendencies. Often it does not matter if details are left out. Our goals are reached even if we have changed reality to fit preconceived notions. We make categories and believe other people's opinions. At other times, however, we need to have a closer match between what is "out there" and the meaning we make of it.

Perception and Sensory Modality

Difficulties in perception at the sensory level may come from individual traits or from changes due to age or injury. Vision includes the ability to judge distance, to see both near and far, and to see colors. Vision may change as a person gets older. Some people have an astigmatism; others are nearsighted or farsighted. As muscles in the eye age, the eyes may focus differently. Some people's ability to see colors decreases with extreme age. Some people are born blind or color blind. Many others lose portions of their sight through injury or disease. Studies of vision and the structure of the eye are important in correcting visual problems.

Some people have a preference (maybe an inherited one) for visual data. They use a **visual modality** for receiving and processing information. They may pay more attention to how something looks than other aspects of the event. When they remember something, it may be done in the form of pictures and visual patterns. You may want to communicate with them or find out what they have experienced. If so, you will be more successful if you use drawings, videos, and other graphic materials. They might prefer reading rather than listening. They may respond strongly to color and visual presentation.

Hearing is the second major sense. People can hear the frequency or pitch and intensity of a sound and can detect the time interval between sounds. People hear within a narrow range of sounds, and continued exposure to a high volume of sound can result in decreased ability to hear. Some pitches and tones may be harder to hear than others. Have you ever had your hearing tested?

Just as some people prefer the visual aspect of an experience, others pay more attention to what they hear. They use the **auditory modality** first. Some people will react to sounds first and attend less to visual and other data. They may be more likely to talk about something rather than read or watch movies or work at a computer. What someone looks like is not so important as is what they say. Response to music will be strong.

Have you ever gotten in trouble because you didn't listen? You may forget to do a chore if you are just told to do it. However, if the same chore is written down on a list, you remember. If so, you probably prefer visual data over auditory stimuli.

■ **Figure 8.16**
Examine this and other advertisements you see and hear. What ideas do you think they are trying to present? Which senses are they trying to please? *(Courtesy of The Coca-Cola Company)*

Some people have a preference for the **kinesthetic modality** when they perceive and process information. Movement and their body sense are their dominant focus. Communication in the form of action, body stance, and other nonverbal data will be most important in sharing information with them. Experiences may be stored in the form of tensions in certain muscle groups. If the experience is upsetting, they are more likely to go running or kick a ball, rather than to yell or write a letter.

When you design for a person or group of people, it is helpful to consider what will and will not catch their attention. How do they process information? What is likely to be interesting, but not so different that they will distort your message? What cultural factors may influence how your design is received and evaluated? Remember that people are unique. Even though they are members of a group, you will have to look deeper to understand them as individuals.

You are familiar with many products of technology developed to enhance human senses. A sense of balance keeps us oriented to the planet. We know the importance of this sense when we try to walk in a carnival fun house, where walls and floors are distorted. This natural sense of balance makes us prefer to live where the floor is level. Interior designers have found that even the color and shape of walls can have an effect on human balance.

Part of this balance is dependent on the inner ear. Have you ever gotten dizzy or nauseous when you have stuffed up ears? People who have frequent ear infections may get tubes implanted in their ears to improve both hearing and balance. Planes, boats, and space vehicles are designed to keep a stable sense of down and up.

The sense of taste is closely associated with the sense of smell. Food may seem relatively tasteless if your nose is stuffy. The sense of taste is often weakened in persons who smoke. Some change in the sense of taste and smell also occurs as people get older. Our preferences for certain foods may change as our senses change. Food technologists use scents and flavorings in formulating a food that is tempting to the taste.

Development of food processes and artificial foods also considers the senses of taste and touch. Finding an oil substitute for butter that melted on the tongue just right was quite a technological feat! Textures must be acceptable or the food will not be accepted. Have you ever watched an infant learning to eat solid foods? They often spit out what feels different on their tongue.

General body sensations include the sense of touch, pain, muscle movement, and vibration. Touch provides important messages between people as well as between people and pets. Pain is a warning that something is wrong inside or outside the body. We are just beginning to understand the nature of pain and to develop methods of controlling it. Humans, in general, try to live their lives in a way that brings as much pleasure, and as little pain, as possible. Can you think of how this has an effect upon the technology we develop?

Some strong memories and emotional reactions are associated with taste and smell. Research in these areas is limited and relatively new. Deodorants to cover up and prevent odors that people find uncomfortable or disgusting have been developed. Perhaps you have the "scent of pine" in your car. Perfumes have been used to cover up natural body odors for centuries. Spices were once a highly prized material. Some were used to cover up the taste of slightly spoiled foods. Often scents and spices are used to enhance natural taste and smell. Some are so valuable they have been used instead of money in trade.

You can enhance your understanding of an event by enriching the amount of information you receive. A situation can be understood best when you have data from as many of your senses as possible. Reading about something is not the same as reading and seeing pictures. Actually being in the situation with the smells, sounds, and sights is quite different from being told a story about it. A recent development is the use of a process called **virtual reality**. In this combination of video and computer, a person can get close to the sensation of "being there." The experience is still far less than reality, more like being an integral part of an animated movie. Virtual reality does provide unique

■ **Figure 8.17**
Virtual reality creates a 3-D image that changes in response to the viewer behavior. This scientist can view the simulated movement of the vehicle through air, from several angles. Information about the design can be obtained before a physical object is made. *(Courtesy of Ford Motor Company)*

possibilities for entertainment and learning. Designers can get a better idea of what an environment might look like before it actually exists. Decisions about placement of knobs, controls, and instruments in a vehicle can be made while it is in the design stage. This approach is helpful in matching the product to the user.

EMOTIONS AND BEHAVIOR

Humans are consumers and creators of technology. Feelings and attitudes are as important as ideas in determining what will be designed and what will be used. Emotions are very important to our survival and learning. They give us the impetus to do something about reaching goals. Human emotions provide the mental energy behind the decision—to act or not act. This is the "engine" that drives technological development.

Events happen to us all the time. Some of them we find pleasant, some are threatening. Remember, however, that what happens to us only starts the process. We bring our expectations, developed over time in our own unique way. These cognitive templates are used to create our joy, sorrow, hate, or jealousy. Our beliefs, attitudes, and expectations influence what we perceive. The words and other symbols we use to categorize the experience and think about it will influence the emotions we feel. Have you ever had a friend get angry at something you thought was funny? The event is the same, but your reactions were very different.

When you want to evoke emotions or ideas in another person, you will select the symbols that are likely to be interpreted as you wish. Understanding the cultural templates of another, the language they use, and the symbols of their familiar groups can be helpful. The choice of labels, colors, forms, shapes, and lines can influence how someone will feel and behave. What colors would you use if you wanted to make someone think that your design would enhance their health? What images would you use if you wanted to make a scary movie for preschool children?

The impact of the messages you receive from someone else can also be understood. As you learn more about yourself, you can begin to predict what reactions you will have. In this way, you can influence your emotional response. You don't have to be influenced in ways that are not best for you. If you stop long enough to check what you are thinking, you can rethink ideas that are not appropriate to the new situation. Just as old gloves may not fit you like they did last year, your ideas may not work anymore either.

When we experience a very strong emotion, the emotion takes priority over logic and thinking. If the experience is extremely threatening and painful, we even shut out the perception. We may tuck away the memory in such a way that we do not remember it. Later, we use a great deal of energy to avoid circumstances like this event. The way an experience makes us feel is essential information for learning and for modifying future behavior.

Another way to respond, however, would be to recognize that when you have a strong emotion, you have been surprised. Something is very different from what you expected! Knowing where you learned those expectations can help you decide what to do. You can take the time to think about the ways you are processing information—what data might you be deleting or distorting? Is this the only way of looking at the event, or are other people reacting differently from you? Do you want to change your ideas or the situation?

Some people avoid computers or other technological products. They may believe they will break the machine or make "stupid" mistakes. If someone thinks that technology is mysterious and overwhelming, they may let other people take responsibility for decisions about technology. These approaches may not be helpful as people try to make their own lives better.

Figure 8.18 ■
How a person feels about a challenging assignment depends on what they think about it.

HUMAN WANTS AND NEEDS

In the chapter on Designing, we indicated that many products are designed to serve a human need. Many other technological products are developed just because people want them. Abraham Maslow and Erik Erikson developed models to explain **human motivation**. Maslow's model can help explain how we approach an experience—which aspects of a situation will be more important at the time. (Refer to Figure 8.19A.) Erikson identified **psycho-social stages** in human development to show which needs are dominant at different ages of life. Erikson's model appears in Figure 8.19B. Many forces operate on humans all at once. However, the ones we attend to most are influenced by the stage of growth we are in. Satisfaction of needs leads to further growth.

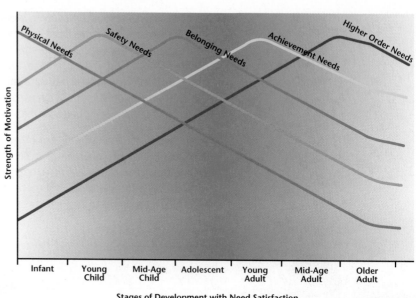

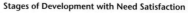

Figure 8.19A ■
Each of us has a set of needs that influences our motivation. According to Maslow's stages of development, each need is satisfied to a sufficient degree and other needs become primary. We can develop higher order needs when we have a history of need satisfaction and when we can count on our resources.

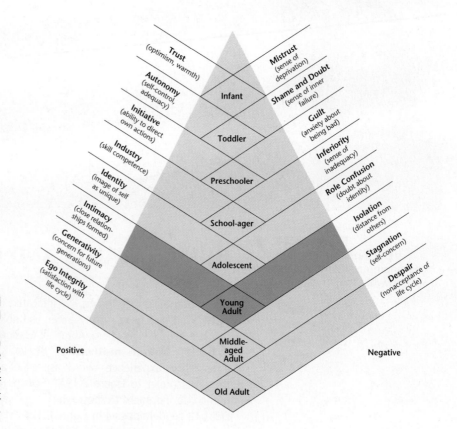

Figure 8.19B ■
This model of the psycho-social tasks, drawn from Erikson's work, shows the age at which a person may be trying to build the resources to meet the task. Anyone can get "stuck" at a given level if the environment does not support healthy growth.

■ **Figure 8.20**
Human infants are born little and
dependent on others for survival
and development. *(Courtesy of
Good Samaritan Hospital,
Cincinnati, Ohio)*

Learning to Trust—Basic Physical Needs

All living beings are vulnerable to their environment. The prenatal environment of humans is critical to their development for the rest of their lives. Some deficiencies may be made up after birth, but the direction of development is set very early in prenatal development. The health of the mother is imperative if the embryo is to develop into a strong and healthy infant. The effects of malnutrition, drugs and alcohol, and disease or injury are often severe enough that the effects remain apparent throughout that person's life. Many technological efforts have focused on correcting and ensuring the healthy development of the fetus and newborn infant.

Human beings are born little and unable to survive alone. Unless there is another person to nurture them and supply what they need on a relatively consistent basis, human infants die. They cannot just walk off and eat some grass, as a calf or chick might. Most animals bond with those who supply their needs. This caregiver does not have to be the biological parent, but the infant's needs must be met on a consistent basis.

Early in life, the infant learns whether to **trust** people and the world. In this early process of feeding, protecting, and nurturing the child, the caregiver helps the infant develop a bond with other people. The caregiver must put aside his or her own needs and be sensitive to the messages sent from the infant. If the major motivation of the caregiver is to nurture the development of that new, vulnerable human being, they both satisfy developmental needs. Child care is demanding, vital work.

An unintended consequence of the technology of producing baby bottles and infant formula has been the ease with which the baby can be left alone to feed. If this happens, the sensory stimulation and socialization from being cuddled and talked to or sung to is lost. This is as important to social and intellectual growth as milk is to physical growth. A positive result of technology is found in the super-absorbent fibers used in diapers. Staying drier has cut down on painful diaper rash. The use of adhesive fasteners means the baby and the caregiver no longer get stuck with diaper pins. It is easier to learn to trust when there are fewer sources of pain.

In cases of mild neglect to extreme abuse, the care of the infant is erratic, inconsistent, or punishing. The infant is hungry, wet, crying, and ignored for too long at a time or treated in a rough manner. These infants learn not to trust—the world, other people, or their own impulses. If the sense of trust is not built in the very early months and years, development at *all* later stages of life will be difficult. These people will be motivated to hoard resources and try to defend them, even when there is plenty. They accumulate more than they can ever use, even though they realize that other people are lacking in basic resources for food and shelter.

However, even the adult who was carefully nurtured as an infant can experience events in which physical needs and survival are threatened, such as war or natural disaster. Old, dependable resources no longer work. In these situations, survival and threat of pain are always the bottom line in deciding how to act. When you are hungry, freezing, sick, exhausted, or suffering from other physical deprivation, all other needs take second place. The person or group who can supply you with food, warmth, medicine, and the like can exercise great influence over you.

The resources of people in many parts of the world are spent on subsistence and survival. People who are poor and homeless spend very little time worrying about whether people like them. They care, yes, but most of their energy must go toward getting warm and less hungry. When food is available, they eat it. There is little concern for the future, when today is so uncertain.

Autonomy—Basic Need for Safety and Predictability

Toddlers who have a sense of trust begin to explore and test things out for themselves. Caretakers now must see to it that they are safe from dangers, poisons, and exposure to harmful materials and events. Childproof containers, playpens, and safety seats are designed with the recognition that healthy toddlers need to move! Toilet training can teach the toddler that his or her body is okay and that even the odors and "toxic" products they produce are acceptable. Sometimes the environment is not childproofed and exploring results in disapproval from the caregiver or in pain and injury. Sometimes caregivers are disgusted with the child's messes or get extremely angry when the child breaks things. The child's developing sense of confidence in self can be damaged.

If the world has been experienced as predictable, you can let go of your concern for pain and move away from home base. You can take risks. You can let go because you trust that you are okay. Hopefully, the child has developed this sense of predictability by the time he or she is three years old. However, events can change.

■ **Figure 8.21**
Learning self-defense techniques is seen as important when people believe their personal environment is not safe. *(© D.C. Riggott/Tony Stone Images)*

Even the most trusting and venturesome adults can be threatened and safety becomes of primary concern. Suppose your neighborhood is threatened by burglaries and the streets are not safe. A great number of tools and devices have been developed to help humans feel more safe. Alarm systems, guns, knives, armor, and other protective clothing are examples.

People and groups who can help you be safe and secure have the power to influence the way you behave. You may promise to do things that you consider wrong in order to get someone to help you in a fight. On a larger scale, countries build up arsenals of weapons and form agreements with other groups in order to deal with a potential threat to their safety.

Initiative—Social Need for Belonging

The next level of development for trusting, confident preschoolers is that of trying to create things and express ideas. Budding language skills, painting, building, and mimicking adults are all part of the child's sense of initiative. It is important that the caretakers approve of the efforts, interest, and curiosity of the child. Showing them the "right" way and punishing them for mistakes can crush their questioning and damage the child's ability to achieve at later stages.

Once people trust that they are physically secure and safe, concern about social issues and belonging become more important. You work to please people. The first people we want to please are our parents and family. At later ages, it is important that teachers and people your age accept you.

Most cultures begin a formal education process once the child has reached about five years of age. Some societies expect children to learn to read, write, draw, calculate, understand their culture, and get along with other people. Some societies expect their children to learn to find and grow food, make cloth and shelters, and care for animals and stores of materials. Rules for playing games, and interacting with elders and with people from other social groups are taught. Habits for caring for health and well-being are developed. At this point in development, all societies have expected standards for the work that children do. They are expected to work—whether it be at formal or informal learning. Children learn the skills needed to be a contributing part of their society. The necessary tools and environments are developed to suit these expectations.

Figure 8.22 ■
All societies expect children to learn the skills and knowledge that the group has accumulated and considers important. How children are educated varies widely. Some schools are well-equipped to support learning from several media sources, while others are not. *(Right: David Hiser/Tony Stone Images)*

When a sense of belonging is a dominant need, you are influenced by what others might think or feel. You try to look right, act right, and do what is expected in order to fit in. You buy things that promise to make you acceptable and lovable. You try to look better, smell better, sound better—just like the group to which you want to belong. You may even dye your hair purple in order to get the attention of the "right" group of schoolmates.

Yet anyone who has past experiences of success in being accepted can lose this sense and be **vulnerable** again. If a change occurs, such as a divorce or remarriage—if the family moves and you change schools, this security changes. There is a new social group in which to negotiate one's place. New rules are needed.

Vulnerability also comes with expected patterns of growth. You may have felt fine about yourself as a child, but then your body changed with puberty and adolescence. You probably spend a fair amount of time looking in a mirror and comparing yourself to others your age. The question of whether you belong is predominant over other concerns. You may even starve yourself in an attempt to be accepted. Products that promise to build muscles, rearrange your muscles and fat, or make you taller or shorter are tempting.

Achievement—Industry and Identity

Suppose our developing person has been successful. A sense of physical well-being, safety and comfort, and love and belonging has been accomplished. Now the person becomes more interested in standing out from the group—in

■ **Figure 8.23**
Uniforms and common dress are one way of gaining a sense of belonging which can contribute to a person excelling at what she or he does. That, in turn, helps the individual move to new levels of independence. *(Top: Courtesy of Queen City Metro; Bottom: Courtesy of Girl Scouts of the USA)*

excelling. Achievement, pride in unique abilities, and judgments now are important. If you are secure in your sense of belonging, you can choose to be different. You can go against the decisions of the group. Your own ideas and values are important in the decisions that you make. You still care about others, but you are not dominated by the need for their approval. People at this level of motivation can be creative and work to change the group.

During the school-age years is the time when children learn what talents and interests they have that set them apart from the group. They may learn to compete and work for better grades, a shorter time in a race, or a larger collection of baseball cards. The dominant motivation now becomes developing and expressing those unique aspects, and gaining recognition as being different.

The task of an adolescent is to get ready to be an adult. This is not easy, however. Often it is hard to predict which adult groups are "right." How will expectations change from now to the future?

Some groups want the adolescent to fit in and not question the rules and patterns that elders have developed. In societies that are rapidly changing, the task of adolescents is to examine what they have been taught and decide which rules and values to keep, modify, or discard. When the society contains many groups who differ in expectations and rules, the adolescent has the dilemma of deciding which seem most accurate and most effective. Sometimes this questioning is welcomed, but often the result is conflict and crisis. Gang warfare, competing cliques in school, and family quarrels are part of this struggle for some adolescents.

Intimacy, Generativity, and Integrity—Truth, Justice, Order, and Beauty

Maslow indicated that after we have built the resources for satisfying survival, safety, belonging, and achievement needs, humans move to a different level of motivation. Higher-order motives require that the person let go of everyday concerns about themselves. People begin to strive for beauty, order, and a sense of justice and meaning. Some may call these luxuries. Even people who are very poor want these conditions—they just don't have the resources to spend on them.

It is true that some people have sacrificed their lives to make a point about justice and fairness. However, these were not people who were deprived of basic needs early in life. Most higher-order motives are based on at least some degree of satisfaction of physical and social needs over a period of time. Poor and starving people are easily led and deceived. If a society is to be orderly and democratic, all its people must be adequately cared for in terms of physical well-being and safety.

Once young adults have developed a strong sense of what is and is not important in terms of their own values, they strive to share those values with other people. The developmental task at this stage is achieving **intimacy** with others. The love and affection that is developed here does not have the component of competition and dependency that is evident in earlier relationships. Adults who have a secure sense of identity can share their ideas and emotions without

■ **Figure 8.24**
People who are ready for intimate relationships are also ready to think through the consequences of their behavior for themselves and their partner. Many technological activities have been developed for the purposes of controlling fertility and birth.

demanding that the other person agree with them. The recognition of differences is viewed as enriching, rather than threatening.

At this stage, conditions are good for choosing and making a commitment to a life-mate. This person can be a companion with whom you can share your most intimate thoughts and dreams. This intimacy may or may not be expressed sexually. The concern is for sharing and the well-being of both people in the long term. Immediate pleasure is balanced against long-term consequences. The long-term best interests of both partners are considered.

Relationships that seem to be intimate often are not. Partners and friends may be chosen because of fear, competition, or to prove something about oneself. You may choose your friends or partner based on how they look or what they can do for you. The development of cosmetics and clothing styles designed to attract the attention of others are products that serve this motivation. Think of other things people do to themselves in order to improve their chances of gaining attention from another person.

People even bear children in order to "prove" that they are adults or to get someone to really love them. As you can see, these are not the conditions for nurturing that infant, who needs a father and mother who can put the child's needs first. Even in ancient societies, methods and devices for preventing and encouraging conception were designed. Recently, products and procedures were made more available and reliable. They still are only as reliable as the user, however.

Once people are secure in having someone to share life with, they often become concerned about future generations. They want to make the world a better place. Sometimes this motivation is satisfied through nurturing children—your own or other people's. Sometimes it is expressed in work that is meaningful and has lasting effects. Concern for the well-being of all people and for the environment, for now and for the future, is felt. This stage of adult development is called **generativity.** "Giving back" to others seems to be just and adds meaning to the person's own life.

If people have reached middle and late adulthood with the previous needs satisfied to an adequate degree, they will be able to face their own mortality with a sense of pride in what they have done with their life. They can recognize their many mistakes and other ways they could have lived. However, their life has an **integrity** and seems to fit in the cycle of life and death. Lacking the conviction that their life has made sense, people try their best to deny aging and their eventual death.

Many resources and technological activities can be viewed as age- and death-denying efforts. Some people are unable to take pride in the quality of the life they have lived. They are especially vulnerable to messages that promise that they can live longer or not die at all. Most societies have developed some rituals that help people express some of their concerns about death and give meaning to the life cycle.

Technological advances in life-support systems and other techniques to provide "death control" have raised many serious questions. Ethical decisions—who should receive these resources, what quality of life can people maintain, and who

decides—are extremely difficult. If you could design and produce something that would make everyone live an extra 100 years, would you? What consequences can you predict?

Much of technological development has been addressed to physical needs, safety, and perhaps to belonging and self-esteem. The rewards and punishments that influence the behavior of well-fed, secure people reflect their values, personal goals, and standards. Concern for the welfare of others and for the planet are a part of **enlightened self-interest**. We want to live with others and few of us can survive alone. Therefore, being concerned for ourselves involves caring for the rights and needs of others and for the environment. Also, it is logical to think through the consequences of what we do for the future as well as the present. These ideas become important when you are trying to get a group of people to work toward a common goal. We will return to these concepts in the chapter on Time, Space, Capital, and Management and the chapter on Consequences and Decisions.

HUMANS AS DECISION-MAKERS

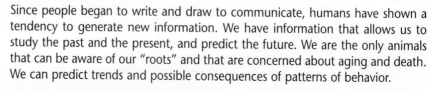

Since people began to write and draw to communicate, humans have shown a tendency to generate new information. We have information that allows us to study the past and the present, and predict the future. We are the only animals that can be aware of our "roots" and that are concerned about aging and death. We can predict trends and possible consequences of patterns of behavior.

The vast availability of information has both positive and negative effects. Once, a well-educated person could read every important book in the world. Now, people have trouble keeping up with monthly magazines of interest to them! It is possible to feel overwhelmed by an information overload. We may decide to let others stay informed and make the decisions. Or, we may make the decisions without even considering the available information. A major source of influence today is access to "knowledge banks" and skill in using that information to persuade others.

With advances in communication and transportation, we have the ability to know what is happening in other parts of the world and even other parts of our universe. Almost instantly, we can learn of events happening to our neighbors on the other side of the globe. We can also communicate directly with someone who lives miles away from us by using computer links and information highways.

The access to information and the ability to use tools, information, energy, and human capabilities gives increased complexity to ethical decision-making. The importance of deciding on values, goals, and definitions of fairness are evident as we think of long-term consequences to our planet and all its people.

■ **Figure 8.25**
In the future, personal computers may be built into our garments and be readily accessible for use. *(Courtesy of NEC Corporation)*

SUMMARY
CHAPTER 8

Technology is a uniquely human activity. There are many ways in which humans interact with and use resources. Since humans are both the producers and consumers of technology, the human factor is critical in any technological event. Humans have developed tools suited to human dimensions and abilities. We have also developed tools to extend our senses and increase our available power. Materials are used for survival and comfort. In general, people have tried to design and use technology to make their lives contain more pleasure and less pain. We attempt to satisfy our curiosity and learn more about ourselves and the environment through the use of machines and processes that create information. Information processing becomes vital to developing useful technology that produces constructive, long-term outcomes.

The processes of accurately perceiving and processing information related to a problem have been made easier by technological advances. It is still a human function, however, to gather and interpret the information, make decisions, and reach conclusions. Tools and machines can be used to correct the way humans organize and process data through deletion, distortion, and generalizing. Emotions are valuable in providing the motivation to decide on a course of action that is constructive.

Knowledge about the stages and requirements for growth and

■ **Figure 8.26**
Changes over time are expected, predictable, and inevitable. Many technological activities have been designed to speed up or slow down these natural changes. Many products are used to expand human capabilities throughout the life cycle. *(Marc Rubin for Human Ecology Forum, Cornell University)*

development can be helpful in understanding wants and needs. Self-understanding can help you avoid pressures from others that are aimed at your own vulnerabilities. Designing goods and services for others will be more successful if outcomes are made user-friendly.

Humans can be nurturing and responsible, or destructive and shortsighted. The degree to which basic needs have been satisfied influences whether the approach will be defensive or trusting. Since humans share the same planet, the consequences of resource use are shared by all. Humans make the decisions about how resources are to be used. Will tools be used to help or hurt? We can weigh present needs and wants against future concerns. These are all part of the complex, difficult, and critical activities that humans must do. Knowledge bases and other results of technology can be of assistance.

ENRICHMENT ACTIVITIES

Look back to the portfolio pages you have developed and identify the tools and materials that are being made to help people survive physically— to keep them safe, to belong, to achieve, or to improve the quality of their life.

- Clothing is often used to convey a message to other humans. In groups, design a wardrobe to convey the following messages. Draw the person in the outfit, find pictures in magazines to illustrate, or actually gather the items and have a fashion show.

    1. I am a down-to-earth, natural person who cares about the environment.

    2. I want to impress others with my success, money, and fame.

    3. I am interested in ideas and the life of the mind and do not care that much about how things look.

    4. I am an active, energetic person who likes to move.

    5. I want to belong and look like the in crowd.

    6. I don't want people to notice me; let me fit into the background.

- As a class, dream up a product that is of little or no use to anyone. In groups, design the product and its packaging and advertisements. Group according to the level of development your customer is likely to have on Maslow/Erikson's stages. (For example, one group will market to people who are concerned with basic physiological needs. Another will have customers who are primarily motivated by safety/autonomy concerns. The next will be concerned with belonging/initiative concerns, and so on.) Compare your advertising campaigns with some of the ads you view every day on radio, TV, and other media.

MATH/SCIENCE/TECHNOLOGY LINK

Select a different part of the world. In groups, do research about the lives of people your age in that region. What technological products and activities are available to them? What types of tools, materials, energy sources, and information do they have access to? Write a letter or contact a class or pen pal by E-mail. Find out as much as you can about what they like most and least to do, what their plans and hopes are, what their greatest concerns are, what they do for fun, etc. Compare these findings with the answers your own classmates give to these questions.

R E V I E W   A N D   A S S E S S M E N T

1. Compare a human to a machine. Identify the different subsystems.

2. Identify five different products that are designed to protect humans from exposure to conditions that might physically damage them.

3. Make a collection of products designed to improve the physical appearance and comfort of people.

4. Compare the "toys" (things they use for enjoyment) of people at different ages. For what purposes were they developed?

5. For your portfolio, make a collage of the different materials that people use in everyday activities.

6. From your portfolio page relating to the energy spectrum, identify the potential hazards to people for each form of energy they use. Indicate the likelihood for harm from each.

7. Talk to a reference librarian or someone in the community who regularly uses various data banks. Predict some of the ways in which people will gain access to information in the future.

8. What is meant by "sensory templates"? Identify tools that have been developed to correct for deletion or distortion for each of the senses.

9. What is meant by "cultural templates"? Identify at least three times when you or someone you know got into conflict with someone who had different expectations because of cultural/family differences.

10. Identify some of the major events in your life that make a difference in what you expect from yourself and other people.

11. What are the stages of psycho-social development? Identify people who are at different ages. Design a survey instrument that will help you learn more about their goals and what they consider their accomplishments so far in life.

12. For each stage of development, identify "traps" to healthy accomplishment of developmental goals.

CHAPTER 9

Time, Space, Capital, and Management

Figure 9.1 ■
Time and space, along with capital and human input, are resources used to implement the design process. (© 1994 Michael Simpson/FPG International)

INTRODUCTION—PUTTING IT ALL TOGETHER

. . . .

You can see the results of centuries of human design and problem solving all around you. Materials, information, and energy have been used along with the required processes and tools. Other, less obvious resources have been utilized

253

KEY TERMS

assets
capital
capital intensive
chain of command
communication
conceptual skills
crowding
deficit spending
economic resources
expert power
family work
hierarchies
human capital
human skill
intellectual property
interest
investment
**investment in human
    capital**
labor intensive
legitimate power
maintenance
monetary system
non-economic resources
organization chart
ownership
personal recognition
plane
**principle of supply and
    demand**
privacy
private property
promissory note
punishment power
quality circles
referent power
resource management
reward power
rules
shortcuts
technical skill
territory

in technological events as well. Time and space, along with capital and human input are used. This chapter examines the resources of time, space, capital, and labor. Methods of combining all the resources in order to implement the design process are explored as well.

An example might help. Suppose you are planning a surprise birthday party for a friend. Many decisions must be made. How much money can you spend? How long do you have to make plans? What stuff will you need? Should you have a picnic or decorate your home and have it there? How many people can you invite before the neighbors begin to complain? Will you buy a cake or make it? How will you invite people so that your secret is kept safe and no one will crash the party? What is the best time to start and how long will the party last? How can you minimize the chances for damage and accidents? Who will do the work—setting up, serving, and cleaning afterwards? These are only a few of the required decisions. We will return to the trade-offs that are needed in the chapter on Decisions.

With any design and technology activity, people must decide (1) what they want/need; (2) the information, materials, tools, energy, and processes to use; (3) the time, space, and money available; (4) the work to be done; and (5) how to tell if the results are satisfactory. This is not enough, however. Processes must be designed for putting resources together in a way that achieves the desired outcomes. You will need to manage the resources for planning the birthday party for your friend. These processes are called **resource management**.

Putting together one party requires less management than if you are to repeat the activity several times. Suppose you are very successful in planning this party. Your friend recommends your work to someone else. You decide to become a social manager and start a party-planning business. In order to be successful, you must consistently deliver a good outcome over time. A plan for time, money, space, and other resources is needed. If you hire people to help, you must communicate your expectations clearly and motivate them to do their jobs well. Will you need to employ people and keep an inventory of supplies so that you can respond quickly when someone wants your services? This involves management—the coordination of the use of all resources in order to create the desired outcome. This, in turn, involves consistently reaching the standards that you specify in your design briefs.

A SENSE OF TIME AND ITS USE

We all get the same number of hours in the day and, as a species, tend to live a maximum number of years. We are probably the only species that concern ourselves with how much time we have. Planning the use of time as a resource means deciding how to spend it to achieve our goals. This leads to the belief that people can waste time or that, if you hurry up or rush, you can save time. You may have heard people say, "Time is money." This concept reflects the view of time as a resource that can be used and traded.

There have been several attempts to change our sense of time—to speed it up, slow it down, or freeze it. Films allow you to speed up, slow down, or freeze the action that has been recorded. Have you ever seen slow-motion photographs? Many photos are taken as the high-speed event happens. These photos are then shown at a standard speed that allows the viewer to study each segment of change and movement. Time-lapse photography allows one to condense a process that takes a great deal of time. The viewer can experience an event in a short period of time, rather than waiting around for the event to happen. Changes that may be overlooked, because of their subtlety, become evident. Do you have the patience to watch the railroad track rust? Or a flower bloom?

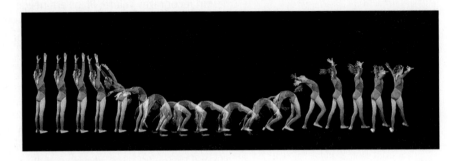

Figure 9.2 ■
A slow-motion, step-action photograph can be used to analyze each move made by an athlete. *(© 1992 Globus Brothers Studios/The Stock Market)*

Time Changes All Things

One thing that never changes is change itself. You can "freeze" segments of events by recording pieces of the event on video, photos, or by saving objects from the event. No one can just stop time, however. Neither can you have immediate change in just the blink of an eye. Processes take time. Have you ever thought about what in your life you might want to preserve, just as it is, forever? What might you like to speed up? Slow down?

One of the effects of the passage of time in nature is that of entropy. We used this term in several of the previous chapters. In the conversion of energy, some energy is no longer usable for the intended work. With information, entropy occurs when there is a loss of a portion of the signal. With materials, including living materials, the term is used for expected changes over time. Systems tend to break down and become disorganized. Materials age. They disintegrate, such as in rusting and corrosion. Mistakes occur in reproducing cells of living materials, including in our own bodies. Signs of aging and some diseases are linked with changes in the body's ability to read the DNA code and reproduce cells accurately. We will return to these ideas in the chapter on Biotechnology.

Time, along with gravity, tends to be a leveler. Mountains erode, valleys fill up with sediment, and muscles sag. Over time, organisms die and decay, and new organisms are formed. Each living organism has its own natural life cycle. Even the most dramatic event becomes a dim memory or a small incident in the larger scheme of history.

■ **Figure 9.3**
Time is one of the factors in human development. Each generation is born and ages. *(© Ken Fisher/Tony Stone Images)*

Have you ever gone into an abandoned house? If so, you can try to imagine the people who lived there and see some of the signs of how they arranged their lives. Dirt and decay will be evident and insects and small animals will probably be living there now.

Humans try to slow down or stop some of this tendency toward disorganization and entropy. Painting, polishing, cleaning, and other **maintenance** efforts are an attempt to keep order in the face of entropy. Cleaning is never "done"; the process begins again as soon as you turn your attention elsewhere.

The life work of many people is that of maintenance. Sometimes elaborate plans are made for keeping machines, systems, spaces, and groups of people functioning as originally intended. Part of the management process is to keep systems functioning, to maintain them in spite of expected entropy of resources. Another management task is to predict how long the product is likely to function before mistakes and failure can be expected. Nothing can last forever, and a good design recognizes failure. Plans are needed for replacing things before they wear out or break down and throw the whole system into chaos. Plans should include methods for reclaiming materials and for recycling resources after the life of a product.

Everything Takes Time

In any technological event, all processes necessary to achieve the desired outcome require the resource of time. In producing a product, you may decide to use a process that takes the least amount of time. For example, some wines are allowed to age for years under carefully controlled conditions. Wine makers, over the years, have learned to control the various ingredients that make up the flavor and aroma of the wine. What once happened over years of time in cellars and caves, now can be done in the factory. Now synthetic ingredients can be made to produce a standardized product in days rather than years.

How people use their time depends on their own values and expectations. You may spend hours with computer games or talking on the phone. Your parents may think you are wasting time. You think your activity is important.

The first step in managing personal time is to record what you currently do. Next, how does your use of time match your goals and objectives? Are you spending your minutes and hours in a way that is likely to reach your goals? If not, which activities could you realistically eliminate? Are there major goals that you are not including in your time use?

Some of your life is already divided into fairly fixed time periods. School classes, meeting times, and TV shows are scheduled on a consistent basis. Some events are not regular and vary in the amount of time required. How good are you at estimating how long it takes to write a paper? Cook a meal? Build a design model?

It is also useful to make a projection about your future years. Imagining your life when you will be 30 years of age may seem difficult. How about when you will be 50? Life projections and planning are methods for taking what is known from the past and using it to achieve goals over time. Some goals take many years of dedicated effort to achieve. Receiving a high school diploma is one. Some goals take prior achievement, so they must be planned in "layers." If you want to be the president of a company, you need to get a good work record first. You also want to avoid any blemishes on your record that may cost you a promotion years later.

Figure 9.4 ■
This employee posts the goals and activities for her staff to accomplish in a sequential order. Thus, all the workers involved can see their tasks and the relationship of their jobs to all the other jobs. (© 1994 Davis Freeman)

When designing a product or process, you must plan how much time to use. Recipes and instructions, whether for baking a cake or firing a ceramic product, may list the optimal time for processing. You may have seen the results of ignoring this processing time from other people's experiences. Have you ever fired your pot for too little time or baked a cake 30 minutes too long? Have you ever underestimated how much time to save to do your homework? Many people base the cost of their services on how much time they estimate a job will take. They need to be very accurate, or they are out of business!

After people have used a given process several times, they become more efficient. A trend in technological development is to develop **shortcuts**. These are processes that take less time. Time/motion and trend analysis studies have been generated to decide which sequences and procedures take the least amount of time to produce the desired result.

DESIGN BRIEF

Time Management

- Choose two days of the week, at least two days apart. Use the chart provided in your *Portfolio and Activities Resource* and complete the following.

- For each half hour of the 24 hours, keep track of how you spend your time. For each activity, categorize as follows:

  B—caring for basic needs   I—self-improvement
  H—helping others   R—recreation and leisure
  P—paid work   F—family work
  O—other (explain)

- Go back over your list of activities and identify whether you are using your time wisely to help you reach your goals. Place a check mark beside those activities that you believe are suited to your goals.

- Next, analyze times that do not match goals. Decide whether you would like to change. Write out your reasoning.

- Create a plan for improving your use of time.

MATH/SCIENCE/TECHNOLOGY LINK

- Calculate the percentage of time that you spent in each of the categories for time use.

- Make a pie chart of your time management.

- Compile class data and make a pie chart of usage for the class.

- Compare the average for the class with your own use of time.

A SENSE OF PLACE AND SPACE

As you read this book, you are probably oriented with your head farther away from the Earth than your feet are. Our sense of direction is based on our orientation to the forces of gravity. This may seem obvious to you. But if you were living on a space station, spinning through space, you would have to design something that would produce an orientation. Actually, the Earth is our familiar massive space station. We are spinning through the universe. As would be true with other large spinning bodies, we are oriented in line with the center of the mass. Down is toward that center, and up is away from it.

Figure 9.5 ■
We all need a sense of direction and orientation to space, whether we are on a planet or flying through space. *(Courtesy of NASA)*

People need a sense of living on a level **plane**. Homes are built over a flat plane or a series of planes. People do not function comfortably on a slope and they get nauseous or dizzy without a dependable orientation to space. One form of torture is to deprive the person of a means to maintain a sense of time and place. Carnival rides spin and move in erratic directions, defying gravity. Losing your orientation may be exciting for a short time, but you would not be able to tolerate it for very long. There is a real sense of comfort in getting back to solid ground and restoring the expected orientation in space.

Territory

Each of us wants some **privacy**, a sense of familiar space, and a place to work and play with others. Just how this is defined, however, is dependent on your cultural templates. Privacy is a matter of perception. Suppose you were reared in a large family or an urban area where many people share small spaces. You may get frightened and feel vulnerable in the wide open spaces of a park or desert. If you grew up in a rural area, you may feel frightened and smothered when you walk the streets of a densely populated city.

A sense of **crowding** is also a matter of perception. At a dance or football game, a large number of people can jostle each other and push into a limited amount of space. This is seen as fun. The mood changes when someone smells smoke or starts a fight. If people panic and push to get out, their number is seen as a source of danger. This crowd is frightening even though the number of people and the space remain the same.

City planners, architects, organizers of entertainment, and interior designers are a few of the professionals who study and plan the use of space in order to achieve desired effects. Densely populated areas can be made comfortable when the space is planned to provide for appropriate functions, privacy, traffic flow, and aesthetic appeal. A truck that sells prepared food at the curb may be able to deliver a large menu if the space is carefully organized and arranged. Optical illusions, mirrors, and tricking the eye involve methods designers use to make space appear larger or smaller.

Familiar Space

All of us wish to have a sense of home. Again, the cognitive templates we use determine our reactions. Home represents our ideas of shelter, security, and belonging. It is also a place to express personal preferences. Some people view the world as their home and move from place to place with their few possessions. Others build elaborate houses, well beyond the structure required for shelter alone.

Houses can be an indication of status and achievement. Space can be used to express the need to compete and impress. Some people spend a great deal of their economic resources to gain and maintain houses. Their home is their castle in which they want to have control over what happens in their private space. Other people work very hard to keep a roof over their head and hope that they can have a small amount of privacy—even though they regularly hear their neighbors' activities.

Some societies have little or no sense of **ownership.** The idea that someone should claim ownership to land, air, or the ocean is incomprehensible. Some Australian aborigines have a sense of space that includes certain paths they travel, rather than a fixed location in which to live. Some Native American tribes view the environment as belonging to nature. Humans who use the products of the environment are expected to take only what they need, to be respectful, and to give back to the Earth. Resources are to be used with a sense of responsible stewardship. We will return to these ideas in the chapter on Consequences.

In other social groups and cultures, elaborate documents for describing ownership have been developed. If you live in the United States, you have already been taught many rules about ownership and **private property**. This includes ideas about private and public space. Your investment and concern about space and its use probably varies with whether you "own" it or not.

When people lose their homes or move to a new home, they often attempt to stake out their claims to the new **territory**. They explore surrounding areas that may affect their new space. Homeless people often choose a carton or doorway that they specify as their space, which they try to protect as theirs. They may even mark the boundary of their space in a way that indicates their claims. Have you ever noticed that the same homeless person is on the same street corner as you go by?

You may have noticed this marking of territory. If you regularly sit in a certain chair in your classroom, do you get angry if someone new sits there? Do some groups of people you know, regularly "hang out" at certain places. Do they

■ Figure 9.6
Wars have been fought and conflicts negotiated to decide on rights and the ownership of land. Maps, deeds, and international agreements have been developed over centuries of exploration and conquest.

defend "their" territory if a person from another social group comes into the space?

Who owns land? Air space? Ocean space? Who decides how any resources in that space can be used? How can you gain ownership and access? These questions have been addressed by numerous technological endeavors. If there are ample resources, people tend to use what they want. If there is a great deal of empty space in a park, you could probably camp out for a night. No one would know or care. If a very large bear exerts a prior claim, however, you will probably leave quietly and quickly. Even a little skunk might be able to persuade you to leave.

The history of human groups is filled with battles over territory. Each warring group believes that its rights to claim the territory are superior to those of the other group. Elaborate agreements are often made. Surveys, maps, and legal statements about boundaries are drawn up. Many people are killed in battles over boundaries, sometimes even boundaries that were decided centuries before.

Agreements are made as to the appropriate use of the space and rights of the owner. Deeds, titles, and charts are drafted. Currently, multinational groups are trying to decide on fishing and mining rights in the ocean.

As population density increases, more and more regulations about the use of space and the resources on that space arise. Conflicting claims and shared consequences make these decisions complex. Zoning regulations for the use of land are in place in many parts of the United States. Regional plans for water and treatment of wastes exist in many areas. Individual rights extend only to the point that they do not overlap and infringe on the rights of others. Concerns about pollution, clutter, and resource use recognize our interdependence with other humans and other species. They also recognize that we are all temporary occupants on the same planet.

Management of Space

In any technological event, there must be some space in which to conduct the activity. How that space is arranged and the flow of one activity into another influences the outcome. Once the sequence of steps is determined, the amount of space, the placement of tools and materials, and the amount of time needed for each step should be organized. Suppose you were designing the space for the surprise party mentioned previously. What steps would you take?

Consider your own private space. List the activities that occur there. How often do you share that space with others? Is there a place that is just yours for your belongings? Even if this is a shoe box stored under a bed, all of us want a place to put those things we find special and personal. How can you arrange your space and decorate it to make it more comfortable, pleasing to you, and expressive of your style and interests?

Previously we talked about spending and wasting time. Time cannot be reclaimed once it is gone. Unlike time, space has some dimensions that can be made tangible. If it is cluttered or used for unimportant activities, it can be renovated. We can change our use and sometimes can "restore" space. The idea of ownership of space gives rise to renting or selling space; that is, space can be treated as a resource that can be exchanged for other resources. The resource can be managed so that we reach our goals.

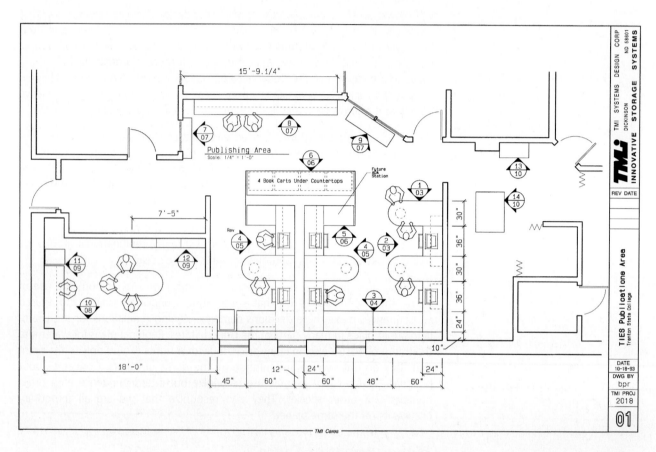

Figure 9.7 ■
This is a plan for part of the office of *TIES Magazine* at Trenton State College. It includes spaces for functions such as the reception of visitors, group meetings, production work for artists and writers, assembly and layout, privacy for work and telephoning, and storage and display areas. *(Courtesy of TMI Systems Design Corporation, Dickinson, ND)*

D E S I G N   B R I E F

Space Management

• Create an improved space management plan for your own bedroom. Be sure to identify all the functions that space should perform. Make an inventory of the tools and materials you currently use or store in that space. Use what you have learned about the principles of visual design and arrangement to make your room more aesthetically pleasing and functional.

• Choose a public space, such as a park, store, cafeteria, hospital, office, or school. Identify the functions that space must serve. Analyze current arrangements and evaluate the traffic flow, task sequences, barriers, and the elements of effective arrangement. Make a floor plan of the current arrangement. Make a floor plan for an improved arrangement.

MATH/SCIENCE/TECHNOLOGY LINK

• Calculate the costs of the improvements you have indicated on your design for improving your personal space.

• Analyze the use of color and lighting to create the effects that support the function of the personal and public spaces in your designs.

• Identify the relative costs for renting and buying an office, factory, or other space in your geographical area. If relevant, compare costs in the business district versus suburban area or a mall.

CAPITAL—HUMAN AND MONETARY RESOURCES

Resources can be categorized in several ways. Regardless of their source, resources that are scarce, transferable, measurable, and usable for production purposes have been called **economic resources**. Other resources are seen as **non-economic**. They are consumed for purposes other than the production of goods and services. Whether the resource is material, time, money, or space, it can be called economic or non-economic, depending on its use.

Materials are consumed for survival and pleasure. Time and money can be spent for activities that are relaxing. Time may be used for daydreaming. Space can be filled with clutter. Each of these resources could be classified as non-economic under these circumstances. However, if you decide to exchange these resources with other people, they can be seen as economic resources. If your job is to produce ideas for new marketable products, the time to daydream is an economic resource. Food in excess of what you need can be traded. The store of food is then an economic resource.

Capital refers to an accumulation of economic resources that can be traded for what you need. Since the earliest time, when people had more of something than they needed, they shared, wasted, stored, or traded it. If you were from a group that believed that goods and services belong to everyone, the fruits of the hunt or harvest were freely exchanged. No one kept track of how much each individual produced or consumed. Preserving and storing what is not needed was a group activity.

When the concept of private property is the dominant idea of the group, the individual produces for his/her own use. Excess goods and services are stored for another time or are traded. If the goods are shared (not traded), even within one's own family or social group, this is considered an act of charity. The person who produces the most, has the most. Any accumulation of stored resources can be used or traded, however, as that person decides.

When the economic resources are in limited supply, those people who control them can ask whatever price they believe they can get. The **principle of supply and demand** governs the exchange.

■ Figure 9.8
When the demand is high and supply is short, people often try to develop a synthetic product that meets the criteria. These synthetic diamonds are as hard as the real ones and are suitable for industrial grinding—maybe even for jewelry. *(Courtesy of GE Corporate Research and Development)*

If the price fixed is too high for other people who want to trade, they often attempt to find or make substitutes. When latex was unavailable because shipping was curtailed during World War II, resources were spent on research and development to produce synthetic rubber. Fake jewels and copies of art objects and period furniture can be so realistic that special equipment is needed to detect the differences. When prices went up for fossil fuels, people became more concerned about conservation. Research into alternate fuel sources was given more attention and funding. Because the price is high, people are willing to **invest** their resources to make the synthetic material or product.

The medium of exchange has changed over time, but the need to trade remains constant. The tendency is for people to specialize and develop skills, talents, and resources in a few specific areas. This makes us dependent on each other for some of our wants and needs. Very few people in the world can or wish to produce all of the goods and services that they, by themselves, need and want. Look around you and identify how many of the items that you use came from your town. Which came from other parts of the country? The world?

What Is This Money Worth?

Currently, most of the exchange of goods and services involves a promise to reimburse the person at a later time. The money that you might have in your pocket is basically a **promissory note**. Anything can symbolize an exchange if people keep their promises. Most societies have developed a **monetary system** that allows exchanges within the entire group.

Even the most established monetary systems fall apart when the group no longer believes that the money is valuable. When there is a threat to the group, such as war or major crisis, people rush to obtain the actual materials they can consume. They may hoard more than they use and basic materials may become the new medium of exchange. In crises, critical resources such as food and water, and basic tools and energy sources may not be available. The worth of the money declines. When trust is there, people would rather put the accumulated resource in a safe place or invest it. Money will not spoil, needs little room for storage, and is flexible in use.

Even as economic capital is stored or used, however, it can grow or shrink. Changes in supply, demand, and trust make the capital worth more or less. When several people get together and pool their resources, each may get a record of the amount they put into the enterprise. Stocks and bonds are such records. These stocks become more or less valuable as the enterprise thrives or falters. Paintings, jewels, and products that are seen as beautiful and rare may be collected as a form of wealth. Again, the important aspect of this wealth is perception.

You or the government can go into **deficit spending**. The United States is currently in debt to the tune of several trillion dollars. This debt continues to grow. How can this happen? People who have capital are willing to loan it. In addition, **interest** is charged as a partial payment over time. In some cases, there is no real expectation that the money will be repaid. Sometimes the person who made the loan would rather the borrower did not repay the loan itself, just the

■ **Figure 9.9**
Using an automatic teller machine (ATM) means that the amount is deducted from your account immediately. (© *Michael Dzaman Photography*)

interest. The interest on the loan is a reflection that the money is being put to "work."

Ben Franklin, one of the early founders of the United States, wrote in *Poor Richard's Almanac* "Neither a borrower nor a lender, be." What would he think of the United States today? This quote seems very outdated in a country where credit cards and massive loans and mortgages are commonplace. Centuries ago, moneylenders were seen as crooks. Charging money as interest was not allowed in good company. If you could not pay for something, you simply could not have it or use it.

Traditionally, wealth and capital consist of **assets**—tangible items, such as gold, minerals and gems, land, raw materials, and an inventory of salable products. Land and space are not portable. They have the advantage of being relatively indestructible. Certainly, the minerals from land can be taken away so the value is lessened. Space can be cluttered. Ownership of both can be stolen or taken away by trickery or force. However, land and buildings (real estate) have historically been viewed as a reliable form of capital. They are potential sources of income.

Currently, a major form of wealth is not tangible at all. This wealth often is in the form of information—records on a computer. In theory, the symbols on the computer file are backed up by a tangible item (such as gold in Fort Knox). Today, many people's jobs consist of trading money! How many jobs can you name that involve buying, selling, and handling money? As countries trade with each other, elaborate plans have been developed for the exchange of goods, services, and information.

Human Talent and Energy as a Form of Capital

A major form of capital that takes a slightly different form is **human capital**. People have abilities that they can develop and exchange with others. Humans vary a great deal in their style, talents, interests, motives, stamina, and capabilities. We each have abilities that can contribute to others through work.

Each of us must work in some way. Maybe you are wealthy enough to hire other people to do most of your work. However, there are some chores and responsibilities that even the most wealthy people must do for themselves. Personal and family work involve making or exchanging for the goods and services required for ourselves and our families. Caring for ourselves and family and maintaining a living space is also included. In the past, most of what a family needed was produced in the home. In the last few decades in the United States, many of the needed goods and services are purchased. Both parents often work outside the home to gain economic resources. Instead of producing goods and services, family work often means resource management and consumer decision-making.

Unlike buying and selling in the marketplace, **family work** is seldom paid for with money. Children may get an allowance, but that is so they can gradually learn to use money wisely. Each family member may have some personal share of resources. Within a family or intimate social group, many of the tasks are accomplished through cooperation and sharing. The tasks are conducted because of a commitment to the care of each other. Dependent children and elders may contribute less work. When there is a special need, one group member may be given more of the resources than others get. This uneven exchange represents a recognition of special needs and unique contributions. This is unlike what would happen to an employee who did not do her/his share of the work at a paid job.

Because family work does not involve a wage or monetary price, this resource is often overlooked when human capital is discussed. Yet, more than one-half of the labor that is done in a society is personal and family work. This contribution is not included in official records of labor and production. Yet, where would we be without it? How much is it worth to have a really good cook in your family? What price would you put on caring for a sick child? How much do you count on having someone arrange family and social gatherings?

FAMILY WORK TASKS FOR MANAGING A HOME

The following is a list of some of the jobs that are done to keep a family and home functioning. Think of others. Identify who does each of these tasks in your home. If your family hires the services of others to do the task, who finds these services, purchases the help, and monitors how well the job is done?

Buyer. Select, purchase, and monitor the supply of food, clothing, equipment, housing, materials, and services.

Health Care Provider. Care for people when sick. Provide preventive care and monitor each family member's physical well-being.

Tutor. Help with homework, arrange educational experiences, and monitor intellectual development of family members.

Manager. Delegate tasks, monitor the quality of performance, and motivate all family members to do their part.

Tailor. Alter clothing, and make and repair textile items.

Laundry Worker. Wash, clean, store, and maintain clothing and other items.

Driver. Take people where they need to go, and arrange for transportation.

Gardener. Establish and maintain garden and lawn, grow food and decorative plants, and monitor the health of all household and outside plantings.

Counselor. Assist with problem solving and monitor the emotional development of family members. Help with social relationships and value formation.

Maintenance Worker. Maintain and repair house, other buildings, vehicles, furniture, and equipment.

Cleaner. Wash, polish, and otherwise keep clean and in functional order all equipment, utensils, and structures, both inside and outside the house.

Cook/Caterer. Plan menus and prepare food for everyday meals and special occasions. Plan and arrange decorations that will enhance the food service.

Waiter. Serve meals, and clean up and store food service equipment.

Errand Runner. Deliver messages, pick up and deliver items, and provide personal services to other family members.

Accountant. Maintain household budgets, pay bills and taxes, develop short- and long-term spending plans, manage investments, and plan and implement actions to provide for fiscal well-being of family.

Interior Decorator. Design and maintain the beauty of the environment in which the family lives. Monitor the aesthetic well-being of family.

Social Worker. Seek out and arrange for community services that will assist the family in providing for the growth and development of each member. Attend to programs in the community that can enrich family functioning.

Dietitian. Ensure that the cook creates healthy meals and snacks, and educate and monitor the eating patterns of each family member so that they obtain needed nutrients at each stage of their life.

Social Secretary. Plan for fun, contact with friends, entertainment, and participation in the social and cultural events. Maintain social calendar and coordinate the activities of all family members.

Caregiver. Be a babysitter/companion for dependent family members. Ensure their safety and monitor their intellectual, emotional, and physical well-being.

■ **Figure 9.10**
Investment in human capital often means education and development of your talents and skills. A high school diploma is information that indicates to an employer that you have successfully completed this level of accomplishment.
(© Myrleen Ferguson Cate/Photo Network)

Brains and brawn. Many people in the world have almost no accumulated wealth. The only resource they can trade is the physical work they can do. With the development of machines, muscle power is replaced with other energy sources. However, even now on a global basis, many people exchange their strength, stamina, and muscular endurance for wages. You could say that they live by their toil and sweat.

Perhaps you have engaged in bodybuilding efforts and have attempted to increase your strength, stamina, and muscle mass. Did you do this with the idea that you would be a more valuable worker? Unless you want to work in entertainment—such as sports, contests of strength, or modeling—the answer is probably, "no." People who do physical labor usually build needed muscles and stamina on the job.

Another form of human capital comes from exchanging knowledge and skill. This labor involves thinking abilities and creation of unique products. Some people are talented in intuition, social relationships, creativity, and problem solving. Just as with other forms of capital, human capital can grow or shrink. In order to have marketable talents and maintain their value, study and practice are needed. Some form of education, along with some supervised practice, can be used to make these talents more valuable to others. An attempt to improve intellectual resources is called an investment in human capital.

Exchanging Human Capital for Economic Capital

A question that has been a source of tension over the ages is, "Just how much is human capital worth?" Again, when the demand is high and the supply is short, people can ask a high price for their work. If there are many people who can do the same job, the amount they can charge for their labor will be low. The employers will shop around for cheap labor.

In most technological endeavors, a group of people are involved. Each person contributes different types and amounts of resources. Some contribute money to buy the necessary materials, machines, and energy; some contribute skills that are intellectual in nature; some contribute their physical labor.

When there is a profit, how should it be divided? Throughout history, those people who physically carry out the tasks to produce the product get only a small portion of the profit. Those who do the planning or contribute their management skills gain more. Those people who contribute the economic capital often take the greatest share of the profit. The people who contribute the money may never see the product or physically participate in its production; yet they are usually the ones who decide how to divide up the profits.

Does the person with the money deserve the largest portion? They take the monetary risks. The laborer would argue that he/she takes the physical risks, however. How much of the profit should be used to ensure the safety and comfort of the workplace? How much should be set aside for workers for their old age or if they are injured or disabled or they are no longer needed to do the job? What resources should be allocated to protect the environment now and in the future? These are all resource management decisions.

In some countries today, physical labor is readily available. Many people want and need to work with their muscles. The money they get in exchange is meager. Intellectual talents and technological skills are in short supply. When this is the case, people with specialized skills can demand a high salary or wage. Little money will be put into machines and labor-saving tools and processes. There is little concern for protecting workers, since the supply is great. If a laborer is injured or worn out, another can be hired. Enterprises that use many people to do the work are called **labor intensive**. Labor is used instead of investing money in costly equipment and processes to do the job.

In other places, very few people want to or are available to do physical labor. Perhaps the population of workers is small or they are engaged in other activities (such as being sent off to war). Perhaps the population has become educated so that they have intellectual skills and demand higher wages for these understandings. Technological endeavors in these countries are likely to use machines and energy sources other than human muscle. These enterprises are called **capital intensive**. Capital is used to buy the materials and machines so that fewer people are involved in the work. Humans may monitor and trouble-shoot, but work is done by high-priced machines. An accumulation of economic capital is required before this kind of production can occur.

Figure 9.11 ■
Farming in some countries is done in a capital-intensive manner. Expensive tools and processes are used and fewer workers are employed. *(Courtesy Allis-Chalmers Corporation)*

Think of jobs that you would not like a human to do—for safety, comfort, or other reasons. Think of jobs you just would not like to turn over to a robot, no matter how well-programmed the machine might be. What new tasks are needed, if you do turn the job over to a robot or computer? The tendency in technological development is to design ways to relieve humans of more and more of the routine and dangerous tasks. A resulting concern, however, is the need for assurance that people will learn new skills and develop their human capital, as physical labor is replaced by machines.

Some say that if a machine can do it, let it. There are many tools and machines that now do what people had to do for themselves in the past. Most homes in the United States and many other countries in the world now have hot and cold running water and sewage. Power sources such as electricity and gas for heat, cooling, lighting, cooking, and refrigeration are available. These systems make it possible for a person today to have a lifestyle equal to one that required many servants a century ago.

Which are your special talents and interests? Are you strong and do you like to work with your muscles? If so, a job in which your physical labor is required may be for you. What types of jobs make use of the abilities you have? Some people are not the right size for some jobs—too large or too small. Body build and type are a factor in choosing the work most suited to you. Are you more suited to be a ballet dancer or a construction worker who uses heavy equipment? Can you sit

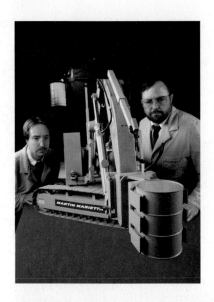

■ **Figure 9.12**
Human vulnerabilities mean that some jobs cannot be done by people without harming them. This robot has been developed to handle toxic materials. The human who controls the robot is able to work in a safe location and does not come in contact with the toxic materials. (*Courtesy of Lockheed/ Martin*)

still for a long time, or do you need to move often? Some people are happy to sit and work at a computer terminal for hours at a time. Others would rather climb on top of buildings and put out fires. Some people want to be alone, others want to work in groups.

Another area of concern is protecting the ownership and rights to the products resulting from working with your brains. This concern has led to various ways of documenting ownership of **intellectual property**. In the chapter on Design, we explored how to document and how these efforts could lead to patents, copyrights, and registration. In research reports on the work of others, it is important to clearly indicate the source of the ideas. This shows that you honor the talent and work of the author. Often, payment in the form of royalties is required when you use a play, song, or other product created by someone else.

MANAGEMENT OF LABOR AND OTHER RESOURCES

Economic resources can be exchanged for wanted goods and services and can be put to work themselves. Money and other capital can be invested so that it is available when needed for exchange. As we indicated before, sometimes it is possible to obtain and use the goods and services on credit and pay later.

Suppose you wanted to purchase a great new bicycle but did not have the money. What are your options? In some cases, the owner will let you make small payments each week or month until you have paid for the bike (and some interest charges). If your reputation for paying on time is good, you may be able to make another, bigger purchase later. In this way, you build a credit rating that can make it easier to buy what you want, even before you have the resources for repayment.

Managing money can help to ensure that your money "works" for you. In order to manage your money, it is useful to keep records. It is helpful to learn by starting out with small amounts. In that way, the cost of mistakes is small. An allowance is one means by which you can have supervised practice with money and learn to manage it to accomplish your purposes. Do you have a job, get an allowance, or get paid for some family chores?

Some people save a part of any money they get. Others have trouble with saving, and spend money as they get it. A balance between savings and spending is more likely to get you what you want in the long term. Of course, this means it is important to decide just what you want, now and for the future.

The first step in the management of any resource is trying to figure out what is important, both for now and the future. Some goals can only be accomplished over time, with the cooperation of others. You may want a car of some kind that is just yours, before you are 20 years old. Maybe not! You may know that you want to study astrophysics and have a Ph.D. before you are 30. Maybe not! You may want to work in a restaurant and have three children before you're 30. Maybe all of these are your goals!

Whatever your goals, it helps to clarify them. In that way, you can think through a plan for reaching those goals. There are certainly no guarantees that you will reach your goals. Many events are not predictable. However, not deciding is to decide; that is, people who do not decide what they want do not turn their talents to accomplishing goals within their reach. Also, they may not recognize an accomplishment when they do achieve it.

The design and problem solving model learned in Chapter 1 can be used to improve your use of time, money, space, and human capital. The outcome may be personal. The results may not be a tangible product, but the procedure is the same as that used in designing a product or solving another type of problem.

D E S I G N   B R I E F

Managing Capital

- Create a display to indicate monetary systems from several countries and parts of the world. On the display indicate the relative worth of the monetary systems.

- Design a form (such as a spreadsheet) to use in identifying the costs of materials and time that are needed for you to carry out your design and technology activities.

MATH/SCIENCE/TECHNOLOGY LINK

- Select a country in which the population is high and technological development is low. Compare their gross national product with that of the United States and/or another technologically developed country.

- Compare at least three banks in your area in terms of the costs and benefits of using their financial services.

- Compare several of the leading credit cards in terms of annual fees, charges for credit, payment schedules, and requirements for obtaining credit.

- Use a computer program to calculate the true interest on a loan to buy a bike that costs $700.

Task Analysis and Scheduling

Some techniques of management can help people as they organize their work. Actually, some of these same principles can be applied to office, factory, or school work. It often pays to observe in the actual situation as a first step. (Remember when Enrique and April first visited Aunt Sarah?) Sometimes taking a video of a person doing the task provides a means for analyzing each element of the expected performance.

■ **Figure 9.13**
When accomplishing a complex task, especially if several workers are involved, a plan for space and sequence of duties is needed. An assembly line allows each worker to do a segment of the total task. The materials and tools that they need are placed in that location. *(Courtesy of NASA)*

Each separate job that must be done can be examined. Which jobs require the most attention? This may be because of safety or because the quality of the outcome is very important. Note the jobs that require the most skill, and those where mistakes would be most costly.

Next, identify routines—jobs that must be done over and over and over. Which jobs are boring? Is the worker likely to lose concentration—maybe even doze off? Which jobs tend to be put off because they are distasteful? Which of the jobs are ones that no one wants to do?

The next step is to organize the jobs so that time, energy, and other resources are used most efficiently. Jobs that are routine can be streamlined. Stresses can be managed.

Tips for Management of a Complex Job

- Place equipment and materials so that they are easy to reach as you do the job.

- Plan the arrangement so that energy is used efficiently.

- Vary the parts of the body that are stressed by the way you sequence the jobs. (Follow a job that requires lifting and moving with one that can be done while seated.)

- Tasks that require little attention can be combined with those that are more engaging, but will tolerate interruptions. (Doing the laundry and your homework at the same time, for example.)

- Jobs that can be expected to be distasteful or boring can be placed before a job that you do like. (Also try to identify why you dislike the job and see if there is something that can be done about that.)

- Schedule jobs that require skill and concentration and those in which mistakes will be costly at the time when you are most alert and full of energy.

- Choose the day of the week that is likely to be matched to the requirements of the task. (Some jobs are best scheduled on Tuesday, rather than just after a weekend. Friday afternoon may not be a good time to try to start something new or start an intricate project.)

- Plan breaks and changes from a task that requires concentration to one that is less demanding.

D E S I G N   B R I E F

Project Management

- Select a product that you or one of your classmates have developed in previous design work. Create a plan for putting it into production. Estimate how much time, space, and capital you would need to start a business in which you make and sell the product.

- In groups, select possible menus for a dinner to be prepared for parents' night/to be sold to make money for a class project/for the school board (you decide as a class). Create a theme for presentation of the food and dining area. Make lists of all the materials, tools, space, labor, money, and tasks needed. Prepare a proposal that is a complete plan. Analyze the tasks required, who will do what, when, etc. If you have the resources, carry out the plan. If you don't, prepare a plan for obtaining the necessary resources.

MATH/SCIENCE/TECHNOLOGY LINK

- Use the examples provided in the *Portfolio and Activities Resource* to develop a GANTT chart or a PERT chart to show the necessary tasks for the dinner, their sequencing, and the approximate times for starting and finishing each. Make a note of the critical times; i.e., the times by which certain tasks must be done without hurting the outcome of the entire job.

- Make the list of costs as accurate as you can, based on actual market research.

- Interview a manager of a food service operation. Ask that your plan be criticized and improve it.

Working Together in Groups

People have always had to work together in order to accomplish most of their goals. Therefore, how to gain the interest, motivation, and cooperation from groups of people has been a central concern throughout history.

Suppose you want to give your mother a really pretty coat for Mother's Day. You don't have the money to buy it or the skills to make it. What could you do? Can you get your brothers or sisters to contribute? Can you all find jobs that will earn enough money so that if you pool resources you can buy the coat? What if your sister wants to give your mother some new curtains instead? How can you convince her? Also, how can you convince the local store owner to give you a job after school?

In any endeavor in which several people work together, there are issues of management. How can a team be built that will put aside individual wants and focus on common goals? What plan will encourage individuals to contribute their best to the group effort? Issues of leadership and relationships between coworkers are important. Analysis of the tasks needed to make the product and allocation of profits and benefits will be needed. Blending individual needs with group needs is critical. Many people have studied the process of getting people to work together in small and large groups.

You have some experience with working in groups in your family, school, and neighborhood. Some of the insights you have learned in living and working with others can help you understand organizations, even those as large as a major corporation or government. Many of the same issues and principles apply.

Who Is Boss?

Power, in terms of humans and group relationships, is the ability to influence the behavior of others. There are five basic types of power that can be used in managing group relationships. These types of power are **reward**, **punishment**, **referent**, **expert**, and **legitimate**.

Even in a group as small as two people, you can expect to have disagreement and conflict. These conflicts get more complex as the group gets larger. How this conflict is settled can either hurt or help the relationship in the long run.

Those people who have the ability to give out **rewards** and **punishments** are able to influence others to act in the way they want. You do many things you might not really like to do because someone rewards you for it. This reward may be such things as good grades, money, smiles, love and affection, saying nice things about you, or being your friend. What rewards will you work for?

Many people work to avoid punishments. The rewards just described can be taken away or withheld. In other cases, punishment means physical, social, emotional, or intellectual damage and pain. If people threaten to steal your bike or beat you up, you let them influence you—at least until you can protect yourself.

In our discussion of motivation in the chapter on Humans, we examined the kinds of rewards people work for at different stages of life. We also indicated the relationships between needs, depending on whether they had been adequately satisfied previously. Power that counts on giving and withholding rewards does not always work. The reward must be matched to the needs that are important to the individual's stage of development and level of need satisfaction.

In a group where one person or just a few people hold most of the rewards, they can dominate others. Those who lack power take a passive position and just go along with whatever the other person decides. History is full of examples where "might makes right." Those with the most money, weapons, or power to intimidate act as boss.

Another source of influence over people that is found in management of humans is **referent power**. Referent power comes not from who you are or what you can do, but from who you know. You follow the directives of one person because he/she represents the person with power. In many work groups, power is delegated. The supervisor's power may come from carrying out the plans of the "big boss." Think of examples of people and groups whose power comes from their association with someone who is really in control.

Two other types of power may or may not be related. **Expert power** refers to the ability to influence someone based on the expert's knowledge and skills. If you want to build a bridge or write a play, you may try to find people who are experts. You listen to what they suggest and let them influence you. The same can be said for listening to a famous athlete who is coaching his/her game.

But suppose you listen to the same athlete about what shoes to wear or what soda to drink. What does the athlete know about shoes or soda? In this case, the power you are giving the athlete comes from the role or position of that person, not from his/her abilities. You might ask a physician about a problem that is outside the field of medicine. The advice you get may not be any better than if you asked your brother. Yet, you let the physician influence you because of the respect you give to the position of doctor. The role is the source of authority, and this is called **legitimate power.** Recognize that you give the power to someone because of his/her legitimate position. You may respect your parent when he/she says to change clothes and dress right, but you may or may not think your parent is an expert on style. Which authorities do you let influence you because of the role or position they hold? Which people do you listen to and follow their directions because you respect both their skills and their position?

When people work together, a leader is usually needed. In small groups, this may be an informal leader. The leader may be one person or another, depending on the task. As organizations become more complex, leadership may be more formal. Few people have all the skills needed to understand the total organization.

Skills Needed in Management

In order to lead a group or organization, three types of skills are required: technical, human, and conceptual. **Technical skill** means that you can do the actual task or make the product. **Human skill** involves communicating, and motivating people to work together. Helping people become and stay interested and effective as workers requires human skill. **Conceptual skills** involve recognizing the bigger picture. How does this task fit with what other groups might be doing? What are the overall requirements? How should the work be divided and delegated? What criteria will be used to decide when the outcome is achieved? What changes in society and the world affect the work of this group? These questions are part of the overall conceptual analysis of the functioning of the group over time.

As an organization gets larger and is divided into subgroups, a leader may need less technical skill and more overall conceptual skill. No matter where in the leadership structure a person works, however, the need for good human skills remains. The actual people and jobs may be different, but the need for interaction and communication skills remains constant.

■ **Figure 9.14**
This athlete is an expert on the importance of good shoes for playing basketball. Would he be an expert on good shoes for dance? *(Courtesy of Nike, Inc.)*

As an organization grows in size, people are divided into subsections and teams. Some say that a supervisor can effectively manage no more than about six people. It is hard to know what is going on with more than that number of people or divisions. In some large agencies and businesses, there are many subdivisions.

Some organizations are developed on a hierarchy, with each level more and more specialized. An **organization chart** is needed to show who is in charge of each division. An example of an organizational chart is shown in Figure 9.15. Those who are in charge of making major decisions at each level are identified. Each person knows who his/her supervisor and colleagues are. They may have almost no contact with workers in other divisions or at other levels of the organization.

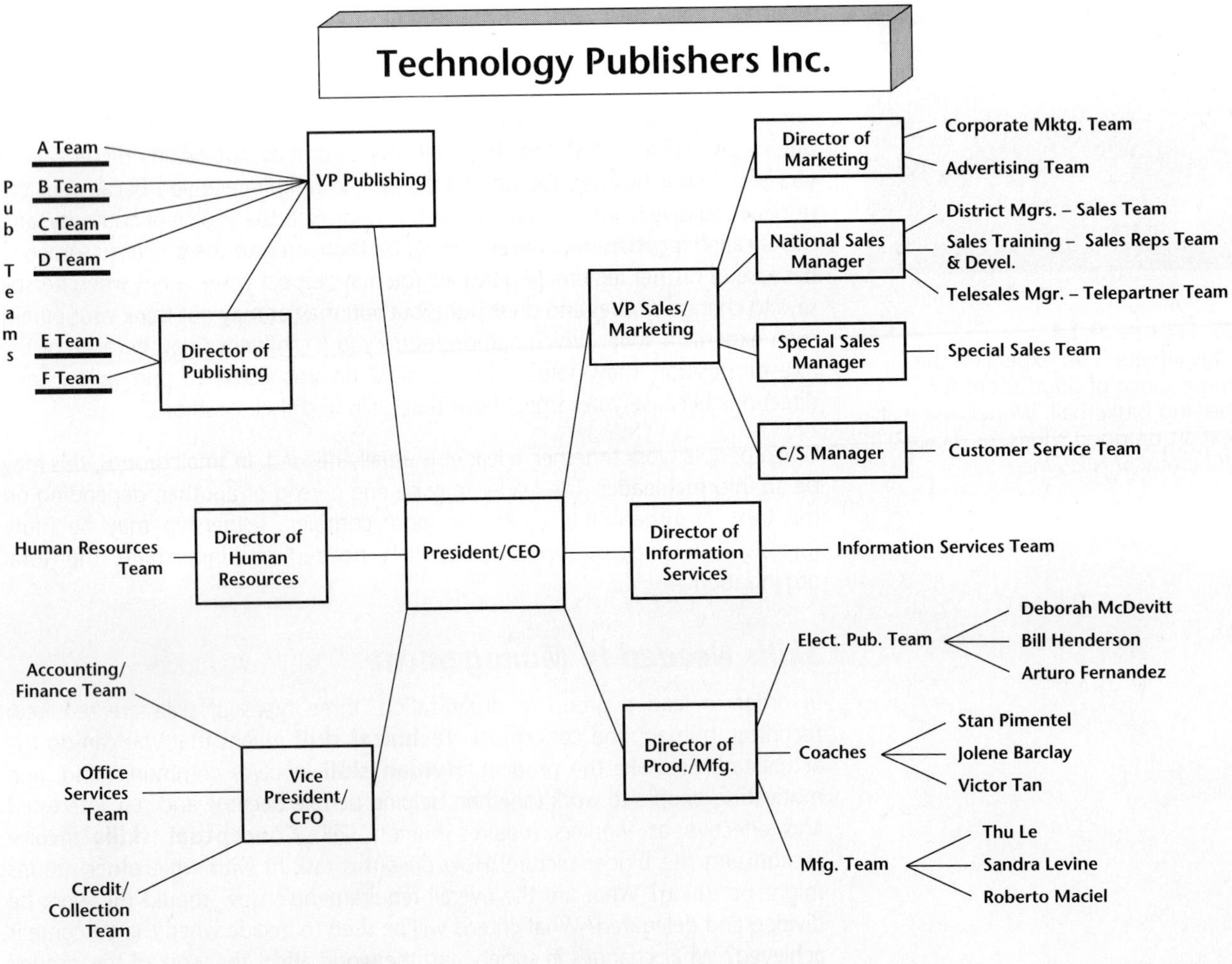

Figure 9.15 ■
An organization chart shows the chain of command. Some organizations have many levels; some have very few.

Leaders in the divisions who actually make the product need good technical skills. Yet, decisions may be made about the product without direct information from them. Morale of workers, loyalty to the group, and pride in the product may deteriorate. Policies may be implemented that workers see as unsafe or unfair. They may conclude that the bosses don't know what is going on.

The Board of Directors may have good conceptual skills and know more about what changes occur in the global marketplace. Their decisions may be made to protect workers and the company in the long run. They may conclude that the workers just want to get a paycheck and are not concerned about the big picture. They think that the technical workers don't know what is going on.

Another approach to management views information from all levels of the organization as important in achieving products of high quality. This approach does not break up a company into specialized subgroups and **hierarchies.** Even as an organization grows in size, the teams that are formed contain workers who contribute at all stages of production of the final product; that is, if the product is a shoe, the subgroup has people who make each part of the final shoe. Workers from marketing, purchasing, consumer affairs, personnel services, and other aspects of staff are also included. This total group of workers communicates with each other and makes suggestions about maintenance and change of the system for making and selling shoes. They also discuss the management of the company itself.

Although each worker's job is specialized, they can see the results of their labor in the final product. In addition, they contribute their understanding of their part of the process to the decisions that are made for the company as a whole. In these management groups (sometimes called **quality circles**), all people gain some sense of the conceptual ideas involved in decisions. They also are reminded of the technical skills and the conditions under which the product is actually made. (Refer to Figure 9.16.)

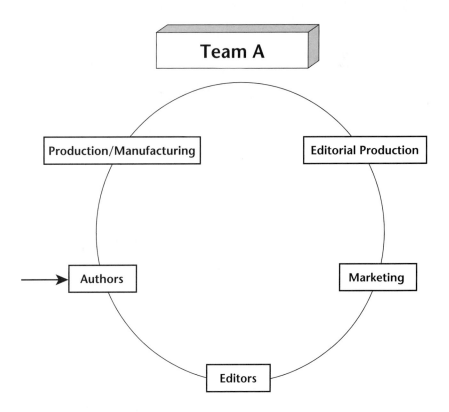

Figure 9.16 ■
An organization, such as a publishing company, can be managed with quality circles, groups that contain workers from all aspects of production and distribution of the product.

What do you think? Which system of organization do you believe will work best to create high-quality products for the least expense of resources? For many years, business in the United States has reflected the belief that people do their best when they compete with each other. The organization rewards those who carry out the tasks as they are told. When the worker is promoted, more creativity may be required, as long as company procedure is followed. Workers are expected to leave the company if dissatisfied, or be fired if they no longer fit company needs. If the corporation needs workers in another part of the country or the world, the worker is expected to move.

Many companies around the world are currently reorganizing so that subdivisions include the components of the whole system. The hope is that worker morale and loyalty will be improved as they participate in decisions that affect them. With the attempts to become more competitive, globally, attention is turned to how to improve the productivity of workers.

Assumptions about Why People Work

Organizations and groups are put together and maintained with certain rules and principles. Underlying the management structures is a theory about the nature of human beings and their motivation to work. Some groups reflect the principles that evolve from the idea that people are basically trustworthy and want to have meaningful work. If behavior does not reflect this trustworthiness, it is assumed that reasons can be found in the person's background or in the workplace. Attempts are made to help people see how their own needs are satisfied through their work with this group. Efforts are made to get each worker to recognize the meaning of his/her work to the whole product. Attention is paid to the relationships between coworkers.

WHY DO PEOPLE WORK?

THEORY X ASSUMPTIONS

1. People are born lazy. They work only if someone makes them do it.

2. People need to be told what to do. They need to be afraid of punishment or of being deprived to keep them doing what the company wants them to do.

3. The average person prefers to be told what to do. They wish to avoid responsibility. People have little ambition and work for the money.

(Based on work by Douglas McGregor, *The Human Side of Enterprise*. McGraw Hill, 1960.)

THEORY Y ASSUMPTIONS

1. For people, working is as natural as playing and resting.

2. It is not necessary to force people to work. When motivated, they are self-regulating and are happy to have a goal to strive for.

3. People are motivated to work when they can see the results of what they do. Results are a form of personal achievement.

4. The average person not only likes, but seeks responsibility.

5. People are highly imaginative and creative when they have the chance to be.

Another contradictory idea about human beings and why they work is that people are basically lazy and will work as little as possible. If this theory underpins the organizational structure, there will be strict supervision of workers. Attention will be paid to the allocation of rewards and punishments. Competition will be encouraged. Efforts to detect cheating and falsifying records will be constant. Systematically, supervisors will listen in on phone conversations. Random checkups of work will be done. The hierarchy of power in the company will serve as an incentive for behaving as the company wants. The rewards of being the boss are evident and only available to those who work according to the company policy.

Which of these ideas are reflected in the organization of your school? Which reflects your experience? What have you seen that motivates the people you know to do their best work?

A Healthy Group

The functioning of an organization or group can be assessed by checking in four areas. These checkpoints are useful, whether they are applied to a family or to a work group. First, people are individuals and have a desire for **personal recognition** and growth. A well-functioning group is managed in a way that the individual/personal skills, talents, and needs are recognized, used, and developed for the good of the larger group. You do not have to "check your personality at the door" if you want to belong.

Secondly, every group needs **rules** that provide the structure and security of knowing what is expected. Without appropriate rules, too much energy is spent in trying to figure out who does what, when, and why. Job descriptions and personnel policy manuals are developed when the organization gets too big for informal communication and negotiation. Rules are enforced fairly and consistently to all workers in the group.

Good organizations recognize that experience counts. A person's status changes as he/she develops a track record as a trustworthy, competent member of the group. These people gain power for changing rules or bending them during unusual circumstances. The respect they have earned puts them in a good position to negotiate major changes in the rules. Their investment in the group is trusted and thus, they are listened to.

The third aspect of a well-functioning group has to do with **communications**. The rules discussed above are clearly stated, are available to all, and are directly applied. In order to utilize individual contributions and ideas, people are encouraged to communicate emotions, ideas, and suggestions freely. Signs, charts, and other methods of communication are used in ways that are effective in getting messages across. The arrangement of space indicates that people are encouraged to interact with each other. Time and schedules indicate that communication is considered valuable.

In large organizations, there is often a **chain of command**; that is, a worker is expected first to speak to a supervisor and try to work out problems as close to their source as possible. In this way, those people who experience the problem are encouraged to work out their differences. In addition, the solution will be reached by those who live with the consequences.

The last aspect of a well-functioning group is the ability to gather resources from outside the group, as appropriate. No organization functions well as a closed system. Changes outside the group have an impact on the effectiveness of functioning inside the group. Some leaders hang on to old practices just because this is the way things have been done in the past. They may lack the conceptual skills to keep their group responding constructively. Yet they have the legitimate power in their role of boss. What long-term results would you predict?

Making Trade-offs Between Resources

Time, space, labor, and capital are interrelated resources involved in technological events. People may decide to spend the money and pay for the labor of someone else in order to save their own time and efforts. Some jobs cannot be accomplished in the way you wish because of the lack of space. You cannot make more space unless you rearrange items, remove extraneous materials, change processes, or rent/buy more room. If you are not busy and have spare time, you exchange your time and labor for money. If you are stressed out and have too much to do, you will look around for someone to take over some of your work in exchange for resources that you have.

As you have conducted your design and problem-solving projects, you have had to make decisions about which materials, energy sources, and tools or machines to use. In addition, as you work to produce the desired outcome, questions about available time, space, and money must be considered. As you analyze each step, the most efficient sequence and placement of tools and materials will be decided. In some cases, you can save money, space, or time without compromising the final outcome. In some instances, you need people to bend, lift, and do physical labor. In others, only a person with a high level of skill and understanding can accomplish the job. As with using any other resources in technological endeavors, decisions about time, space, labor, and capital are not easy. Criteria for desired outcomes for both now and the future must be considered.

Figure 9.17 ■
Aunt Sarah, as the person who actually uses the new storage and retrieval system, is the best source of information for suggesting problems that need further attention. Because Aunt Sarah has a visual handicap, someone who is an expert in braille or other symbolic languages may be a valuable consultant.

DESIGN BRIEF

Managing Groups

- Create a display on how to negotiate conflict. Role-play and video examples of fighting unfairly versus negotiating. Include the video in your display.

- A biotechnologist has just been commissioned to make a clone that represents the best traits and skills of you and your classmates. She wants to put information in her database. Decide on the list that you will give her. Using your ergonomic figures from previous design briefs, decorate and/or dress the figure to reflect some of these traits. Write a marketing statement that will encourage an employer to hire your clone.

MATH/SCIENCE/TECHNOLOGY LINK

- From Family Work Tasks listed in Chart 1 (see pages 266-267), calculate how much it would cost to hire all family work done by an outsider. Compare your figures with any figures given in Help Wanted ads in your local paper.

- Identify a job that you or your family needs to have done, such as painting, repairing, cleaning, doing taxes, etc. (Do not choose jobs you do for pleasure or would not trust others to do.) Estimate how much the job will cost. Give a realistic monetary value to your time and that of your parent(s). How much can you pay for the labor of another person to make it cost-effective to hire someone rather than do it yourself? Find the estimates others have made and compare them with your own.

SUMMARY
CHAPTER 9

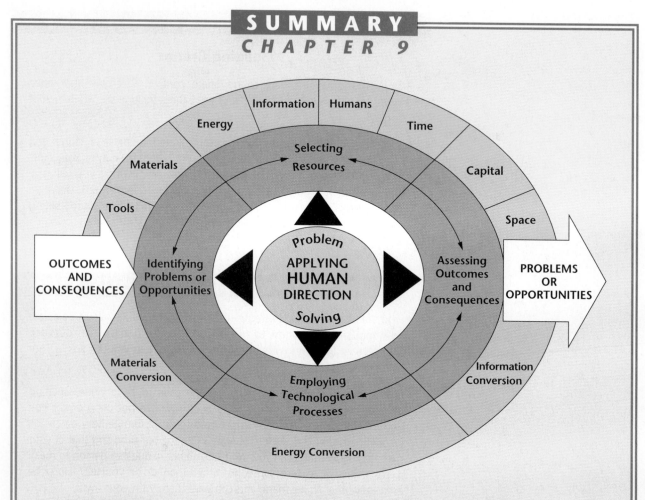

■ **Figure 9.18**
The management of resources involves making decisions about many things at the same time.

Any technological activity involves materials, tools, energy, information, and humans. Special attention was given in this chapter to the information systems of time and money. Space and its relationship to human needs and wants was explored as well. The work that people contribute to technology can be in the form of energy or information. They provide the muscles and the brains.

All of the resources that are used need to be arranged and man-

aged in order to produce the desired outcome effectively and efficiently. The goal of management is to produce the best outcome with the least expenditure of resources. When your design is successful enough that you decide to produce it again and again, the issues of management become more complex.

When groups of people work together to accomplish a goal, leaders are needed. The skills needed for such leadership are

technical, conceptual, and human. The need for the skills varies, depending on the type of management and the level within the organization. Healthy groups utilize individual talents, have rules, and communicate clearly the expectations of the group and individuals. Outside resources are used when the group does not have what it needs to solve problems and change over time.

Trade-offs between resource types is sometimes required. Some tasks that are labor intensive can only be done well by people. Others can be accomplished by the use of capital to replace human efforts.

As with any other resource that is involved in technological endeavors, decisions about time, space, labor, and capital are not easy.

Criteria for desired outcomes for both now and the future must be considered.

ENRICHMENT ACTIVITIES

- Gather data on the stock market for the same day, for the last five years. What social and economic events were happening at the time?

- Brainstorm the qualities of a valuable employee. Interview a manager of a small and a large corporation or business. Compare your list with that of the manager. Ask the managers what skills they find most important for their success as managers. Ask for examples in which they have had to use technical, human, and conceptual skills.

- Look in the Help Wanted ads of your local paper for a job that interests you. Write a statement of your skills, goals, experience, and special reasons why you should be considered for the job.

REVIEW AND ASSESSMENT

- Identify the resources that must be managed in any design and problem-solving activity.

- Describe three products you use that are designed to save time.

- In your own neighborhood, school, and home, identify at least five examples of places in which the space is reserved for a specific function.

- Define "capital," "capital intensive," and "labor intensive." Give examples.

- What is the relationship between psychological aspects, such as trust, and the economic value of different forms of capital?

- Give examples in which you use money, time, and space for economic reasons and for non-economic reasons.

- What are the types and sources of power?

- Describe the skills needed by a manager/leader at different levels of management.

- What are the assumptions behind hierarchical organization versus flat organization of workers?

Systems of Technology

Introduction to Unit III

Next, we turn to the ways in which technological activities fit together. In Unit I, we discussed the process by which technology is developed—design. In Unit II, the resources used in any technological event were examined. Then, we explored the management of these resources to effectively and efficiently carry out the design. Your choice of resources is limited to the products and processes of a number of other technological activities, some that are occurring at the time and some that are the results of development in the past. Certain major related activities have evolved over the years as humans tried to develop the products and environments they wanted.

These interrelated sets of technological activities are often considered the "systems of technology." This Unit will identify some of these complex systems and indicate how different aspects of a given system are interrelated. In addition, we will see how a network of relationships between the systems themselves forms the structure of our current technological environment.

A system includes a set of interrelated parts, working together for a purpose. Each part is important to the success of the whole system. Each is affected by the characteristics and functioning of all other parts of the system. Together, more can be accomplished than if individual components operate on their own. In order to work effectively and efficiently, a consistent order of operation is needed. This set of expectations (rules) describes how each part must work together with others if the task is to be accomplished. Systems vary from very simple and straightforward in operation to extremely complex and variable.

Perhaps an example will help. A loaf of bread could be called a system. Each ingredient is important to the outcome, and the combination of the ingredients must be made in a given way in order to get an effective blend to make the product you want. The recipe and production plan contain the rules for the system. The production of a loaf of bread depends on a number of other larger systems. Some are organized for producing the materials and tools needed. Still other systems are involved in getting all of the resources required together at the right place and right time. Even more systems must be operational in order to get the bread to the consumer while it is at peak quality. These larger technology systems are addressed in Unit III and include Physical Systems, Bio-Related Systems, Communication Systems, and Control Systems.

Chapter 10, Physical Systems, includes three types of interrelated systems—construction, transportation, and production. Some of the basic components of construction systems are examined, including structures built for support, shelter, containing materials, and directing transport. The importance of compression and tension in creating structures that will achieve their function as well as support their load is examined. Types of structures—rolling, floating, flying, and transporting—are described. Common features of transportation systems are addressed and include the degrees of freedom of movement, pathways, and purposes. The types of production systems include extraction, harvesting, and refining. Ways for extracting materials through mining and pumping; harvesting of foods and fibers; and refining metals, foods, and fuels are other aspects included in the chapter. Conservation, reuse, and recycling systems have been developed. We will examine how physical systems can be designed to manage resources for now and for the future.

Chapter 11 addresses one of the oldest, as well as the newest, types of systems related to using and improving living organisms. Stages of development of biotechnology are described, beginning with early attempts to improve the food, fuel, and fiber supply. Later, technologists turned to isolating and manufacturing the organisms that would provide the desired effects. The most recent stage, modern biotechnology, involves engineering changes in the basic genetic structure of materials. Biotechnology systems in agriculture, health, waste management, and environmental issues are discussed. Some of the emerging bioethical decisions are treated as well.

In Chapter 12, Communication Systems, we build on some of the concepts introduced in the chapter on Information. The components of sender, receiver, transmission media, and feedback are essential aspects of the communication process. Open and closed systems, entropy, and developments in graphics and telecommunications are considered.

Unit III ends with a treatment of the control of systems. Chapter 13 treats inputs, controls, and outputs as they are designed to maintain a system over time. Closed loop, self-regulating systems require feedback of information and the use of machine logic in their decision-making. These developments are particularly important in complicated systems, where the human turns over most of the guidance and control to machines.

After studying this unit, we hope that you will have a better understanding of how many of the resources that you use have been produced and made available to you. In addition, we hope that by understanding the interaction between systems, you recognize how they can be designed, controlled, and changed.

10 Physical Systems

Figure 10.1 ■
Systems for transport have evolved over time as humans develop ways to travel farther and farther from home. Transportation requires vehicles and pathways that are the results of construction and production systems.
(© Boden/Ledingham/Masterfile)

INTRODUCTION—SYSTEMS OF CONSTRUCTING, TRANSPORTING, AND PRODUCING

. . . .

Where will you live 20 years from now? Perhaps your home will be suspended from a mountain or buried beneath the ground. You might choose to live in environments such as the sea or in space. Wherever you live, you

289

KEY TERMS

containing structures
custom manufacturing
directing structures
dynamic load
extracting
floating structures
flying structure
fractional distillation
geothermal energy
harvesting
heavier-than-air vehicles
hydrocarbons
impact structure
intermittent
  manufacturing
lighter-than-air vehicles
load
manufacturing
mass production
nuclear fission/fusion
pathways
polymerization process
purpose
recycle
reduction
refining
reforming
rolling structure
sheltering structures
static load
support structure
transport unit
transporting structures
truss
vehicle
ways

will need some form of shelter to protect you and make you comfortable. You will need methods of processing and storing the resources and the materials you require. You will also need to transport those goods to you, and transport yourself to other places.

Whatever products and materials you use must come from somewhere. Very few of these will actually be produced by you and your family. You will rely on production systems to grow, harvest, mine, refine, process, manufacture, or otherwise gather and produce the materials and products that you use.

In this chapter, you will read about the physical systems that are required for you to obtain the goods and services you want and need. You probably take for granted many of these systems. You know that milk does not just magically appear in a carton in the grocery store. Neither does the shampoo you use, the airplanes that fly, nor the buildings in which you live, work, and play. But have you ever explored how many processes, how much labor and capital, how much material, energy, and information go into the delivery of those goods and services to you and your neighborhood?

We will explore the common aspects of constructing, transporting, and producing. As you apply these ideas, you will recognize the technological systems that people have developed. You can consider how you might apply the resources available to improve those systems.

CONSTRUCTION SYSTEMS

The technological activity of building various structures is called **constructing.** Over time, people have built a variety of structures for a variety of functions. Each of these structures has parts that must be held together in such a way that they do what they are intended to do. Once you study the common features of structures, you can look around you and recognize how structures are built, how they are alike, and how they are different. With the knowledge and skills gained, you will be able to design a variety of structures, perhaps some that are more effective than those around you now.

The Types of Structures

Suppose you are one of the first humans to decide to live on the moon. The first thing you will need is some type of shelter that will protect you from the environment and allow you to organize and protect your equipment. You will also need pathways to get from one shelter to another, and so on. Each function will require different forms of construction.

All structures can be classified into these five basic types: **supporting, sheltering, containing, directing,** and **transporting**. Many structures are combinations of these types. The function of the structure is important in deciding where the strength is needed and how the parts are put together.

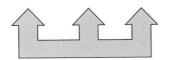

Supporting

Sheltering

Containing

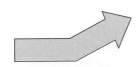

Directing

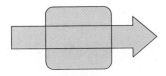

Transporting

■ **Figure 10.2**
All structures fall into five basic types: supporting, sheltering, containing, directing, and transporting.

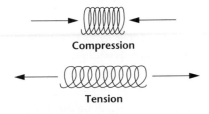

Compression

Tension

■ **Figure 10.3A**
The two major forces on structures are compression and tension.

THE FIVE DIFFERENT TYPES OF STRUCTURES

Supporting structures hold themselves and other things up.

Examples: bridges, lighthouses, antennae, chairs, tables, and shelves

Sheltering structures hold themselves up and something else out.

Examples: houses, castles, forts, barns, and bus stops

Containing structures hold themselves up and something else in.

Examples: Corrals to restrain animals; containers to store food, fuel, medicines, and other materials

Directing structures control the flow of materials and humans going from one place to another.

Examples: pipelines, canals, levees, mazes, stairs, tunnels, and viaducts

Transporting structures provide support and move themselves and something else.

Examples: Vehicles that fly, hover, roll, or float

Examples of the five structures are shown in Figure 10.2. There are several key ideas to explore before considering each of the five types of structures. These include compression and tension forces, static and dynamic loads, design, and materials characteristics. These are discussed in the following pages.

Compression and Tension

The forces of push and pull. Any structure must hold up its own weight and the weight of the materials it holds in, holds out, or carries away (usable load or "payload"). Some of this weight (load) acts on the structure with a force called compression. This force causes the molecules of the material to squeeze together. Other parts of the load act with a force called tension. These stresses pull apart the molecules of the materials in the structure. Most of the time, the structure must handle both types of forces at once. A structure often is designed to balance these forces to achieve equilibrium. (These forces were discussed previously in the chapter on Materials.)

DESIGN ACTIVITIES

Place a thick sheet of Styrofoam on the floor and walk across it. It will support you as you walk. You could use the sheet of foam as a supportive structure to help you float in water. Your weight is a compression force, and you can see the effects of compression as the foam becomes dented and thinner in places where your weight compresses it. Can you use this same

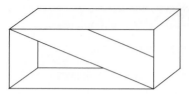

(A) Add solid brace.

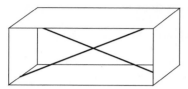

(B) Add internal flexible braces.

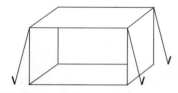

(C) Use a minimum of two external braces.

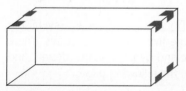

(D) Use a solid joint to provide a rigid frame.

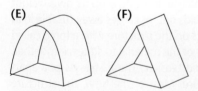

Use a more stable form—(E) the arch, (F) the triangle.

■ **Figure 10.3B**
Structures can be made more stable by using one or more of the techniques shown above.

sheet of Styrofoam to build a bridge? The sheet can be laid across a small stream of water with each end resting on a bank. If you attempt to walk across, the sheet will be subjected to compression and tension. The force exerted by your weight pushes the material together on the top side and pulls the material apart on the bottom. Make cuts in the top and bottom sides of a smaller piece of the same foam so that the effects of the compression and tension forces become very apparent. Place a load on the top of the sheet and observe what happens to the cuts on the top and bottom of the foam sheet.

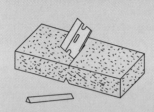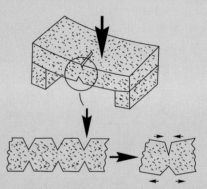

■ **Figure 10.4**
The forces that act on a structure cause stress on its parts. The two principle types of stress are tension and compression. Tension tries to stretch or pull materials apart. Compression acts to squeeze or force materials together. A simplified plastic foam model of a bridge illustrates compression and tension forces.

DESIGN BRIEF

Design a means to counteract and balance the forces, pitting the compression force against the tension force. Design and develop a model of your bridge. Use a can of sand with a weight equivalent to one-tenth of your weight. Can you use braces on the bottom and/or on the top of the bridge so that the Styrofoam sheet would support the structure itself and your weight as you walk across it?

MATH/SCIENCE/TECHNOLOGY LINK

Examine some bridges and bridge-like structures. What types and shapes of braces are used on the bridges? Which do you think would be the best to use on your bridge problem? What makes these braces strong? What contributes to making these braces stable? How much will the "can of sand" model weigh in pounds and newtons if it is one-tenth of your weight? Is there a difference when the object that represents your weight is placed gently on the bridge model as compared to dropping the object on the model?

The triangular shape, called a truss, that is used to brace a bridge can also be used in other structures. The **truss** is a special kind of triangular brace that takes advantage of this stable geometric form. Usually, you cannot see the way in which the forces of compression and tension have been considered in the structural design. All structures must be built to withstand forces of compression and tension. Structures must withstand several kinds of weight. If the material itself cannot support the load, a triangle is often added.

Static and Dynamic Loads

There are two types of load that any structure must support. The materials and design of the structure must work with both types. The first type of load is called **static load.** Static load is weight that does not move. Your body on a mattress and snow on the roof are examples of static load. The weight of a sidewalk may not be very important when it is distributed evenly over a large area of ground. Its weight will be more crucial, however, if that sidewalk is built over a road and its weight is concentrated on only a few supportive columns. A refrigerator shelf may be able to stand by itself, but it may collapse under the weight of the food you put on it.

The second type of load is **dynamic load.** Dynamic load is weight that moves around. When you were little, you were probably warned not to jump on your bed. Your weight while sleeping exerts much less force than when you jump up and down. In fact, if you jump on the bed, your dynamic load may be double or triple that of your regular weight. Obviously, such an increase could cause the bed to collapse. Walking or running on a footbridge exerts a different load than just standing on the bridge. Hopefully, when you are walking across, no one will decide to shake the bridge or jump around. From experience, you may have already sensed that this increased movement makes the bridge more vulnerable. Now you know it was the effects of the dynamic load that you were experiencing.

■ **Figure 10.5**
These students attach support lines to the cables and roadbed of their prototype of a suspension bridge. They have placed weights on the roadbed to represent a stactic load. What might happen if a moving (dynamic) load was placed on the bridge? *(Courtesy of Salvadori Educational Center)*

Designing Structures for Loads

In the past, people typically placed one piece of material on top of another to build a structure. This building technique counted on the mass of the materials to provide necessary strength. Mass structures, such as dams, rely on their own weight to provide the strength needed to hold back the force created by the water it restrains.

Historic Greek buildings were structures that required transporting, lifting, and placing huge stones. This kind of mass structure is called the post and lintel structure. The support column is called the post and the piece across the top is called the lintel. The distance between the posts was determined by the amount of tension the lintel could withstand before it would break. This approach is still used in many buildings today.

(A)

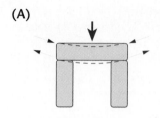

(B)

(C)

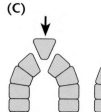 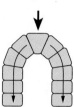

Figure 10.6A, B, C ■
Three different approaches to structures are (A) post and lintel, (B) cantilever, and (C) arch.

■ **Figure 10.7**
The Firth of Forth bridge in Scotland is a good example of forming arches through the use of cantilever structures.

The second approach, shown in Figure 10.6B, is the cantilever. The cantilever is similar to an extended post and lintel where a side extends out beyond the support. The bridge shown in Figure 10.7 was an innovative use of the cantilever design feature to achieve arches that spanned large distances.

Later, the arch was introduced as a design feature. The arch takes advantage of the compressive strength of the materials used for the structure. Arches support loads far greater than could be supported by the same materials in a post and lintel structure. Can you identify modern buildings that use arches?

Early structures were made of stone, wood, and other natural materials. As materials with greater compressive strength were developed, more ambitious structures could be built. Some arch bridges are built of steel and concrete.

The arch also served as the forerunner to the flying buttress. The flying buttress is an external support to a building—much like an extended arch—that was used to build inspiring cathedrals. (Refer to Figure 10.8.) The flying buttress provided a mass to anchor or support the walls, arches, and the roof. Massive walls were no longer needed to support the weight of the roof. Therefore, walls could be made higher, thinner, and open for glass windows. This change stimulated the creation of the beautiful stained glass windows found in cathedrals such as the Notre Dame in Paris.

■ **Figure 10.8**
The builders of the Middle Ages controlled the compression forces to construct cathedrals of truly magnificent size and beauty. Use of the flying buttress on the outside of the cathedral allowed lofty, open space inside.

D E S I G N   A C T I V I T I E S

A lumber import company has a lumberyard adjacent to a dock along a river. They require a crane that can off-load the large bundles of lumber from the boats lying alongside the dock. The crane must be able to take the bundles from the hull of a boat and place them in an orderly fashion anywhere on the ground within the yard. The lumberyard is 200 yards long parallel to the dock and 20 yards wide from the edge of the dock. (The bundles of lumber weigh less than 2 tons and represent little problem in terms of the weight to be lifted.)

D E S I G N B R I E F

- Design a crane that is capable of lifting the loads and placing them anywhere within the area described. (NOTE: You may want to provide a fixed crane with a long overhead boom or arm, or you may want to provide a crane that can move.)

- Develop a working model that includes the structures and mechanisms necessary for its operation.

Choosing Materials to Withstand Loads

Stones in an arch pack together under load. They have high compression strength. On the other hand, rope tends to fold over itself when pushed together. Ropes have low compression strength, but they are very strong under tension. Early shipbuilders took advantage of this knowledge as they chose materials for their moving structures.

Shipbuilders matched materials with the kind of force that each part of the structure had to withstand. The intent was to reduce the weight of the structure. Sails, with ropes attached for control, were combined with heavy, wooden masts. The solid, heavier materials are used where compression is expected. Light, flexible materials that are very strong when pulled are used for parts that must withstand tension. Light line, strong wire, cables, and cords are examples of materials used for tension structures. Applications of these can be seen in sailing ships, circus tents, suspended bridges, and hang gliders.

Sometimes materials are changed or developed to achieve the design purpose. New synthetic materials are light, but very strong. Even some of the conventional construction materials have been improved. Cement was known to the Romans, but was probably discovered in its natural form by accident. The Romans apparently mined the crude cement they used as mortar in their construction. The "recipe" for cement and concrete was not developed until much later.

The Romans would be amazed at today's uses of concrete. Concrete, like stone, compresses well but does not withstand tension. For example, highway slabs crack under the weight of traffic if steel reinforcing bars are not added for tensile strength. Another technique developed to overcome this weakness uses wires and tension for strength. Wires are placed into a form and pulled so that the wires are under high stress or tension. The form is then filled with wet concrete. The concrete hardens around the prestressed wires. When the form is removed, the tension on the wires packs the concrete even more tightly. The material system of the concrete, with its compressive strength, and the wire, with its tensile strength, allows the prestressed concrete to withstand heavy tensile and compressive loads.

There are many other recent examples of material systems used for construction. Plastic and other materials can be reinforced with fibers during production to provide strength. Pieces of lumber are sometimes glued together to produce strong, laminated beams. These material systems take advantage of the characteristics of each material to improve the ability to balance tension and compression forces.

Supporting Structures—Holding Things Up

Support structures are usually stable and firmly anchored in one place. They are often built to provide a flat, usable surface, such as a floor. Examples are highways and building foundations.

Most roads, other than those made of dirt, are composed of materials that make a strong, smooth surface to support vehicle traffic. Modern highways use concrete or asphalt to form a surface that withstands heavy traffic and the changes caused by fluctuations in the weather. Asphalt can be found in a natural state or can be obtained by evaporating petroleum. Currently, some highways are being built that use recycled glass and plastics mixed with asphalt.

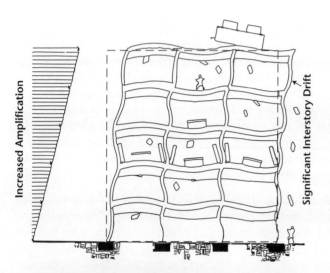

Buildings with conventional structural systems bend and deform during earthquakes. Accelerations of the ground are amplified on the higher floors. Deformation and distortions could be permanent.

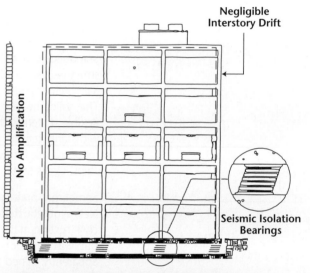

Buildings on base isolators do not deform. Movement takes place at the level of the isolators. Floor accelerations are low.

Figure 10.9 ■
Conventional design for construction considers some shifts in the load with minor movements caused by wind or settling of the foundation. Special design is needed, however, to withstand the erratic movements of an earthquake. This design isolates parts of the building from each other. Movement in the foundation and bottom floors is contained and not amplified in the upper stories of the building.

A short section of road may look very similar to a foundation slab for a building. But a foundation has to hold up another structure. Additional support may be needed in the form of footings, piles, or piers. Footings are solid pieces of material at the bottom of a structure. Piles are solid pieces of strong material that are driven into the earth to hold the structure in a more stable position. Piers are supports that are constructed of reinforced concrete, wood, or other strong materials.

Poured footings, usually of concrete and metal, are used when the earth can support the weight of that building fairly easily. If a structure must be erected on earth that cannot support the structure, such as clay or sand, piles are driven through the soil until they reach bedrock. Bedrock is a layer of rock deep under the surface that does not shift as soil does. A foundation is poured on top of the piles. Sometimes, bedrock or suitable load-bearing soil is too far below the ground surface. Holes must be dug deep in the ground and concrete poured into them. These concrete piers then provide support for the building.

Support structures may also hold materials high in the air. Towers are often truss structures that are built for height. They are similar to bridges except that they are designed to support their own weight plus side pressure from the wind. Some towers are used for communication, such as TV broadcasting. Towers may be stabilized by the use of guy wires or cables. These wires provide tensile strength and stability, yet they add little to the weight of the structure.

Some support structures must withstand the dynamic load of the wind. Some must carry heavy moving loads far above the earth or water. Others support the structures that move over them (such as trains). Still others stand up under static loads of snow, earth, or water.

Sheltering Structures—Holding Things Out

Sheltering structures often need a little more strength than that needed to hold themselves up. Most sheltering structures have roofs. Some add walls, floors, and foundations. This all depends on what the material inside is being sheltered from.

Early wooden bridges were covered to protect the bridges from the effects of the weather. The covered bridge is a simple illustration for examining other sheltering structures. Most shelters have a roof to protect the structure itself. There are many different kinds of roofs with many shapes and styles. All roofs, however, are designed to provide a watertight covering, primarily to hold out the weather.

In geographic areas where there is a great deal of snowfall, roofs are constructed to withstand the extra weight. Likewise, roofs on shelters where severe storms and strong winds often occur must be built to withstand those forces.

Walls are not always required. Some shelters need only be strong enough to support the roof. In those cases, posts may be used to hold up the roof. The sides of the building remain open. Floors and foundations may or may not be required. Portable shelters, such as tents, need little foundation for stability when erected. Tents are sheltering structures that integrate the roof and walls into one unit. The floor may be the earth itself. Mats may be added for comfort, or a waterproof floor may be added.

Inflatable structures are essentially large balloons of durable materials held up by air pressure. The inflatable building is a type of flexible shell that is relatively inexpensive and requires almost no support. The structure stays up because of the difference between the air pressure inside and the pressure outside the balloon. Inflatable structures are often used to cover large athletic fields because they require no posts or columns. You want to be sure that the materials used for the structure can resist any forces that make holes large enough to let out the air pressure, however! Actually, one such structure was slightly deflated when a snowstorm occurred. The weight of the snow was enough to make a large rip in the structure and the entire roof caved in. Luckily, no one was inside watching a game.

■ **Figure 10.10A**
The roof of this large airport in Saudi Arabia provides shelter from the sun, but must also support its weight and resist the forces of wind and weather. *(© 1993 Kimball Hall/The Image Bank)*

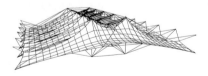

■ **Figure 10.10B**
This computer image of the San Diego Convention Center shows how the forces of compression and tension are balanced. *(© 1986 Horst Berger Partners)*

Containing Structures—Holding Things In

Many structures hold or store things. Containing or storage structures serve not only to store materials, but also to restrain materials. Containing structures also help control the environment in which we live. Usually, the walls are the most important part of containing structures.

There are many different kinds of storage structures. Water towers store water for use in homes and businesses. Oil and gas tanks store fuels. Desks and lockers hold your school supplies and books. Grain elevators and silos store feed for animals. Jails and corrals are containing structures. Levees are used to keep river water within a passageway. Ocean and river waters can be restrained with a dam. The most extensive and famous system of dams and levees is found in Holland. Two-thirds of the people there live on land reclaimed from the sea.

Look around you and identify the storage and restraining structures that you see. For what functions were they designed?

Directing Structures—Providing Passageways

Directing structures provide passageways by which vehicles, people, and materials are transported. The uses of these pathways are important in the technological activity of transporting. Directing structures contain materials, usually liquids, and control the movement of those materials. Pipelines have been built in many parts of the world. Pipelines are designed to carry gases, liquids, and some solids. Solids, such as powdered coal, can be mixed in water or oil to make a slurry. The slurry, something like runny pancake batter, can be pumped through a pipeline. The shape of a pipeline and the path it takes directs and controls the movement of the materials to the desired destination.

■ Figure 10.11
The pipeline from Alaska, through Canada, to mainland United States uses modern materials to support and direct the flow of oil. *(Courtesy of American Petroleum Institute)*

The Alaskan pipeline, shown in Figure 10.11, is a sophisticated offspring of a similar structure nearly two thousand years older. The Romans built directing structures to move water to their cities. They used gravity to move materials, a principle that still makes sense today. All directing structures are much more efficient when they use gravity. Much can still be learned from the work done by early aqueduct and canal builders.

Transporting Structures—Moving Things

Transporting structures are portable containing structures. They usually have a floor and walls. They may also have a roof. Flat railway cars can carry more cargo if they have containing walls. A roof may be needed to shelter the cargo from the environment. Transporting structures for gases need a floor, walls, and a roof to keep the vapors from escaping. Often, the structure is built in such a way that the roof and walls are in one piece, like a bottle.

There are three basic types of transporting structures: **floating**, **rolling**, and **flying**. **Floating structures** can be tiny or awesome. From the stern or back end of a modern supertanker, the front of the ship is nearly a quarter of a mile or more away. On the other hand, a kayak is so small and light that it can be lifted with little difficulty. Despite its size, the kayak can be used in rough water by those skilled and brave enough to venture out on the ocean or on the white-water sections of a river.

There is much diversity in the types of floating structures. Many different materials were available to the early shipbuilders. In some places, reed and papyrus were the only materials that were plentiful enough to use for boat building. These boats had limitations due to the delicate nature of the reed and papyrus. All materials have limitations, however. Reed, hide, wood, iron, concrete, and steel all have distinctive characteristics that influence the design of floating structures. All structures are influenced directly by the technology available to the builders. Boats built today are very different from the boats built 500 years ago because of the tools and materials available today that were not available to the earlier builders.

The form of a **rolling structure** is significantly different from a floating structure. Floating structures have large surface areas over which the weight is distributed. Rolling structures focus their load on very small points of contact between the wheels and the road or other surface. This reduces friction and makes the vehicle move efficiently. The frame of a rolling structure, such as an automobile or a bicycle, must be able to support the total weight of the vehicle and its load.

Rolling structures employ a rigid frame for the passenger. In many cases, the frame forms or holds a container or payload compartment. Another approach provides a container by using a shell structure that integrates the frame into the shell. The main frame or shell transmits the weight of the vehicle and its payload to the wheels. The rest of the structure supports the necessary parts of the vehicle, such as the engine, and serves as a buffer for the payload in case of impact (as in a crash).

The development of **flying structures** was different from that of floating and rolling structures. A different kind of knowledge had to be accumulated before people could design a structure to fly. Boats could be made to float and carts could be made to roll with little need to understand why they worked. Airships and airplanes were considerably more difficult to design, however.

Airships are lighter than air. They are similar to boats in that they must follow the principles of specific gravity. That is, the weight of the structure and its load must be equal to or less than the weight of an equal volume of the water, air, or other material in which it travels. A blimp is an example of an airship that floats in the denser air near the Earth's surface. Weather balloons and other structures that are meant to fly at higher altitudes must be made much lighter to compensate for the low density of the air at that level.

Airplanes do not work as airships do. Airplanes are lifted by the flow of air over their wings. This flow creates a partial vacuum that pulls the wing and the airplane up. The wings of an airplane are very strong beams that are specifically designed to hold up the weight of the plane during flight. The airplane structure must be able to withstand the static and dynamic loads experienced in flight.

■ Figure 10.12
The kayak, a very lightweight structure of considerable strength, has evolved in design over hundreds of years. (© Michael Kevin Daly/The Stock Market)

■ Figure 10.13
This modern mountain bike uses some of the principles learned in aircraft design to increase the ability of the tires to absorb the bumps along the road. This makes for a much smoother and safer ride. (Courtesy of Giant Bicycle)

D E S I G N   B R I E F

You are challenged to build a glider to meet one of the two following briefs:

• Design a glider capable of flying along a straight path over a distance of at least 10 meters. The glider is to be made from thick paper or card and constructed by cutting, folding, and joining not more than two pieces of material.

• Design a glider capable of flying along a straight path over a distance of at least 30 meters. The glider may be made from a framework which supports a membrane covering.

In both the above activities, record your design on paper and include notes beside your drawings which explain your design decisions. Evaluate the performance of your glider in terms of ease of construction, robustness of the glider, how well it meets the required performance, and the reliability of the flight path over a series of launches.

MATH/SCIENCE/TECHNOLOGY LINK

An airplane's dynamic load and stress increase greatly as the plane moves through turns, dives, and pullouts. These movements must be made against the pull of gravity (g). These maneuvers increase the g-load of the plane. The g-load is the weight that must be pulled or lifted against Earth's gravity. A g-load of one (1 g) is equal to the plane's weight at rest or flying on the level. A plane pulling out of a dive has a g-load much greater than one. You may have felt these g-load forces if you were in an airplane as it takes off and turns.

You may also feel the opposite sensation when you jump off a diving board, come down in an elevator really fast, or speed down on a roller coaster. These are examples of negative g-loads, where the forces acting on you are less than the weight of gravity.

• How does this relate back to dynamic load?

• What does increasing the g-load do to an airplane in flight?

• Can you relate this to what happens to you when you ride a roller coaster?

• How would positive and negative g-loads affect the design of an airplane?

• How might you test the models you build for g-loading?

Figure 10.14 ■
An unmanned power glider is designed to measure the deterioration of ozone at altitudes up to 80,000 feet. *(Courtesy of NASA)*

When an airplane is flying straight and level, the weight that the structure must support is very predictable.

However, there are immense structural loads on an airplane pulling out of a steep dive. Lifting force must be developed to keep the plane from continuing the dive—and crashing! The lift must be developed by the wings and may have to exceed two, three, or four times the plane's weight. This is equivalent to 2 g's, 3 g's, or 4 g's of load. Only acrobatic airplanes and military fighter planes are built to handle loads of more than 4 g's. Steep dives at high speeds can snap the wings off a standard airplane.

Spacecraft also present many construction problems. They must be built to withstand the large g-loading required to get them out of the Earth's gravitational force. They must withstand gravitational forces again during landing. Yet, the demands on spacecraft are not as severe as one might think. The major problem is to contain the required environment so that equipment can operate properly and humans can live in safety and reasonable comfort.

Moving vehicles often become **impact structures.** A car, plane, rocket, or boat must be designed as a high-impact, collapsible structure. The materials and design of the vehicle allow for the loss of mechanical energy if it should collide with another vehicle or a nonmoving object, such as a wall or mountain.

■ **Figure 10.15**
This car-crash testing device can help determine if the passenger compartment is protected from the force of the crash. *(Courtesy of Ford Motor Company)*

D E S I G N   B R I E F

Use the drawing of the model car provided in the *Portfolio and Activities Resource.* Cut out and assemble the model. Design and develop a structure that will absorb the greatest amount of energy.

• Test the energy absorption using a stationary car-crash testing device similar to the one shown in Figure 10.15.

- Test the car and its structure to see how well it works and how the design might be improved. When you are ready, check with your teacher to test your car and structure in the moving car-crash device.

MATH/SCIENCE/TECHNOLOGY LINK

- Identify the science concepts that are related to the vehicle crash-testing problem above.

- Identify which of the concepts you have applied properly and which you have applied improperly.

- Describe which of the concepts are most important in making your solution(s) to the above problem more effective.

- Describe how you can translate these concepts and apply them to improve your solution(s).

TRANSPORTING SYSTEMS
. . . .

Imagine going from California to New York by subway in 21 minutes. One design for such a subway would only use wheels for slower speeds. The cars would be suspended by a magnetic field and would run in tunnels melted by a nuclear-powered machine called a Subterrene. The Subterrene would melt rock to form a glass-lined tunnel in which the VHST (Very High Speed Transit) vehicles would run, or more accurately fly. The air in the tunnels would be evacuated so that the projectile-shaped tubecraft could achieve speeds approaching 14,000 miles per hour floating on an invisible magnetic field.

Figure 10.16 ■
High-speed transport systems in Japan and Germany use magnetic forces for levitation and propulsion of the passenger vehicles. This is called a mag-lev system. *(Courtesy of Popular Science Magazine)*

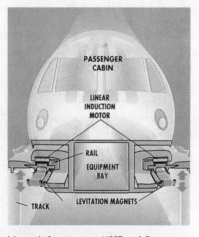

Magnetic forces attract HSST and German Transrapid trains up toward the rails. A linear motor (above) propels Transrapid.

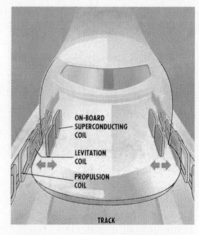

A new test track for Japan's repulsion-levitation maglev train has lift and propulsion coils on guideway walls.

Currently, major research and development activity is attempting to make magnetic levitation (mag-lev) usable on a large-scale basis. Mag-lev transportation systems are possible today but are still very expensive if implemented. Supporters of the proposed systems claim they would be considerably cleaner and more efficient to operate than many of the existing forms of transportation.

Transportation as Moving

Very seldom do you stay in one place for very long. Humans tend to be on the go, taking with them their materials and devices. Sometimes moving from place to place is very complicated and requires intricate support systems. In other instances, the transport may be as simple as carrying your things in a backpack as you wade across a stream. People send items, animals, couriers, and other people from one place to another. The destination can be as close as across the street or as far away as another planet. Transporting is the technological activity in which people, machines, energy, materials, or other substances are moved from one place to another.

MATH/SCIENCE/TECHNOLOGY LINK

Make a list of the major products you used in your design and technology activity. Find out where they were made originally. Where did the ingredients for the product come from? How did they get to your home or school?

DESIGN BRIEF

Design and develop a map that shows the source of the tools, materials, and products you have used in your design and technology work. Show the most appropriate means by which these materials could have been transported. For each of the different methods of transport, calculate how long it would take for the materials to reach you. Determine which materials require quick delivery and which are less time-sensitive in their transport.

PATHWAYS, VEHICLES, LOADS, AND PURPOSES

All transporting systems include the use of resources and some form of construction system. This may be to contain or shelter, or to constrain and direct the payload. The structure may be a backpack, basket, pocket, automobile, submarine, skates, pipeline, or rocket.

Many transportation systems utilize some form of vehicle. This vehicle travels along a pathway. The travel is for some definite purpose. The key ideas in transporting are **pathways, vehicles, loads,** and **purposes**.

The simplest transporting systems may have the most controlled pathways. The vehicle travels in only one direction. Some transport paths are more restricted than others. For example, compare the VHST tunnel with the airways used by commercial aircraft.

Pathways: One-way, Two-way, and All-ways

The concept of **ways** is a key to understanding the differences between types of transportation. Ways are the pathways from one place to another. Ways include roadways, highways, seaways, railways, waterways, and airways. The type of pathway used depends on the number of directions in which the vehicle is to move. Transport in one direction usually requires the most expensive forms of ways, since a great deal of control must be built into the system. The vehicle must be more confined and controlled in its movement. Examples are one-way streets, railroad tracks, high-speed tunnels, subways, elevated rails, conveyors, and long pipelines. These pathways all cost more money to build and maintain since a separate path is needed for flow back to the source.

One-way transport allows movement in only one direction. Ski tows, escalators, slides, and firefighters' poles support travel in only one direction. Materials moved by such devices as pipelines, aqueducts, and some conveyors are also restricted to one direction of travel.

Figure 10.17
The tunnel that has been bored under the English Channel, the "Chunnel," used three bores for digging. Two, 24 feet in diameter, are for the auto-trains that go in each direction. The third is a service tunnel for maintenance work and the like. (© David Teich)

Two-way transport allows travel in two directions. Examples are railroad tracks, some conveyors, and office building elevators. The travel of a railroad train is restricted. It can only move forward and backward on the tracks. The train has no freedom to move anywhere but along those tracks. Travel in two directions is not limited to trains, however. Some ferry boats move along a fixed rope or chain as their guide track. A primitive bridge used a supported rope as its track. Some people use clotheslines that work in the same fashion. Self-propelled people movers also operate on this principle.

Three-way transport allows movement forward and backward and to one side. Examples of this type of transport are seen in a riverboat that moves with the current or a sailboat that uses square sails. A boat that uses square sails can sail in all directions except directly into the wind. It was not until the invention of the lateen sail and the stern rudder that sailing ships could tack, or zigzag, into the wind. Earlier boats with square sails and side-steering oars were, to a great extent, blown with the wind.

Four-way transport is the most widely used of any of the forms. Travel by foot allows four directions of movement. The early transport of goods and materials was done by people carrying the burden on their backs or heads. Pack animals made the transport easier.

■ Figure 10.18A
The addition of a sail to the surfboard allows the surfer to move from side to side and forward by catching the wind. The sail is similar to the lateen sail that improved early sailing ships.

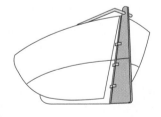

■ Figure 10.18B
The invention and use of the lateen sail and stern rudder made boats much easier to control.

THE DEVELOPMENT OF ROADS

The introduction of conveyances for dragging, such as the sledge and the travois, also influenced four-way transport. They were used on well-worn pathways or trails. This approach became common in land transport and the beginnings of roads were seen. When the wheel was used as part of a wagon, the need for roads was already established. Roads, however, were slow in coming. The entire Greek empire had only one paved road. During the time of the Roman empire, roads were improved. You may have heard the expression, "All roads lead to Rome." This saying reflects the attention that was paid to building roads for the transport of Roman armies and their supplies. These same roads were used to transport captured materials and people back to Rome.

In most places, however, roads were primitive pathways until the time of the Industrial Revolution. The lack of good roads, however, did not interfere with the significance of the wheel in the growth of transportation.

The relationship of the road and the wheel in moving heavy loads provided the basis of the land transport system that continues today. Roads restrict four-way movement and change it to two-way movement. With heavy loads, the wheel requires the smooth surface of a road. The vehicle then has to follow the direction set by the people who built the road.

Five-way transport, or travel in five directions, can be done in the air and beneath the surface of the ocean. Fixed-wing aircraft can move in five directions, every way but in reverse. Some submarines, such as those that tow divers under water, travel in five directions. Airplanes that go up will come down, and, hopefully, submarines that go down will come up. In both cases, however, there are set limits in which the travel can be accomplished.

Lighter-than-air vehicles and submarines both float in their own oceans. Lighter-than-air craft, such as hot-air balloons and blimps, depend on the air to hold them up as they move. A submarine moves under the surface of the water. The movement of a balloon up and down in the air and a submarine up and down in the water is accomplished by changing their buoyancy, or ability to float. A hot-air balloon can be made lighter by dropping weight out of its carrier basket or by heating the air in the balloon. It becomes heavier, in a manner of speaking, by letting off small amounts of its lifting gas or by reducing its hot air source.

Heavier-than-air vehicles operate in a different manner. They move in an ocean of air. A great deal of lifting energy is required to get a heavy airplane or helicopter off the ground and into the air. The aircraft must continue to move through the air rapidly enough to create a low pressure area above its wings. This low pressure zone creates a vacuum, a form of suction, that holds the craft up. (See Figure 10.19.)

Vehicles capable of traveling in five directions must have methods of propulsion, navigation, and control. All must maintain an established relationship with the force of gravity. The propulsion, navigation, and control of these craft are accomplished in relation to the closest horizontal plane of the Earth's surface.

Six-way transport is possible by helicopters, some submarines, and spacecraft. True sixth direction of movement removes the restriction of gravity. Consequently, the spatial reference that gravity provides is also removed. Right and left, up and down, and back and forth can no longer be determined by referring to a ground surface. Perhaps you have seen pictures of astronauts as they bounce around on the ceilings and walls of their spacecraft. This freedom gives no control of movement. Therefore, the astronaut attaches him/herself to some reference point, such as a chair. Magnets on shoes and gloves are needed in order to control movement from place to place. When work is done outside the spacecraft, cables are run from the vehicle to the astronauts and the objects they are working on, to keep the people and things from floating away.

More freedom—less control. The mag-lev vehicle discussed earlier in this chapter is an advanced form of two-directional transport. Subways and trains are other more conventional means of transportation. The uniqueness of these systems is that part of the controls are built into the path. The flow of vehicles or materials is controlled by the pathway itself. Obviously, many more controls must be built into the tracks of a train or of a high-speed tubecraft than into a slower pipeline or canal. The potential of one- and two-directional movement is great because the control can be built in. Moving materials within a pipeline or moving a vehicle on tracks can be made easier through the automatic control of that movement. Travel at thousands of miles per hour in a tunnel below the surface of the Earth can be made safe. It can be safer than travel at the same speed in aircraft flying above the surface of the Earth. But building such transport systems requires a large amount of resources such as labor and capital.

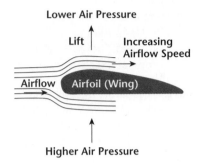

■ **Figure 10.19**
The shape of the airfoil (wing) and the flow of the air causes a difference in pressures that lift the wing and airplane.

Figure 10.20 ■
Within its environment, the submarine has six degrees of freedom of movement. *(Courtesy of General Dynamics)*

Whether we are transporting eggs, bullets, or people, freedom of movement has both advantages and disadvantages. In transportation technology, more freedom means less control. Safety of the payload (including passengers) is a significant factor to consider. One solution is to turn the navigation and guidance of a vehicle over to an automatic system. Automatic guidance and control becomes more difficult and costly as the number of potential directions increases.

Vehicles

Containing and propelling.   When people think of transportation, they often think of trucks, cars, and planes. But water is transported in pipeways, normally called pipelines, as are oil, natural gas, and other materials. These pathways also provide the method of containing the material as it is transported. There may be devices built into the pathway to keep the flow distributed evenly. Other devices may measure the amount of flow.

■ **Figure 10.21**
Trains often transport goods, using specialized cars that can be loaded directly onto trucks, ships, or other trains for the next stage of their trip. *(© Wendt Worldwide Photography)*

Some types of transportation require separate structures for containing the goods, materials, or people. These containers are often called the **transport unit**. Examples are railroad cars and modular containers loaded on the bed of a truck, as shown in Figure 10.21. Some transportation units have their own power or propulsion, while others must be attached to a power vehicle. Others are propelled by devices within the pathway. Tractors, trailer trucks, railroad engines, and tugboats are examples of vehicles that provide the basic power for moving the transport units.

Vehicles that combine a propulsion unit with a container unit are relatively common. Such a vehicle is the automobile. The automobile unites the engine and the cargo space, or passenger compartment in this case, into one combined unit.

There is now a rapidly growing trend toward the use of standardized containers that can be used in more than one type of transport. For example, there are containers that can be carried in a trailer truck for land transport and then loaded onto a ship for sea transport. At the next port, the container is again loaded onto a trailer truck for delivery by land to its final destination. Use of standardized containers requires agreement and cooperation among transport companies. Trucks, trains, and ocean liners must all have similar spacing attachments and loading equipment.

Loads

Feathers versus bricks.   Any vehicle will have a limit to the amount of cargo it can transport. In some cases, such as the hauling of rolls of fiberglass insulation, a vehicle may be filled in terms of space but carry only a fraction of its weight potential. In other instances, such as hauling bricks, there may be plenty of space left in the vehicle after the maximum weight has been reached. At this point, two related ideas enter the picture: useful load and operational load. The useful load, in the brick hauling example, is the number of bricks that we are able to haul. The operational load is the total weight of the truck, driver, driver's helper, fuel, tie-downs, pallets, spare tire, and the like. All vehicles have load limits in terms of their strength and power. It is possible to reduce the operational load so that the useful load can be increased. For example, by reducing the amount of fuel in the tanks, leaving the driver's helper at home, and reducing the number of tie-downs and pallets, the total useful load could be increased. Lighter materials in the structure of the vehicle can also increase its efficiency, as long as the strength of the vehicle is not decreased.

Purposes

Why move? The reason for transporting a cargo of people or material is a major element of transportation. The transportation goal is to use as few resources as possible to achieve the purposes. In other words, we want to move a cargo of people, goods, or materials in the shortest possible time at the lowest possible cost.

The speed of travel may be of primary importance. Transporting people to work places an emphasis on speed. Time lost in travel is time not available for use in working, playing, or relaxing. Transporting strawberries to market requires a slightly different interpretation of time. The important thing is not how fast the fruit is moved but whether it reaches the market at the point of ripeness. The fruit should also reach the market when the demand for strawberries is high. Speed is important, but timing, in this case, is even more important.

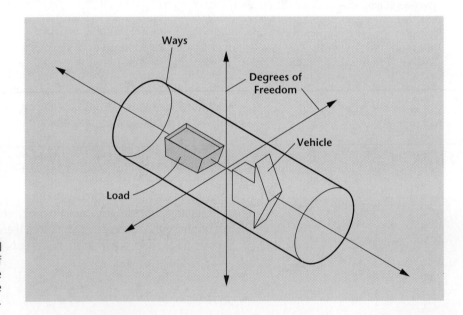

Figure 10.22 ■
This conceptual model of transportation systems includes the elements of pathways, vehicle load, and degrees of freedom.

Sometimes, the high cost of fast transport may outweigh the advantages of speed. Here, a cost/benefit analysis may be appropriate. Sometimes, when you want to get to another country, you prefer a leisurely ocean liner over a high-speed jet transport. Interstate buses continue to attract passengers, even though they may be slower than automobiles. Decisions about transporting systems weigh the benefits of one type of system over another. If resources were unlimited, we could develop many types of transport. We could give people the choice of which to use. Resources are not unlimited, however, and choices must be made. How many resources can we dedicate to which systems? Which will provide the desired outcomes with the least amount of cost, for now and for the future?

MATH/SCIENCE/TECHNOLOGY LINK

The freedom of movement of a space vehicle is possible until another planet catches the spacecraft in its gravitational field. Then, a new reference plane is established. "Up" and "down" are in relation to the surface of that planet. The freedom of movement possible in outer space can pose some problems for space travelers. Without reference points, it is difficult to keep track of where you are going. It is also difficult to keep a sense of time. The sun does not rise and set in the 24-hour cycle of Earth. Artificial means must be designed to prevent disorientation.

All spacecraft have definite limitations in the amount of energy they can carry. Because the planets themselves move, nearby planets are the easiest targets and will be explored first. If we are to go into deep space, the exploratory craft cannot count on only the fuel it can transport. Energy from mechanical and electrical sources will be exhausted long before reaching the target. Breakthroughs in the utilization of energy from different portions of the energy spectrum expand the possibility of deep space travel. Devices for converting energy from solar power already look promising.

People will not be confined to the Earth for long. Time, speed, and available energy are critical in deciding where and for how long we might travel. As we leave our planet, the challenge of maintaining control along with freedom of movement will require new "ways."

DESIGN BRIEF

People who will travel long distance in space or who will live and work in space for long periods of time will be more comfortable if they are in an environment that has a gravitational effect similar to that on Earth.

- Design and develop a working model that illustrates how a gravity of 1 g can be maintained in a space colony.

PRODUCTION SYSTEMS

The materials, vehicles, ways, and structures that we use are all produced. Very few of the items that we take for granted as a part of our everyday lives exist in the original state, or were gathered by us. Most of the goods and services we use are produced by others for our choice and consumption. The illustration in Figure 10.23 shows one of the first applications of the moving assembly line for the magneto assembly operation at the Ford Motor Company Highland Park Plant in 1913. Magnetos were pushed from one worker to the next, cutting production time in half.

Figure 10.23 ■
The late Henry Ford implemented the moving assembly line to produce the Model-T, which resulted in a dramatic increase in the number of automobiles produced. *(Courtesy of Ford Motor Company)*

Production includes any activity that yields or creates a product. Let's consider an example. Perhaps you have given little thought to the clothes you wear to school every day. They have a long and complex history, however. All over the world, people wear the denim pants from a design that originated in the United States as work clothes. Several different technological activities are involved in producing jeans.

All jeans start in a cotton field. This field may have been in the warm regions of the United States, or perhaps in another cotton-producing country. The cotton may have been planted, hoed, and harvested completely by human labor. If the cotton came from the United States, several machines were probably used. The cotton was cleaned, ginned, dyed, and otherwise prepared for use. The fasteners were made from metals or plastic. All of the required materials were transported to one place by the most economical means. This may have been by ship, train, or plane. Weaving machines made the cloth. The fabric was cut according to patterns that have been developed over many years. These patterns are based on the body measurements of many people. These measurements were averaged and patterns were made in several different sizes and shapes.

The cut pieces of cloth were machine-stitched, probably on an assembly line. Some people put in zippers while others put on pockets. The process continued with each machine or person following a standard part of the process. The sewn garments were then riveted at stress points to make them wear longer. The garment was then pressed. The production system started with raw materials and ended with a finished product.

Extracting, Harvesting, and Refining

Extracting minerals. Our planet contains the materials and energy needed to produce the products we use. The process of taking those natural materials from the earth, air, and sea is called **extracting.** Extracting includes mining materials such as coal, salt, and diamonds. It also includes drilling and pumping for materials such as gas, oil, water, and sulfur.

MATH/SCIENCE/TECHNOLOGY LINK

The scarcity of resources has created a need for technological innovations. Before the sixteenth century, the major source of fuel was wood. In England, however, wood became scarce because it was used faster than it could be replenished. Wood became scarce once trees were cut down for fuel. You may recognize a similarity between the sixteenth century dilemmas and those currently faced in some parts of the world. Now, forests are being cut down in order to supply the energy and material needs of the population. As the population becomes larger, forests are cut down faster than trees can grow back. The amount of forestland is depleted and the ecological balance is threatened.

In the sixteenth century, before wood became scarce, mining was considered an affront to nature. When the need for an alternative fuel source became strong enough, mining coal became an acceptable method of extracting fuel.

Early mines were relatively small, shallow, and dangerous. Mines were made deeper after new ways of building support structures were developed. But deep mines had problems with ventilation, water, and transportation. The problem of removing water and getting sufficient air to the workers had to be addressed. Lifting the materials to the surface of the ground also required the development of new machines and structures. This new economic need contributed significantly to the development of the steam engine—a machine of major historical importance.

Modern mines may reach several thousand feet into the earth. The deepest mines are the gold mines in South Africa that go down more than twelve thousand feet. This is deeper than humans have ever been before.

One method of mining is long wall mining. In this way, coal can be removed mechanically in large cuts, allowing the earth to settle and fill the open space. There is no need to build piers and supports in the mined sections, so mining costs are reduced. The process is safer because there is less danger of cave-ins.

Open pit or strip mining is used to obtain gravel, coal, copper, diamonds, and other materials. A great deal of topsoil is moved and huge ugly holes are often made. Open pit and long wall mining require less complicated support structures. Air shafts and protection from cave-ins are not needed as in deep mining.

New laws in some states require that mining companies restore the land before they move on to another site. Trees and grass must be replanted and the land must be restored for other use.

Drilling and Pumping.   Far beneath the surface of the Earth are deposits of the hydrocarbon remains of plants and animals. These are in the form of oil. Over millions of years, the remains of animals and plants piled up on prehistoric forest floors. Eventually, these remains were covered by soil during the later stages of

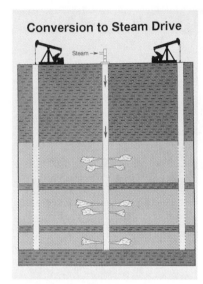

Conversion to Steam Drive

Steam →

■ **Figure 10.24**
Conventional pumping techniques can remove only about one quarter of the available oil from a well. But if steam is injected into the mine, oil can be forced out of pores and cracks of the rock. This helps increase crude oil production. *(Courtesy of ARCO)*

Earth's development. In time, pressure and heat formed a new form of carbon (oil) from these organic materials. Drillers try to reach the oil and pump it to the surface so that it can be refined for use. This is often a difficult effort. It requires costly structures and high-level technology.

Just as the scarcity of wood in sixteenth century England stimulated mining activities, the scarcity of oil has caused new developments in drilling and pumping. New methods have been developed to get the maximum amount of oil from wells that were previously abandoned, such as pumping water, heated liquids, or solvents into the wells to liquefy oils that are too thick to pump. (See Figure 10.24.)

A very new approach to increasing the yield of oil from wells requires the use of microscopic plants. A form of algae is grown in ponds or in a factory. This algae produces a slime around its cells. This slime can be inserted into the oil well, where it attaches itself to the oil. Removal of the oil-laden slime makes it possible to pump normally inaccessible oil from the underground rock particles.

As you can see, when people's demand exceeds the supply of fuel that they could easily obtain, they either develop ways of using another fuel or they develop new techniques for squeezing more fuel from the Earth. The population of the world continues to grow and the wants of people increase. As we begin to recognize the limitations of resources, many people begin to question how these resources should be distributed. We can expect that this conflict will grow until certain critical problems are faced. As long as we depend on materials that cannot be replaced, people will be competing for the limited supply that exists. We will return to this issue in the chapter on Decisions.

Growing and harvesting. Growing and **harvesting** is required to produce many of the materials used for food, shelter, clothing, and fuel. In early times, people hunted and gathered the materials they needed. They noted the locations where certain plants and animals could usually be found and harvested what they could find. People moved with the herds of wild animals and foraged for what they needed. Later, plants and animals were domesticated and grown in constructed locations. Learning about the conditions for germination and fertilization for reproduction, requirements for nutrients, water, and light, and the application of this knowledge, led to improved production yields. Systems for making and using tools were developed, such as the plow, fertilizers, improved animal foods, and controlled irrigation.

Over time, people have learned more about other living organisms that affect the growth of their crops. Attempts to shield domestic plants and animals from their enemies led to techniques for weeding, trapping, and otherwise removing and resisting unwanted plants and pests. Herbicides, insecticides, and organic growing systems have been developed.

The desire for more efficient production techniques has led to larger and larger systems, similar to factories. Large fields and huge enclosures on land and in the water are used to grow large quantities of plants and animals. Some attempts to improve the production of foods and fibers change the plants and animals themselves. An understanding of the reproductive cycle, cross-pollination, grafting, and cross-breeding are used to produce desired changes. Modern biotechnology changes the basic genetic material, so that significantly altered plants and animals and new varieties can be developed.

Recent developments involve growing and harvesting plants and animals that are too small to be seen without a microscope. Micro-algae are grown for such products as fish food, high-protein/low-cholesterol human dietary supplements, food colorings, thickeners, and vitamins.

Since the beginning of time, people have tried to predict the best time for planting and harvesting and have tried to protect against natural events that damage crops. Usually, all farmers can do is plant their seeds and hope for the best. Hydroponics is a method for growing plants, inside a structure, without soil. The nutrient balance needed by the plants is in a solution that circulates over the plant roots. Light, temperature, and humidity are controlled, along with a careful watch for any signs of unwanted insects or disease-causing microbes. Foods can be grown year-round and in many locations—even space stations and caves.

Large-scale production often requires adding agricultural chemicals for fertilizing the crops and for controlling weeds, disease, and insects. These chemicals increase the need for capital investment during planting and harvesting. The development of no-till methods of planting and harvesting illustrates a modern answer to problems of erosion and energy use. No-till means that the farmer does not use a plow or other tools to break the ground except to insert the seed when planting. Using chemicals to control weeds and insects eliminates the need to plow the land while the plant is growing. Because the land is covered, this method reduces soil erosion from wind and rain. It also reduces the amount of labor and energy costs, since the farmer travels over the land fewer times. Use of a variety and large volume of chemicals may increase the cost of production. The increases may be offset by improved quantity and quality of harvest. However, questions about the effects of the chemicals as their residues go into the food chain and groundwater must be considered.

One answer is to improve the chemicals that are used. New herbicides and pesticides have been developed from natural sources or have been made to degrade quickly into harmless by-products. Precise timing of the application of these chemicals is needed so that they will be effective before they biodegrade. Another modern approach is to make the plants themselves resistant to given diseases and pests. Therefore, the more "ecologically friendly" herbicides and pesticides can be used in smaller amounts.

Previously in this chapter, we discussed the need for wood as fuel and building material. Large-scale tree harvesting is relatively new. Trees have been cut and used for many years. As the demand for wood threatens to exceed the supply, farming techniques of planting and tending have been changed. One company conducted a 20-year effort to develop "super seedlings" of softwoods such as the Douglas fir. About one dozen of the finest trees were selected from nearly 100,000 acres of forestland. A few ounces of seeds from these trees were planted in a greenhouse and carefully nurtured. Each year the best seedlings were saved. Finally, after many years of work, enough new seedlings were ready. The company used these super seedlings in areas that were being harvested for

■ **Figure 10.25**
Modern, capital-intensive farming methods require large investments in equipment that often consumes large amounts of energy. Crops can be irrigated, fertilized, and sprayed with herbicides and insecticides mechanically. *(USDA Photo by Fred Witte)*

lumber. Super-seedling trees grow seven times faster than naturally grown trees. This process will help, but it will not be enough to take care of the demand for wood that leads to deforestation.

Harvesting is not limited to farming land crops. Fish farming is also a harvesting activity. Artificial structures are placed underwater to provide surfaces on which algae can grow. The algae, in turn, attract certain fish populations that can be harvested in highly efficient ways.

Sophisticated harvesting techniques, especially those used by floating fish factories, threaten some species of fish. Floating fish factories are ships designed to collect the fish, and process and preserve them in one continuous operation. They catch large volumes of fish, some of which are plentiful species. Other rare or endangered species may be caught and killed in the same process.

Plants, such as kelp, are also harvested from the sea. An abundant material, chitin, comes from seashells and is being used for a variety of purposes, such as cleaning up oil spills. Oysters are "seeded" by adding sand, then the cultured pearls are collected and made into jewelry.

Many natural materials must be processed before they can be used for human needs and wants. That process is called **refining.** Foodstuffs, fuels, and many other materials can be refined to make them more useful.

Refining metals. Most metals that are used in manufacturing and construction are refined. Except for a few select metals, refining is essentially the process of driving out excess oxygen. This kind of refining is called **reduction.** The metal is reduced to a more pure form. The reduction process is illustrated in Figure 10.26.

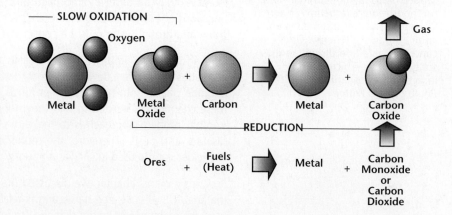

Figure 10.26 ■
Many metallic ores are metals that have been oxidized (taken on oxygen). The metal is extracted from the ore by a process of reduction. The ores are heated by a carbon-based fuel. The oxygen combines with the carbon from the fuel to form carbon dioxide, leaving the molten metal behind.

All metal ores that contain oxygen must go through some form of reduction process. Some reduction processes are quite simple, such as reducing iron ore to iron and tin oxide to tin. Other metals, such as uranium and cobalt, are more complex. All of the processes, however, need large quantities of heat to complete the materials conversion process from the ore to the metal.

Iron is one of the most useful and important metals. It is the basic ingredient of steel. Without steel, industrialized societies could not function.

A blast furnace is used to drive off oxygen from iron ore. This converts the ore to iron. Such a furnace can be as tall as 100 feet high. To melt the iron in the furnace, temperatures have to be in excess of 1537°C. Once a blast furnace has been started, it must be operated continuously to conserve energy costs. The materials are fed in at the top and the molten iron is drawn off at the bottom. If the furnace is shut down, time and a great deal of energy is needed to heat everything back up to the proper temperature.

Metals are not the only materials that are refined for later use. Ceramic materials, such as Portland cement and limestone, are also reduced or refined. Limestone is an important ingredient for concrete. To make Portland cement, clay and limestone are mixed together and heated in a kiln. These materials are then ground to a fine powder and bagged for shipment to the job site.

Refining aluminum requires a different process. Bauxite, the ore of aluminum, shows little effect from direct application of heat. The ore melts but the aluminum does not separate. After years of research, Charles Hall found a practical way to obtain aluminum from bauxite. He passed electricity through the melted ore. This process is outlined in Figure 10.27. The melted aluminum settled to the bottom of the furnace where it could be drawn off. Before Hall's process was used, aluminum cost more than $500 per pound. Following this improvement, aluminum costs dropped to about $1 per pound.

■ **Figure 10.27**
The preparation of aluminum ore for producing aluminum is more complicated than that of most other ores. The product of this phase of producing aluminum is a white powder called "alumina." Electrical current is then used to melt the alumina, which can be siphoned off to be used for producing aluminum products. *(© 1982 Aluminum Association/ Lifetime Learning Systems, Inc.)*

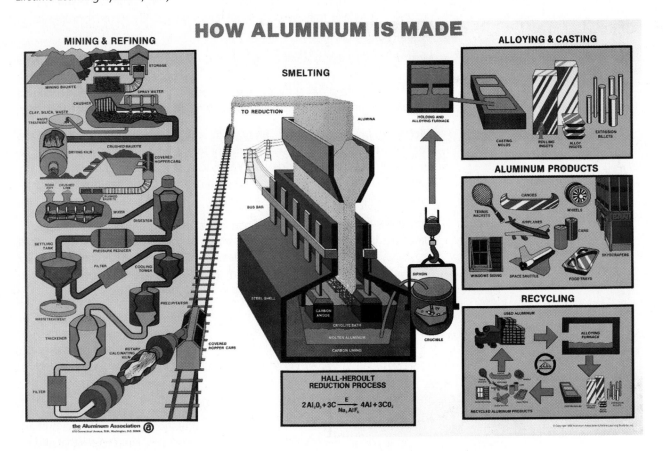

HALL-HEROULT REDUCTION PROCESS

$$2\,Al_2O_3 + 3C \xrightarrow[Na_3AlF_6]{E} 4Al + 3CO_2$$

Refining fuels. Just as the use of coal changed many of the processes for making products and started what is called the Industrial Revolution, the use of petroleum had a major impact. How do you heat your home? What fuel is used to transport you to school? Do you use any plastic tools or synthetic fibers?

Refining petroleum produces a range of usable fuels, as well as materials for manufacturing. The crude oils consist primarily of **hydrocarbons**. The three main groups of hydrocarbons are the paraffins, the naphthenes, and the aromatics. Oil refining requires two types of processes. Fractional distillation is the physical separation of the oil into several useful forms. The second type of process of reforming includes cracking and **polymerization**.

In **fractional distillation**, the crude oil is heated to boil off the many liquids. The tall columns at an oil refining plant are sophisticated distilling vats in which the oil is heated. The lighter parts of the oil, called fractions, boil off first. These vapors rise in the column and are drained off into holding tanks. Liquids that boil off next are extracted at different heights in the column. The oil is usually sent through several columns. Sometimes it is recycled more than once before the distillation process is complete.

Refining processes also change low octane fuel to higher octane fuel. This process is called reforming. The polymerization process joins molecules to form long, more complex molecules. Many plastics are formed through polymerization processes. Additives may be used to increase the stability or other properties of the material.

Most of the coal that is mined today is used in its natural state. Coal can also be refined to produce synthetic natural gas. This synthetic gas has many of the same properties as natural gas and can be sent through the same pipelines. Through a coal gasification process, coal and hydrogen react together to produce methane.

What we call waste today is beginning to be recognized as a resource. Animal and plant wastes, as well as refuse from the products we use and consume, can be reclaimed or recycled. In addition, energy can be captured and refined as the materials biodegrade. Waste can also be burned to provide heat for buildings or to generate electricity.

Refining foods. Foods are refined to remove impurities or to change the food to be more desirable. Sometimes, substances are added to improve the shelf life, the nutritional value, and the appearance of the food. Milling grain into flour is an example of food refining. In many parts of the world, flour is made from wheat, but it can also be made from rice, rye, or barley.

Different flours may be blended to provide the desired characteristics. Flours for bread, pizza, and cake are all different. In addition to blending, flours may be enriched by adding vitamins that may have been lost during the refining process. Rolled oats, cornmeal, and even sugar are refined in essentially the same way.

Producing and Manufacturing

Producing energy. In Chapter 6, special attention was given to the primary sources of energy. We mentioned the sources, processes, and devices that can be used to generate useful forms of energy.

Selected energy devices are used to convert prime energy into secondary sources that are readily usable. These energy conversion devices include solar energy units, geothermal devices, windmills, tidal-powered generators, hydroelectric power plants, nuclear reactors, fossil-fuel power plants, steam turbine generators, magnetohydrodynamic generators, and fuel cells. Several of these systems are illustrated in Figures 10.28–10.31. All represent important means of producing energy. Solar, geothermal, and fuel cell energy remain largely undeveloped.

Technological systems for producing enough energy to meet the demands of our world population will be severely challenged in the next few years. We will return to this issue in a later chapter on Consequences and Decisions.

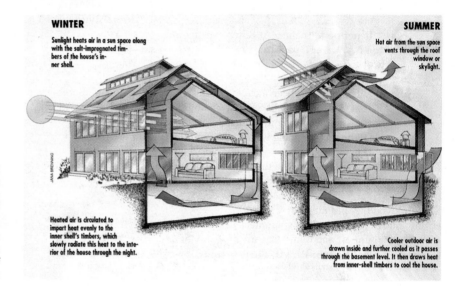

Figure 10.28 ■
The temperature of sunlight is capable of heating and cooling homes. Such systems are popular in climates with numerous "sun days" (southern climates), but can also be effective in climates where "sun days" are limited. *(Courtesy of Popular Science Magazine)*

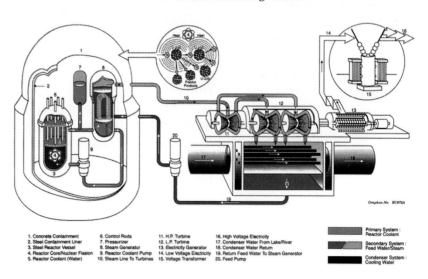

Figure 10.29 ■
Nuclear energy providing heat to generate electrical energy. *(Courtesy of Virginia Power)*

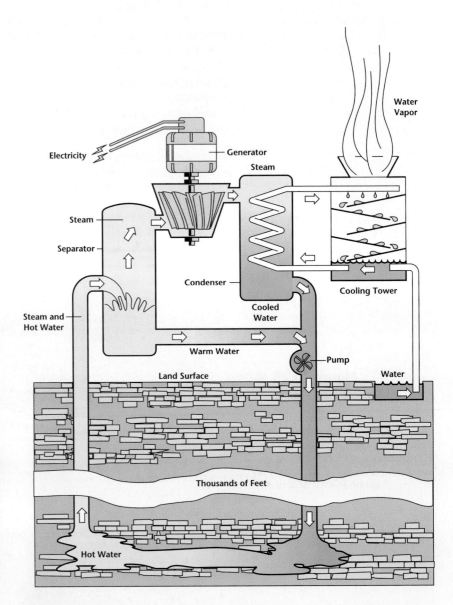

Land Surface

Thousands of Feet

Figure 10.30 ■
Because the Earth's crust is so thick, only limited use of geothermal energy has been made for producing electricity. If new techniques can be developed to reach and tap the thermal energy of the molten rock far beneath the crust, the problems of fuel shortage would be solved.

Manufacturing

Manufacturing can be thought of as the activities through which useful and salable products are made. The end result of manufacturing is a product that is ready to be used. Obviously, some manufacturing processes produce only certain materials and parts of a final product. For example, automobile tires are manufactured to be used as parts of a completed automobile. The preparation of the materials for the tires may include extraction and refining as well as manufacturing.

Manufacturing falls into three types: (1) the custom producing of items, (2) the intermittent producing of "batches" of products, or (3) the continuous production of products in mass. Custom manufacturing can operate on a small, medium, or large scale. Intermittent manufacturing usually operates on a

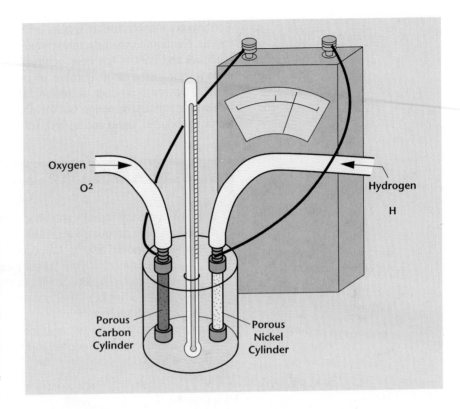

Figure 10.31 ▪
Fuel cells, originally designed over a hundred years ago, now take many different forms. All fuel cells capture energy in the form of oxygen ions released in the oxidation of conventional fuels. Fuel cells provide a direct conversion of energy and avoid the wasteful detour of first generating heat to be converted into useful energy. Direct energy conversion (DC current) reduces the loss of usable energy, or entropy.

Within the figure: Oxygen → O^2; Hydrogen H; Porous Carbon Cylinder; Porous Nickel Cylinder

medium or large scale. Continuous or mass production can operate only on a large scale when a great quantity of a given item is produced.

Custom manufacturing may range from small activities to very large ones. An individual making a one-of-a-kind piece of jewelry is an example of a small custom activity. The production of a large airplane or ocean liner is custom manufacturing on a grand scale. In the custom manufacture of individual items, there may be concern to make no two items exactly alike. Often, the uniqueness of an item is valued.

Intermittent manufacturing is used when there are too many products to be made by the custom method. The products are processed in quantities called "lots" or "batches." Manufactured lots refer to groups of individual items that are made at one time. For example, the tires mentioned previously would be produced in lots. Batches are quantities of materials or products made all at once and then separated into individual products. For example, paint is mixed by the batch and then poured into appropriate-sized containers for marketing. The label on a can of paint will often have the identification number of the batch from which the paint was taken. Bakery cookies and bread are usually mixed by the batch and then portioned out for baking as individual items.

Continuous or **mass production** is used when we need a very large number of identical parts. Mass production by continuous means is possible only if there is a mass market for the product. Products such as automobiles, refrigerators, radios, toasters, and telephones are mass produced on assembly lines.

Continuous manufacturing is also used to process oil, chemicals, foods, and metals. Continuous manufacturing is possible only if a fairly sophisticated level of feedback and control has been achieved. A production line must be a working system, not just a large number of unrelated operations. For example, it is possible to mass produce a million loaves of bread with a thousand bakers working at a thousand ovens, but the process is not continuous. Ordinarily, bread is mass-produced, using automated, self-monitoring and correcting systems.

DESIGN ACTIVITIES

A new company recently moved into your area and is very interested in developing new products for children. Wanting to take advantage of the creativity of young adults, the company approached the technology education department of your school with a proposal. Specifically, they are interested in producing kits, similar to the one illustrated here, for young children (ages 5 to 11). The kits are intended to help young children understand and apply some of the key concepts of design and technology at a level appropriate to the age and development of the user. Three kits will be developed and produced, one for ages 5–7, one for ages 7–9, and one for ages 9–11.

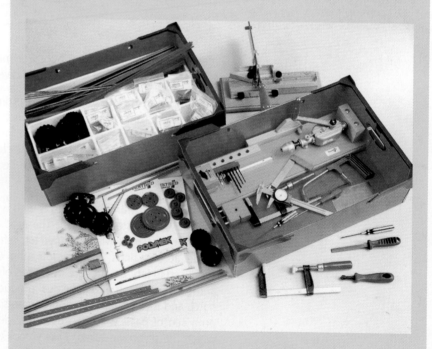

(Courtesy of Unilab's Polymer Integrated Modeling System)

Design Brief

Select one of the age ranges above. Using the **Design Loop** as a guide, begin the process of clarifying what you plan to accomplish and how you will implement that plan. Some of the activities for each of the phases of the Design Loop are identified below:

Identify and define the problem.
- Clarify and investigate the needs and opportunities presented by the problem.
- Conduct interviews with one or more interested teachers and with children of that age.

Develop the design brief.
- Clarify the results you want to achieve.
- Describe and plan the types of activities that you think the children can do.
- Write the design brief that will guide your work.

Explore possible alternatives.
- Conduct a search for solutions and information.
- Consider several alternative experiences that children might use to learn about design and technology. Write out, illustrate, and print instructions for the activities.

Accumulate and assess the alternatives.
- Identify a range of alternatives.
- Consider each alternative and choose the best solutions.

Try out the best solution.
- Experiment with the solution and translate it to a model and prototype.
- Determine what items to include in the kit. Identify which of the items you would buy and which you would make or assemble.
- List the sources and cost of each of the items you would buy.
- For those items you would manufacture, divide up the list among the class. Individually or in groups of two or three, do the following:
- Develop a plan for producing each of the items on your list using the technology facility in your school.
- Compile the information from the plans of the class groups. Determine the required tools, materials, energy, information, people, space, time, and capital required for producing 100 of your kits.
- Design a production system that your class can actually carry out.
- Develop a package that will contain all the items and your instructions.
- Make the package safe, attractive, and efficient for use and storage.
- Conduct a pilot run and produce multiple sets (perhaps 100 kits).

Evaluate the results.
- Test the solution and assess the process that was followed.
- Evaluate the overall effort. Include the assessment of (a) the finished items and kits, (b) the methods used to produce the product, (c) the work of the production team, and (d) your own participation and work as a team member.

- Test the kit by having children of the appropriate age, individually or in groups, use the product. Observe their behavior and interview them to get their reactions and suggestions. Be particularly alert to any safety hazards.
- Develop a report and presentation on the ways the product and production system can be improved.

MATH/SCIENCE/TECHNOLOGY LINK

- Identify the developmental characteristics of children who are potential users for your Design and Technology Kit.

- Determine methods for testing materials and tools that will be included. How much strength of grip do children this age usually have? What tolerance limits will you build in? How strong should materials be? Devise tests for strength and durability.

- How much do kits such as this usually cost? Make a spreadsheet to show the costs of each item in the kit and the total amount for each kit. Calculate the differences in cost for a custom-made kit versus those produced in batches of 20. Do the same for costs of batches of 100.

- Develop time lines and PERT charts for the production of the items, as well as for the assembly and production of the kits.

- Compare ergonomic statistics for the children ages 5–7 versus ages 7–9 versus ages 9–11, and determine how a specific product might differ for these age groups.

- Identify differences in safety concerns for each age group of users. Design symbols that will warn the children of potential hazards. Identify what items will be excluded because of potential safety factors.

- Create instructions that can be understood by children who cannot read and those who can read at third-grade level. Count the number of syllables in the words you have used and the number of words per sentence. Simplify the instructions to create a clear, accurate message that will most likely be understood by children with limited reading skill.

Conservation and Reuse

Some of us worry that if we continue to use the Earth's materials for production and construction, the supply of materials will soon be exhausted. Buckminster Fuller, a famous futurist, disagreed. He saw the Earth as a spaceship on which all the materials here when the Earth was brand new are still here, but in different forms. In Fuller's view, the problem is not in running out of resources, but to **reuse** them by learning how to reclaim and **recycle** the resources that are available. Some of the by-products of modern industry and current consumption patterns have made this simple goal a complex dilemma, however. Toxic materials, wastes from nuclear reactors, and the use of materials at a pace far faster than the Earth's ability to recover or absorb pollutants are issues that must be considered, even at the point of the original design.

We have already developed systems by which many of the materials that are used for production can be recycled, reclaimed, and reused for future use. Even after products wear out, many of them can be used again. Glass bottles, for example, can be reclaimed for their glass. The glass can be used to make new bottles or can be mixed with road-paving materials. Even fewer resources are needed when the glass bottle is washed and reused, eliminating the need for reprocessing.

Often, the issue is one of attitudes and values. Your jeans might have holes and be frayed. You may just throw them away or give them to a charity. When frayed jeans are in fashion, used jeans often sell for more money than the original price. The idea "new is better" leads to a tendency to use something and then throw it away. If we put more value on reusing products, we can save valuable resources.

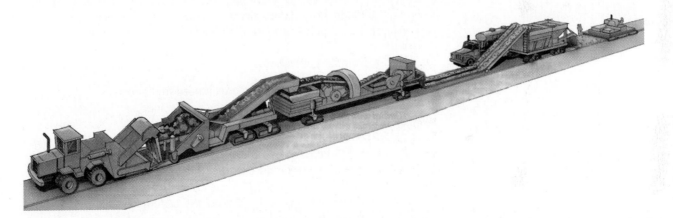

Figure 10.32 ■
This road rebuilding system uses a conveyor to transport materials throughout each stage of the system. First, slabs of pavement are removed, crushed, and separated from the steel reinforcement bars. Next, the powdered concrete is fed through a mixer to make new paving materials. Last, the material is poured onto the roadway and smoothed to make a "new" paved surface. *(Courtesy of Popular Science Magazine)*

Yet, reclaiming materials such as fibers, glass, aluminum, iron, and paper can save money and other resources as well as the environment. For example, it takes much less energy to make new cans out of melted ones than it does to make more aluminum from aluminum ores. The problems that come from energy allocation and use will make reclaiming of materials more and more necessary in the future.

Another aspect of the recycling process is the use of the reclaimed materials in making new products. Currently, many people recycle old newspapers. Relatively few printing companies use recycled paper, however, and we now have an excess supply of recycled paper. If the system is to work effectively, all parts of the system must contribute. Consumers need to be convinced that the effort they make when they sort and recycle is worth it. Effective processes for reuse and reclaiming must be developed. Manufacturers need to be convinced that the use of reclaimed materials makes good business sense. Packaging can be designed to contribute as little waste as possible. Products can be designed so that when their useful life is ended, they can be taken apart and the materials can be reclaimed.

Everyone complains about taking out the garbage. Garbage is something we would like to do without. It takes money to get rid of it and it smells. But even garbage can be considered as a resource. It is possible to design incinerators so that the garbage that is thrown away each year could generate heat that is equivalent to burning 400 million barrels of oil. One out of every ten barrels used in the United States could be replaced with "garbage oil."

Kitchen and lawn wastes can be composted and used for mulch and fertilizer. Currently, much of our garbage goes into dumps called landfills. We are running out of space for new landfills, and dumps are closing every day. The ultimate goal of reclaiming and reuse is to reduce the amount of waste that goes into the landfills.

The concept behind recycling is to "close" the system. The scrap, waste, or refuse from the system can be reintroduced and used again and again. Some creative ways of using reclaimed materials have been invented. Worn-out automobile and truck tires, ground up into pieces, have been used with asphalt to build roadways and sidewalks. The rubber has a nice bounce when you walk on it and rain water drains right through, leaving the walkway free of puddles. The focus on recycling is of relatively recent origin. If we turn our talents to designing ways for reusing the scrap, waste, and refuse that come from our industries, farms, and homes, we can reduce the amount of materials and energy needed to make products.

As we discussed previously, there is an urgent need to use sources of energy that are renewable. Research has shown that the weeds, food, and garbage thrown away by most families and the human waste that is regularly flushed down the toilet can be converted into methane gas. The gas can be produced in sufficient quantities to reduce each family's energy bill. Your home could be designed to capture your family's waste products so that they could be used in a methane converter. This converter would produce gas to supplement the fuel you burn for heat. Such an approach would be closing the system of your home waste environment so that the refuse could be used as a raw material. The idea is not new. Farmers have used animal wastes to fertilize crops for years. Animal dung is a major source of fuel in many developing countries.

Wood is a major fuel in many areas of the world and attention is required to find more efficient ways of using it. In some countries, large amounts of time are spent gathering and transporting wood for fuel. This time will be even longer in areas where the demand becomes greater than the supply. Where wood is the only available cooking fuel, people could benefit from having more efficient stoves. The destruction of forests and often of the species of plants and animals that live in these forests has produced major controversies. What are the arguments, pro and con, over laws and regulations restricting the right to harvest lumber in certain areas of the United States? What are some of the global issues that are of concern in this issue?

In all of the previous examples of recycling, two central ideas occurred again and again. Waste and refuse can be considered as a potential resource. A system of operation, whether large or small, can be closed by feeding back into the system any part of the garbage, scrap, waste, or refuse that can be put to productive use. Recycling, therefore, is another practical case of establishing a self-sustaining technological system.

SUMMARY
CHAPTER 10

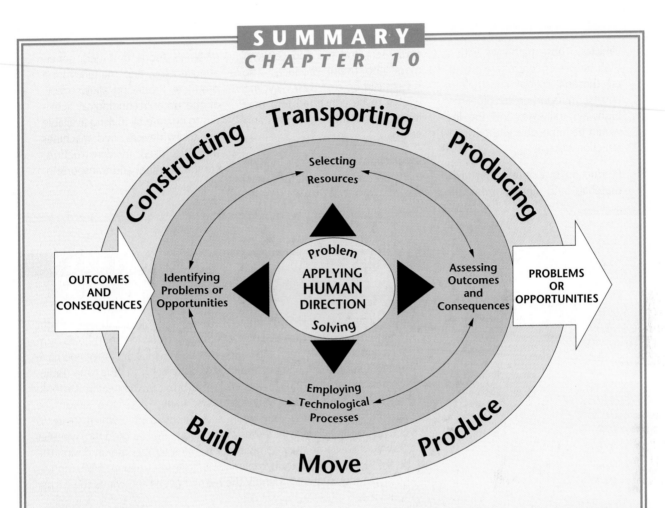

■ Figure 10.33
Contructing, transporting, and producing represent the physical systems of technology that are used to build, move, and produce the structures, products, and environments that people require.

Structures play an important role in the technological activity of constructing. All structures must be designed to hold themselves up under moving and stable loads. This is accomplished by the choice of materials and the structural design. There are supporting structures that hold something up, sheltering structures that hold something out, containing structures that hold something in, directing structures that move something away, and transporting structures that move themselves and their loads away.

Constructing provides necessary environments for shelter, movement, and work. The relationship of constructing to transporting is complex and varied. There are directing structures, rolling structures, and flying structures. Each has unique construction problems.

Transporting is an activity of moving materials or people from one place to another. It requires the use of paths or ways, and some means of propulsion. The energy may be provided by gravity, but is most often generated by motors

or engines attached to stationary machines or mobile vehicles. The paths used in transporting vary from the high control/low freedom of one-directional (or one-way) movement to the low control/high freedom of six-directional (or six-way) movement. New hybrid systems that integrate different levels of freedom and control are being developed.

The close relationship of transporting and constructing activities is apparent in the pathways required for floating, flying, and rolling vehicles. Those pathways with a high degree of control are provided through special constructing efforts. Transporting by pipelines, highways, railways, and the like would be impossible without construction systems.

Production is a system that takes materials and energy from the earth, sea, and sun and converts them into useful products. The system can be closed by recycling wastes to be converted into new forms of usable materials and energy.

Production activities include extracting materials and fuels, harvesting food and fuels, refining materials, foods, and fuels, generating energy, and manufacturing products. Producing allows many of the other technological activities to operate by making available the tools, devices, and machines that are used in constructing, communicating, and transporting.

ENRICHMENT ACTIVITIES

Design Briefs

• Design a shelter to take with you to the beach. The shelter should be light in weight so you can carry it easily. It should help you transport items from your home to the place on the beach that you will locate. Once you have opened it up, it should be big enough to contain you and the tools (radio, etc.) and materials (food, towels, etc.) you bring to the beach. It should shelter you from excess ultraviolet rays, heat, wind, and sand. The structure(s) should help your family and friends identify which shelter is yours. Develop a means of letting them know, from a distance, whether you would welcome their company or not. Remember as you design the shelter, to consider the tools, materials, energy, information, time, money, and space requirements. Identify the human needs and wants the shelter should meet.

• Design a model of an electric, motor-powered vehicle that can operate on a sand table that simulates the surface of the Moon. The vehicle must be able to cross the sand table with a gross weight (vehicle, load, and driver) that is equal to 50 times the weight of the motor that is used to drive the vehicle. (The small electric motors available to you in school are usually small and lack significant torque. However, they do rotate at a high speed. This can be used to your advantage in your design, through speed reduction arrangements that use a belt and pulley or a gear train for the drive system.)

• Design a meal that can be mass-produced and stored without refrigeration for use on an extended biking trip. List any instructions, including any required tools and materials, for making the meal more palatable before eating.

MATH/SCIENCE/TECHNOLOGY LINK

- Calculate the weight that your vehicle would have if you were actually on the Moon.

- Survey various batteries and motors to determine the power (wattage) each can generate. How much does each cost? How much does each cost to operate per hour? Determine the expected durability of each in terms of how long it could power the vehicle. If you were to add a solar cell collector, what change will be added to the vehicle's weight? Calculate whether the added weight will be offset by the added energy boost.

- Determine the energy needed for bike riders traveling 50 miles a day. What foods provide high energy and are digested efficiently? Which foods supply quick energy and which supply the body with energy slowly, for a longer duration?

REVIEW AND ASSESSMENT

- Identify the different types of structures and provide several examples of each.

- Illustrate how static and dynamic loads affect structures differently.

- Demonstrate how compression and tension are different in post and lintel, arch, and cantilever structures.

- Identify the advantages and disadvantages of each of the different types of structures.

- What are the four key ideas in a transport system and how are they related?

- Identify the advantages and disadvantages of each of the transport systems that have a high degree of freedom as compared to those that have a high degree of control.

- Provide some practical examples of extracting, harvesting, and refining.

- Select a means of producing energy that you find interesting and develop an illustration of how the system operates.

- Select an interesting product and develop a graphic presentation of how it is manufactured.

- Identify as many examples as you can of how recycling could be put into operation.

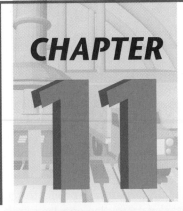

Bio-related Technology Systems

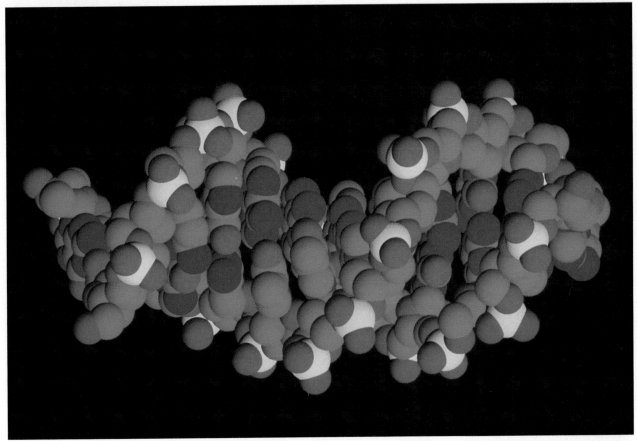

Figure 11.1 ■
Bio-related technologies evolved when the structure of DNA was identified in the early 1950s. Modern biotechnology can change some very fundamental aspects of foods, medicine, and the life cycle of plants and animals.
(© Ken Edwards/Science Source/Photo Researchers, Inc.)

INTRODUCTION—HARNESSING LIVING ORGANISMS

"Transgenic Animals," "Recombinant DNA," "Gene Splicing," "Cloning," "Bioprocessing," "Genetic Engineering." Big titles like these are often seen as headlines in newspapers and magazines. These terms refer to the new

KEY TERMS

absorption
adsorption
algae
allergens
antibiotics
bacteria
biocatalysts
bioconversion
bioethics
bioprocessing
bioreactors
biosensors
callus
cells
chemical bonding
chromosomes
clone
DNA
E. coli
entrapment
enzymes
fermentation
filtration
fructose
genes
genetic engineering
growth chamber
growth medium
high fructose corn syrup
   (HFCS)
hormones
Human Genome Project
immobilization
immune
insulin
microbes
microorganism
mold
mycoprotein
mutation
nucleus
nutrient agar
patents
pharmaceuticals
pollutants
recombinant DNA (rDNA)

technological revolution that is becoming more and more important to each of us—the biotechnology revolution.

Revolution? Actually, for centuries people have tried to learn more and more about the living organisms around them, as well as themselves. They built various systems based on this knowledge in order to enhance their lives. Sometimes their very survival depended on these systems. Through observation and selection, people improved the quality of many useful materials and avoided outcomes that were destructive.

Within this century, advances in knowledge and tools allowed people to take what happens in nature into factories for controlled production on a large scale. However, in only the past few decades, increased understanding of living organisms, along with new tools and materials, has made some revolutionary solutions possible. The ability to intervene and alter some basic life processes is what makes this modern biotechnology a revolution.

In the most basic sense, biotechnology uses living organisms, or a part of an organism, to make some product or perform some service. In the old sense of the term, this meant using the whole organism, as it appears in nature. With modern biotechnology, we are able to use parts of the living organism. We can also combine parts of organisms to produce brand new organisms, not naturally found in nature. It means that sometimes we do not have to wait for the natural life cycle processes of the organism to occur before it reproduces. And the offspring does not necessarily have the same characteristics as the parent. Sound revolutionary?

How much does biotechnology influence you on a daily basis? Can you recognize its processes and outcomes?

Take a moment and complete this short survey. From the following list of activities, decide if there might be a biotechnology connection. Answer "yes" if you think biotechnology might be involved, or "no" if it is not.

1. _____ Wash the dishes?

2. _____ Drink from a water fountain?

3. _____ Glue a plastic model together?

4. _____ Eat a steak?

5. _____ Throw away trash?

6. _____ Eat a cheeseburger?

7. _____ Drink a soft drink?

KEY TERMS

sedimentation
selective breeding
sensors
single cell protein (SCP)
sludge digester
tissue culturing
toxins
transgenic
vaccines
yeast

Actually, the answer to all of the questions is "Yes!" Each of the activities listed has a possible connection to biotechnology. These connections are briefly described below:

1. **Wash the dishes? Enzymes** may be used in detergent to break down protein stains from food on dishes. These enzymes can be made more efficient by using **genetic engineering** techniques. (Some are already found in laundry detergents.)

2. **Drink from a water fountain?** Harmful substances in drinking water might someday be removed using genetically engineered **microorganisms**. The water may have already been treated with biotechnological sanitation procedures before it is "reused" for drinking purposes.

3. **Glue a plastic model together?** Biotechnology can be used to improve the adhesive strength of some glues.

4. **Eat a steak?** Using new genetic engineering techniques, improvements on vaccines that fight cattle diseases are possible. You have more healthy cattle and more good steaks!

5. **Throw away trash?** Termites and similar animals have certain bacteria in their stomachs that allow them to digest paper or wood products. A **bacteria** found in the stomach of a cow may be genetically engineered to be similar in function. Then, cows could digest wood products. Cows would be able to eat the trash you throw out. This would help to solve a waste problem.

6. **Eat a cheeseburger?** The piece of cheese on your cheeseburger gets its smell, flavor, and texture as a result of bacteria. Cheese makers look to biotechnology for new strains of bacteria that will consistently give the characteristics they want. Some processed cheeses already contain enzymes. The meat in your cheeseburger may someday also be made by bacteria. This "meat" will taste just about the same and possibly give you better nutrition. Making the bun for your cheeseburger involves an old biotechnology process, where yeast is used to make the bread dough rise.

7. **Drink a soft drink?** Most soft drinks today are sweetened by a product of enzyme technology called **high fructose corn syrup (HFCS)**. HFCS is made by converting the **sucrose** molecules in cornstarch to fructose, using enzymes. Fructose is twice as sweet as sucrose, so it takes less to make your drink sweet. The enzymes that change the sucrose come from bacteria. Genetically engineering these bacteria can produce more and improved enzymes that reduce the cost of making fructose. Your soft drinks will cost less!

Did you get all the answers correct? Were you surprised how biotechnology plays a part in many of the things you do?

Biotechnology is currently being applied to such developments as (a) increasing the quality and output of food, both from plants and animals, (b) providing diagnostic tools for health care, (c) producing alternative energy sources, and (d) tackling some environmental cleanup tasks.

Biotechnology is defined as "any technique that uses living organisms, or parts of organisms, to make or modify products, to improve plants or animals, or to develop microorganisms for specific uses."

Because biotechnology can have such a large impact on our lives, it is important to understand it. Then we will be better equipped to use it in ways that are beneficial both now and in the future.

In this chapter, you will gain a better understanding of what biotechnology is and how the processes have changed over time. We will explore some traditional and modern technological systems that are based on an understanding of biological processes. We will study how technology is coupled with biology in the areas of agriculture, energy, health care, the environment, and bioprocessing. Decisions and consequences, along with rules that are needed to govern the safe use of these new skills, will be introduced.

Figure 11.2 ■
New, more effective medicines and health care products have come from the application of bio-related technologies to the medical field. (© Lonny Kalfus/Tony Stone Images)

STAGES OF DEVELOPMENT

Biotechnology applications were part of human history long before they were completely understood. People were limited to the tools that they had and the observations they could make. Thus, it was not until the development of more and more powerful microscopes that people could identify that there were plants and animals too small to be seen by the unaided eye. (These tiny organisms are called microorganisms; they can only be seen with microscopes.)

Once they could be seen, microorganisms could be identified and their by-products could be examined. Some microorganisms could be used in ways that were beneficial. People also found out what would stimulate or prevent the growth of these tiny beings.

Recently, tools have been developed that allow us to see into the structure of the cell itself. We can also manipulate and make changes on parts of extremely small organisms. We can use parts of the organism to change the rate and type of growth.

Stage One

Traditional biotechnology makes use of organisms, plant or animal, just the way they are found in nature, without altering them. Since early history, people noticed that changes occur over time in materials. Changes in smell, texture, flavor, and appearance of foods, fibers, and other materials were sometimes desirable. If so, people tried to repeat the process. Other changes made them sick or the materials less useful. Methods for avoiding these changes were developed.

In addition, people noticed that, from one generation to another, changes occurred. Although offspring of both plants and animals share many characteristics with their parents, there are differences. Sometimes, the differences are great. In other cases, the contributions of each parent are obvious.

■ **Figure 11.3**
These cattle have the right characteristics for high-quality beef. *(USDA photo by Fred Witte)*

Animals were bred with other animals in hopes that combining the characteristics of each would improve the offspring. Selection of seeds for the next season's planting was based on characteristics that were wanted in the next crop. Improvements were made that increased the yield, as well as the quality of the food, fiber, or other material.

Attempts to control spoilage and illness were based on the knowledge of the time. People did not know about **bacteria, yeasts, molds,** and **viruses**. They could see bugs, rodents, and weeds. For example, although people could not identify just what caused the Bubonic plague, they could see that there was less disease where there were fewer rats. Malaria comes from a parasite that is carried by the mosquito. Killing and preventing the growth of mosquitoes decreased the incidence of malaria. In both cases, the control of the disease meant controlling the animals that carried the microorganism. The most effective means of disease control, even now, is preventing exposure to the disease.

In the middle of the nineteenth century, Pasteur suggested that fermentation and some diseases were caused by tiny, **one-celled** animals. Lenses that were made at the time allowed some investigation of the growth of these **microbes**. Pasteur also showed that these microbes grow in colonies, from a single ancestor. This discovery led to much more precise control of fermentation processes. It also led to better control of some diseases. Food preservation techniques to kill the microbes led to canning.

Preservation of foods through drying, salting, pickling, canning, refrigeration, and freezing are based on knowledge of the conditions needed for the growth of organisms that cause food to spoil. Attempts are made to control moisture, oxygen, and temperature so as to retard their growth. Recognition of the growth patterns of organisms that are toxic to humans can be used in food processing, storage, preparation, and serving. (Refer to Figure 11.4.)

Food is often cooked to a high enough temperature to kill many of the microorganisms that cause spoilage and disease. Food-borne illnesses have always been a problem for people. Today, as people eat from food service systems or rely on processed food, we are dependent on those who cook and handle the food. Hopefully, they bring the food to a high enough temperature to kill or retard the growth of the bacteria. Hopefully, also, if the food is held hot for a long time, it is kept hot enough. (The same concern goes for keeping foods cold enough. Have you ever wondered whether to throw away that leftover that has been sitting in the back of the refrigerator for who knows how long?)

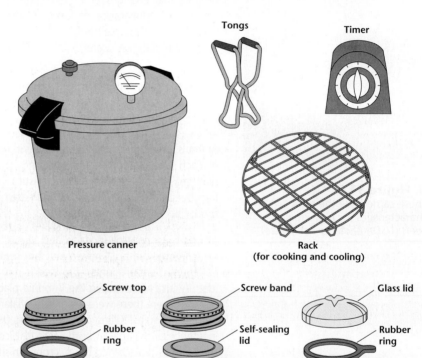

Figure 11.4 ■
The discovery of bacteria and its effects on the spoilage of foods led to several approaches for preserving foods. Special tools for home canning were developed so that bacteria could be destroyed and the food could be maintained in conditions that retarded or prevented further bacterial growth.

People have always tried to explain and treat illness. Early healers gathered herbs for various injuries, infections, and other maladies. Why some people were hurt, ill, or died was explained as their fate. (Current reflections of this belief are bad luck, bad karma, and the wrong sign.) Attempts to appease the gods were made and those who were sick were often sent away or isolated. Later, the explanation was bad blood, or humours. In this case, bloodletting and using leeches to suck out the bad blood were typical remedies.

After germs were discovered, procedures for the care of the sick began to change. Joseph Lister tried to show that germs (microorganisms) could be spread through contact with an infected person. He advocated that physicians wash their hands as they turn from treating one patient to another. He wanted hospitals to be

■ **Figure 11.5**
For centuries, people have deliberately altered their foods by encouraging the growth of microorganisms. The flavor and other characteristics depend on regulating the growth of those microbes that are wanted and preventing contamination from other damaging microbes. In the past few decades, processors have introduced the desired microbes and maintained growing conditions under tight controls in factory conditions. *(Middle: © Garry Gay/The Image Bank)*

cleaned. These ideas were met with resistance at first. However, improved sanitation and sterilization of instruments meant that far fewer patients died. Today, places where many sick people gather must take care to maintain sanitary conditions. Where the diseases are particularly dangerous, sterile conditions must be maintained and quarantines are imposed.

Human history is full of epidemics, from plagues to the current crisis with AIDS. Understanding of how disease organisms grow and are spread has led to improved prevention programs, sanitation systems, drugs, vaccinations, and attempts to control disease transmission. Traditional biotechnology has been increasingly effective against bacterial and yeast infections. Control of viruses required modern biotechnology, however.

At approximately the same time that microbes were first seen, Gregor Mendel wrote about his systematic experiments and observations of genetic transmission through **chromosomes**. He identified the principles of how characteristics of each parent are combined in the offspring. This knowledge improved the ability to predict and control the outcome of reproduction. These understandings made the breeding of plants and animals for specialized purposes more predictable. Mendel also identified that, every once in awhile, a very new characteristic is found. **Mutations** are these surprises that occur in reproduction. They are often called errors in reproduction; however, these errors are often quite useful. We will return to this concept later in this chapter.

Examples of very early biotechnology processes are found in **fermentation**, which was used to make bread, cheese, beer, and wine. The Babylonians made use of fermentation as far back as 6000 B.C., even though they had no idea how the process actually worked. By 3000 B.C., the Sumerians were using it to brew as many as 20 different types of beer.

Fermentation is a method of growing microorganisms that use complex organic molecules as their food. As they digest the food, they naturally break these molecules down into simpler substances that we can use. An example of this is when yeast cells convert sugar to carbon dioxide and alcohol in the absence of oxygen. This is the process used to make beer and wine.

Another yeast-fermentation process that you are familiar with requires that the carbon dioxide be trapped in a flour mixture to make bread. This traditional biotechnology, fermentation, relies on the natural biological processes to improve a product.

Early cheese making probably started when someone stored milk in a container made from the stomach of a calf or goat. Although they did not know it at the time, the stomach lining contains an **enzyme** that caused the milk to form curds and whey, rather than spoil. Someone liked the results and saw that the cheese did not make people sick. They repeated the process next time they had some milk that they wanted to preserve. Using a goat's stomach produced different results than a cow's stomach. Storing containers of cheese in a cave that had blue mold growing in it produced even different results. For centuries, people developed different flavors by introducing different molds into the cheese. However, they did not isolate the specific ingredients that produced the results until they had the tools to observe at the microscopic level.

Stage Two

Traditional biotechnology uses what is known about biological processes and materials as found in nature. Through these approaches, such as those just described, we have been able to control some organisms and natural processes. In addition, we have found and used some new organisms that improved agriculture, brewing, and the production of some **pharmaceutical** drugs. Within the last century, more and more microorganisms have been identified. There is better understanding of naturally occurring **toxins**, **allergens**, and **nutrients**. Enzymes and specific microorganisms can be isolated, selected, and screened for very specific use.

In several cases, production of the selected microbe was brought from the farm or individual enterprise to a factory. In this way, greater quantities could be made in more controlled circumstances. Naturally occurring mistakes could be controlled. Contaminants could be eliminated. Precise measures of ingredients, temperature, acidity, and other conditions could be managed. Manufacturing of enzymes and other microbes made them available in large quantities and in purer, easier-to-use forms.

Exploration throughout the world found even more microorganisms for development of desired products. Examples include stable molds and bacteria that grow in hot springs such as in Yellowstone National Park, other molds that can survive and grow in arctic temperatures, and microalgae that can grow in the Dead Sea. Even more surprising are the bacteria found in the thermal vents on the ocean floor. The fact that they can grow at such high temperatures and pressures makes them suited to uses where ordinary bacteria would be killed. Isolating and stimulating the growth of a wide variety of existing microorganisms is an ongoing research activity.

With improved techniques for production, natural processes could be extended. Parmesan and Romano cheeses could be made by adding the stomach of a goat kid to a vat of milk in a cheese factory. Other cheese was made by extracting and adding the enzymes from the fourth stomach of unweaned calves. (Note that producers still depended on the goats and calves for the enzymes.) Beer, wine, vitamins, soy sauce, vinegar, mushrooms, sprouts, citric acid, yogurt, and food gums are regularly made by traditional biotechnological processes.

During World War II, medicine was critically needed to treat the wounded and stop bacterial diseases. By this time, many of the principles that were useful for screening and selecting microorganisms were understood. It was found that a mold produced penicillin, a very good **antibiotic**. (An antibiotic kills bacterial growth and is an effective medicine.) What was needed was improved production methods to produce large volumes of the medicine. As you may recall, over time mutations occur naturally as plants and animals reproduce. The focus of the work in producing penicillin was to create desirable mutations. Naturally occurring "mutants" or "mutated strains" that grew faster and thus produced more penicillin were selected. Over 30 years, increased yield was achieved by stimulating mutation of the mold, using ultraviolet light and chemicals that caused mutation. The molds were developed that produced much greater quantities of penicillin than the original molds could make.

DESIGN ACTIVITIES

- Select a plant that is grown for food, fuel, or fiber. Read about the changes in varieties over the last few centuries. Identify ways in which the selection of seeds and grafting have been used in order to improve the desired characteristics of the variety.

- Attend a pet show or stage one for the school. Identify the different characteristics to use in judging the "Best of the Breed."

- Make yeast bread. With a microscope, observe and record the changes in the dough during the stages of fermentation and rising. Vary the conditions for growth and measure the differences in characteristics of the bread (E.g., add more sugar, fat, and/or other ingredients; use different types of flour; add more/no yeast; vary the temperature for rising; knead/only stir until ingredients are mixed; cover/leave uncovered). Develop a plan for ensuring the quality of the next bread that you bake.

- Make cheese/grow mushrooms. Visit or write to a cheese manufacturing/ mushroom production company to identify the kinds of microbes (and their sources) currently used in production. Identify the conditions needed to ensure that contaminants from the air, utensils, ingredients, etc., are controlled.

Stage Three

In the early 1950s, the structure of **DNA** was identified. This discovery was essential to the beginning of what is called modern biotechnology. At this stage, people looked for ways to alter the internal structure and functions of the organism itself to produce something new. Over the next 20 years, many new processes were developed and applied to problems in several arenas. Technologists no longer had to rely on creating mutations or on searching out new varieties of plants and animals. They could engineer the changes that were desired.

Over the next few decades, modern biotechnology will begin to challenge some basic assumptions people make about nature and natural processes. It is important for you to have some basic understanding of this so you can make informed decisions.

The expectations for modern biotechnology are very high. There is hope that it can provide the tools needed to fix many of our current environmental problems. Polluted water and air, depleted natural resources, and an ever-growing amount of garbage may be addressed. Modern biotechnology may also be able to improve health care, provide new sources of energy, and agriculture methods to solve world food problems. These hopes stem from the realization that the new skills in biotechnology can produce a wide variety of materials. This might include everything from liquid and gaseous fuels, plastics, and dyes, to antibiotics, synthetic growth hormones, and drugs to treat burns and strokes.

Yet, the majority of the work being done in biotechnology today is an extension of natural processes. What has brought about the drastic changes are the new techniques for manipulating the genetic material inside cells. These new techniques resulted from a greater understanding of the processes that take place inside the cell. It is the understanding of cell structure and function that serves as the basis of the biotechnology industry. Some attempts are made to speed up the natural life cycle. Other efforts will produce brand new varieties of life.

CELLS AND GENETIC ENGINEERING

Cells are the basic building blocks of plants and animals. All living things are made up of one or more cells. The average human body has more than one hundred billion individual cells, all living and working together as one unit. There are liver cells, heart cells, bone cells, skin cells, cells that sense light, cells that sense odors, and many others. Cells that make up multicellular organisms, such as humans, all depend on each other for their survival.

As indicated earlier, there are also other complete organisms made up of only a single cell, called microbes. These tiny beings do not depend on other cells for their survival. Each cell is able to perform all the functions necessary for its life and reproduction. Most important to the biotechnology field is that these kinds of single-cell organisms can reproduce at a very rapid rate. Much of modern biotechnology is based on the use of this type of cell. A model of a cell is shown in Figure 11.6.

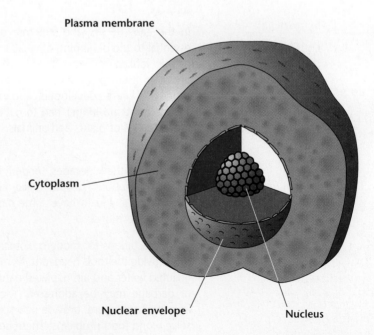

Figure 11.6 ■
The typical human cell, microscopic in size, operates like a small factory that converts materials, energy, and information into new forms. Information is stored in the genetic material (DNA) of the nucleus. Within the cytoplasm, small biological "mechanisms" convert selected materials into energy and other materials into useful components for the cell and the human.

Plasma membrane

Cytoplasm

Nuclear envelope

Nucleus

Cells are tiny, living units that make up all plants and animals. Each is a little package of life with a membrane that keeps all the contents inside. A very important component is the nucleus, a hollow shell containing many strings of a chemical called deoxyribonucleic acid (DNA). Cells reproduce and replace

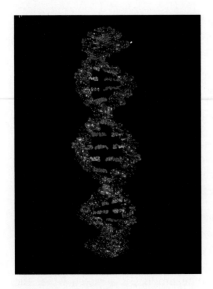

■ **Figure 11.7A**
The discovery of deoxyribonucleic acid (DNA) led to the development of modern biotechnology. Segments of DNA, the genes, include specific information about how the cell should function and reproduce. (© *Douglas Struthers/ Tony Stone Images*)

themselves by dividing. When a cell divides, the strings of DNA bunch together into **chromosomes**, which then make identical copies of themselves. The chromosomes pass on vital information to the new cells. The DNA strings contain many short segments known as **genes**. Genes are the sections of DNA that carry the information cells need to make an enzyme or protein.

In humans and other complex organisms, cells are highly specialized to form different body parts and functions. In bacteria and other single-cell organisms, genes are contained on a single genome, or circular strand of DNA. (Refer to Figure 11.7A.) During reproduction of complex organisms, each parent contributes one-half of the genetic material for the new offspring. The next generation takes on characteristics that are a combination of each parent. For single-cell organisms, all of the genetic material for the DNA is duplicated, and before the cell divides, a complete, identical strand is passed on to the new daughter cell. Therefore, all offspring are identical to the parent cell.

The enzymes and proteins manufactured by cells are the products biotechnologists often produce in large quantities. Modern biotechnology also genetically alters microorganisms so they can be used to produce very specific products, or perform very specific services.

Genetically altering an organism involves moving the genes from one organism into the cell of another organism. This process is known as genetic engineering. Modern biotechnology began with the development of genetic engineering, which became possible only by learning how cells store and read genetic information.

Figure 11.7B ■
Using a "cut and paste" technique, bonds between the segments of DNA are weakened. The segments are altered or a piece from another strand of DNA is selected. The new segment is inserted and fixed in place. Specific enzymes are the tools for weakening and strengthening the bonds. The new combination is called a recombinant DNA (rDNA).

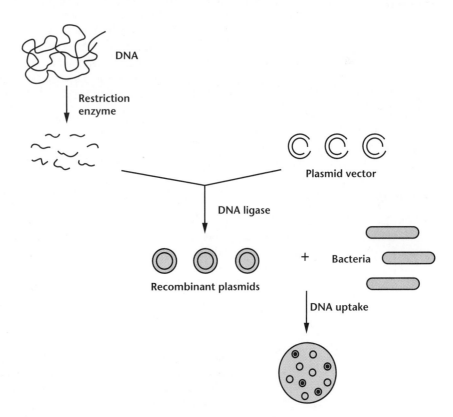

■ Figure 11.8
Products that have been created through the use of rDNA include anticancer and antiviral drugs, foods and fibers, and insecticides and herbicides that are safer for the environment. (© Barry Bomzer/ Tony Stone Images)

A piece of DNA containing the desired gene is taken from one organism and inserted into the DNA of a different organism (usually a bacterium). The organisms used for genetic engineering can come from humans, plants, and animals. The new strand of DNA that was made by combining DNA from both organisms is called **recombinant DNA**. (Refer to Figure 11.7B.) As this name implies, recombinant DNA (rDNA) is made by recombining pieces of DNA. The organism containing the rDNA now has new genetic information. The next generations will also exhibit characteristics of the new gene.

Bacteria resulting from further growth are called **clones** because each cell has identical DNA in its genome. This is desired by the technologist who wishes that each cell produce the same enzyme or product.

Seldom are we contented with just one new organism. Genetically-engineered bacteria are often placed in special growing tanks known as **bioreactors** or fermentors. There they can multiply at a very rapid pace and produce the needed product. Sometimes the product is a large quantity of the new organism that can be used itself for a given purpose. Often, the product is a material that the organism makes as part of its life cycle of using nutrients and eliminating wastes.

Bioreactors are huge tanks where growing conditions such as temperature, nutrients, pH level, etc., are carefully maintained. Actually, the real factories in a biotechnology production facility are the cells themselves. Bioreactors are designed to create the best conditions for growing genetically engineered cells. The design must be based on how cells grow, divide, and pass on the traits stored in the genes of their DNA.

Currently, only relatively simple organisms can be altered through the rDNA process. Cells with a short life span tend to reproduce rapidly. Thus, changes can be produced over many generations in a short period of time. Because the cell can reproduce without combining with another cell, it is possible to keep the changes without the contamination of other genetic material. We will discuss possible uses of these batches of new bacteria later.

Splicing Genes

The genetic makeup of living things has been traditionally altered by humans through a process known as **selective breeding**. Organisms exhibiting beneficial traits are selected and crossed to bring about offspring that display a blend of those traits. This blend of traits comes from the combining of genes from both parents in the chromosomes of the offspring. This is a faster way of doing what nature does through natural selection. But selecting and breeding organisms with desired characteristics still takes a long time. Genetic engineering, using rDNA, speeds up this process dramatically. Bottlenecks that limit the rate of change are bypassed by inserting the gene with the specific trait directly into the chromosome of another organism. **Gene splicing** first requires the isolation of the chromosomal DNA that has the desired gene. Then, a method for loosening the bond between the strands of DNA must be found.

Biotechnologists now use enzymes to separate out a gene and splice in a new one. These enzymes are called **restriction enzymes**. They weaken the bonds between specific strands of the DNA. (Think of a string of pearls, with each pearl as a DNA strand.) The restriction enzyme is used to separate the targeted portion from the rest of its DNA. This gene is moved and combined with the DNA from another organism. The bonds within the receiving organism have also been weakened so that it will accept the new strand. A second enzyme is then used to glue, or bond, the new combination into place. The result of this splice is that the foreign gene has now become part of the organism's genetic makeup. A slightly different organism has been created. The new organism is now capable of exhibiting the characteristics of the new gene.

Splicing genes is a quick and powerful way of improving the traits of plants and animals. For example, farmers in countries where it is dry and hot have trouble growing large potatoes. Their potatoes have the genes that allow them to survive in a harsh climate. However, they lack a gene that would make them grow large. Splicing the gene for "largeness" or more growth from another species of potato into the species that has a gene for "hot and dry," produces a potato that is slightly different. The new potato not only grows in the harsh climate, but grows large as well.

Interesting approaches have been devised to insert genes into plants. One technique is the particle-gun method. Metal particles are coated with selected DNA and shot into the plant. The cell walls heal up and the tiny particles remain in the cell. They are too small to interfere with natural cell functioning, but they give the plant the characteristics from the new DNA. Another method of introducing new genetic material is through using electric currents. The plant cell wall must be removed first, however.

The process of splicing genes, or transferring genes from one species or variety to another, has many interesting possibilities. The genetic makeup of organisms can be altered and new species of animals and plants can be created. These new techniques have provided a wide variety of benefits. Plants and animals that are healthier, with greater yield, may be developed. Organisms can be made that will digest waste products and help to clean up the environment.

Knowledge about DNA and genes has been expanded by a project called the **Human Genome Project**. The goal is to identify the function of each segment of human DNA, and to produce a map that will decode all of the information contained in the DNA. Such a map will increase the potential to repair or inactivate genes associated with handicapping conditions and potential disease. The location of factors associated with gender, race, and sexual orientation can also be located. Changes there might be ultimately possible. What will be considered a handicapping condition? Who will make the decisions? What are the advantages and disadvantages of diversity, even those which we often see as presenting problems?

Increased understanding of the role of **hormones** has led to the development of products that stimulate growth of muscle mass. Other hormone medications can be used to regulate ovulation, fertilization, and the implantation of fertilized ovum. Still other products can induce gender-related behaviors and appearance. Others impede the signs of aging. Basic questions about ethical and social consequences are and will continue to be debated.

DESIGN ACTIVITIES

- Read about the development of the engineered tomato that was planted in open fields in 1993. Identify the reasons for the development and the controls that were placed on the growth of the tomatoes.

- Find out the standards that are used by the United States Food and Drug Administration (FDA) in deciding the list of additives and ingredients that must go on the label of products consumed by people. (This will be the GRAS—generally accepted as safe—list.)

- Take a position and prepare a debate. Tomato soup companies should/should not be able to use engineered tomatoes in their soups without any additional warnings on their labels.

MATH/SCIENCE/TECHNOLOGY LINK

- Grow a tomato plant from a seed. "Engineer" the direction in which the plant grows by placing it in a box with a hole cut in the top or side. Record the length and direction of the growth daily. Chart the growth of the plant. Contrast this experiment with genetic engineering.

- Investigate the research on a specific food-related effort. Determine what results have been reported. Compile your notes for participation in class discussions.

BIOPROCESSING

Much of current biotechnology, whether traditional or modern, occurs in industry. Modern production systems use the method known as **bioprocessing**, applied to a variety of different microorganisms. Bioprocessing methods are used to make food, fuels, chemicals, fibers, medicine, and products for environmental management.

In each of these industries, the two major procedures used are fermentation and **bioconversion**. These two procedures require supporting technologies, such as fermenters, bioreactors, sensors, and electronic control mechanisms, to name just a few. The need for these supporting technologies creates the strong bond between biology and technology.

Fermentation

Growing microorganisms in a vessel such as a fermentation vat to get them to produce a chemical or pharmaceutical product is known as fermentation. Biotechnology uses microbes in this way to make many different chemicals. The microbes are fed and grown on specially prepared food called a **growth**

medium. The growth medium is usually prepared using a nutrient, or perhaps a waste chemical from other industries, mixed with water. A large container called a fermentor, or bioreactor, is used for the **growth chamber**.

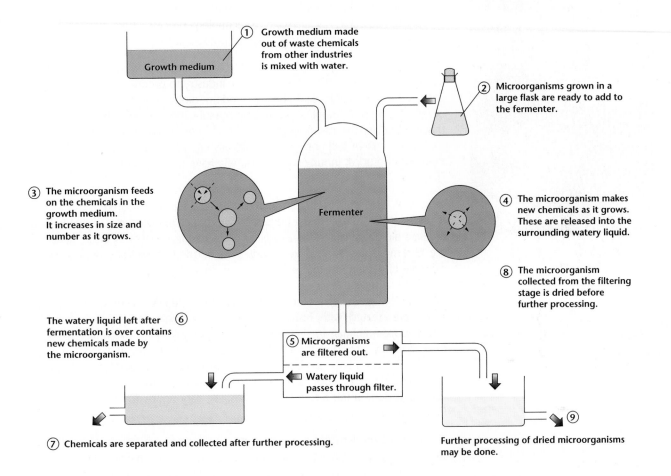

① Growth medium made out of waste chemicals from other industries is mixed with water.

② Microorganisms grown in a large flask are ready to add to the fermenter.

③ The microorganism feeds on the chemicals in the growth medium. It increases in size and number as it grows.

④ The microorganism makes new chemicals as it grows. These are released into the surrounding watery liquid.

⑧ The microorganism collected from the filtering stage is dried before further processing.

⑥ The watery liquid left after fermentation is over contains new chemicals made by the microorganism.

⑤ Microorganisms are filtered out.

Watery liquid passes through filter.

⑦ Chemicals are separated and collected after further processing.

Further processing of dried microorganisms may be done. ⑨

Figure 11.9 ■
Once the microbe has been altered to contain the characteristics that are desired, the problem becomes how to grow many of the new organisms. Fermentation vats and other types of bioprocessors control the growing conditions so that this reproduction occurs as rapidly as possible. In some cases, it is the new organism itself that is the product. Other times, the chemicals that are produced as a by-product of the organisms growth are the desired product.

As the microbes feed on the growth medium, they grow and rapidly increase in numbers. During this growing period, useful products are made and released by the microbes into the surrounding water. These products are collected by running the liquid through a number of filtration steps.

Fermentation has been traditionally used in the food industry for making beer and wine. Beer is produced from the fermentation of malted barley by yeast cells, which are single-cell, microscopic fungi.

Control mechanisms are extremely important for making sure the microorganisms grow well and produce the right product in the greatest amount. They are also used to detect if the fermentation solution becomes contaminated by unwanted microbes. These controls are generally electronic sensors built into the fermentor. The sensors monitor and regulate growing conditions such as temperature, growth medium, and the amount being produced. Industrial fermentors use computers to keep track of the information coming from the sensors. Operators can determine when to make adjustments in the fermentation process.

■ Figure 11.10
This fermentation tank is used for growing rDNA proteins for use in the pharmaceutical industry.
(Photo by Ted Horowitz/Courtesy of Schering-Plough Corporation)

Although industrial fermentors are quite sophisticated and expensive, simple fermentors can be constructed. Common items such as plastic soda bottles, aquarium tubing, a thermometer, and some yeast and sugar water are all that is needed. Using these materials, the basic control and operation of a fermentation process can be done. A small-scale biotechnology process where microorganisms are used to produce a product can be demonstrated.

The food and beverage industry is not the only industry in which fermentation is used. It is also used in the pharmaceutical industry to produce therapeutic drugs needed in the health field. Before modern biotechnology techniques were developed, most therapeutic drugs were manufactured using a chemical process that was costly. They could only produce the drugs in very small amounts. Today, recombinant DNA technology has drastically changed what can be produced in the pharmaceutical industry. Here, bacteria have been turned into microscopic "factories."

A successful example of modern biotechnological efforts is related to the disease diabetes. Different cells in the human body produce many chemicals that help to regulate biological processes, as well as protect against disease. Certain genes on the chromosomes control the production of these chemicals. Insulin is one example of a chemical that is made by animal cells. People with chromosomes that have a defective insulin gene cannot make insulin, which they need to metabolize sugar in their blood stream. Such people have diabetes. In the past, they were treated with insulin produced by animals. Insulin could only be obtained from animals in very small amounts. Some patients were allergic to it. Biotechnology provided a way to produce human insulin in large amounts, quickly. This was made possible by transferring the human gene for insulin into a bacterium called **E. coli**. Growing large amounts of insulin-producing E. coli bacteria is done using a fermentation process, very similar to the way it is done for brewing beer or making wine.

The tools of modern biotechnology are also focused on problems caused by viruses. Viruses, whether they are part of a plant or an animal are quite different from bacteria and yeasts. Viruses incorporate themselves into the DNA of the host organism and use the host's processes for their own reproduction. A virus can remain dormant in the host for some time and only show its presence after years of living in the host genes. Mutations can occur as the virus adapts to its host and learns to resist attacks by the host **immune** system or medication.

Figure 11.11 ■
Viruses present a special problem for biotechnology. The virus attaches itself to the DNA of the host. Thus, the host's natural immune system has trouble identifying the virus as alien.
(John Karapelou/© 1993 Discover Magazine)

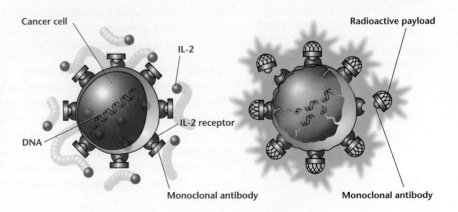

Have you ever wondered why the development of a **vaccine** or effective medicine for the common cold is so difficult? Or why finding a cure for AIDS or cancer is so elusive? Part of the problem is in how the virus works. If it is using the host's own genetic material for its mutations, any medicine is likely to kill host cells as it kills the virus. In addition, there are several viruses involved, as well as the series of mutations that each virus makes. A great deal of work is needed here.

To use this fermentation process, bacteria must first be genetically altered by splicing the gene for making human insulin into the DNA of the bacteria. The genetically engineered bacteria are grown in a fermentor that contains the correct nutrients and environmental conditions they need to grow and multiply. Over a relatively short period of time, the bacteria multiply into very large numbers. At that point, they are removed from the fermentor and broken open. The insulin proteins contained inside them are separated out and purified. The insulin that was made from human insulin is much easier for humans to use. Thus, it can be administered to diabetics who might have had allergic reactions before. Using this same fermentation process, many other important therapeutic chemicals and proteins can be produced.

DESIGN ACTIVITIES

- AIDS, cancer, and the common cold have eluded the pharmaceutical industry in coming up with a cure or effective treatment. In groups, select one of these, or another disease that currently poses a major challenge to medicine and health management. Trace the history of what has been discovered and treatments that have been tried. Write to or interview someone who is engaged in research and development. Identify barriers they face and successes they have encountered. Prepare a display for your school bulletin board that explains your findings, along with the most recent recommendations on how to protect oneself from exposure to the disease and how to keep the immune system healthy in order to avoid or delay the progress of the disease if exposed.

- Make high fructose sugar. Prepare a collage of labels from foods in which this product is used. Identify the costs/benefits of its use. Be sure to include nutritional, as well as economic, criteria.

- A current controversy is the use of Bovine Stimulator T hormone to increase milk production in dairy cows. Identify how the hormone works. Why are some farmers, as well as general citizens, so opposed to its recent approval by the FDA? Why are other people so in favor of its use?

- It is possible to increase the body's production of muscle through the use of growth hormones. Identify some of the concerns that have been expressed if the hormones are used in production of animals for food. Athletes who compete must undergo tests for evidence of the use of steroids for bodybuilding. Why? Research the effects of such use on the human body.

MATH/SCIENCE/TECHNOLOGY LINK

Conduct a study of the reported cases and mortality due to HIV-AIDS during the last decade in the United States. Chart these data and discuss, noting cases per one million people/year, gender difference of afflicted people, and other pertinent data. Project the rate of increase for 10, 20, and 30 years into the future, assuming that no more effective means for the prevention/cure are employed in the next 30 years than are currently being used.

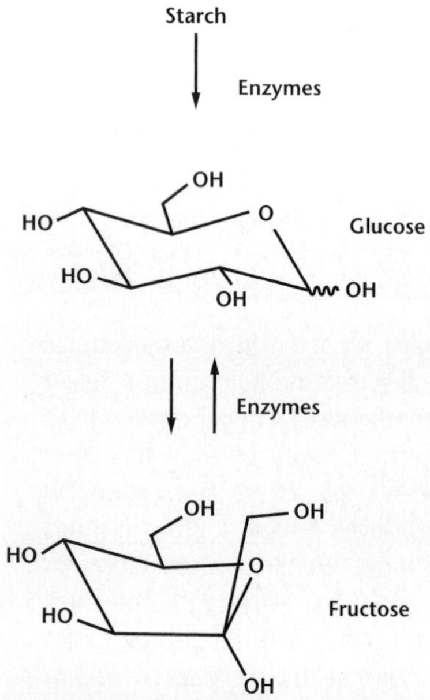

Starch

Enzymes

OH

HO

O

Glucose

HO

OH

OH

Enzymes

OH OH

HO

O

HO

Fructose

OH

■ Figure 11.12
High fructose corn syrup is a sweetener used in many foods. An enzyme is used to convert starch into the mixture of glucose and fructose.

Bioconversion

The other major component of bioprocessing is bioconversion. Unlike fermentation where a product is actually made by a cell, bioconversion uses enzymes or cells to cause a biological reaction to take place. The reaction occurs faster than it normally would. Enzyme molecules or cells used in bioconversion processes are known as **biocatalysts**. They speed up biological reactions, though they do not become part of the reaction itself. Such biocatalysts are used in industry. They are also found in a number of household items used every day, such as laundry detergents, meat tenderizers, and contact lens cleaners.

Enzymes can also be used to improve fuels for our cars. To do this, enzymes are first used to change starch into simple sugars. Yeasts are next used to change the sugars into alcohol, which can then be added to gasoline. This creates an improved fuel for cars.

In the health care field, enzymes are used to help diagnose patient illness. Test strips have been specially designed to hold enzymes in a small square of paper at the tip of the strip. Testing body fluids, such as urine, with these test strips can help to determine what might be making a patient ill or if a woman might be pregnant. Antibodies for certain illnesses can be used to detect the presence of disease-causing organisms, such as salmonella. This method increases the speed and accuracy with which dangerous foods and other products can be detected and taken off the market.

Immobilization. Bioprocessing methods address the problem of separating the product made by the microorganism from the medium in which it was grown. Typically, the separation of the organism and growth medium from the product requires a series of **filtration** and purification steps. These steps take a great deal of time and are expensive. They make use of the enzyme or cell only one time. This factor adds to the cost of manufacturing. An excellent technique found to reduce the separation process, as well as use the cell or enzyme repeatedly, is **immobilization**. This process is shown in Figure 11.13.

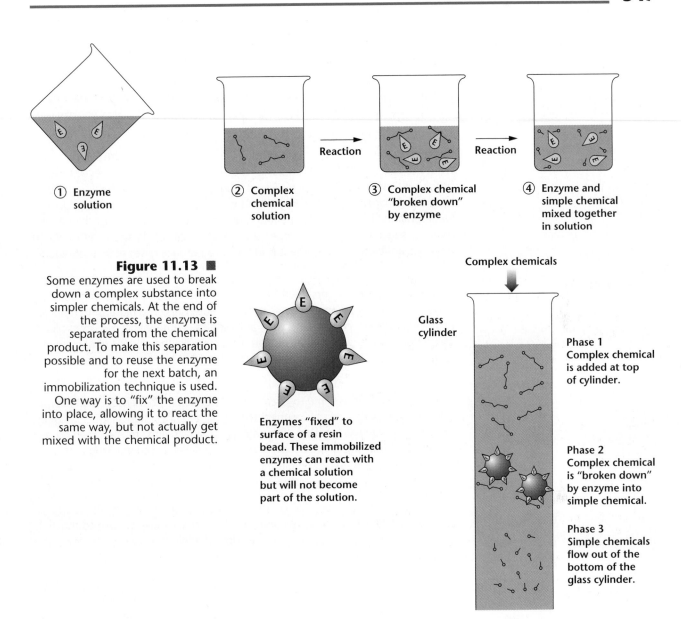

① Enzyme solution

② Complex chemical solution

Reaction

③ Complex chemical "broken down" by enzyme

Reaction

④ Enzyme and simple chemical mixed together in solution

Figure 11.13 ■

Some enzymes are used to break down a complex substance into simpler chemicals. At the end of the process, the enzyme is separated from the chemical product. To make this separation possible and to reuse the enzyme for the next batch, an immobilization technique is used. One way is to "fix" the enzyme into place, allowing it to react the same way, but not actually get mixed with the chemical product.

Enzymes "fixed" to surface of a resin bead. These immobilized enzymes can react with a chemical solution but will not become part of the solution.

Complex chemicals

Glass cylinder

Phase 1
Complex chemical is added at top of cylinder.

Phase 2
Complex chemical is "broken down" by enzyme into simple chemical.

Phase 3
Simple chemicals flow out of the bottom of the glass cylinder.

Simple chemicals

Immobilization can be done in three ways: **chemical bonding, adsorption, and entrapment**. Chemical bonding means that some enzymes can be linked chemically to other molecules. This method often damages enzymes. Immobilizing enzymes or cells physically can be done by adsorbing them onto the surface of different things. Adsorbing basically means that things stick to the surface of the material but do not combine with it. Enzymes and cells can be adsorbed onto such things as glass or resin beads, carbon particles, and even pieces of ceramic.

The easiest and most gentle immobilizing method is entrapment. This is where the cell or enzyme is held inside a porous, gelatin-like substance. Because the gelatin is porous, it allows the chemicals to reach the cell or enzyme inside the gelatin that is surrounding them. It also makes it possible to use the same cells or enzymes repeatedly. Remember, all this activity is going on at a level you can only see with a microscope!

In each of these methods, the goal is to "fix" the cell so that, although it mixes with the chemicals it is growing in or the product it is making, the cell remains insoluble. When the cell is in this "fixed" form, it is said to be immobilized. Yet, it can still function as if it were free. The use of immobilization techniques is widespread in industry. It can be found used in the food industry to change one type of sugar into a sweeter one (sucrose to fructose), in the health field to produce antibiotics, and in the fuel industry to produce gases such as butanol.

Immobilization techniques are also becoming very important for the development of biosensors. Research is attempting to find ways for immobilizing enzymes onto organic membranes to make biosensors. The biosensor combines a microelectronic component with a biological component. For example, a biosensor could measure many human processes, like the amount of glucose in a person's blood stream. This kind of information would be very useful to someone who is diabetic.

There are several problems to overcome in developing biosensors of this type. Mini-electrodes that can be coupled to specific enzymes must first be designed so they can work together. Biosensor research seems to be opening up possibilities that were once only possible in science fiction. Programs now underway in Japan are developing a "biomicrochip" and a "biocomputer" that have organic molecules taking the place of semiconductors.

The goal is to eventually place microbiosensors inside a person's body to constantly check their physical condition. The sensor must be able to function without becoming an actual part of the body. In addition, it cannot be viewed by the body as a foreign substance, or else your immune system would attack it.

Such a system might eventually be made to send a signal to a hospital computer. A physician could use the information to make a diagnosis. Wouldn't it be interesting if the physician could send a repair order to that microchip in your body, so that adjustments could be made?

Biosensors may also have applications in industry. Combining biosensors with machines could lead to designing robots that could actually see, hear, and touch. Today, ideas like this seem to be more dream than reality. But over the past 30 or 40 years, the dreams of many scientists have led to some amazing realities by engineers and biotechnologists.

Currently, some dreaded diseases have been all but eliminated. Smallpox killed thousands of people before a vaccination was developed. Ultimately, the disease was eradicated worldwide. There is a controversy now about whether the samples of the smallpox virus, stored for research purposes, should be destroyed. Or should they be kept as representatives of that life form?

Figure 11.14 ■
There is a growing recognition of the need to maintain genetic diversity. When biotechnologists want a certain characteristic, they may be able to find it in a rare variety or breed of plant or animal. Yet, many varieties and breeds, and sometimes even an entire species, die out because of natural or techno-logical outcomes. One attempt to guard against the loss of diversity has been the National Seed Storage Laboratory of the United States Department of Agriculture. There are also international efforts similar to this attempt to preserve the variety of life.

BIOTECHNOLOGY IN AGRICULTURE

Technological efforts to decrease death have been more readily used than efforts to decrease birth. Therefore, the world population continues to grow and the demand for resources increases. Improved agricultural techniques become more important if all people are to have necessary materials for food, clothing, shelter, and energy use. Farming methods, whether for food, fiber, or forest products, need to be made more efficient.

Plant and Animal Biotechnology

To increase the productivity of farm crops, biotechnologists look for ways that help reduce the risks and dangers of farming. One of the largest sources of farming risks involves the weather. Sudden drops in the temperature often do great damage to young crops early in the spring growing season by freezing their leaves. Frost can kill all plants, especially young ones.

Recently, bacteria on the leaves of some plants were found to have a special protein in their coat that causes ice crystals to form. To prevent frost from forming on the young plants, biotechnologists used genetic engineering to remove the specific gene in this bacteria. Without this gene, the bacteria, nicknamed "ice-minus," was no longer able to form ice crystals on the leaves of the plants. Spraying these genetically altered bacteria onto young plants provides a "shield." No ice crystals form on the plants when the temperature drops slightly below freezing. This altered bacteria makes it possible to protect young crops and prevent this source of financial disaster for farmers.

In addition to creating new bacteria that prevent plants from freezing, genetic engineering can also help reduce the damage done to crops by insects. Research is underway to develop a bacteria that will grow on corn and other similar farm

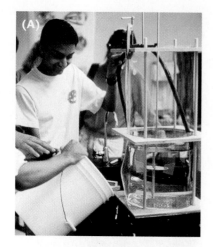

■ **Figure 11.15A, B, C**
(A) This student is conducting field tests of a bioreactor he and his fellow students have designed and developed. (B) This algae pool was constructed of simple modeling materials. (C) Students make alginate beads for use in testing an impeller system. *(A & C: Courtesy of TIES Magazine, B: Courtesy of John Wells)*

crops to produce a natural insecticide. This bacteria is not harmful to animals or humans. However, the natural insecticide the bacteria makes will kill insects such as cutworms which eat and destroy the crops. By using this genetically altered organism, farmers can greatly reduce the amount of chemical **pesticides** they normally use to control insect pests that feed on their crops. Fewer chemicals used on the crops reduces the damage to our environment.

Another way to control pests without the use of poisons and insecticides involves an understanding of the reproductive cycle. Males of the species that are harmful to crops and humans are captured. After being sterilized, so that they cannot fertilize eggs, they are released into the environment. If the sterile male mates with the female, there will be no offspring. The population of the unwanted animals will be decreased. The amount of chemical needed to control the effects of these pests is less.

Genetically engineered bacteria may further help to reduce the amount of chemicals needed to grow crops. Fertilizers are typically added to the soil to give plants the nutrients they need to grow. One of the most important elements in fertilizer is nitrogen. Adding nitrogen to the soil provides plants with a ready supply of an important chemical they need in order to grow. But only a small portion of the nitrogen actually gets used. The rest often gets carried away in the runoff that occurs when it rains.

This nitrogen rich water causes problems when it finally reaches a stream or pond. The excess nitrogen causes **algae** to grow so fast that they cover the surface of the water. When the surface of the stream or pond is covered, they block the sunlight. Without the sunlight, the plants at the bottom cannot grow. The algae's natural enemy, bacteria, invades but must compete for the short supply of oxygen. Also, fish eat them since the plants they usually eat are in short supply. The decay of the plants also uses up the oxygen in the water. Eventually, fish die along with other organisms that live in the stream or pond. Have you ever seen or smelled a stagnant pond? The smell is the decay of plants and fish that are killed.

One way of changing this process would be to introduce more of the bacteria that naturally eat the algae for food. Another is to decrease the need for fertilizers. Some microbes are capable of changing the nitrogen in the air into fertilizer in the soil for plants. Biotechnology has given scientists the tools needed to develop a bacteria that lives in the roots of plants and collects nitrogen from the air. This nitrogen is available for use by the plant. The bacteria are like tiny natural fertilizer factories. This eliminates the need for adding chemically manufactured fertilizer to the soil. Using biotechnology in this way helps to keep the environment healthy and safe.

Several development efforts are currently focused on identifying the genes that will help to improve plant crop production. The next step is finding a way to splice these genes into the plant's own DNA. This use of biotechnology gives farmers new varieties of plants and animals. Transferring traits between plants and animals normally took years using traditional crossbreeding methods. But so far, only a small number of genes have been successfully transferred into a limited variety of plants. Yet, there is great expectation for the application of biotechnology in plant agriculture. A small sample of the improvements biotechnologists are currently looking to develop include the following:

- Growth, even in extreme heat or cold, without damage

- Growth where there is too much or too little water, salt, and acid

- Crops for animal food that have high nutritional value and are easy to digest

- Immunity and resistance to certain diseases, pests, and weeds

- Extended growing seasons, so that harvest can occur over a longer period

- Controlled ripening, so the produce can be stored longer

- Stronger walls on fragile, watery fruits and vegetables, thus allowing shipping with less damage

- Cheaper methods for producing flavorings and ingredients that are currently very expensive

- Changes in the type of fats and oils produced

- Increased yield of the produce

- Improved amount and quality of nutrients

- Produce antibiotics which kill pathogenic fungi

■ **Figure 11.16**
Biotechnological solutions have been developed to resist disease and improve the yield and the crop's ability to withstand mechanical handling in harvest and transport. Tomatoes have been altered to have thicker skins that resist bruising. Rather than picking tomatoes green and artificially ripening them when they reach the market, growers can now allow them to ripen longer on the vine. Improved taste and nutritional value are the result.

The application of biotechnology methods can also be used to control animal diseases. Using gene-splicing techniques similar to those applied to plants, bacteria can again be used like tiny factories. This time they produce vaccines and antibodies that protect animals against diseases such as scours (a diarrhea-causing disease that kills young farm animals) and hoof-and-mouth disease. The use of genetic engineering has also led to improvements in increasing animal productivity.

Animal productivity refers to such things as growing larger farm animals in shorter periods of time. It may include increasing the amount of wool or milk an animal makes and the general ability of the farm animal to produce useful products. The productivity of an animal is tied to certain substances in the body, known as hormones, that can regulate growth. Hormones are substances naturally produced by animals, but generally in very small amounts. Increasing the amount of hormones produced would improve the productivity of the farm animal. Bacteria can be made to produce growth hormones that are normally made by an animal. Hormones produced this way are called **synthetic hormones** and can be given to animals through injection or in their feed.

Increasing the production of growth hormones can also be done by genetically engineering the genes of the farm animal itself. This is done by splicing the specific genes that cause extra growth hormone production into the cells of growing animals. With this gene inserted, the animal now makes more growth hormone naturally. Research to improve the productivity of farm animals by means of growth hormones and vaccines continues at an increased pace today. This may provide solutions to future food production problems.

It is easy to see how genetic engineering can be a very powerful tool. It gives the biotechnologist a way to make a small change in an animal or plant in order to give it a new beneficial trait. Once that change has been made, the next step is to grow the altered animal or plant to produce large numbers of the new strain or crop. An additional disadvantage of traditional techniques such as grafting (see Figure 11.17A) is that the effects are not passed on to the next generation. What is needed are identical copies of the altered animal or plant.

Bud Grafting

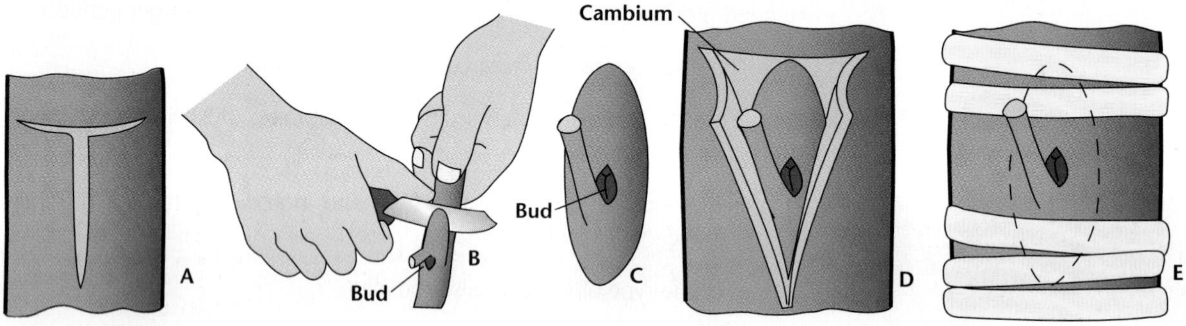

■ Figure 11.17A
Methods for combining the characteristics of one plant with another have traditionally been accomplished through grafting. A twig with a bud that will produce new growth is attached under the bark of a plant that is already rooted and growing. The bud is held into place until it attaches to the rooted plant and the two grow together.

Tissue Culturing and Cloning

For plants, identical copies can be made using a technique known as **tissue culturing.** This technique allows the biotechnologist to grow an entire plant by using as few as one cell from the original plant. The one cell taken from the parent plant can be made to grow in a small test tube on a special food mixture called a **nutrient agar**. A small lump of cells, called a **callus**, soon forms on the nutrient agar. A specific mixture of nutrients produces certain types of growth. The callus can then be cut into many small sections and each piece placed in another test tube with nutrient agar. Each piece of callus will then grow into an identical plant, or clone, of the original parent plant. The callus can be made to grow into a small plant with leaves, stem, and young root system by changing the nutrient mixture of the agar they are feeding on. Each of the plants grown in this way came from the one cell of the original plant. Therefore, all are exactly identical to the original plant. The process of tissue culturing provides a way to make exact copies of valuable crop plants in large numbers.

The purpose of cloning and culturing plant cells is basically the same as those of traditional breeding approaches. We want to increase the crop yield, improve the beneficial traits, and reduce the labor and production costs. A great deal of work still remains to be done in perfecting the technique of tissue culturing. (Refer to Figure 11.17B.) Improvements must also still be made in the methods used for cloning major crops such as soybeans, maize, and wheat. In addition, better ways need to be developed for finding and removing from plants the genes controlling the undesired characteristics. However, in spite of the work yet to be done, the combination of genetic engineering and tissue culturing can be expected to have a great impact on crops over the next decade.

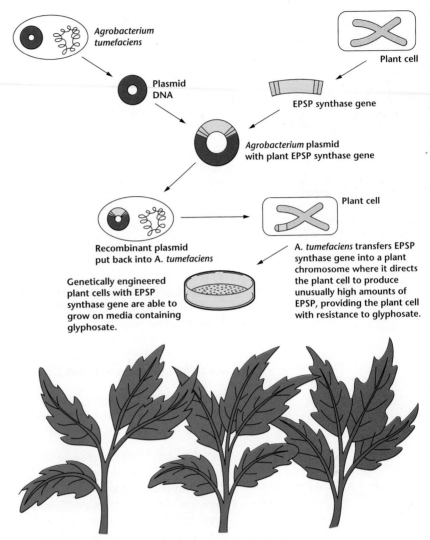

Agrobacterium
tumefaciens

Plant cell

Plasmid
DNA

EPSP synthase gene

Agrobacterium plasmid
with plant EPSP synthase gene

Plant cell

Recombinant plasmid
put back into A. tumefaciens

A. tumefaciens transfers EPSP
synthase gene into a plant
chromosome where it directs
the plant cell to produce
unusually high amounts of
EPSP, providing the plant cell
with resistance to glyphosate.

Genetically engineered
plant cells with EPSP
synthase gene are able to
grow on media containing
glyphosate.

Figure 11.17B ■
Tissue culture can now stimulate
the growth of a whole plant from
a single cell of the plant. The
advantage of this process is that all
of the plants that are grown from
this cell have exactly the same
characteristics.

Whole plants are regenerated from the single cells and are able
to withstand spraying with glyphosate.

Food from Microscopic Plants or Animals

In an ever-growing world population, biotechnology professionals strive to provide enough necessary food and other materials. In the not-too-distant future, microorganisms themselves might well be a major source of nourishment. The thought of eating microbes may seem strange, but in fact we already do. There are many microbes in such foods as yeast extract (Marmite), cheese, and yogurt. Look on the label to see whether "active cultures" is listed as an ingredient.

Eating microorganisms is not new. In fact, cheeses have been made and eaten for many hundreds of years. There are even records from the early 1500s of people in Mexico making small cakes with a cheese taste from the slime found in lakes. This was probably algae known as *Spirulina maxima*. There is also a tribe in Africa, the Kenembu tribe, which eats a microbe called *Spirulina platensis*. These

■ Figure 11.18
Cheese has been made for centuries, using bio-related technologies. Currently, however, cheese is being changed again through biotechnology. In this process, the fat in the cheese is replaced with a cellulose-based ingredient. *(Courtesy of FMC Corporation)*

microalgae are often high in protein, have cholesterol-lowering effects, and are easily digested.

The Japanese currently market products called "functional foods." These foods are designed to include many valuable nutrients. They also include helpful bacteria and other microorganisms that will live in the human intestine and aid in digestion and use of nutrients.

Protein produced by bacteria, known as **single cell proteins (SCPs)**, was first used on a large scale as a food for human consumption in Germany during World War I. During the war, the large shortage of food caused scientists to work toward the development of a process to grow microbes because of their high protein content.

Today in many parts of the world, there is still a shortage of food and people are starving. It is very important that foods high in protein are eaten because they are needed for growing body tissues. Using microbes to make food could help reduce world hunger. SCP is a valuable food because it has a very high protein content and contains vitamins and other essential chemicals the body needs. Another advantage in using SCPs as food is that they can be grown quickly. They do not have the same large growth requirements needed for raising farm animals. The ability to grow them quickly makes them cheaper to make and less expensive to buy. Much of the SCPs are used to make animal feeds, but more and more they are being used to produce foods that look and taste like meat for people to eat.

In the chapter on Materials, we indicated that the human need for nutrients can be satisfied by a wide range of sources. No one said the animal had to be big! Another food source is a tiny plant—mold.

Certain molds contain high levels of protein and can also be used as a food source instead of the usual animal or plant crops. The protein made by these molds is called **mycoprotein**, which can be produced by having the microorganism grow on cheap waste materials such as wastepaper, straw, or molasses. After growing for a time on the waste materials, the microorganism is filtered off and put through a drying process. The dried microbes can then be used as food. This type of food can be made to taste so much like steak, meatloaf, or vegetables that you really cannot tell the difference. (Have you ever accused the cook of making a steak taste like a paper plate? Well, this is just the opposite!)

Most of the biotechnology processes you have read about so far are aimed at producing a specific product. But some of the biotechnological processes developed to control the pollution in the environment work in the opposite way. They are designed to destroy the specific raw materials of pollution.

The ecology of the Earth is a complex system designed to break down natural wastes and to reuse them. It is the natural recycling process of the planet. But the new kinds of pollution produced by modern industry differ greatly from those of nature. In addition, in parts of the world where many people live close together, they produce waste in quantities too great to be naturally processed. The capabilities of the ecosystem are severely strained and cannot keep up. Cleaning up previous damage, as well as preventing future ecological harm, presents quite a challenge to technology.

One approach under development for controlling waste is to use or develop special microbes that will "eat" unwanted waste. Some microbes even produce something useful during the process. Actually, this is being done today with microbes that can produce methane gas while breaking down household sewage. Nearly every city and town in the developed world has invested in the type of biotechnology known as sewage processing.

Clean water is something we all need to remain healthy. Our cities and towns pour household wastes into the sewer system daily. The sewer water must be purified before it can be put into rivers or be reused. Exposing wastes to sunlight, heat, and air cannot kill all of the harmful organisms. Most sewage purification processes depend on the work of tiny microbes to break down domestic and agricultural waste products. The first part of the process uses a **sedimentation** tank to allow suspended solids to settle out. Liquid on top of the solids is drawn off and transferred to an aeration tank where microorganisms grow and feed off the waste in the water. Afterwards, this water can be placed into a river or stream.

The solids, however, must be passed on to a **sludge digester**. Here an assortment of microorganisms are introduced to break down the sludge into a form that can be used as a fertilizer. How each microbe actually functions is still not fully understood. So far, biotechnologists are content not to try and improve the method through some form of genetic manipulation.

A second way of reducing pollution in the environment is to use biotechnology as part of the production process itself. An example would be using microbes that feed on sugar to produce the chemicals needed to make plastics. This process is far less hazardous to the environment than making plastic products the traditional way, using oil-based raw materials. The oil-based raw materials must be piped or shipped to the plastic factory. Transporting increases the possibility that oil might be spilled or leaked into the environment and cause damage to wildlife and beaches. Additionally, starches can be added to plastic so that it will biodegrade more quickly than traditionally made plastics.

The use of biotechnology to reduce the effects of oil spills and other similar pollution emergencies is quite valuable. In some cases, nutrients are added to the oil-contaminated water or soil to enhance the ability of naturally occurring bacteria to degrade the oil. Another method makes use of biological materials that will soak up or absorb a wide range of unwanted pollutants. These types of materials are known as **biosorbants.**

A good example is the material chitosan, which is made from the shells of crustaceans such as shrimp or prawns. The hard outer shells are made of a substance called chitin that has recently been found to have some useful **absorption** properties. Oil and similar substances cling tightly to chitin when they are mixed. Because of this, systems can be designed that use chitin to remove pollutants, such as oil, that have washed into drainage culverts or similar runoff areas. Chitosan is extremely plentiful in many parts of the world. In fact, people have been looking for a use for this cheap material.

Biotechnologists are hopeful that the ideas being developed today on waste management will go much further in time. It is important that efforts continue toward identifying those microorganisms that can naturally break down

■ **Figure 11.19**
Absorbant chitosan from the shells of crustaceans may prove to be very useful in reducing the effects of oil spills and other similar pollution emergencies.

pollutants and industrial wastes. In the future, new microorganisms can be genetically designed to remove various pollutants. The more hazardous pollutants, such as pesticides and herbicides, remain in the environment for long periods of time and are a challenge to modern biotechnology.

BIOETHICS

Modern biotechnology is considered a technological revolution because of its immediate and profound effects on society. In the past ten years alone, biotechnology has had an enormous impact in the areas of agriculture, medicine, and environmental control. The advances in biotechnology clearly hold great promise for improvements in these areas. Serious questions concerning its regulation and control arise, however. The benefits of this new technology will depend largely on its effectiveness, cost efficiency, and the assurance of health and safety for people and the environment we share with other living beings. **Bioethics** is concerned with understanding all the legal, social, cultural, ethical, and economic implications of conducting biotechnology research.

Genetic engineering has given the biotechnologists the means to change certain specific traits of microorganisms, plants, and animals. Those microorganisms that produce new antibiotics or proteins have been patented for years. In 1987, the United States Patent Office approved the patenting of modified or changed, higher, more complex life forms created by scientists in the laboratory. **Patents** give the inventor (or the company for which he/she works) the right to sell their new design or to sell the rights of the new design to someone else. Patents also prevent others from using the invention for 17 years in exchange for not keeping the work secret.

The approval to patent new life forms has brought about a great deal of controversy over ethical issues and environmental concerns. Questions are being raised about whether it is wise to allow individuals to own new or modified life forms. Answers to these questions are not easily found. And even though the new patent law does not apply to humans, it does apply to a variety of other organisms such as genetically engineered cows, pigs, mice, plants, and bacteria. Recently, however, there has also been discussion about whether information from the Human Genome Project can be patented or registered.

In the chapter on Physical Systems, we explored how many years ago there was controversy over mining techniques. When people began to do deep shaft mining, there was a great deal of resistance and controversy about whether it was right to violate the Earth. Think of the potential for controversy as we begin to use the tools to change the processes by which life and death occur! The issue of ownership is only a part of the greater issue of creating new life forms and altering the process by which individual traits are determined. We will return to these and similar questions in the later chapter on Consequences and Decisions.

Other areas of biotechnology need monitoring and control to ensure the safe and appropriate use of this new technology. Gene therapy, DNA fingerprinting for forensic purposes, vaccines from recombinant DNA, and the Human Genome

Project all require regulation. Research is currently being done to find the cause and possible treatment for many genetic disorders that affect millions of people. Also, mutations have produced widely diverse plants and animals. Which varieties will people a few centuries from now consider valuable? What happens when a genetically engineered organism is released into the environment?

Microorganisms are being genetically altered to clean up oil spills and help solve waste problems by breaking down toxic wastes and garbage. Altered microorganisms are also being used to make hormones and proteins that are naturally produced, like antibodies and growth hormones, in large quantities for use in the health field.

Biotechnology will affect everyone in some way, from the food they eat each day to the types of jobs that are available to them. A good basic understanding of this new technology will lead to a more practical assessment of what future benefits are possible.

Methods of regulation are being prepared by a number of government agencies. The Department of Agriculture, Environmental Protection Agency, National Institutes of Health, and Food and Drug Administration are some of the agencies in the United States that are responsible for monitoring new products. Many nations have comparable agencies and there are some worldwide "watchdog" groups.

There is overall agreement among biotechnologists and regulatory agencies that regulation of biotechnology research and development is needed. The methods used to regulate this type of research are costly because a number of regulatory agencies are involved in making policy. Consumer groups are also working to ensure that critical information is available in a form that people can use for decision-making. The key to making correct decisions about biotechnology issues is educating people so they can differentiate between fact and fear. People also need to question the source of information and the intent of the sender. Raising the scientific literacy and social awareness of people helps them to make sound decisions on matters concerning ethics and public policy.

SUMMARY
CHAPTER 11

Biotechnology is the use of living organisms, or some part of them, to make a product or provide a service. It has been a part of human life for thousands of years. The earliest uses centered around the making of bread and fermented products such as beer and wine. In these processes, yeast and bacteria were used to make a product, even though the people using these organisms did not actually know how the process worked. In addition, the organisms were used just as they were found in nature, with no alteration to their genetic make up. These earlier methods are known as traditional biotechnology.

Later, more was understood about microscopic plants and animals. The production of the plants and animals, regardless of size, became more efficient as more precise control over the circumstances for their growth was possible. Some of this production was moved from the farm to the factory for large-scale production.

Recently, scientists have learned about how genetic information is stored and copied, and how to change that genetic information. This new knowledge, combined with the ability to alter the genetic information, has enabled the biotechnologist to change the "natural" abilities of an organism. This is known as modern biotechnology.

Altering the genetic material of an organism requires an understand-

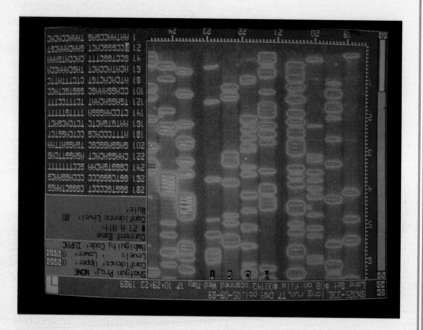

■ **Figure 11.20**
This computer screen displays DNA sequencing information.
(© Peter Menzel/Stock, Boston)

ing of the processes that take place inside the cell. Discovery of DNA, specifically the genes, led to developing the tools to look for and make use of certain genes. This allowed the transfer of genes between organisms, a technique called genetic engineering. Genetic engineering provides a way to quickly do what crossbreeding would take years to accomplish.

Gene splicing makes possible the design of plants and animals that can be grown or raised under a wider range of conditions. Throughout the world, hunger could be reduced and foods could be created with better nutrition, taste, and shipping qualities.

Growing microorganisms to produce a product or perform a service is known as bioprocessing. Industrial bioprocessing methods involve fermentation and bioconversion procedures. Fermentation involves growing vast numbers of microbes in a bioreactor, a large environmentally controlled chamber. Bioconversion uses enzymes or whole cells to make a biological reaction occur at a faster rate.

Biotechnology offers help in overcoming some of the many weather- and health-related risks. Bacteria have been used to shield plants or assist them in naturally acquiring nutrients. Research in this area of biotechnology has led to a reduced amount of chemicals needed for raising crops. Similar research has also led to improvements for increasing animal productivity. Today microbes can even be harnessed to grow valuable foods that are inexpensive to produce and high in protein. Single-cell proteins are one of these foods that can be made quickly using microorganisms and then processed into food products.

Aside from producing products, naturally occurring microbes or genetically altered ones are being put to use in protecting and cleaning up the environment. This part of biotechnology will most certainly become very important to humankind in the future.

Technology solves problems and also creates them. The increased understanding and use of biotechnology has great potential for

making life better, but involves an equally large amount of risk. It is very important that all people who will be affected by new breakthroughs in biotechnology, be knowledgeable about this new technology. Many social and ethical issues will be raised, but responsible decisions can only be reached through an informed public.

ENRICHMENT ACTIVITIES

Prepare a display showing products that have been developed for use in growing plants and animals. Display in the following categories:

1. Resistance to damage, even in conditions of extreme heat or cold

2. Growth in conditions where there are high (or low) levels of acid, salt, and water

3. Resistance to certain diseases, pests, or weeds

4. Controlled ripening

5. Better yield

6. Increased tolerance for rough handling by harvesting and processing machines

7. Cheaper methods of production

8. Improved nutritional value

REVIEW AND ASSESSMENT

- What is meant by the term "bio-related technologies"? Identify some technological systems that have been affected by this effort.

- Distinguish between Stages 1, 2, and 3 of biotechnological development.

- How did the development of the microscope influence the development of biotechnology?

- What is DNA?

- What is rDNA and how is it done? Name some purposes for its use.

- What is the role of the bioreactor or fermenter in modern biotechnology?

- Describe some controversies unique to the biotechnology field. Why are there such divergent, emotionally charged views on what is ethical and safe?

- Identify three ways for immobilization of an enzyme or cell in production.

- Why might manufacturers be so concerned with immobilizing the enzyme?

- Name three ways in which the results of biotechnology have decreased the amount of economic risk for farmers.

- Identify how biotechnological approaches might be used to address some current environmental concerns.

- What is cloning or tissue culture? What are its advantages over ordinary reproduction processes?

CHAPTER

12

Communication Systems

Figure 12.1 ■
Fiber optics is vastly improving our capacity to communicate quickly and effectively. (© *Dominique Sarraute/The Image Bank*)

INTRODUCTION—SENDING MESSAGES TO HUMANS AND MACHINES

● ●

Today, the Earth is ringed with communication satellites that bring us in direct contact with other people and events through telephone, radio, and television. We are able to communicate with ships at sea and spacecraft through

361

KEY TERMS

binary code
carrier
closed-loop process
code
communication
communication medium
communication system
communication vehicles
compatibility
entropy
feedback
fiber optics
frequency
graphic communication
language systems
media
noise
open-control loop
open-loop process
receiver
redundancy
replication
sender
signal entropy
signals
symbols
synchronous satellite
synthesizers
telecommunication
transmit
visible light waves
wavelengths

the use of satellites as relay stations. We have also learned to communicate with machines and even with some of the higher level animals, such as dolphins and chimpanzees.

The use of machines to make communicating easier, faster, and cheaper is a relatively recent development. Until recently, people did not know much about what was happening in the rest of the world. When they did receive information, it was often well after the event had occurred. The importance of information processing machines was made evident in the National Census of 1880. Without the new punch card system, developed by Herman Hollerith, the analysis of the 1880 census would still have been in progress ten years later when the next census began.

Communicating is obviously a major activity of humans. This chapter will give only passing attention to communications as a social, psychological, and anthropological activity. This chapter looks primarily at communication in a technological context. A model of the communication process can help you understand communication among animals, humans, and machines. The discussion here addresses **communication** involving humans and machines.

SENDERS AND RECEIVERS

Communication involves a sender, someone who wants to communicate, and a receiver, someone who is willing and able to listen. A **sender** and **receiver** pass information between them. Humans can communicate with other humans and with machines. Machines can also communicate among themselves. For both humans and machines, communication is possible only when the message takes on some meaning for the participants. Communication can take place when the sender and receiver speak the same language, but meaning is achieved only when the participants have a common base of understanding.

The information to be transmitted must first be changed into a **code**. With human-to-human communication, words are often used as code. On the receiving end, you hear the sound, decode it into words, and puzzle out the meaning. When the message reaches your brain, the transmission process is complete, but as the receiver you must interpret the meaning. In order to interpret the message that has been transmitted, you and the sender must have several things in common.

Compatibility Between Sender and Receiver

There is no **compatibility** between sender and receiver unless they use a common code of **signals** and **symbols** that has been agreed upon in advance. If the message is to be understood, the code must be understood by both the sender and the receiver. For example, if the message sender speaks English but the receiver understands nothing but Spanish, then the message will not be understood. The languages or codes are not compatible. What other types of symbols or signals could be used to allow communication to occur? How about sign language? Or pictures? Would it work to find someone who speaks both languages?

■ Figure 12.2
The control of aircraft as they fly from one destination to another involves communicating between humans and machines. (© Richard Pasley/Stock, Boston)

With a common system of symbols, usually called a language, humans have the foundation for communication. Equally important, however, is the need for a common area of interest and some degree of common understanding. Without a common area of interest, the sender and receiver have little to talk about or to share. Without some common understanding, there is little likelihood that the messages that are transmitted will have any real meaning. As an example, consider the dance performed by honey bees to communicate to other bees where they can find flowers. The bees have a common area of interest, the finding of the flowers as the source of nectar and the raw materials for their honey. The bees also have some understanding as to what the movements of the dance mean. (Refer to Figure 12.3.)

Communication, therefore, requires a common language, common interest, and common understanding. The need to communicate arises when something unguessable must be shared with others. As with the honey bees, we must provide our receivers with some additional information so they can gain some new understanding or modify what they might otherwise do. This aspect of the unguessable represents the difference of what is known by the sender and what is unknown by the receiver. The act of communicating attempts to make the unguessable a matter of common understanding between the two participants. We will revisit the idea of the unguessable information later when a key concept of **signal entropy** is introduced.

Figure 12.3 ■
Bees use a unique dance to communicate to each other where good supplies of nectar are located. The dance tells the bees what direction and how far they should travel to get there. (Left: © Peter Poulides/Tony Stone Images; Right: Reprinted with permission from NSTA Publications, ©1994, from Science and Children, National Science Teachers Association, 1840 Wilson Blvd., Arlington, VA 22201-3000)

Machine Communication

As has been shown in the chapter on Information, early machines took very simple forms. Machines used cams, gears, and linkages to send signals to other machines. Later, the use of holes punched in paper cards or paper tape set the stage for even more sophisticated communication through electronic means.

Now we see some of the old and new methods being integrated in new and interesting ways. For example, the labeling of goods in stores has been a common practice. As the stores got bigger and the number of products got larger, it became very difficult to keep track of what was being sold the fastest and, consequently, what would need to be replenished on the shelves the quickest. Another problem was the time required at the supermarket checkout counter to identify and tally each item to provide a final bill. Part of the answer to this problem is called the Universal Product Code (UPC), which most of us know as "bar codes." Each product has its own bar code which is printed on the package. The bar codes can be read as described below, and shown in Figure 12.4.

READING THE UPC OR BAR CODES

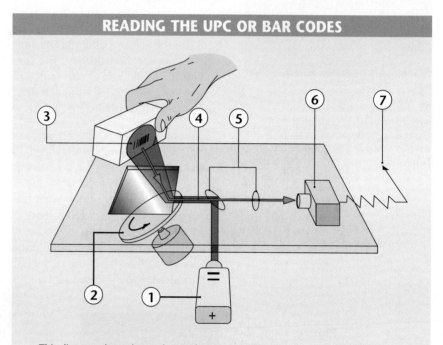

This diagram shows how a bar code system works.

1. Low-power laser sends out a beam, reflecting off a mirror.
2. Beam passes through scanning disc and glass window.
3. Beam shines on bar code.
4. Back scatter reflected back.
5. Mirror and lens.
6. Detector.
7. Signal to store computer.

■ Figure 12.4
This illustration shows how a bar code reader at a checkout counter operates.

The process starts when you bring your purchases to the register to pay for them.

1. The person at the checkout counter passes the bar code over a glass window and the bar code is read by a scanner.

2. A low–power laser projects a beam that reflects off a mirror to the scanning disk.

3. The laser beam is deflected by the disk to scan the space above the window.

4. The beam shines on the bar code and is reflected back (called "back scatter") through the window and scanning disk.

5. The reflected beam, or back scatter, travels past the mirror and is directed by the lens to a detector.

6. The detector converts the laser reflection to electronic pulses which are flashed to the computer.

7. The computer compares the bar code pattern with those in its memory (taking a fraction of a second) and signals the cash register with the price.

The computer can also print a receipt and change the store's inventory to keep it up-to-date. In large supermarket chains, the store computer can communicate automatically with a computer at a supply warehouse to order more products when the supply of a specific product gets low.

Recent developments in communication between machines and humans allow many people to use computers. Many **language systems** have been developed so that people with little knowledge of the computer itself can communicate with it, using familiar symbols. Only a programmer or a computer technician needs to know the highly complex set of signals triggered by the symbols on the keyboard. In other words, when you type "RUN" into a keyboard, the computer does not respond to the inferred meaning, "go ahead and do your work." Instead, the computer has been programmed to turn on specific sets of circuits when the "R" is pressed. Another set of switches is turned on when the "U" is entered into the machine. Still another set operates when the "N" is entered; and still another set of signals is received when the "RUN" appears in combination.

In order for computers to communicate and share information, they must use the same code. One code that is widely accepted internationally is called the ASCII code. ASCII is an acronym for American Standard Code for Information Interchange. It is a binary code that can be used by computers. When computers were first being developed, all commands had to be programmed in using a machine language. It was quickly apparent that an easier means of programming was needed. There are now hundreds of different programming languages available. Now that computers are more sophisticated, they are able to interpret input programming that comes closer to human language than the "0's" and "1's" of machine language. Programming of this sort uses what is called "higher level" language. Some of the early languages that are still commonly used include BASIC, FORTRAN, COBOL, and PASCAL. Other more recent languages include LOGO, FORTH, and C. All higher level languages permit the user to communicate with a computer through a more natural language. The computer is designed to take over the task of translating the higher level language into the machine language required to operate the computer.

Computers made by different manufacturers interpret only the languages compatible with them. The human must use the proper system of symbols and signals to be able to communicate with that particular brand of computer. The processes of communication are the same, however. The human may use keyboards, tape, cards, and the like. It is the arrangement of the symbols that differs. There must be compatibility between what is sent and what the receiver can "read." Even very complex information machines, such as the **synthesizer** and the computer, are still only machines. They can respond only to signals.

At Bell Laboratories in New Jersey, the first machine that was "taught" to translate signals into simulated speech that humans can understand was called a synthesizer. This was the first real "talking machine." Advances since then have led to machines that can translate spoken words into the signals that machines can understand. Now machines are able to receive, as well as send, spoken words. This represents a giant step in communication between humans and machines. Synthesizers are also used for musical purposes. You may have heard music that simulates the sound of many different musical instruments through one synthesizer.

It is very exciting to have computers or robots speak to people and carry out spoken orders. It is easy to forget that even though the robot seems to understand, it is actually responding to the individual sound waves of the words. These are the signals that, when decoded, can trigger specifically programmed events inside the machine.

TRANSMITTING THROUGH MEDIA

Information in the form of a code or message is **transmitted** from the sender to the receiver in different forms of **media**. Each medium uses a vehicle as the **carrier** of information. It is the link between sender and receiver through or on which a message can travel. Suppose you write a letter to a friend. You need paper to carry the words and ideas you want to send. Sending the letter through the mail involves transporting the letter and the information represented in the written symbols. If, however, you decide to call on the telephone, you might be communicating over the wires between the telephones. The wires will serve as the carrier in this instance. If your friend comes to visit, you can talk to each other directly. The carrier will be the air between you. Examples of other **communication vehicles** include radio waves, newspapers, tapes, cables, and laser beams.

On paper, a message is sent by the graphic images. Over wire, a message is sent by electrical pulses. When talking, a message is sent by sound vibrations in the air. A **communication medium** requires that a message be paired with an appropriate vehicle.

It is very important to match the carrier and the form of the message. Each medium has advantages and disadvantages regarding the amount of information that is communicated. Talking directly to a person allows you to see, hear, smell, touch, and move as you communicate. The messages returned to you can be received through several senses. You can ask questions or hear each other's tone of voice. Communication by way of two-way television contact is different, however.

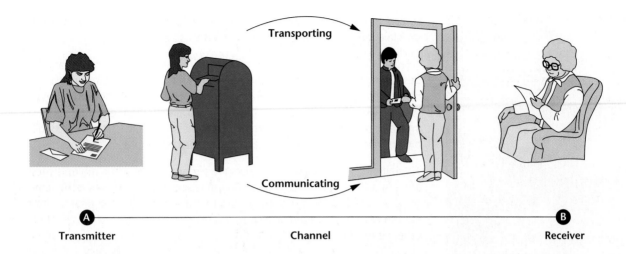

Transporting

Communicating

A Transmitter   Channel   **B** Receiver

Figure 12.5 ■
A common form of one-way communication is shown here. The sender (transmitter) writes a letter. The paper is the medium for communicating. The letter is transported by truck, rail, or plane. The one-way transmission is complete after the receiver has read the letter.

Medium	Message Symbols/Signals	Carrier Vehicle
Print	Words	Paper
Television	Electronic signals	Radio waves, cable
Telephone	Electrical signals	Wire, optic cable, satellite
Facsimile (Fax)	Images, words, electronic signals	Wire, optic cable, radio waves
Voice	Auditory symbols	Sound waves
Recordings	Magnetic or optic signals	Tape, disks
Data processing	Holes or pins to make patterns or magnetic signals	Cards, tapes, drums, or disks
Painting	Figures in symbolic patterns	Canvas and paint
Photographs	Figures in symbolic patterns	Paper, chemicals

Figure 12.6 ■
Selected media, messages, and carriers.

Communication also changes with the use of a CB (citizen band) radio, tape recording, facsimile (fax) transmission, or newspaper advertisement. The effect of the total message and the amount of information that can be sent is different in each case. Some media are static or frozen. They can be stored and replayed. More dynamic media lose the message once it is sent. Some media allow the message to be transmitted extremely rapidly. The receiver can either receive the message rapidly or store it. If stored, the message can be replayed later at a pace appropriate to the receiver.

Some media allow easy **replication**. Millions of copies of a newspaper can be made and transported. This process is impossible for some media. An example is a famous painting. There can only be one original. The image may be reproduced or transmitted by television, but much of the impact of the original, symbolic message may be lost. The use of electronic mail (E-mail) allows the user to send messages to many different receivers at the same time. One message can be

■ Figure 12.7
Fiber optics using light for communicating information provide a very efficient means of transmitting large quantities of data at very high speeds. *(© Doug Armand/Tony Stone Images)*

routed to everyone using the computer-based **communication system** by a simple command to send the information to all participants.

Television is also effective in reaching very large numbers of receivers and is often described as "mass media." Early television reception was quite poor and much of the quality of the original image was lost. Enormous strides have been made in improving the quality of television. With high-definition television, the received image comes closer to the original. The message that is received is still two-dimensional and limited to visual and auditory sensations.

The differences in media and carriers are also important when multiple messages are transmitted. Many telephone calls are sent at the same time on the same wire or cable. The telephone company tries to make sure that the signal transmission at both ends is optimized, similar to a radio station. You sometimes hear parts of other people's messages when you are on the phone. More messages can be sent at one time when radio waves are used. Even more can be sent by microwave communication, and still more on cables using **fiber optics**. (Refer to Figure 12.7.) Carriers have different capacities for transmitting signals clearly and accurately. They also differ in their ability to maintain separate channels for each message.

As more and more information is gained, different carriers are needed to provide more ways of communication. Vast amounts of information need to be stored and decoded later. This needs to be done efficiently and with as little loss of information as possible. Symbolic messages, sent on solid carriers of paper, canvas, and the like, must be transported. However, messages that can be sent as signals on a band of energy can be transmitted instantly and flexibly to remote locations.

Light—The Fantastic Carrier

The role and importance of carriers can be illustrated by comparing **visible light waves** to radio waves. Both of these wave forms are on the electromagnetic spectrum shown in Figure 12.8. Visible light is further along the spectrum in the direction of speed. Light waves, therefore, have a higher **frequency** than radio waves.

Radio, television, microwaves, infrared, and visible light are radiated waves on the electromagnetic energy spectrum. They all travel at the same speed of 186,000 miles per second. However, they differ in **wavelength**, the number of times they vibrate per second. Messages are carried on the wavelength, from one vibration to another. The more vibrations per second, the more information that can be carried.

Visible light vibrates at frequencies up to a million times faster than AM radio waves. The radio part of the energy spectrum is crowded, because it is narrow and many people want to use it. In some frequency ranges, the stations may seem packed one on top of the other. The visible light spectrum has a million times more room than all the radio, TV, shortwave, and microwave bands that we now use for communication. Light waves have great potential as carriers.

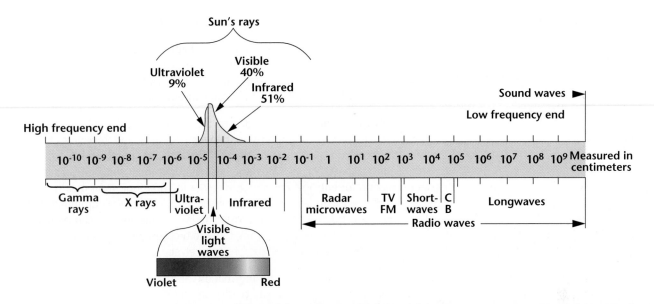

Figure 12.8 ■
The electromagnetic spectrum and its use for transmitting information.

MATH/SCIENCE/TECHNOLOGY LINK

Information Highways

A great deal has been written regarding the development of new "information highways" to improve long-distance communication. The use of information highways is expected to change significantly how we access data and interact with people around the world.

Information highway is a general idea that is used to represent the capability of "moving" large amounts of data easily. Information highway is being used as a conceptual model just as we used the word "tree" as a conceptual model in Chapter 3. Information highway is a good analogy, but of course no one will be traveling on those roads. (All analogies and models have their shortcomings and limitations.)

D E S I G N   B R I E F

Collect materials on the new means of moving information locally, nationally, and internationally. Determine what is needed to access information through an electronic information network. Design and develop a display that will help your classmates understand how that network operates. Design and develop a set of instructions that communicates to others how they can use that network to get access to information related to problems of interest to them.

Entropy and Noise—Losing Information

The concept of entropy was introduced in the chapters on materials, energy, and humans. In the energy chapter, for example, entropy refers to the disorder of energy that results in the loss of usable energy. In a similar way, entropy in this chapter refers to the disorder or loss of usable information.

In this chapter, we introduce the idea of the unguessable information as the essential aspect to be communicated between senders and receivers. In his important work on information theory, Claude Shannon considered the unguessable aspect of a message as synonymous with disorder in communication, or entropy. Entropy as disorder in communication means some of the message is not understood. If there is still disorder after the communication act is completed, that information is lost. The loss of information can be linked directly to the lack of meaning. Entropy can also be caused by sending a message that is incompatible with the abilities of the receiver.

Entropy, as the loss or disorder of information, also occurs during transmission. Transmission of information by any media on any carrier is never 100 percent efficient. The entire signal never gets through. The transmission of information is improved and entropy can be reduced by building in **redundancy.** This means sending more signals than are actually needed. Television stations send excess messages so that even TV sets with poor reception can receive a picture. The television picture itself is redundant. Much of the image could be lost and still understood. The quality would be reduced but the message would still be understandable.

Loss of information may result in partial loss of a signal, over time. Ink fades with time; a tape may lose some of its magnetism; or the carrier may become worn and tattered. In these cases, the entropy increases as the information becomes more and more unclear. This loss is called **signal entropy**. Signal entropy is the loss of usable information as it is changed from one form to another.

Symbols may also lose some of their meaning over time or with translation. We have only a few recorded messages made by ancient people. We have trouble decoding the messages for three reasons: The materials have worn away or decayed; we only have pieces or parts of the object on which the message was recorded; or we are unable to break the code. Our experiences are not the same as the experiences of people many centuries ago. Some languages and symbolic systems have evolved into different forms, and the old codes have been forgotten.

Signal loss is also caused by noise. **Noise** is a special term that indicates interference during transmission. Shannon's model in Figure 12.9 shows the point in communication at which noise interferes. Information entropy and the loss of signal becomes more of a problem when interference or noise increases.

All communication systems have some noise that interferes with information transmission. Radios and telephones have static. A computer may have stray voltage. A photocopier may reproduce dirt on the lens. All of these interfere with transmission. If the static becomes too strong, the voltage too high, or the copier too dirty, the transmission of the information becomes impossible. The signal is lost when the noise level becomes too high. Therefore, the signal must remain higher than the noise level to be received. To maintain good transmission of information, a high signal-to-noise ratio must be maintained.

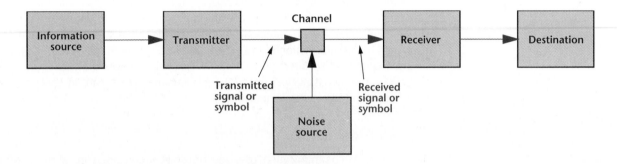

Figure 12.9 ■
This general model of communication shows the major parts of a one-way communication system. Noise interferes with the signals and symbols as they are transmitted on the channel. Entropy takes place during each step from the information source to the destination.

FEEDBACK

Talking Back to the Sender

Have you ever called someone on the telephone and gotten their answering machine? You leave your message, but you have no idea if or when it will be received. This lack of contact with the receiver is a special situation in communicating. This situation is called an **open-loop process** of communication. The sender gets no information back about the transmission. When your message is intended for a machine, the same situation is called an **open-control loop**. Return information from the receiver to the sender is missing.

Most communications have some return contact between receiver and sender. This contact closes the communication circle. These situations are called **closed-loop processes**. Return information from a receiver to a transmitter may be in many forms. A conversation with your friend, a thermostat control on a furnace, the water level control on a toilet tank, and the display screen on a computer terminal are all examples of closed-loop communication.

Feedback is the term used to describe the contact from receiver to sender. Feedback closes the loop of a communication or control process. It is simply information provided by the action of the receiver. This information indicates that the message was received. The sender can use the feedback to learn how to communicate better.

People's facial expressions and questions provide feedback on communications. We can then restate things or give examples to clarify our messages. Feedback happens only if some change, like the change in expression, takes place. Until then, we can only guess at the success of our explanation.

Feedback from and to machines is also based upon change. This change is actually a deviation from an existing situation to a new situation. Change or deviation from the set temperature in a kitchen range provides feedback to the heating system of the stove. A temperature drop in a room provides feedback to the thermostat as a change from the set temperature. The feedback causes a furnace to turn on.

In a later section in this chapter, feedback and two-way sharing of information are combined to provide the user with increased ability to interact with the communication system. Interactive systems enable the user to direct or query the system to get information of interest. An interactive system allows for the flow of information in two directions. Interactive television moves the user away from being a strictly passive viewer to being an engaged participant in choosing the information to be accessed and shared.

Communication by Signal

Some visual communication messages must be transported in tangible form. Other messages can be transmitted on energy waves. The transport of a videotape is a tangible form of visual communication. The broadcast of a television program is visual or symbolic communication transmitted on specific waves from the energy spectrum. Because of the importance of the transmission of information, a study of communication would be incomplete without some attention to what is called **telecommunication.** The prefix "tele," as used in telephone and television, means operating at a long distance. Telephone really translates to mean "long talk" and television means "long see" or "long sight." Radio could have been called "teleaudio" or "long hear."

In 1800, Alessandro Volta invented the electric battery. Volta's battery laid the groundwork for Samuel Morse's invention of the telegraph. The importance of Morse's work was not immediately apparent. This is often the case with new ideas. Other people use the breakthrough and find important applications that were not recognized by the inventor.

The obvious result of Morse's work was the telegraph itself, with its capability of sending coded messages over metal wire at the speed of light. But the telegraph's symbols also changed written letters of the alphabet to signal energy. These dots and dashes, along with the electrical pulses that caused them, were reexamined and are still used, one hundred years later. This **binary code** was crucial for developing systems of information and communication. One such system was used in 1974 to send a radio space message from the observatory in Arecibo, Puerto Rico. This huge radio transmitter stretches between mountains and sends some of the strongest signals leaving the Earth. Its message leaves our solar system less than six hours after transmission. As you will see in the *Portfolio and Activities Resource*, the electronic blips, sent in blocks, can be translated into a visual checkerboard message.

EXAMPLES OF TELECOMMUNICATION

- **Telegraph**—An apparatus or system for transmitting signals in code by electric impulses sent by wire or radio.

- **Telemetry**—The process of measuring and recording television images and readings of instruments at long distances.

- **Telephone**—A system or instrument used to convert speech to electrical impulses to be sent long distances by wire or other carrier.

- **Telescope**—A device for viewing distant objects by means of lenses.

- **Teleshopping**—A network of computer systems that allows shopping for a wide variety of items, from your home.

- **Television**—The process of transmitting visual images by electric impulses superimposed on radio waves to be projected on a luminescent screen (TV set).

- **Teletext**—The process by which users can call up pages of text and graphics and receive the information, superimposed on their regular TV screen.

Systems of Communicating

Technological activities tend to evolve into systems and then into still larger systems. A communication system may be very simple and involve very few participants. An example is the buzzer or doorbell that many of us have in our homes. A visitor at the front door communicates to people in the house through a push button connected to a bell. A more sophisticated system is the telephone system that links a large portion of the people of the world through an intricate information network.

Communication systems may also be established between machines. A simple example is the system between the furnace and the control thermostat in a building. Automated production is a more sophisticated system.

There are also communication systems that use **synchronous satellites**. The satellites are placed in an orbit whose speed matches the speed of the turning Earth. The satellites appear to hang motionless over fixed points on the Earth. Three of these satellites can provide contact with nearly every point on the planet.

Another instance of a communication system is found in the growing number of computer networks. With the growing support of information highways, the linking of such networks can be expected. Although the networks use an alphabet soup of names and technical jargon, the systems are not terribly difficult to use. It is helpful to look at the networks and the process of networking as simply using computers and telephone lines to link people with each other.

Information communicated by satellite may also be printed rapidly in thousands of newspapers. For this process, paper, ink, and other materials must be produced and transported to the printing presses. A network for distributing the papers, handling money, and even recycling the newspapers is needed. All aspects of the system are interdependent. Each element needs all the other elements to work. This interdependency may be evident only when the process stops. A strike affects the paper carrier, workers in paper mills, truck drivers, and people who depend on the newspaper for information.

■ **Figure 12.10**
This communication system shows the use of telephones and satellites, a practice that has become common in telecommunication. *(Top: Courtesy of MCI Communications Corporation; Bottom: Courtesy of Motorola Corporation)*

The exchange of money is another example of a communication system. The trade of surplus goods stimulated transportation systems. It also fostered changes in methods of recording the value of the goods traded. In earlier times, as people traded farther away from home, they needed a common medium of exchange. Salt was an early form of money, as were shells and sand dollars. Coins were made to represent the value of the objects traded. Later, paper money was used to make the exchange of goods and services easier.

The use of symbols rather than physical objects increased the possibility of complicated transactions. A system of credit and loans allowed people to buy goods and services easily, on the promise that they would pay later. Credit and debit cards can now be used to obtain cash, as well as goods and services. Credit cards and cash are being replaced by electronic "blips." This electronic "money" allows machines to communicate with machines and make the money transfers. No physical information in the form of coins or paper will change hands.

Before worldwide electronic banking and marketing could happen, vast changes in communication systems were required. New machines were needed to send and receive messages wherever transactions occur. Common standards for a monetary system were needed. Money matters are often considered critical—worth starting battles and wars over. The development of sophisticated security exchange systems must occur before global monetary transactions can become routine.

FORMS OF COMMUNICATING

Developing a Portfolio: A Communication System

As indicated several times throughout this book, portfolio work is important. It helps you work through ideas that you may have regarding the design and development of a new design. It also provides a means of documenting your thinking as you collect and save your ongoing work. This will be very important as you and your teacher look back on the progress made and problems encountered in your work. The following section is intended to help you improve your portfolio and its presentation.

Symbolic Communication

Printing, photography, lithography, and electrostatic reproduction are a few of the advances in graphic communication. The general term **graphic communication** refers to the visual forms of sharing information. Graphic communication includes the printing technologies. It also includes visual communication, such as sign language for the hearing-impaired. Language forms developed from sign language can even be used for communication with selected animals. More commonly, graphic communication efforts result in some form of printed or visual display of ideas. One form of graphic communication that you have been introduced to earlier is the portfolio.

Developing a Portfolio

A commonly used form of graphic communication in design and technology is the portfolio introduced in Unit I. The portfolios use different symbol systems to convey meaning, such as drawings, illustrations, writing, models, and presentations. As you develop a portfolio, you will attempt to convey to others your design and how you went about getting and developing your ideas. Your message might include text, drawings, photographs, models, sounds, and demonstrations. You will need to choose appropriate media to carry your message, and input the information for storage, processing, and output. If you use a computer to prepare your message, the material will be stored and processed in binary or digital code.

Input

The term input has been used earlier in this book to indicate that information as a "signal" was being fed to the process part of a device or system. Input is being used here to indicate that data are being entered into a computer. Input can be text or images in audio or video forms.

Text. If your portfolio includes text (your name, titles of your work, written descriptions, lists, references, etc.), you will probably input that information by typing it in a word processing program.

Images. Images might include sketches, renderings, illustrations, technical drawings, CAD drawings, charts, graphs, photographs, or animations. Images created on the computer are represented in a digital form. Other images, printed photographs for example, must be digitized. A scanner digitizes a photograph by creating a grid of squares that make up the image. Each square, or pixel, is assigned a number that represents its color. For each pixel, the scanner flashes a light at the photograph and measures the light reflected back to determine the color. The commercial scanner shown in Figure 12.13 and the simplified scanner illustrated in Figure 12.11A work in a similar manner. This similarity should be no surprise since both machines evolved from early facsimile transmission machines that even looked like heavier versions of the simplified scanner.

After images are captured through scanning, they can be manipulated in a variety of ways. For example, the images can be modified to take an animated form. An animation is a series of still images that show or suggest motion. Now these images are often created on the computer using two-dimensional or three-dimensional drawing software. It may be used to illustrate a process, an assembly, or various views of a three-dimensional object. This technique can be quite helpful in illustrating a process that is difficult or impossible to see in a single image. A variation of animation is called "morphing" that starts with one image and ends up with another quite different one. The process of morphing fills in between the two images with a gradual transition from one to the other.

(A)

(B)

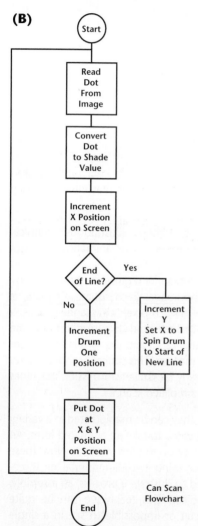

Start

Read
Dot
From
Image

Convert
Dot
to Shade
Value

Increment
X Position
on Screen

End
of Line?   Yes

No

Increment
Drum
One
Position

Increment
Y
Set X to 1
Spin Drum
to Start of
New Line

Put Dot
at
X & Y
Position
on Screen

Can Scan
Flowchart

End

■ **Figure 12.11A, B**
(A) A rotary scanner developed to
help students understand the
process of scanning. (B) The
flowchart identifying the sequence
for the computer program used
with the scanner. *(Courtesy of Frank
Capelle, photo by Michael Dzaman)*

DESIGN ACTIVITIES

Problem: Scanning of graphic images is an important part of capturing and modifying images to create a visual message. Scanners, like most technological devices, often hide the process and consequently make it difficult to understand what is actually taking place.

Design Brief: Develop a "technologically transparent" device that will provide a working scanner. This can serve as a starting point for further investigation into other aspects of technology related to image scanning and facsimile transmission.

Design Solution: The scanner, shown in Figure 12.11A, represents the work of a team of students who developed the device out of relatively inexpensive materials. Turning the scanner moves the image under the photo sensor. Each rotation of the drum results in one scanned line of information. The computer then captures and processes the information to produce the same image on the computer screen.

In order to accomplish the above, software was designed to accomplish a series of tasks. First, it was necessary to take the information from the optical sensor and convert it to a value that could be used by the computer to display the image on the screen. Second, it was essential that the software put a dot on the screen corresponding to the dot just read by the sensor. Third, the drum needed to advance to the next dot to be read. Fourth, the computer needed to advance a line on the screen each time a scan line was completed on the image.

The flowchart shown in Figure 12.11B indicates the flow of the operations to be accomplished. The flowchart helps to identify what the computer program is to accomplish. Each step included in the flowchart could represent several lines of programming.

The sensor for the scanner includes a light source and photo transistor mounted in the same housing. The signal returned by the sensor varies in current, depending upon the amount of light that is reflected—the darker the image, the less light reflected. This signal is then fed into the game port of an IBM-compatible computer. The game port is used to tell the computer the position of a joystick. A joystick is capable of returning a signal that varies in current, depending upon its position. Basically, the joystick has been replaced with the optical sensor.

Once a signal is received, the software determines the location on the drum that it is reading, and determines the equivalent location on the computer screen. The signal from the sensor is converted into a dot of one of 16 levels of grey. The next step is for the computer to rotate the drum another increment and read the next position. The computer keeps constant track of both the drum position and the position of the image on the screen. When the drum reaches the edge of the image, the computer rotates the drum to the next starting position. Thus, row by row, an image is developed on the screen.

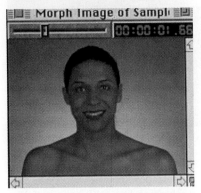

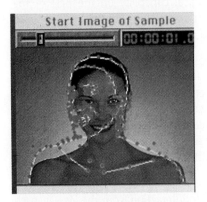

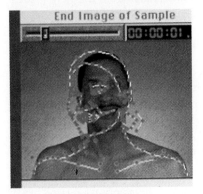

■ **Figure 12.12**
Multiple frames of process
animation, showing "morphing".
(Courtesy of Gryphon Software)

Audio and video. Within the past few years, audio and video production have been influenced significantly by advances in electronics. Digital audio and video devices are available at consumer prices that offer great flexibility and extremely high fidelity. Because these devices record digitally, the information can easily be stored and manipulated by desktop computers. Conventional analog audio and video recordings can be converted into digital code. Video recordings for a portfolio could be used for three-dimensional models and "live" demonstrations.

Processing

Combining text and images for print. For print media, text and images are combined and prepared for output. Page layout (desktop publishing) software can understand and manipulate the machine coding for both text and image information. Generally, the user designs pages by creating boxes in which to place text or images. These boxes become objects on the page and can be manipulated in various ways. For example, they might be relocated, resized, rotated, framed, or given a background color. Text size, font, style, line spacing, and color can be altered. Images can be cropped, scaled, or rotated. They can be converted from color to black and white, brightened, darkened, or have their contrast adjusted. Virtually all page layout programs enable users to create rules (lines) of different weights, patterns, and colors anywhere on the page. In addition, some programs include tools to create other graphic elements such as starbursts, tinted panels, or blended backgrounds. The design decisions you make can enhance and emphasize, or muddle and lessen, your intended communication.

Desktop software is available to help assemble presentations. Choices start with a simple process of sequencing a series of still images to form a slide show. You can also produce multimedia presentations including text, images, sound, and video. More sophisticated approaches can use virtual reality that allows the user to enter a computer-generated world. Regardless of the venue, the designer (or message sender) must make decisions regarding timing, sequencing, and transitions between elements. She may also decide whether the receiver (the person watching the presentation) will be a passive recipient or interact with and control the presentation.

Output

In order to share your portfolio with others, you will publish it, or "make it public." This is the output from the system. Output from a computer usually takes the form of print or graphic representations.

Print. The most common form of graphic output is print. We see it everywhere and have almost reached the point of not even noticing it. As mentioned above, today's desktop publishing systems give the user complete control over page design. These systems also enable easy preparation for print, whether your portfolio will be a one-of-a-kind document or reproduced in quantity. Reproduction may be electrostatic in black and white or color for a limited number, or by one of the traditional printing processes—silk screen, lithography, gravure—for a greater number.

Figure 12.13 ■
This woman is creating a brochure using a desktop publishing system. *(Courtesy of Hewlett-Packard Company)*

Print as output from your communication system has some advantages. It can be inexpensive to reproduce, is easily transported, is commonly understood, and requires no special equipment or technological capability on the part of the receiver. However, print also has limitations. It does not include sound or motion, and generally offers little interaction beyond reading.

Presentations. Presentations are now possible in many different forms. These possibilities offer designers more channels for communicating their messages. The simplest example might be an on-screen slide show accompanying spoken narration. The output may be on the screen of the computer where the presentation was created, or saved to some media (floppy disk, hard disk, CD-ROM) for display on a different computer in a different location. Alternately, such still images can be output to 35mm film with the use of a film recorder.

A slightly more complex example would be an information kiosk displaying a series of images at timed intervals or based on interaction from the viewer, such as pushing a button to advance to the next image. The addition of sounds, narrations, animations, movies, and special effects in multimedia presentations requires more sophisticated output options. Output may be on computer systems or videotapes.

Over the past few years, video production has been influenced significantly by advances in electronics. A range of more powerful and less expensive video equipment has been integrated with personal computers to create what has become known as "desktop video." Figure 12.14 shows examples of equipment available to those interested in developing multimedia programs through this exciting means of communication.

As you output presentations from the computer to slide or videotape, you continue to make choices. For example, you will determine if the advantages of transportability and simpler equipment requirements outweigh the disadvantage of limited potential for interaction by the viewers.

■ **Figure 12.14**
This equipment is used for scanning images for desktop publishing and desktop video production. *(Courtesy of Polaroid Corporation)*

Interactive Systems

As indicated earlier, feedback coupled with two-way exchange of information results in dynamic, interactive communication systems. These systems have evolved and grown more sophisticated, and the user and the equipment work closely together. Interactive systems move users from normally passive to more active roles in communication. The use of interactive systems continues to increase in business, industry, and education.

Interactive communication systems are now commonly used on personal computers (PCs). In addition to user-friendly systems for word processing and data processing, PCs provide programs for desktop publishing, desktop video, and computer-aided design (CAD). New systems that create more productive and exciting interaction between computers and their human users will continue to be developed as the capacity of PCs goes up and the size and cost of the hardware go down.

The combination of television and other media with computers continues to result in new forms of communication. Virtual reality, already mentioned, provides a unique interactive capability that allows the user to manipulate a simulated "micro-world." Users are outfitted with a variety of sensors that allow the computer to determine where the users are looking and what physical moves they are making. The position and moves of the users are then translated into program sequences that generate images as though the users were "inside" the simulation. Virtual reality is already being used for training and holds promise for helping to design and develop new environments. By allowing the designer to get inside the proposed environment before it is actually modeled or built, virtual reality can make it easier and less costly to make changes and improvements to that environment.

Providing interactive communication through the use of simulations is becoming rather common. Programs are used in schools and industry for skill and concept development. As you may have experienced, computer-based games are being developed constantly. Simulations, as serious games, are growing in acceptance. For some types of skill and concept development, simulations provide a power and productivity that is difficult to accomplish through other media.

Simulation and communication have also merged in the daily reporting of the weather. Information gained from satellites provides a view of weather patterns around the world from space. (Refer to Figure 12.15.) The weather we experience on any day is influenced by where and how that weather was shaped in preceding days. It is helpful, therefore, to monitor what is happening to the weather on a day-by-day basis.

The communication system that allows us to monitor the world's weather also provides maps of weather systems as they develop and travel. The path of a hurricane after it has finished its devastating journey can be used to assess damage. Although weather reporting has improved considerably over the past decade to include sharing weather information on a worldwide basis, there is much to be accomplished.

Figure 12.15 ■
This map from a weather satellite shows a hurricane as it moves across the Gulf of Mexico.

Forecasting weather would be improved significantly with a reliable, large-scale computer model of the world's weather. Developing such computer programs, capable of dealing with the millions of bits of data in hundreds of different forms, is a task of enormous complexity. Such a model would require computers more powerful and faster than anything currently available. The model would also require access to greater amounts of information than can now be transmitted through worldwide telecommunication systems.

All instances of two-dimensional means of communicating do not require sophisticated equipment. For example, the use of drawings for communication is very common. We find them in plans for the assembly of products we buy and in the documentation for how a product should be operated and maintained. Drawings are also useful to show what might happen in events that are too small or too far away to observe. Drawings will be particularly helpful to develop your plans and activities before they happen. One example of a different graphic symbol system is represented in the plaques placed aboard the Pioneer 10 and 11 spacecraft before they were launched into space, shown in Figure 12.16. Their visual message provides information about the place, time, and origin of the makers of the craft. The designers of the message were aware they had to send the message in graphic language. Why?

Drawings in the form of charts and graphs can also be very helpful in presenting complex data and information. Charts and graphs provide accepted means of showing relationships that at times may be fairly abstract. For example, the graphs in Figure 12.17 illustrate the findings that emerged from the research and development efforts of a team of students. Three young women were interested in the effect of applying a ceramic coating to the moving parts of a gas-powered engine. The ceramic coating enabled the engine to operate at higher

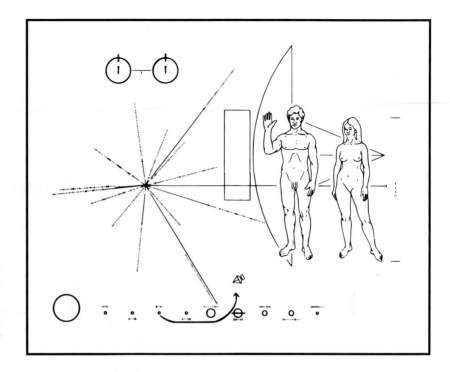

Figure 12.16 ■
This message about Earth, its location, and its people was placed on 6″ x 9″ gold-anodized aluminum plaques. These were attached to Pioneer 10 and 11 for their journeys into space.

temperatures. The team was particularly interested in the effect of this change in temperature on the efficiency of the engine.

The first graph shows the thermal efficiency of the engine with the ceramic-coated parts, compared to an unmodified control engine. The graph indicates that the ceramic engine operated at temperatures ranging from 25°C to 80°C higher than the control engine. The second and third graphs show the mechanical efficiency of the two engines in terms of their horsepower and torque. Information of this sort is difficult, if not impossible, to convey in words alone. Charts and graphs become valuable tools in describing relationships of important variables being studied or manipulated in a research project.

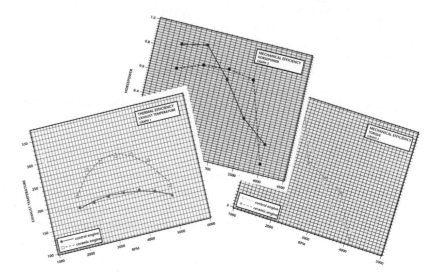

Figure 12.17 ■
These graphs record the findings of a team of young women in their research and development project on the effects of coating an internal combustion engine with high-temperature ceramics. *(Courtesy of Robert Gauger and TIES Magazine)*

THREE-DIMENSIONAL FORMS OF COMMUNICATING

Visual communication can also take three-dimensional forms. Models, displays, and exhibits are three examples that can be very effective in sharing information with others. In all cases, it is important to consider the audience that is to be reached and the message you are attempting to convey to that audience.

Packaging As Three-Dimensional Visual Communication

One common form of communicating by using three-dimensional objects is seen in packaging. As consumers, we come into contact regularly with packages and packaging of all sorts. The package or container may be a permanent one that forms the covering for the product itself. When you made the working model of a design, you may have just included the working parts. However, if the product were to be used over time, some covering or containment is often needed. Often the package is designed with careful attention to the safety of the user, as well as protection for the parts of the product. The structure may be designed to communicate to the user and make operation of the product as easy as possible. Shape, color, choice of covering materials, and other aspects may be chosen to convey a message about the product contained within. The intent is the ease and satisfaction of the user and the durability and functioning of the product.

Another form of packaging is the container in which the product is marketed. This packaging protects the product in transport and in the store. However, the manufacturer usually has more than protection in mind when this packaging is designed. Communication is its major purpose.

Hopefully, the package gives the consumer the information needed to decide whether this item is suited to the consumer's wants and needs. Another intent of the packaging is to persuade the consumer. In many cases, we can be influenced to buy a product because of its package. Since the package is often the first thing we see when looking at a product, the designer gives a great deal of attention to the development of the message carried by that package. There is much that you can learn from a careful study and analysis of commercially made packages.

Packages may appear to be simple, but they are actually rather sophisticated forms of communication. In order to make a sale, the attention of a prospective buyer must be captured because the package is most often seen before the product. If you are to design an effective package, you will need to consider its use, the potential audience or buyers, the image that you want to create, and the information you want to convey.

Once the potential consumer considers the product, more information is needed. Producers are required by law to identify the ingredients of some products, especially those ingested by people or used on their bodies. Products that pose potential risks to safety, health, or the environment must include clear warnings. Perhaps one product has been found much more efficient in energy use than

■ **Figure 12.18**
Examples of different shapes of packaging and the information included on the packages.
(© Michael Dzaman Photography)

GUIDELINES FOR EFFECTIVE MARKETING PACKAGES

- Packages need not be complicated. Keep the design simple. Do not attempt to crowd too much information onto the package. Simple drawings and graphics are often the most effective.

- Packages have only a moment to catch someone's attention. Packages, therefore, must communicate quickly. The package must help the consumer understand what is in the package as quickly as possible.

- The package should be compatible with the product. The package must reflect the image that the designer was trying to achieve with the product.

- Packages must have an aesthetic appeal to the consumer. Effective packages must be more eye-catching than packages of competing products.

■ Figure 12.19
A three-dimensional exhibit developed for student recruitment. The internal display made creative use of movement to create more interest and capture the attention of onlookers. *(Courtesy of Department of Technological Studies, Trenton State College)*

competing products. The package designer includes this required information, along with special features that will enhance sales. In the design activities at the end of this chapter, you will have the opportunity to look at several different types of products, their ingredients, and product claims.

Other means of three-dimensional communication include models, displays, and exhibits. Considerable attention has been given to the development of models. As indicated, not all models are three-dimensional in nature, but when three-dimensional models are used, they can be an effective means of communicating ideas to others. Similarly, displays and exhibits provide effective means of attracting the attention and engaging an audience in the activities designed to transmit the desired message.

The exhibit shown in Figure 12.19 was designed by students and faculty as a part of a recruitment effort to attract students into the teaching of technology as a career. The exhibit was designed to include a number of moving parts that included a device for lifting large ball bearings to the top of the exhibit. The balls then rolled down through a number of gates that directed each ball in a direction different than the ball before it. This exhibit involved the use of motion to attract attention and can be considered a dynamic system of communicating.

SUMMARY
CHAPTER 12

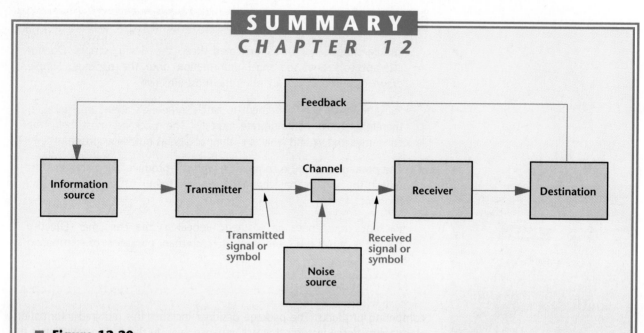

■ **Figure 12.20**
This illustration of the more complete communication model includes feedback as a key part of the process.

The sending of symbols and signals occurs between two or more participants, machines, humans, or both. Characteristics of the carrier, such as the frequency of an energy wave, determine how much information can be sent in a given length of time and the amount of redundancy needed. Transmitted information goes through a natural process of deterioration called entropy. The loss of symbols/signals due to entropy affects the efficiency of the communication. The effect of signal loss is reduced by sending more than is needed or by choosing a

more efficient carrier. Communication systems require feedback before communication ends.

Communication through the use of visual symbols follows the same rules and principles of signal transmissions. Symbols are very important as a connection between machines and humans. Signals are used in connections between machines. Communication systems are an integral part of the other technological activities. Information must flow among the participants involved in all technological activities. Without a sophis-

ticated communication system, many activities would come to a standstill.

All communication can be described through the general model in Figure 12.20. Information moves from a source to a destination through a transmitter to a receiver. The signals and symbols as they are carried on a channel will lose their fidelity because of some source(s) of noise. Feedback, as we will explore in more depth in the chapter on Control, may also be a part of communicating.

D E S I G N   B R I E F

- In preparation for creating a package for one of the products that you or your classmates have designed, collect sample packages for study. Carefully take the packages apart so they can be laid out flat. The shape of the box laid out flat is called the "pattern" or "development." The term development has the same meaning as used in Chapter 2 when you created a two-dimensional development of the electric car. Packages are designed and developed in a similar manner.

- Design a package for a toy product developed for young children or a food product developed for teenagers. Determine the size and shape of package needed. Develop a two-dimensional pattern for the package. Conduct tests to see if the pattern provides you with a package of appropriate size and shape. Plan the graphic message you want to use on the package to draw attention to the product. Determine what information about the product must be included by law on the package. Determine how you will incorporate that information on the package. Develop and render the final package design on card stock. Cut out and shape the package in its final three-dimensional size and form.

- Design and develop an exhibit that will help younger students understand more about design and technology. The students will be attending your school next year and will be involved in design and technology activities. What information will you need about the students? What information will you plan to share with them?

MATH/SCIENCE/TECHNOLOGY LINK

- Determine how math is related to the design and development of packages that are developed from flat sheets of materials into boxes.

- What are some standard sizes of cans and boxes used for food?

- How would you change the size of a No. 10 can to optimize the quantity of contents that could be stored in the can?

- What would be the optimal box size to store 20 ounces of granola, using the least amount of cardboard?

- Are cereal boxes made to optimize their volume while minimizing the materials used in them? Why are boxes for cereal made the sizes they are?

- What advantages are there for forming containers of plastic that are square instead of round?

- Why are tin cans made round rather than square?

- What are the daily nutritional requirements for a teenager? Compare this with the percentage of RDA (recommended daily allowance) listed on a box of cereal or cookies.

- What does the required labeling of food include? Why?

- What is required in the labeling of a product for young children? Is the labeling adequate? Why? What would you change?

ENRICHMENT ACTIVITIES

1. Describe the model for communication. Include the sources of problems in the communication process.

2. Identify at least three instances that illustrate incompatibility between sender and receiver.

3. Explain how a bar code reader can be used to communicate information to a computer.

4. Compare and contrast the effectiveness of different carrier vehicles.

5. Describe a message for which the following medium would be best: print, television, videotape, face-to-face conversation, telephone, still photograph, or audio recording.

6. Define entropy and identify methods for reducing entropy in communication.

7. What is the function of feedback in personal communication? In communication between humans and machines? In communication between machines?

8. Identify some of the factors to consider if you were designing the permanent cover of a radio you designed and made. What factors would you consider in creating a package in which to sell the radio?

CHAPTER 13

Control Technology Systems

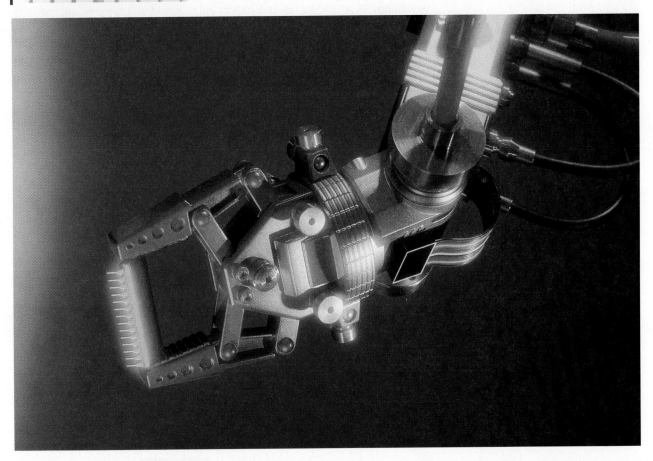

Figure 13.1 ■
Many different devices and approaches, including robotics, have been developed to provide the means for controlling various technological systems. *(© Joseph Drivas/The Image Bank)*

INTRODUCTION—DIRECTING AND REGULATING SYSTEMS

. . . .

Control is an important part of our lives, and you engage in control activities many times every day. Your control activities may include instances of picking up and using a pencil, turning on a fan, unlocking and opening a door, adjusting

387

KEY TERMS

adjusting
analog devices
automation
binary system
bit
byte
chip
control
cybernetic
dedicated devices
external program
feedback
gyroscopic action
handling
integration
internal program
linear
logic
manufacturing
memory
process
program
steady state
system

the temperature in a room, turning on and adjusting a television set, and using a computer.

Within this chapter, we will focus our attention on control technology as a part of the larger concept of control. Control technology is defined as the technical means by which we direct and regulate different systems. Control technology can be as simple as a single switch in an electrical or pneumatic circuit or as complex as a computer network.

The idea of control technology fits within the concept of the systems approach that has been used throughout this book. As you will see in this chapter, control technology is an important and logical extension of the systems approach to technology. The ideas, concepts, and activities included in this chapter are intended to help you become skilled in the design, development, and use of control systems. Through the activities that follow, you will have the opportunity to design and build devices for controlling a wide range of technological systems.

AN EARLY EXAMPLE OF CONTROL TECHNOLOGY

Machines that operate automatically had been around long before James Watt developed his improved version of the steam engine. One improvement that Watt made on the steam engine was to add an automatic control device that could regulate the speed of the engine.

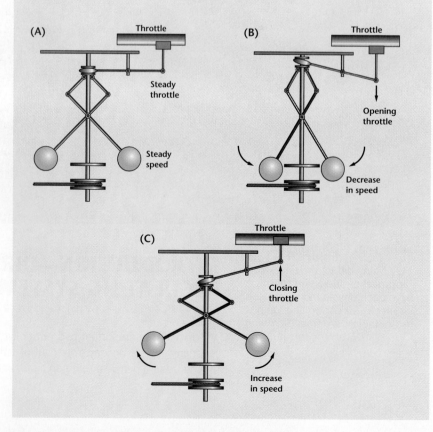

Figure 13.2A, B, C ■
The three illustrations in Figure 13.2 show (A) the flyball governor maintaining the steam engine at the desired speed, (B) the governor opening the throttle in response to the engine slowing down, and (C) the engine running too fast and the governor closing the throttle to slow it down.

CONTROL SYSTEMS

Systems Approach to Control Systems

Initially, most of us see control technology as a rather complex and difficult aspect of technology. But even very complicated instances of control can be made relatively easy to understand if we look at the essential aspects of control. The systems approach, as introduced in the chapter on Information and the preceding chapters in this section, involves looking at the major parts or building blocks that make up any **system**. The systems approach, including the components of input, process, and output was introduced in Figure 7.5. This model represents change whether in the operation of a simple tool, converting materials to new forms, or transporting people or goods from one place to another. All of these instances involve the use of control.

Several examples of control devices used for the basic input, process, and output are illustrated in Figure 13.3B. The examples represent the types of control systems that you can use in designing solutions to control problems that you will encounter. In some instances, modular, prefabricated systems can be used to develop control circuits. In other instances, you may find it more appropriate to use discrete components and fabricate your own control system. Let us take a closer look at the three aspects of control as they are applied in different systems and circuits.

Systems inputs. As indicated in the chapter on Information, inputs can take many different forms. For control systems, all inputs involve energy—energy that is converted into information. The example input devices shown in Figure 13.3B, and those introduced earlier in the chapter on Information, are all available as you design specific control systems. All the input devices provide a means of sensing a condition you want to monitor. For example, you may want to monitor conditions related to the temperature, humidity, or light level of an environment, or monitor the movement of something—an object, wind, or water. You will make a number of key decisions in the design of a control system. One of these key decisions is choosing the "system inputs" in the form of the sensor or sensors that can monitor the condition you wish to control.

System process. There are a variety of devices that can take the input signals and use them to control a system. The control of the system involves using the input to provide a more usable output. This could take the form of **adjusting** the volume of an amplifier on a stereo or setting the level of light by adjusting the variable light control.

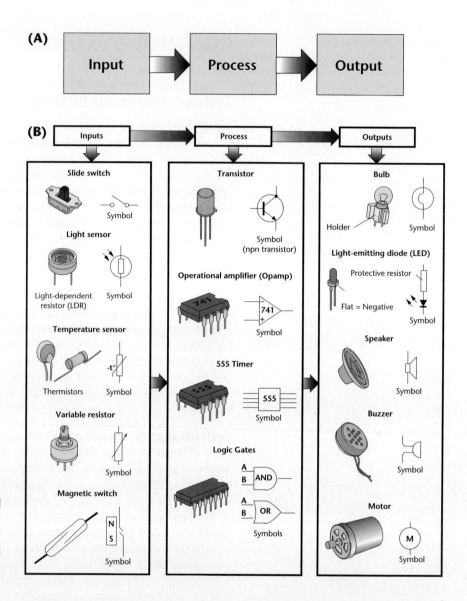

Figure 13.3A, B ■
(A) Control systems can be shown using block diagrams. (B) Examples of input, process, and output devices showing the use of discrete components.

The process aspect of a control system may also involve the changing of one kind of input into a different kind of output. Examples of this, shown in Figure 13.3B, include sensing sound to turn on an outside security light, or sensing temperature change to turn on a cooling fan. In many cases, the input provided by the sensors is not important in itself but as a means to interpreting the condition to be monitored and changed. For example, a control system could use a variety of signals to turn on a security light. The input could come from a light-dependent resistor to monitor the light level, a pressure switch to monitor if someone is standing on the welcome mat at the door, or a movement sensor to determine if someone is approaching the house. A control system could use any or all of these to trigger a buzzer, light, or other output devices to alert people in the house or to initiate another function, such as dialing a predetermined telephone number.

Your second key decision in designing control systems will be to determine the "system process." You will have to choose the method of converting the input signal into a form suitable to operate the output devices. This will require that you make some decisions about the output device that you want to use and then determine how you can control that device.

System outputs. Figure 13.3B illustrates several potential outputs of a control system. All of these outputs, however, involve changing energy from one form to another more useful form of energy. For example, as you design a control system you may want the output to be in the form of light, sound, heat, or movement. If the output is to be sound created by a buzzer, you will need to design a control system that can provide the energy needed to operate the buzzer. The energy requirements for a buzzer are relatively small. The same is true if you plan to light a small bulb or an LED.

If, however, you need to light a larger bulb or turn on a heating device, the power requirements are much greater and the control devices must be heavier duty. You can control devices that have large power requirements by introducing an intermediate control device. You can use the "output signal" of a light-duty control system to provide an "input signal" to a second heavy-duty control system.

This illustrates the third of the key decisions you will make in designing control systems—determining what the "system output" will be. The output of the system should have the capability of changing the operation of the device, system, or environment you are trying to control. With all this in mind, it is time to turn to designing and developing control systems.

DESIGN AND DEVELOPMENT OF CONTROL SYSTEMS

Elements of Control Design

Designing of control systems is best accomplished by focusing on the end result. This means you will be most concerned with the outputs of the control system and how they will be used to control another system. In most instances, it will be very helpful if you make this decision early on. This will allow you to determine what specific output device your system will have.

In a similar manner, you will need to choose the input device. If you plan to use a sensor, you will need to choose it almost at the same time you are deciding on an output device. Knowing what specific sensor and output devices you will use will allow you to determine the input and output requirements of your control system. All of this sets the stage for you to design the process part of your control system.

Designing Circuits

In Chapter 7, two approaches to developing electronic circuits were introduced. (See Figures 7.10A and B.) These two standard circuits are used here to serve as processors. The circuits use a change in resistance to operate—one uses a sensor that causes an increase in resistance, and the other uses a sensor that causes a decrease.

Throughout the book you have seen and used the concepts of input, process, and output. These are three of the basic parts required in control circuits. Now let's see what this would look like as you apply the concept of "change in resistance" and what you have learned about input, process, and output as you design a control circuit. Consider this problem.

D E S I G N   A C T I V I T I E S

Problem: You have a summer job at the beach distributing leaflets to swimmers and sunbathers urging them to try out a new product called "SunBlocR." The job requires you to be at work very early, but only on days when the sun is shining. On days when the weather is overcast, you can sleep in.

Design Brief: Design a circuit that will provide a wake-up alarm if the sun is shining in the morning.

Proposed Solution: Because all control circuits have three basic sections, you can use a detailed block diagram to work out which components you will need. (See Figure 13.4A.)

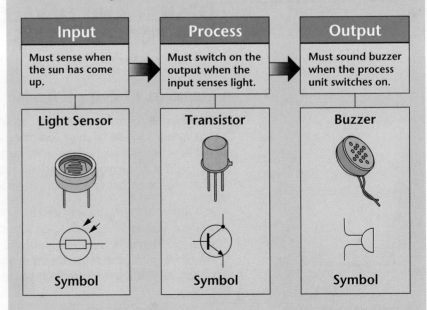

Input	Process	Output
Must sense when the sun has come up.	Must switch on the output when the input senses light.	Must sound buzzer when the process unit switches on.
Light Sensor	Transistor	Buzzer
Symbol	Symbol	Symbol

■ **Figure 13.4A**
Using a block diagram and the systems approach to identify inputs, process, and control outputs of a system and identifying possible components.

Choosing Components: Referring to the components in Figure 7.5, you will find the components needed for this job. You may use modular component boards for the job. (See Figure 13.4B.)

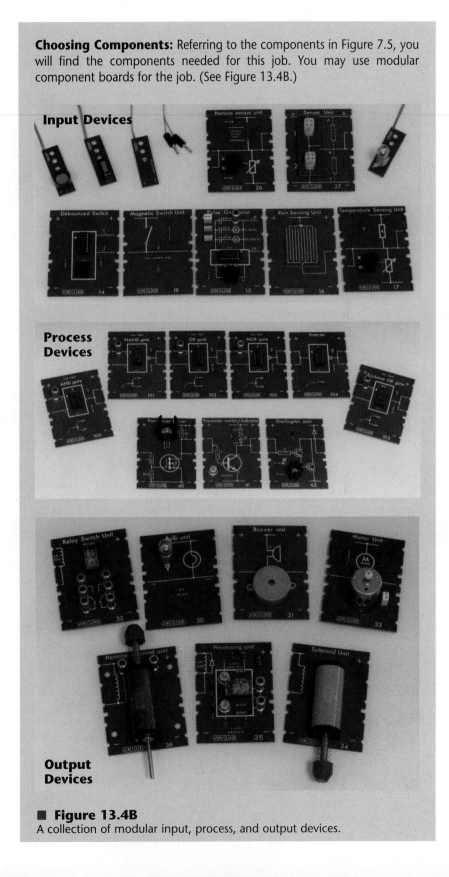

■ **Figure 13.4B**
A collection of modular input, process, and output devices.

Circuit Diagram: Using the components you have selected, you need to draw a circuit diagram showing how the components fit together. (See Figure 13.4C.) For more information on circuit design, refer to your *Portfolio and Activities Resource.*

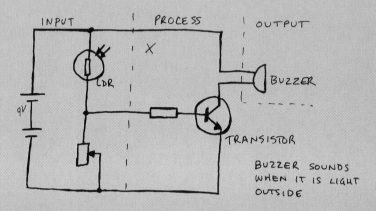

■ **Figure 13.4C**
Selecting appropriate components.

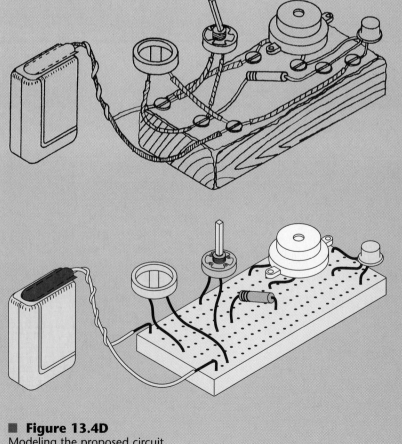

■ **Figure 13.4D**
Modeling the proposed circuit.

How does the circuit work? The sensor provides the *input* part of the circuit shown in Figure 13.4C. The light-dependent resistor (LDR) reacts to changes in light level. The resistance of the LDR gets low when in the light. Its resistance gets high when in the dark. As the light gets brighter, the resistance of the LDR drops (and the voltage drop across the LDR is less). This means that a larger proportion of the resistance is dropped across the variable resistor. (These changes in voltage cause the voltage on the base of the transistor to become more positive.) When the voltage increases by 0.6 volts, the transistor switches on and current begins to flow in the base circuit (see blue section). This in turn causes a larger current to flow in the collector circuit (see red section) that makes the buzzer operate. (The two circuits overlap on the top leg labeled "X.")

The variable resistor is used in the *input* part of the circuit so that you can adjust when the *process* part of the circuit will switch on the *output*. You can adjust the resistor so that the buzzer will sound when dawn is breaking or when the sun is up and it is bright outside. From experience, you will find that it is common to use a variable resistor in the input part of a control circuit.

Modeling

Circuits seldom seem to work on the first try. It is important to "model" the circuit early on in the building process. Modeling provides a quick way to test if the circuits operate correctly. Illustrated in Figure 13.4D are two forms of modeling. The top example uses a "home-built" approach, while the other uses a commercial "prototyping-board" approach. The home-built one is less expensive, but the prototyping board is faster. Either will allow you to work out any problems before the circuit is converted to a more permanent form.

Developing Circuits

After the circuit has been tested through the modeling process, you are ready to build the circuit in a more permanent form. The most-used permanent circuit is called a printed circuit and involves making a printed circuit board (PCB). Before you can make a PCB, you will need to lay out and draw the circuit to the proper size to fit the board that you have.

Planning the circuit layout. In the next step in making a printed circuit, you will need to use the earlier sketches and your circuit model. These products of thinking on paper and modeling of solutions will help you avoid mistakes in your fabrication work. As you prepare your layout, you will need to provide adequate space for each of the components and the proper number of connections set at the right distances apart. Figure 13.5 illustrates a layout for the parts used in the circuit.

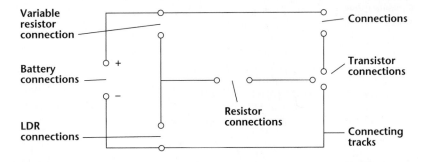

Figure 13.5 ■
Planning the circuit layout.

1 Trace a copy of the circuit onto acetate sheet to make a mask. Transfer sheets will help you to draw accurate connecting points for the components.

2 Place the mask, face up, on the ultra-violet light box. Peel the backing paper off the laminate and place it on top of the mask. Close the box and expose the laminate for about four minutes.

3 Put on goggles and an apron. Pick up the laminate with tongs and place it carefully into a solution of sodium hydroxide which will remove the exposed etch-resistant paint. Use a brush to gently agitate the solution.

4 When you can see the copper clearly, use tongs to remove the laminate from the solution and wash it thoroughly under running water.

5 Put the laminate into the etching tank basket with the tongs. The tank contains ferric chloride which will dissolve all the uncovered copper.

6 After a few minutes all the unwanted copper will have dissolved. Remove the laminate and wash it thoroughly with runnning water.

7 Use a miniature drill to make holes for the legs of the components. Wear eye protection and make sure that the laminate is held firmly on a piece of wood.

8 Push the legs of the components from the insulated side of the laminate through to the copper strips. Solder them in place.

9 Trim the legs of the components level with the base of the laminate using wire cutters.

Figure 13.6 ■
The above figure illustrates the steps involved in making a printed circuit board (PCB). *(Photo by Derek Henderson Photography)*

All of the above work brings you to the final full-size layout of the control circuit shown in Figure 13.5. The full-size layout will provide the basis for the artwork that is required for making a PCB. For detailed instructions for making a PCB, see Figure 13.6.

There are many other circuits that you could use in your design and development work. Some of these are included in the activities of the *Portfolio and Activities Resource* that you can develop and use when appropriate.

CONTROLLING TECHNOLOGY SYSTEMS

In this section, three types of systems will be explored in more depth. Starting from the simplest and moving on to the more sophisticated, they are open-loop systems, closed-loop systems, and self-regulating systems.

Open-Loop Systems

Of the different types of control systems, open-loop systems are the simplest. An open-loop system links the basic building blocks in a **linear** fashion with the input leading to the control block, which in turn links to the output block. The illustration in Figure 13.7 shows a toaster as an example of an open-loop system. The control input to the system allows you to adjust how dark the toast will be. As shown in the illustration, you actually adjust the distance a bimetallic device must move before the toaster is switched off. The sensor part of the device operates from the heat of the toaster. The process part of the control system actually moves as it is heated and the control arm comes closer and closer to the output switch. When this mechanical switch is triggered, a spring causes the toast to pop up while at the same time turning off the electricity to the heating coils. Once you have made your adjustment and placed the bread in the toaster, the process runs its course. That is what makes it an open-loop system.

■ **Figure 13.7**
A toaster as an example of an open-loop system, showing its adjustable control.

Dark Toast Setting

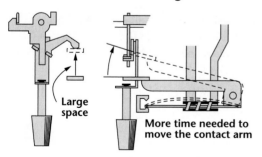

Large space

More time needed to move the contact arm

Light Toast Setting

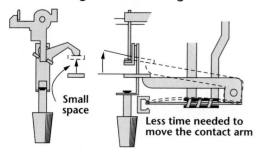

Small space

Less time needed to move the contact arm

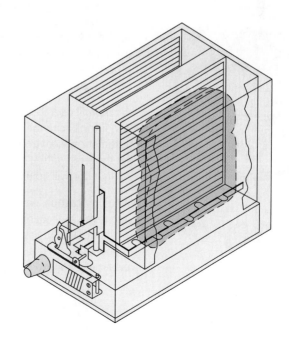

The awake-alarm system described in the preceding section on designing circuits is another example of an open-loop system. In the awake-alarm circuit, the system turns on and sounds the buzzer. You can adjust the resistance to determine when the buzzer comes on, but the system cannot do this itself. Neither can the system sense whether you are in your room. It will operate anyway. Unless you are there to turn the system off, it will sound the buzzer until it gets dark or until the battery runs down.

Closed-Loop Systems

When a closed-loop system is used, the intent is to initiate a desired action, and to provide some way of checking to see if that action has been carried out correctly. The means of checking the action of a system is referred to as **feedback**. As suggested above, the operator of the sound system makes use of feedback to determine if the sound is of appropriate loudness and quality. Figure 13.8 shows a sound system as an example of a closed-loop system.

In this instance, the feedback is possible as the operator senses the sound output of the system and makes changes in the control settings. Other instances where the radio provides the feedback are found in the scan function that automatically searches out the next station, and in the automatic volume control that keeps the sound level within an acceptable range. These instances of control do not rely on humans but are designed into the system so that the control becomes self-correcting.

Self-Regulating Systems

Control is very important in technology. Devices and systems must provide for human regulation from outside the system. In some instances, it is also useful for devices and systems to be self-regulating, or **cybernetic**. The steam engine controlled by a flyball governor, shown earlier in Figure 13.2, is an example of such a system. Another example of a self-regulating system is a windmill. As the wind shifts, it exerts force on the tail of the windmill. This swings the windmill blades to face into the breeze. The windmill has taken full advantage of the wind velocity and the blades continue to rotate briskly.

A windmill operates as a cybernetic device because it adjusts and controls itself as conditions change. A refrigerator, as another example of cybernetics, automatically controls its operation. Plug the refrigerator into an electrical outlet, set the temperature adjustment, and the system will be on "automatic pilot." The refrigerator will adjust itself to maintain the proper degree of coldness.

A toy gyroscope, shown in Figure 13.9A, spins on its tip without falling over. It also resists any change in direction. This gyroscopic action of any spinning wheel is what makes it possible to ride a bicycle without holding onto the handlebars, though this is not recommended. **Gyroscopic action** is important in the cybernetic control of airplanes.

■ **Figure 13.8**
This sound system is an example of a closed-loop system that can be adjusted by the operator and automatic controls. *(Courtesy of Churchill and Klehr)*

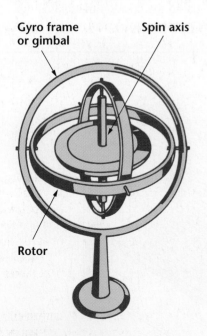

Gyro frame or gimbal   Spin axis

Rotor

■ **Figure 13.9A**
This toy gyroscope is very similar to those used to keep airplanes level in flight.

MATH/SCIENCE/TECHNOLOGY LINK

An Early Self-Regulating System

In the early days of the airplane, an inventor named Lawrence Sperry entered an airplane in an international competition held by the Aero Club of France. A prize of 50,000 francs, a lot of money at the time, was to be awarded for the safest airplane in the competition. Sperry's entry had a gyroscope attached to the airplane's controls. Remember that the gyroscope is a rapidly spinning device that does not like to have its direction changed. Sperry fastened wires from the gyroscope to the controls of his airplane. If the airplane tipped to the left or right, the gyroscope would pull the controls to the right or left and cause the craft to become level again. By connecting the gyroscope to the airplane controls, Sperry was able to make an airplane that was self-correcting.

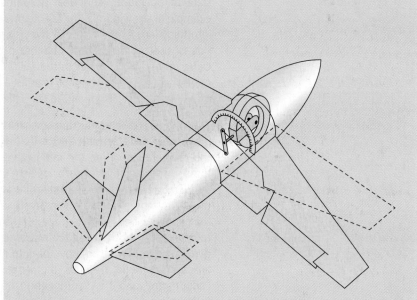

■ Figure 13.9B
The gyroscope in an airplane is able to control the plane and keep it level in flight. As the airplane tips to the right, the gyroscope continues to spin in a level position. The control stick connected to the gyroscope pulls the left aileron up and the right one down. This causes the airplane to straighten up and fly with its wings level again.

To demonstrate his confidence in the new system, Sperry had his mechanic crawl out on one wing during the flight. As he and the mechanic sped by in the airplane, Sperry removed his hands from the controls and stood up in the cockpit. He waved with both hands at the judges. The gyroscope prevented the plane from tipping under the weight of the mechanic. History did not record what Sperry's mechanic thought about the stunt, but Sperry won the prize.

Cybernetics is defined as "the automatic control of a machine toward a desired goal." The tail orients the windmill to face the wind. The refrigerator controls cause it to work only enough to reach the adjusted temperature setting. Sperry's gyroscope keeps airplanes flying straight and level. When we can set a machine to accomplish a task on its own, without further human adjustment, we have a cybernetic, or self-regulating, system.

Feedback in Control Systems

Some important questions must be answered before automatic machines are possible. How does the machine "know" when it has reached the desired goal? For example, how does the windmill know when it points directly into the wind? How does the refrigerator know when the correct temperature has been reached? How does the airplane know when it is flying straight and level? In all cases, the answer is in the concept of feedback.

In our three examples, adjustments continue until the machines reach the desired state of operation. This goal is sometimes called the **steady state**. In the example of the windmill, the tail continues to turn until the windmill faces directly into the wind. As the windmill swings around, the force of the wind against one side of the tail is reduced. Changes caused in the orientation of the windmill cause immediate changes in the tail. These tail changes are feedback from the original change. The windmill reaches a steady state when the force of the wind is the same on each side of the tail. This steady state represents the desired goal.

The refrigerator and the airplane also attempt to reach a steady state. The refrigerator has a sensing device that reacts as the temperature changes. When the temperature inside the refrigerator is warmer than the setting, the device turns on the cooling motor which drives the compressor and lowers the temperature. When the temperature is cold enough, the device turns off the motor. In Sperry's airplane, the steady state is established by the level position of the airplane when the gyroscope is first turned on. The gyroscope does not want to change its position, so each movement of the airplane is counteracted by the gyroscope so that it can move back to its original level position. All of the above are accomplished because the systems have the capability of setting an "adjustment" and then "comparing" the output of the system to see if it matches that setting.

DESIGN ACTIVITIES

The application of feedback for control involves using the compare/adjust function of the systems model shown in Figure 13.10.

One of the members of your family has become interested in gardening and wants to use a small greenhouse to give plants an early start while the weather is still too cool for planting outside. There are a number of interesting control problems related to even a small greenhouse. Three possibilities include providing water to the plants, providing adequate lighting, and controlling the temperature of the greenhouse. Each of these control functions will require the capability of comparing/adjusting the system. For example, in a large greenhouse a thermostat might be used to control a furnace that would be used to heat the greenhouse. As shown in Figure 13.10, the adjustment and comparison process will determine if the furnace needs to be turned on or not.

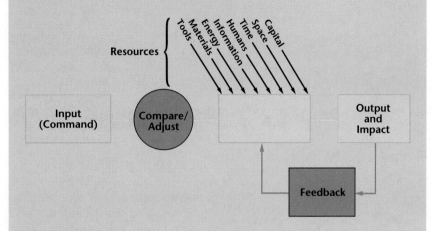

■ **Figure 13.10**
The systems model is shown here with the feedback and compare/adjust components highlighted.

Similarly, a moisture indicator would be necessary for sensing when the soil for the plants had become too dry. The control of ventilation might require monitoring the light levels provided for the plants.

• Design and develop one of these, or other appropriate controls, for use in a small greenhouse. Be sure to determine the setting that is most appropriate and the means you will use to compare the output of your system to that setting.

• Design and develop a hydroponics system that involves monitoring such conditions as nutrient levels, pH factors, temperature, humidity, light levels, and time of day.

MATH/SCIENCE/TECHNOLOGY LINK

Consider the following concerns about the greenhouse and the plants that will be placed in it. What other ideas or concerns do you think are important for improving the greenhouse?

What is a suitable range of temperatures for the plants in the greenhouse?

What nutrients do the plants require?

What influence does the chemistry of the soil have on the plants?

How do plants grow in a hydroponic greenhouse without soil?

How can you determine what size the greenhouse should be?

How often should the air be exchanged?

What effect does the temperature of the soil have on plant growth?

What effect does artificial light versus natural light have on the plants?

DEVELOPING SMART MACHINES

A great deal of attention has been given to the idea of "smart" machines. As a result, some machines have been developed to perform jobs requiring some decision-making. Some of the discussion about smart machines has been turned into entertaining but misleading ideas of machines taking over and controlling humans. The following section is intended to introduce the basic means that can be used to develop machines that "think."

Logic and Decisions

Automatic control is not possible unless the machine is able to determine when the changes it makes are enough to reach the goal. Machines cannot think as people do. Cybernetic machines do, however, operate on the principle of machine logic.

Machine logic is simply the "unthinking" response to controls. The tail of the windmill will unthinkingly turn the blades directly into the high winds of a storm. Although it was a logical move, the mill might be destroyed by the force of the wind. The feedback control system simply does what it has been designed to do.

Using AND Logic Gates

Simple logic gates that use common electrical switches were introduced in the chapter on Information. The same concepts can be applied in digital electronics by using electronic devices called integrated circuits (ICs). ICs, often called "chips," can have hundreds, even thousands, of electronic components in a very small device. ICs use microminiature solid-state switches to accomplish the same functions as the electrical switches.

One commonly used IC that provides the AND function is the 7408. The 7408 device, shown in Figure 13.11C, contains four AND gates, each with two inputs and one output. Integrated circuits can be obtained with three inputs or more, but still will have only one output.

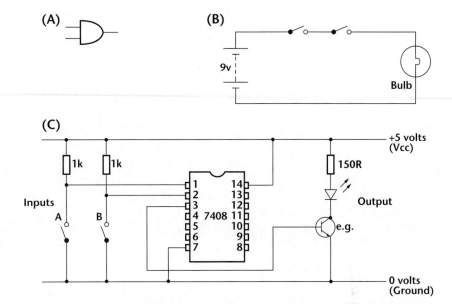

Figure 13.11A, B, C ■
(A) An AND logic gate. (B) An AND circuit that controls a light-bulb. (C) A 7408 AND logic gate.

The input for the 7408 and other similar ICs is described as a "1" or a "0," often called logic "1" or logic "0." An input of logic "1" means a switch or other device is providing an input, usually a signal of five volts. If the switches are left open (turned off) and there is no input, or logic "0," the voltage will be at 0. If both inputs to the logic gate are connected to five volts, there will be an output. However, if either or both inputs are connected to 0 volts, there will be no output.

(At this point, you may want to refer to the *Portfolio and Activities Resource* to see outline diagrams of the more commonly used integrated circuits.)

Using OR Logic Gates

A second commonly used IC that provides the OR logic function is the 7432. The 7432 device contains four OR gates, each with two inputs and one output.

Consider, for example, a system that will sound a doorbell if a person presses the button at either the front or back door of a house. This system has two inputs, one at each door, that represent an OR logic gate function.

In the OR circuit shown in Figure 13.12B, activating either switch "A" or switch "B" will sound the doorbell. Figure 13.12C illustrates what the circuit might look like if developed using an IC. The same IC circuit could be used to activate a relay to switch on (control) a circuit that has larger power requirements than the nine volts used to power the IC circuit.

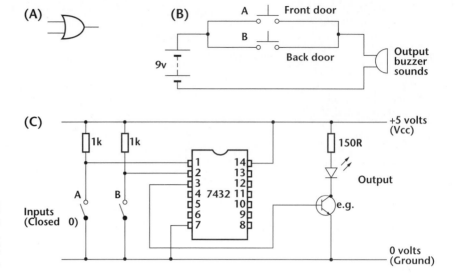

Figure 13.12A, B, C ■
(A) An OR logic gate. (B) Doorbells representing an OR logic circuit. (C) A 7432 OR logic gate in an IC control circuit.

Using NOT Logic Gates

Another commonly used IC is the NOT logic gate. An IC that provides the NOT logic function is the 7404. The 7404 device contains six NOT gates, each with one input and one output.

The function of the NOT gate is to invert the signal. If there is an input signal (on=1), there will be no output signal (off=0). Similarly, if there is no input (0), the inverter provides an output signal (1).

In the circuits shown in Figure 13.13, (A) represents the NOT symbol. In Figure 13.13B, the bulb will light if the switch is open (off). Activating the switch will cause the safety light to go out. Figure 13.13C illustrates what the circuit might look like if developed using an IC. In this circuit, closing the input switch creates a current flow through resistor R1. This also creates a positive input (on) at pin "1." This signal is inverted to no (off) signal from pin "2" to the transistor, causing the light to go off.

It is important to note that all of the above IC circuits have limitations as to the amount of electrical power they can supply as an output. If these circuits are to be used to control devices with higher power requirements, it will be necessary to use more robust control devices. This is accomplished by using the IC circuit to control other circuits. This can be as simple as activating a relay to switch on (control) a circuit that has larger power requirements than the nine volts used to power the IC circuit. The output signal from an IC circuit can also be used to trigger a transistor capable of operating as a switch. Additional information on these uses of IC is provided in your *Portfolio and Activities Resource*.

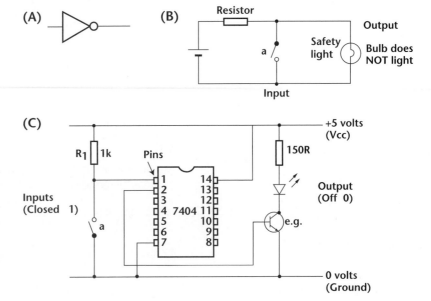

Figure 13.13A, B, C ■
(A) The symbol for a NOT circuit.
(B) A simple NOT circuit for controlling a safety light.
(C) A 7404 NOT logic gate in an IC control circuit.

Production and Controls

An early advance in technology was the change from muscle power to other forms of energy. Humans still had to sense the information, monitor the processes, and provide the guidance and control of the machines and tools, however. As long as the production was the custom-making of individual items, it was all right for people to supply the guidance and control. But as demand grew for products, attention turned to developing more precise machines and more efficient energy production and distribution systems.

As people wanted more and more technological products, key mass production techniques such as interchangeable parts emerged. Mass production required high-quality outputs, which in turn required information (feedback) on how accurately the items and parts were being made. As more and more complicated products were made, the manufacturing and quality-control processes became more intricate.

DESIGN ACTIVITIES

There has been a lot of talk about "smart machines" and the role they can play in taking over responsibility for specific tasks. Some of these machines have become very sophisticated; many have been used for some time and operate on a simpler yet similar basis. Consider the following as example problems that could be approached by using logic control devices.

• Provide a safety guard for an automatic stamping machine that will ensure the operator will have both hands out of the way when the machine goes through its stamping process.

- Provide an automatic door opener that can be operated by individuals approaching the door from either side.

- Provide a means for helping an attendant in a parking lot monitor exits that are out of sight and determine if any gate has been left open or if the wooden arm has been broken.

- Provide a means to ensure that both sets of doors on an elevator are closed before the elevator car is moved to a different floor.

- Provide a means of turning off the doorbell on your home so that you will not be bothered at night.

- Design and develop a logic system that can be integrated into your project work.

- Identify several circumstances where logic systems might be used in your home, at school, or in the community.

■ Figure 13.14
Large numbers of automobiles, all exactly the same, can be produced by automated production systems.
(Courtesy of Nissan)

AUTOMATION AND CONTROL IN PRODUCTION

Automation has had a major impact on production. One of the first automated industries was set up by Oliver Evans, an American millwright. Evans used conveyors driven by water to move grain and other materials in a flour mill.

Evans' conveyors transported the materials to the machines that ground the grain and sorted the product. Simple automated devices such as those used by Evans have been available for hundreds of years, but completely automated manufacturing has emerged only within this century. Within the past few decades, automation procedures have been applied to many different types of production systems. Automobile assembly plants and oil refineries are examples.

Automation has three parts that work in close-knit harmony with each other: the **process** by which the product is being manufactured; the **handling** by which the materials are being transported to, through, and from the machines performing the process; and the **control** by which the process and handling are kept coordinated with each other. These three parts comprise a **manufacturing** system. The controls provide the feedback so that the system can sense changes in the handling and processing and can bring them back to a desired state of operation. Automation, therefore, is a very practical use of the principle of cybernetic control.

(A)

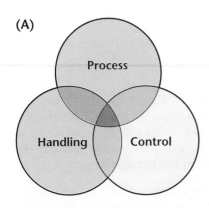

■ **Figure 13.15A, B**

(A) The integration of process, handling, and control. (B) Not all instances of automation and control are necessarily complicated. The mechanism illustrated here is almost elegant in its simplicity. A device such as this fills the millions of boxes of cereal that are consumed at breakfast each day.

In a production line that fills cereal boxes, as shown in Figure 13.15B, it is important that all three concepts of process, handling, and control operate as one related system. When this happens the operation becomes automatic. The control parts of the system include the sensing devices that provide information. These are used to control and to change the processes. Other sensing devices provide information to direct the handling of the materials and parts. The materials are then converted through the intended materials conversion processes. Thus, process, handling, and control can be united into one functioning system as represented in Figure 13.15A. As such systems become more sophisticated, control systems of computers and microprocessors become essential to monitoring and processing all the information that is needed and used to operate and control the system.

Over the last several decades, there has been increased use of devices and machines that can make quality-control checks on the production processes. These machines send information as feedback to the production machines which in turn make the required adjustments. This improves the processes while the operation is going on. At first, the quality checks were made and feedback was provided through mechanical means. Improvements were made using electrical and then electronic devices. As electronic sensing, logic, and memory devices have become more sophisticated, the control of complicated processes by machines was possible. With the advent of robots, fully automated industries are now a reality.

(B)

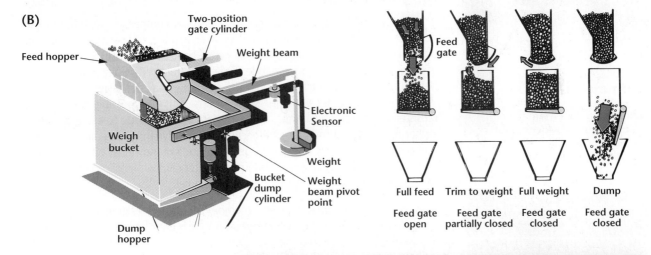

D E S I G N   A C T I V I T I E S

There are many instances in industry that use automation and control in production. This could include filling food containers, stacking or sorting packages for storage, moving parts on a production line, or the control of the nutrients, light, and heat for a hydroponics greenhouse.

The drawings in Figure 13.15B represent a cereal box filling operation on a production line. Clarify the problems related to the automatic operation and design a system that would control the operation that is illustrated.

COMPUTERS AND CONTROL TECHNOLOGY

Analog and Digital Devices

The first computers were analog in nature and were developed to measure processes. The speedometer in an automobile is an analog device that allows for measuring speed. Other examples of analog devices are the gas gauge in cars, a clock with hands, and a pressure gauge in a boiler. **Analog devices** do continuous sensing and measuring of something that is happening.

Digital measurements are discrete. In fact, digital means "unit." Digital systems can be developed on any number "base," but the most common digital system we use works on a "base of 10." Digital computers work on a digital system of "base 2." The numbers in a base 2 system are "0" and "1." The common name for this system is **binary system,** because there are only two numbers used. The binary "0" and "1" are often represented by switches that are turned on (1) or off (0). Although the switches could be mechanical or electrical—as they were in the early stages of computers—they are now almost solely electronic. Electronic switches are extremely small, fast, and inexpensive.

Computer Memory

As indicated above, early computers were huge in comparison to those in use today. Each memory **bit** was in the form of a relay and vacuum tube that together were about the size of a small cereal box. Later, transistors provided a bit of **memory** in the size of a marble. They were hundreds of times smaller than the vacuum tubes. The transistor did the same thing as the relays which made on/off connections, but with no moving parts. This was possible because the transistor was made of dynamic materials that could be made to act as a switch. Further reductions of memory size occurred as the dynamic materials continued to improve. Soon it was possible to get thousands of individual bits of memory fitted onto a small **chip**. Now a modern computer chip with millions of electronic circuits can easily pass through the eye of a needle!

Many small home computers have a memory capability measured in megabytes. A megabyte is equal to one million **bytes**. A byte is equal to eight bits, with each bit representing a micro-miniature electronic switch in an off (0) or on (1) position. A megabyte, therefore, is roughly about eight million bits. If even a small computer now available had to be built using the original vacuum tube and electrical relay system used 50 years ago, it would fill a large building. Small computers that you can easily carry in your shirt pocket have many thousands of times more power than the first computers, and they cost less than a single part of the original computers that provided a single bit of memory.

Computer Programs

Memory is required to get any computer to perform a complex set of actions over and over again. It is also necessary to call up the proper sequence of actions at the proper time. The series of commands that tells a computer what to do is

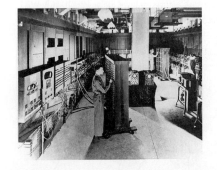

■ **Figure 13.16A**
ENIAC, the first electronic computer, used a room full of radio tubes to provide modest calculation capability. *(Photo courtesy of Unisys Corporation)*

■ **Figure 13.16B**
The later generation of computers became progressively smaller with the introduction of the transistor and then the microprocessor chip. *(Photo courtesy of International Business Machines Corporation)*

called a **program.** All programs may be thought of as having two parts: an internal program to control the operation of the computer and an external program to do the actual work. The **internal program** is determined when the computer is built. The specific procedures needed to control the internal workings of the computer are established by setting up the thousands and thousands of switches in specific patterns. As described earlier, this is accomplished by talking to the computer in a "machine language" of "0's" and "1's." The bar code reader introduced in the preceding chapter is another example.

In this example, the simplified bar codes have only five bars rather than the 30 or more that are commonly used. The five bars provide code numbers for about 100 items—the 30 bars provide for millions of items. When the simplified bar code for a product is scanned, the combination of black bars and white spaces are sensed by passing the reader over the bar code. The information captured in the scan is then sent as a series of signals to the computer. The computer interprets the signals and translates them into the number and name of the coded product. This information can be used to create a receipt for the buyer, to update the inventory for that product, and to order more products if appropriate.

■ **Figure 13.16C**
The continued reduction in the size of processor chips has resulted in laptop computers thousands of times more powerful than the original ENIAC. *(Photo courtesy of Hewlett-Packard Company)*

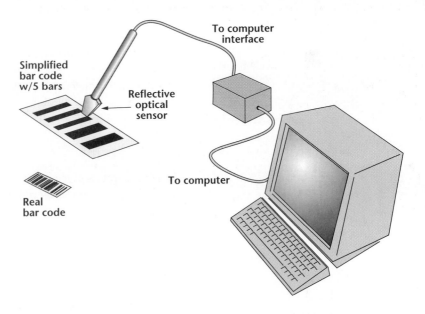

Figure 13.17 ■
A simple bar code reader system with simplified codes and hand-operated scanner.

Commands are sent to a computer through an **external program**. The commands can be as simple as touching the numbers on a microwave oven. The usual method of sending commands to a computer is by typing on a keyboard. A person can work at the keyboard and use the computer without knowing the machine language being used inside the computer. Gears in a standard car transmission also represent an internal program that is built into the machine. When you shift gears, you apply the external program by choosing the gear that you want, but you can only choose the gears that were provided in the internal program. The same is true of a computer. You can only command at the keyboard what was provided for in the internal programming.

The computer programming for the bar code reader introduced in Figure 13.17 could be done in a variety of languages. Because each of the languages is different from the others, it is helpful to develop a general flow diagram that can be used to guide the development of the programming in all the languages. The flowchart provides a means of planning and describing the software systems that can be used to read the bar codes.

The flowchart shown in Figure 13.18A is quite similar to the one used with the scanning device described earlier in Figure 12-11B. The flowchart in Figure 13.18B is more specific and a step closer to the machine language programming required by the computer, shown in Figure 13.18C.

Some microprocessors (small computers on a chip) are built so that their internal programs cannot be changed. These are **dedicated devices.** They only perform specific functions. Dedicated chips are used in microwave ovens, electronic ignitions for cars, computer watches, musical birthday cards, video games, and pinball machines. You cannot program these chips to do any other tasks. The performance of these devices appears to be quite complicated because of their large memory. The machine must be programmed to understand that when "X" happens, it is to do "Y." When you play a game and make move "X," the dedicated chip takes predetermined action "Y."

Computers can control machines to do complicated procedures because of their fast memory and decision-making capabilities. Onboard computers controlled the landing of spacecraft on distant planets. Computers control sophisticated robots that perform amazing feats in **manufacturing**, including the assembly of parts for other robots. The robot exemplifies the **integration** of the three elements of automation—process, handling, and control.

Integrated Systems

When all three concepts of process, handling, and control operate in one related system, automatic operation is achieved. The control parts of the system include the sensing devices that provide information. These are used to control and to change the processes. Other sensing devices provide information to direct the handling of the materials and parts. The materials are then converted through the intended materials conversion processes. Thus, process, handling, and control can be united into one functioning system.

The work of humans in designing and controlling large systems is made easier by the computer in two ways. First, the computer is used to construct a simulation of the system during the design and improvement phases. Second, the computer is used to record the data during the operation of the system. Through simulation, the system can be modeled and experimented upon safely and economically. It would be unwise and sometimes impossible to try out a new approach without first conducting a simulated trial run. How would you like to be the first passenger on a new train that had never been tested? Designs of large complex systems may also be simulated and tested. For example, the illustration in Figure 13.19 shows the operation of a 19-state, 17,000-mile railroad system.

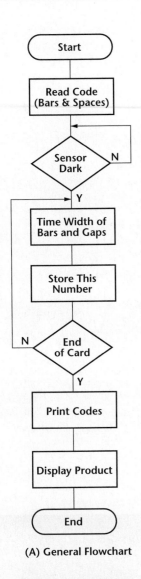

(A) General Flowchart

Figure 13.18A, B, C ■
These flowcharts describe: (A) the general operation of the bar code reader, (B) more specific steps involved in reading a product code, and (C) a few lines of programming for that product. *(Courtesy of Barbara Stala and Cindy Yayac)*

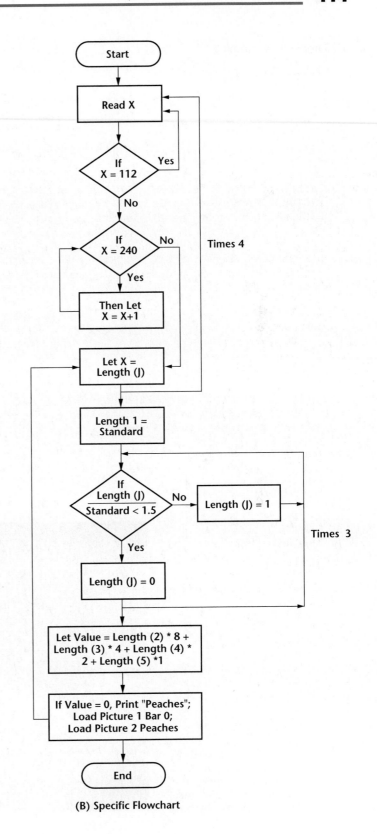

(B) Specific Flowchart

(C) Sample Programming

```
Let STANDARD = length
For J = 2 to 5
If length(J) / STANDARD < 1.5 Then length(J) = 0 Else length(J) = 1

Next J
Let value = (length(2) * 8) + (length(3) * 4) + (length(4) * 2) + (length(5) *1)
Select Case value
    Case 0
    LABEL1.Caption = "PEACHES: GEORGIA - $0.69/LB."
    PICTURE1.Picture = LoadPicture("c:\vb\graph\BAR0.BMP")
    PICTURE2.Picture = LoadPicture("c:\vb\graph\PEACHES.BMP")

    Case 1
    LABEL1.Caption = "CARROTS: FLORIDA - $0.39/BUNCH"
    PICTURE1.Picture = LoadPicture("c:\vb\graph\BAR1.BMP")
    PICTURE2.Picture = LoadPicture("c:\vb\graph\CARROTS.BMP")

    Case 2
    .
    .
    .

    Case 15
    LABEL1.Caption = "ONION: BERMUDA - $0.39/LB."
    PICTURE1.Picture = LoadPicture("c:\vb\graph\BAR15.BMP")
    PICTURE2.Picture = LoadPicture("c:\vb\graph\ONION.BMP")

End Select

End Sub
```

Figure 13.19 ■
This train dispatcher uses computers to control the operations of a 17,000-mile railroad system. *(© Randall Hyman)*

The significance of the computer as a part of larger integrated systems lies in its ability to juggle many decisions without confusion. The larger a system becomes, the more information that must be handled to monitor and control that system. In reality, the handling of the data can be managed only with the aid of powerful computers. The recent development of new and smaller computers promises to make the task even easier and cheaper.

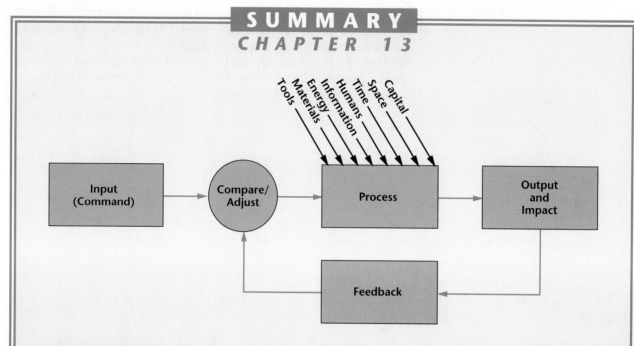

■ **Figure 13.20**
Control plays a central role in any self-correcting system.

The concepts of control and cybernetics relate closely to concepts of communication. The feedback aspect, so important in communication systems, often takes the form of mechanical devices. Communication between two or more people does not necessarily require a machine, but the principles are the same. Cybernetics refers to communication and control as applied to machines and human systems.

Transportation, production, and communication can fit together to form a cybernetic manufacturing system. A cybernetic system is possible only if there are channels through which feedback can flow to the decision points in the system so that the necessary changes can be made.

Finally, the advances that have been made in control technology give the impression that machines are becoming smarter and smarter. Advances continue to be made in machine logic, through smaller, more powerful microchips and new approaches to computer programming. These are leading to computers with processing capabilities hardly considered possible a decade ago.

In the early stages of technology, tools and machines were developed that allowed humans to turn part of the heavy work and drudgery over to machines. This released humans to use their creative and critical abilities for more interesting and productive ends. The advances being made in the development of computer hardware and software promises to relieve humans of decision-making and information handling. The potential impact of this on the work and productivity of humans remains unclear and will only be seen sometime in the future.

ENRICHMENT ACTIVITIES

- Design and develop a means of signaling when the kitty litter box needs to be emptied.

- Design and develop a means of automatically dispensing cream, sugar, artificial sweetener, and lemon into coffee or tea.

- Develop a flowchart that indicates the sequence and decisions involved in one of your daily routines, such as getting ready to go to school.

- Develop a flowchart that you can use to prepare a computer program to control a motorized device that you built in an earlier activity.

- Identify a problem in your home or at school for which a control system is appropriate. Design a potential solution to the problem.

- Identify a problem related to your studies in science for which a control system would be appropriate. Work with your science and technology teachers to design and develop a system that could be used in your school.

MATH/SCIENCE/TECHNOLOGY LINK

- Determine and describe the math and science that is related to the operation of the flywheel governor illustrated in Figure 13.2.

- Determine the high and low end of sensitivity of the common sensors used to provide input for a control system.

- Determine the chemical processes involved in making a PCB and what takes place during those processes. Identify what safety precautions are necessary in making a PCB.

- Explain how the toy gyroscope shown in Figure 13.9A actually operates and how a gyroscope is used as part of a control system for an airplane.

- Describe how a hydroponics system operates. Describe the means you would use to monitor and control conditions related to the nutrient levels, pH levels, temperature, humidity, and light levels for the system.

- Identify and describe the math and science concepts that are related to logic gates and the three logic circuits illustrated in Figures 13.11, 13.12, and 13.13.

- Explain how memory devices used in current microcomputers operate and how so much memory capability can be squeezed into such a small space.

- What changes do you expect might happen in the near future in computer use and methods of input and programming?

ENRICHMENT ACTIVITIES

- Identify the different components of a control system and the role that each of them plays in its operation.

- Explain what is meant by the statement, "The output of one system can serve as the input for another."

- Identify the different sensors and other devices that can be used for the input aspect of a control system. Describe how they function.

- Identify the different devices that can be used for the output aspect of a control system. Describe how they function.

- Explain how the process circuit of a control system can be controlled by a change of resistance in the input device.

- Describe the differences of open-loop, closed-loop, and self-regulating systems.

- Describe the role that the "adjust/compare" aspect of a control system plays and how it relates to the "feedback" aspect of the system.

- Describe how AND, OR, and NOT logic circuits operate and provide some examples of how they can by applied.

- Explain the significance of integrating "process, handling, and control" in an automated manufacturing operation.

- Describe the role that flowcharting can play in planning for computer programming.

- Identify and describe some of the large systems that require the use of computers in order to operate.

UNIT *FOUR*

The Impact of Technology

Introduction to Unit IV

As you come to the final unit of this book, you bring with you some knowledge of design and problem solving, the resources of technology, and the way in which technological activities are interrelated into systems. In this section, we attend to some historic and future developments in technology. As we wrap up the study of design and technology, we hope that you conclude that you *can* make a difference in your own life and in that of others, even people you do not know.

Chapter 14, Investigating, Developing, and Improving, identifies some of the concerns that people have had as they traveled and explored the planet, outer space, and the oceans. Tools and devices were developed in order to investigate the world beyond our senses, the very large, the ancient, and the very small. Research and development efforts and the approaches used by each are described. In order to gather more information and investigate future possibilities, several approaches can be used to forecast the future. These ideas provide insight as we place information from the past, present, and future together.

In Chapter 15, Consequences and Decisions, we examine the many possible consequences of technological development. These effects may be intended and unintended, immediate and delayed, and desirable as well as undesirable. By assessing possible consequences, it is hoped that some of the unintended and undesirable effects can be controlled or prevented. One of the major purposes of the study of technology is to provide the basis for informed decision-making. Aspects of decision-making, alternative choices, and sources of influence in making choices are explored in this chapter. Goals, actions, resources, and outcomes are examined in light of values. The complex nature of choices that have personal, social, and ecological effects points out how crucial it is to give careful thought before taking action. We examine how decisions made by a few have an impact on many others around the world and in the future. The importance of becoming aware of interdependence of the people who share this planet is emphasized.

The Epilogue contains some suggestions as to how an individual can make a difference. Even though many day-to-day behaviors do not have obvious effects on others, we examine how that might be so. We also discuss how to act locally, in your own neighborhood, for the long-term benefit of others, even those in other nations and those not yet born.

CHAPTER 14

Investigating, Developing, and Improving

Figure 14.1 ■
Radio telescopes such as this one have reached enormous proportions. They can be linked together with others around the world and can operate as though they were one device. (© 1991 Jack Zehrt/FPG International)

INTRODUCTION—HOW TECHNOLOGY GROWS

We began this book with an examination of the design and problem solving process, and by now you have completed several designs of your own. This chapter looks at how this design process has been applied, over time, to a

419

KEY TERMS

acceptable risk
analogies
applied research
artifacts
Delphi survey
developments
discovery
forecasting
future histories
futurists
half-life
innovation
invention
iterating
location
magnetic North
maximizing the criterion
microscope
modeling
optimization
predictions
pure research
radio telescope
radioactive carbon
reflector
refractor
research and development
rotarios
scenarios
scientific method
stick chart
trade-off process
trend analysis
true North

number of different issues. This is an examination of the manner and processes by which technology has grown and continues to develop. People's desire to know more and improve products and processes has expanded our horizons. Products, systems, and environments have been improved through investigation, as well as design and development.

The desire to know is a major characteristic of the human species. Throughout this book you have read of instances where humans have been dissatisfied with things as they were and wanted to explore beyond what was currently known. New tools and instruments had to be developed to support this investigation.

Advances in manufacturing and production led to automated, large-scale technology systems. These systems can be made self-directing and self-regulating in order to meet the demands of mass production. The continued development of technological activities has been supported by the development of the computer, miniaturization, and improved materials and tools.

INVESTIGATING THROUGH EXPLORING

Direction—Which Way Are We Going?

Early in human history, people noted that the sun and stars moved in predictable ways. The sun would rise and set in the same general pattern day after day and year after year. When the seasonal variations were taken into account, the sun became a very predictable and dependable tool.

Many of the early direction-finding devices used the sun. A sundial uses the shadow cast by the sun shining on a vertical pointer to indicate the time of day. Although they are not as accurate as modern-day digital clocks, sundials were very valuable at the time. The sundial also helped tell early explorers which direction they were heading. The Vikings took advantage of this knowledge and developed a device called a ship's bearing dial. (Refer to Figure 14.2.) The device provided orientation to the north. It had a moveable pointer that enabled sailors to know in what direction the ship was sailing and the bearing (direction) that the ship should maintain.

Except for some accomplishments of the Phoenicians, Polynesians, and Vikings, exploration of the oceans continued without major changes for two thousand years. The invention of the hinged rudder, between 1000 and 1200 A.D., opened the way for safe ocean travel for all people. Prior to that time, all ships were steered with one or two oars. This arrangement made it very difficult to maintain direction on the open seas. The hinged rudder was attached to the boat below the surface of the water. The control of even large ships was now within the power of one person, the pilot or helmsman using a long lever or other means of mechanical advantage. The ship could also perform the difficult maneuvers needed to sail or tack into the wind. Its direction could be better controlled in the changing currents of the ocean. Eventually, bigger and stronger ships were built, with better masts, sails, and rigging.

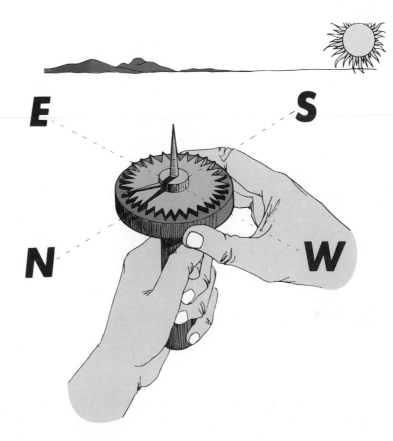

Figure 14.2 ■
The Vikings used a simple but reliable navigational device. The ship's "bearing dial" allowed long ocean voyages like those from Norway to Greenland. The shadow of the sun when directly overhead pointed to the north. Other directions could then be determined by rotating the pointer around the 32-point scale. The ship could be steered by using a preset bearing or direction.

The magnetic compass, which was developed by either the Arabs or the Chinese, provided a much needed direction indicator for sailing on the open seas. Before the compass, sailors had to depend upon sightings on the shore or locations of the sun and stars to keep track of the ship's location and direction. The compass works because the Earth acts as a large magnet. The compass has a magnetic pointer that is suspended so that it can turn freely. Unless it is interfered with by large metal objects or other sources of magnetism, the compass will align itself with the Earth's magnetic field.

The compass needle will always point north and south. But the north on a compass is different from the north on a ship's bearing dial. The compass points to **magnetic North** and the ship's bearing dial points to **"true" North**. There is a difference of about 1,000 miles between the two "Norths." This problem plagued early explorers, because they did not understand what caused the difference in the two Norths. The underlying scientific explanation came later, but in the interim, technological applications were not only possible, but they were also quite successful.

Position—Where Are We?

Devices that determine direction are very important tools of exploration and discovery. But as early explorers sailed the oceans, they found that merely knowing their bearing or direction was not enough. In time, a series of devices was developed that helped to determine location. These included the cross staff, the astrolabe, the quadrant, and the sextant. All four devices work on the same principle. They measure the angle or height of the sun or stars above the horizon.

The compass was far more useful when teamed with an astrolabe or sextant. The compass indicated the ship's direction and the astrolabe found its position. But the relative position of the sun or stars at given times changed as ships proceeded east or west. Because the sundial was not an accurate timekeeper, another device was needed.

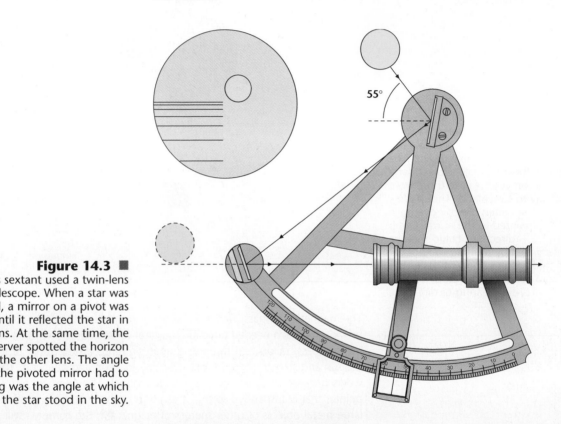

Figure 14.3 ■

This sextant used a twin-lens telescope. When a star was sighted, a mirror on a pivot was swung until it reflected the star in one lens. At the same time, the observer spotted the horizon through the other lens. The angle to which the pivoted mirror had to be swung was the angle at which the star stood in the sky.

The development of the clock provided the necessary information. The clock allowed explorers to measure how far they were from their starting point. Before leaving their home port, sailors set a clock at 12:00 when the midday sun was directly overhead. A sighting was taken the following day when the sun was directly overhead. If the clock said 1:00, it meant that the ship's relative time was one hour different from the time at home base. Because the sun takes 24 hours to go around the Earth and the ship sailed westward, the sailors could determine that they had gone 1/24 of the way around the Earth. By combining this information with a reading of an astrolabe or sextant, the ship's position could be accurately located on a map. Centuries later, mapmakers designated that home port to be Greenwich, England. The calculations of time, day, and location continue to be designated with this location as the "starting point" in much of the world today.

The three related ideas of direction, orientation, and time serve as the basis for navigation, even today. Passenger planes, oil tankers, spacecraft, and satellites all use these three kinds of information. But modern, more accurate measuring instruments have been developed.

Location—Where Have We Been?

A logical outgrowth of early navigation was the improvement of simple hand-drawn maps into accurate maps and charts. These early maps and charts, which took several forms, were the first documents of space. One of the first was the **stick chart**, shown in Figure 14.4A, used in the South Pacific to fix the position and distance between the islands. Some of the sticks also interpreted the wave pattern and currents of a given area of the ocean. The Portuguese developed another type of chart called **rotarios**. The rotarios listed the directions, distances, and hazards of a particular course. They were much like the flight maps used by present-day pilots. Although these maps were not accurate by today's standards, they were helpful to the early explorers.

As navigation instruments improved, more accurate and useful navigation charts and maps were needed. The first maps were drawn on flat paper. However, it was necessary to develop ways by which the features of a round world could be illustrated on a flat surface. The method most commonly used today is a Mercator **projection**, named after the person who developed the technique in the 1500s.

Space—What Is Out There?

Throughout your lifetime, the exploration of space has been an ongoing technological activity. Until the development of the rocket, such exploration was limited to observation and analysis through land-based telescopes and radio equipment. Early observers of the heavens were limited to the use of the unaided eye as a receiver until the invention of the telescope. The earliest telescopes, such as the kind that Galileo used, were of the **refractor** type. Approximately 50 years later, Sir Isaac Newton used the **reflector** telescope. It was a great improvement. Most astronomers still use the reflector telescope which, for nearly three hundred years, was the major device used to explore outer space.

In 1931, the first **radio telescope** was built by Karl G. Jansky in order to ascertain why there was static on the radio. (The radio was in its early development at the time.) Although he did not intend to explore outer space, Jansky's approach led to the radio telescopes that now can receive and collect energy from pulsars, quasars, radio galaxies, and hydrogen clouds. Some even pick up energy that may have been generated in the creation of the universe, billions of years ago.

Oceans—What Is Down There?

Some of the earliest exploring activities required travel on the oceans. Eventually, people became curious about the seas themselves and what might lie below their surface. Undersea exploration was held back for years, due largely to the lack of suitable technology to support human life in that environment. The weight of the water would simply crush the vehicles if they went deeper than their tolerance level.

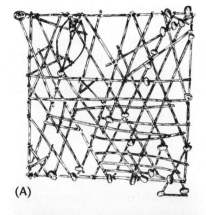

(A)

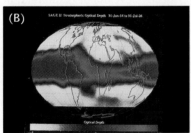

(B)

■ **Figure 14.4A, B**
(A) This is not an abstract sculpture. It is a map used by early Polynesians to sail among the islands of the South Pacific. (B) This map has been developed from data obtained from satellite photos. The colors indicate the depth of debris that circled the Earth following eruptions of a large volcano. *(B: Courtesy of NASA)*

Figure 14.5
Exploration under the sea was made possible only after materials and vehicles were developed to withstand the pressures. This diver is recovering materials from a shipwreck.

Only in the recent past has exploration of the seas begun to blossom. Because stronger metals have been developed, it is now possible to construct deep-sea exploration vehicles that can withstand the extreme pressure to certain depths. They can contain the environment needed to support humans, as well as their instruments.

Maps of the ocean floor and currents have been drawn, much like the maps that were developed when the Earth's surface was first explored. The hills and valleys of the Earth's crust are fairly well-known at this time. However, exploring under the Earth's crust still poses many problems. Seismographs have been used to chart the Earth's structure. Attempts have been made to chart movement of the Earth's layers (tectonic plates) and to predict earthquakes. Right now, we have a more accurate understanding of the makeup of the solar system than we do of the Earth itself.

Microspace—What Is in There?

The **microscope** is the counterpart of the telescope. Telescopes are used to explore outer space. Microscopes are used to investigate micro (small) space. Early microscopes were optical, making use of lenses and mirrors. In recent years, powerful electron-type microscopes have been developed. Explorers of microspace now have powerful new tools of exploration. Electron and X-ray microscopes enable us to explore the structure of molecules themselves. Recent developments in the use of lasers indicate that it may soon be possible to see even more clearly into the microspace of our world. Figure 14.6 shows some of this development.

Some of the most interesting, current microspace exploration is done by medical researchers. New developments in electronic instruments have made it possible to learn much more about how the body functions. Ultrasonic beams, X-ray scanners, and infrared photography can now be used instead of surgery to determine the causes of health problems.

History—How Old Is It?

The exploration of previous years is a search for evidence. Historians search for evidence in the technological remains of past activities. These **artifacts** are products that give us clues about how humans lived centuries ago.

Historians and archaeologists often use special instruments to help determine the age of artifacts. One such instrument measures the amount of radioactive carbon that remains in the object. **Radioactive carbon** atoms decay at a predictable rate, called a **half-life**. Archaeologists determine where the radioactive carbon atoms are in their decay sequence. This information provides a fairly accurate index of the length of time that the material has been decaying. For example, it is relatively easy to determine the age of a prehistoric campsite by sampling the remains of the charred wood. The wood would have been dead at the time of the fire or slightly before then. Thus, we can determine the date of that fire by finding the age of the carbon in the wood.

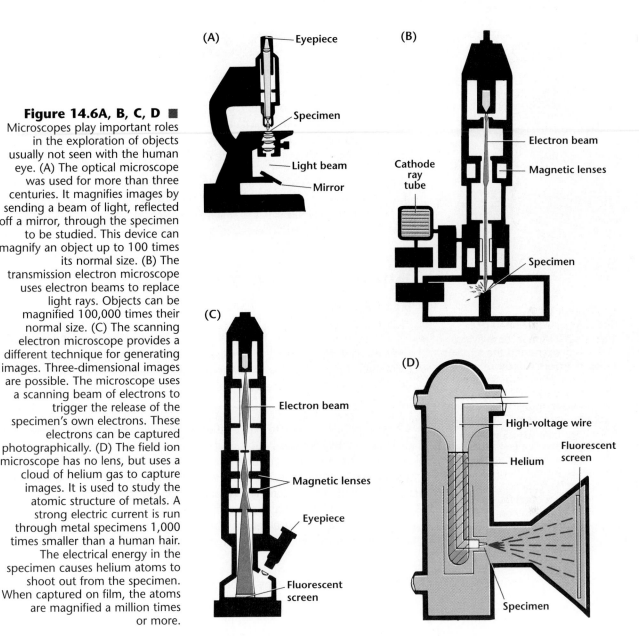

Figure 14.6A, B, C, D ■ Microscopes play important roles in the exploration of objects usually not seen with the human eye. (A) The optical microscope was used for more than three centuries. It magnifies images by sending a beam of light, reflected off a mirror, through the specimen to be studied. This device can magnify an object up to 100 times its normal size. (B) The transmission electron microscope uses electron beams to replace light rays. Objects can be magnified 100,000 times their normal size. (C) The scanning electron microscope provides a different technique for generating images. Three-dimensional images are possible. The microscope uses a scanning beam of electrons to trigger the release of the specimen's own electrons. These electrons can be captured photographically. (D) The field ion microscope has no lens, but uses a cloud of helium gas to capture images. It is used to study the atomic structure of metals. A strong electric current is run through metal specimens 1,000 times smaller than a human hair. The electrical energy in the specimen causes helium atoms to shoot out from the specimen. When captured on film, the atoms are magnified a million times or more.

The archaeologist is a central figure in historical research and exploration. But archaeology has evolved into a scientific profession only in the last 100 years. Before then, technology had not developed enough to provide the important tools needed in this field. The archaeologist now has many technological tools. The airplane is used to find the best possible site for a dig. The camera is used to document the dig. The electronic mine detector is used to find metal objects. The X-ray spectrum analyzer can determine the chemical composition of an object without harming it. The radioactive sampler can determine the age of objects through radioactive carbon dating.

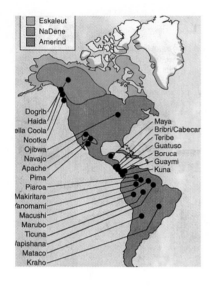

■ **Figure 14.7**
This map was developed from studies of DNA from both living Native Americans and the remains found in archaeological exploration. These artifacts are from tribes who built their civilizations long before European explorers "discovered" America. From this map, investigators concluded that the first Americans came in two waves of immigration from Asia. *(Reprinted with permission from Science. © 1993 American Association for the Advancement of Science. Illustration by D. Defrancesco.)*

DESIGN ACTIVITIES

• Select a tool or device that has been necessary for one of the types of investigation discussed in the previous pages of this chapter. Trace the history of the development of the device that you selected. Write an illustrated paper on the changes in the device over time. Include the contributions it made to the knowledge, capabilities, and expectations that people had as a result of the development of the tool.

• A student from China is enrolling in your school next week. In groups of three to four, create a map of your neighborhood or town for this new student to show aspects that will help him/her explore.

INVESTIGATING THROUGH FORECASTING

Humans have always tried to predict the future. History shows that in each civilization there were a few people considered to have an extraordinary ability to "see" into the future. Sometimes their **predictions** came true and sometimes they did not.

A more sophisticated way to explore the future is called **forecasting**. Forecasting depends on technological tools, but differs somewhat from other exploring activities. The weather forecaster, for example, begins each day by gathering information about local conditions. Barometric pressure, wind direction, humidity, and temperature are recorded. Satellite photos are examined for evidence of upper wind patterns and major storm areas. Data from other stations indicate what kind of weather may be moving into the area. Based on all the available information, the forecaster makes a prediction of what will probably happen.

The key processes of this type of forecasting include the following:

(1) collect information

(2) analyze data

(3) identify probabilities

(4) forecast the possibilities

Most future-oriented exploration follows a similar sequence of action. The weather forecast uses mostly current information. If the forecaster needed to predict the weather months in advance, records from past weather events would be used. Weather seems to occur in cycles. Long-range weather forecasters try to determine these cycles, or trends, to help them in their predictions. This method is called **trend analysis.**

Figure 14.8 ▪
(1) We know there is something called the past and a future. We see neither of them, since we exist between the past and future in a slice of time called the present. (2) You gain knowledge from past experiences and use that knowledge to gain insights into the future. Understanding the future is very difficult, however. (3) Sometimes there seems to be a prankster who lives in the future, who tends to make things happen that we do not expect. We will call that prankster "Murphy." You may have heard Murphy's most famous law—"Whatever can go wrong, will go wrong." (4) Study of the future attempts to identify possible events. Short-range events (one–five years) are somewhat easier to project and have a higher possibility of being forecast than do long-range events.

The computer is an important tool in a variety of trend analysis activities. Satellite, seismographic, and production data can be combined in the computer to predict likely areas for drilling new oil wells. Data from accidents and breakdowns of vehicles, structures, or other products can be used to identify patterns and help predict potential trouble spots. These patterns or trends are important as feedback for making improvements and reducing the likelihood of future accidents.

In recent years, the art of forecasting the future has developed as a specialized area of study. People who specialize in studying the future are called **futurists.** Modern futurists do not try to predict what the future and its consequences will be. Instead, they project the future, by describing possibilities and "alternative futures." Once the possible changes and consequences that lie ahead have been determined, futurists hope that people will use this information to choose wise courses of action.

Projecting Trends

As we just indicated, one method of studying the future requires information about trends. For example, suppose you want to guess about the future developments in electronic devices. You read about a number of electronic inventions that have been created each year over the past century. You find that computers are being made smaller and more powerful each year. Similar changes have occurred in radios and saws with electronic controls. By analyzing the data from several sources, you predict that "smaller and more powerful" may be the wave of the future.

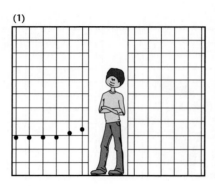

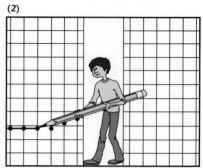

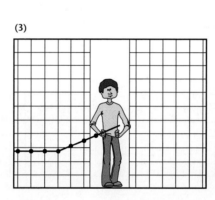

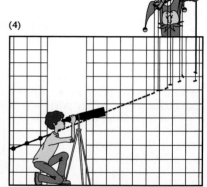

Figure 14.9A ■
(1) First, study related events from the past, such as the growth and development of the telephone. Identify any increase in the number and sophistication of the telephone. (2) Second, attempt to identify if there was a pattern of growth. If so, plot out that pattern of historical growth up to the present. (3) Third, identify if that pattern sets a trend or direction that may extend into the future.
(4) Finally, if there is a trend, project where it might lead in the future. Remember that accurate long-range projections are much more difficult to make than those that are short-range.

■ **Figure 14.9B**
The first commercial telephone.
(Courtesy of A.T. & T. Archives)

Suppose you decided to conduct a trend analysis on one of the resources you have studied (for example, materials). You can start by studying material-related articles about **research and development** in science and industry for the last few decades. You notice improved plastics, ceramics, aluminum, steel, and other materials. In addition, you see that materials are often made on a custom basis to fit the needs of the manufacturer and new products. Based on these trends, you project that there will be an increased demand for custom-designed materials in manufacturing.

DESIGN ACTIVITIES

- You are considering a career using computers. Conduct research and develop a report to use in your career planning.

- Conduct research on the development of an appliance or machine you use at home or in school. Project developments that are possible and probable in the future.

Using Analogies

A second method of studying the future involves the use of **analogies**. Analogies are things and events that have similarities between them. Because they are similar, whatever happens in one event may also happen in the other.

(1)
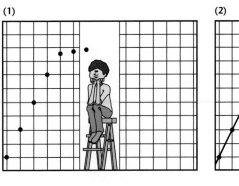

(2)
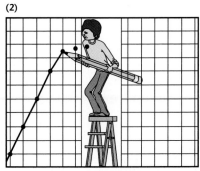

(3)
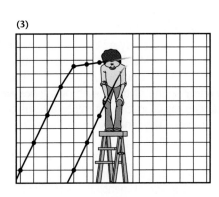

(4)
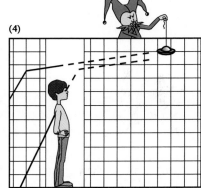

Figure 14.10A ■
(1) After you have identified a trend, you can use it as an analogy to look at something else of interest. Trends in experimental aircraft that are custom-made by individuals can be used as an analogy for projecting potential trends in commercial aircraft design. (2) Once you study the important developments in experimental aircraft design, you identify the directions and patterns. Next, chart out the trends to use as an analogy. (3) The trend of past events can serve as an indicator of a trend for later events. The second trend of events should show a similar pattern of change. (4) If the analogy works, the earlier trend will provide a pattern for projecting the events yet to come. Remember that the future, like Murphy, can be fickle. What you see is not necessarily what you get.

■ **Figure 14.10B**
Burt Rutan, along with a small group of other people, produced experimental aircraft designs that are revolutionizing commercial aircraft design. (*Courtesy of Scaled Composites, Inc.*)

Some futurists, for example, believe that an inexpensive form of synthetic oil will be developed. If so, countries with few oil reserves or resources will not be forced to pay high prices for oil imported from oil-rich countries. This prediction is based upon similar events that happened during World War II. At that time, shipments from rubber-producing countries were cut off and the raw materials for making rubber were in short supply. As a result, natural rubber made from rubber trees became very expensive. In response to this need, scientists developed a synthetic rubber from available chemicals. The price of tires and other rubber products soon dropped.

Suppose you use an analogy to consider the future developments of processes. You study the new developments used to change materials, energy, and information. The developments of today may well be the basis for the innovative processes of tomorrow. Suppose you find articles about the work of scientists who isolated microbes that can live in liquid plastics and epoxies. The microbes appear to multiply slowly and attach themselves to the molecules of the plastics and other organic materials they touch. Using this instance as an analogy, you project that a new biological glue will be developed that makes pieces of wood grow together. Or, by taking a reverse analogy, you project that other microbes will be used for making plastics and other materials more biodegradable.

DESIGN ACTIVITIES

- Based on trends for purchasing computer-related materials (games, software, hardware) nationally, project the amount of time and the type of activities individuals will pursue using their computers. Determine what products have the most potential for sales based on the projected activities.

- Look into the development of such products as records, tapes, and other devices for recording music. Project developments in compact disc production and consumption for the next ten years.

Conducting Surveys

A third future study method relies on conducting surveys or polls. These are not the kind of political opinion polls so often reported in the news. Instead, they are surveys of the opinions of experts in the area under consideration. The **Delphi survey** is one of these methods. If we wish to know when to expect a cure for cancer, for example, we might ask the top ten cancer researchers for their best estimates. The findings and ideas from each researcher are sent to all the members of the group with a request that they reconsider their statements in light of what the other experts have said. This process continues as long as the participants are willing to change their position. The process stops when a consensus (agreement) is reached or when the participants do not wish to reconsider their stand on the subject.

Figure 14.11A ■

(1) The Delphi study starts by conducting a survey with experts in the field. If you wanted to know what the car of the future might look like, you would study people who have knowledge about automobiles and aspects of technology that have impact on autos. (2) After gathering information from your survey, you share the statements made by the large group with all the participants. Based on how others responded, the participants may change their ideas and give you a new set of data. (3) The revised ideas of the total group are then shared with everyone else. This, in turn, may cause some participants to reconsider what they said and give you new responses. (4) This cycle can be repeated until there is no significant change in the positions taken on the ideas. The large part of your sample will now share some agreement as to what the car of the future might be. A smaller group may still disagree. Only time, and Murphy, will tell which group was correct. Or maybe they are all wrong.

(1)
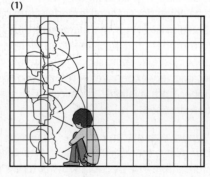

(2)
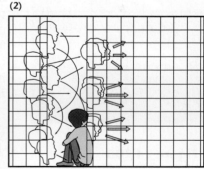

(3)
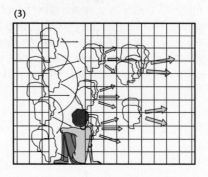

(4)
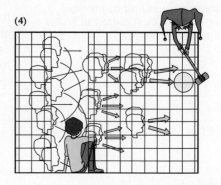

■ **Figure 14.11B**
This concept car, developed by General Motors, weighs 1,400 pounds, carries four people, and will get 100 mpg on the highway, driving at 50 mph. Experts from companies that make materials for the aerospace industry were surveyed for ideas on the use of lightweight materials and designs to make the vehicle light, strong, and fuel-efficient. (© 1992 GM Corp., GM Media Archives)

You can use a survey to predict what new types of tools and machines might evolve in the next few years. You could talk to or write to hardware store owners, engineers, and researchers to find out what new tools and machines they think will be on the market soon. As your experts see what others think, you could give them the chance to revise their ideas. By cycling through the process several times, you obtain better information for your projections. For example, one of your experts might propose that a laser-operated multipurpose cutter soon will be developed. If the idea appears sound, the other participants in your poll will support it. If they think the development is unlikely, they will reject the idea.

D E S I G N   A C T I V I T I E S

• You and a group of your classmates would like some alternative to eating lunch in the cafeteria at school. You have talked with the local vendor who sells snacks and other foods from his truck parked near the school. He said he would offer a wider selection if he knew what students and teachers would buy. Your class has agreed to determine what students and teachers would be most likely to buy for lunch from the vendor.

Conduct a library search for guidelines to design and conduct a Delphi study. Gather the information for creating the most nutritious and appealing lunch that could be safely and economically sold by a vendor. The survey should include students and teachers, as well as experts on food processing and safety, business, and nutrition. Design your survey to tap the expertise of each participant.

• Choose one of the most promising designs that has been developed by students in your class, as you have carried out the design activities from previous chapters. Conduct a marketing survey to determine the likelihood that the product would be salable if you decided to mass-produce it. Identify which groups of buyers would be most likely to want it. Also, in your survey, gather opinions on whether modifications in color, shape, and size would make your design more attractive to a buyer.

Developing Future Histories

Future histories are another means of studying the future. The process begins by pretending to be alive at a time far in the future. By using their imaginations, futurists can then look back, so to speak, and write a "history." The future history describes the events that lead up to that imaginary time in the future. Many science fiction writers use this method to develop their stories. The chief value in this method is the directed use of imagination, somewhat freed from the predictions of experts. An example of a future history about humans and machines is presented here.

CONTEXT—FUTURE HISTORY: HUMANS AND MACHINES

The year is 2050 A.D. Breakthroughs in research about the structure and function of the brain and body were made in the early years of the twenty-first century. This knowledge, combined with what was learned in the 1990s, made possible new and intricate relationships between humans and machines. Machines that assume full responsibility for some functions of humans are now commonplace. Many of these first developments grew out of the use of computers, microprocessors, and robots in the 1980s and 1990s.

The responsibility for producing goods and products has fallen to a very small number of individuals who appear to thrive on the decision-making responsibilities required in large cybernetic "productories" (called "factories" back in the twentieth century). Many other people have become involved in individual and team efforts of another sort. They design and fabricate unique creations that do not lend themselves to mass consumption and production.

The new "Cybuters" now supplement most of human rational decision-making. These machines are helpful in making ethical and moral decisions, although only at the simplest levels.

Most notable at this time is the sophisticated relationship between humans and machines. It is commonplace to find a human and a machine that have been designed as a one-of-a-kind system. These were once called androids but are now termed "humachines." The machine component amplifies the human's physical and mental abilities. It also corrects for the human tendency to delete and distort information. The human capacity to feel, to make new, spontaneous connections, and to make ethical decisions corrects some of the strictly logical tendencies of the machine. Together they can accomplish tasks that neither the human nor the machine could accomplish alone.

This approach to machine and system design came after many years of trying to build computers as "intelligent machines." The practice finally gave way to the development of Cybuters as extensions of intelligent people.

DESIGN ACTIVITIES

- Conduct the necessary research on how "humachine" systems might be developed for work or travel. Plot out the history that would be needed in order for this system to evolve.

- Write a description of your life in 30 years, as you live in your home under the ocean. Be sure that you describe your everyday behaviors and the goods and services you use. Your description should be a projection of scientifically feasible events. Plot out the history of the 30 years of your life to describe how you got to this point.

Figure 14.12 ■
(1) A future history requires that you journey in your imagination to a future time and place, then look back to the present. Consider, for example, what information devices might be like in the year 2075. (2) From your vantage point in the future, write a history of the events that will lead up to that future time. If you imagine a machine that monitors human functioning, you may think of how tiny the energy cells would have to be to allow such a device to be portable, yet comfortable. What events would be needed for that to evolve? (3) In our future history, we link the events together like links of a chain. When we finally reach the present, we can check out whether the chain of events is logical. (4) It helps to write several future histories that might lead to the final future event. Several different perspectives can be taken. As usual, our projections can be drawn off course or even blown apart by events that we did not consider.

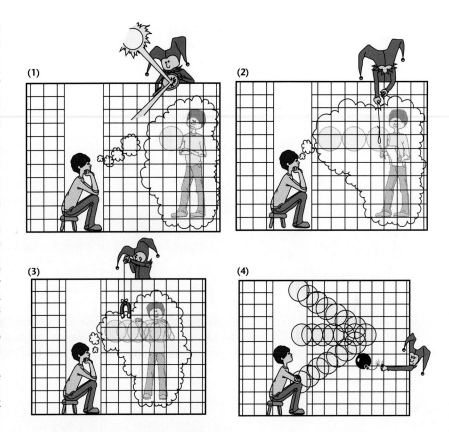

Developing Future Scenarios

Scenarios are imaginative "pictures" of what could happen in the future. Usually, several scenarios are developed for a single problem. Scenarios ordinarily identify a set of events and they lead to other events. You might, for example, develop a scenario describing what would happen if a nearby river were dammed up to form a large lake. Think of the boating, fishing, and other recreational possibilities that the lake would provide. You might describe the new houses that would be constructed along the lakeshore. There might be many new jobs created. Needed electric power might be provided by the dam.

You might picture another scenario for the same event. In that scenario, good farmland is flooded over. Some species of fish may die out and the plant life may change. People are forced to move from their family homesteads and their houses are torn down. Tourists move in. The quiet village is changed by the influx of people who have little or no long-term interest in the ecology of your homeland. This results in traffic jams, vandalism and litter, and the need for a costly new sewage treatment plant. This scenario is different from the first. It predicts many unwelcome consequences if the dam were built.

Many possible scenarios can be developed for a single event. If several are used, better and more reliable results are likely. What actually will occur is usually a combination of predicted effects. Awareness of possible bad results is very important if steps are to be taken to avoid them.

A scenario on information in the future is presented in very positive form below. What are some of the undesired consequences you think might be described in a related scenario?

CONTEXT—SCENARIO: INSTANT INFORMATION

The year is 2021 A.D. Access to instant information has its roots in nineteenth- and twentieth-century developments, such as the telegraph, telephone, television, computer, electronic networks, and the information highway system. Advances in communications increased the range and capability of satellites that orbit the Earth and other planets in the solar system. Through this network, people have access to libraries, data banks, and people with similar interests from all over the world.

Everyone is equipped with a personalized information machine. This portable, high-yield device is given the same social security number as its owner. The device is provided shortly after the human is born and is used in the early stages of life to monitor the intrusion of disease-carrying bacteria and viruses. The device can remind the person that they are low on exercise, that they should take some medication, and warn of potentially hazardous physical conditions. Experts from any location in the world can be consulted when the person needs answers to questions or advice on courses of action.

In addition, the device is a "teacher" for the human. There is no need to remember many facts, since they become out of date so quickly anyway. The communication system gives access to the current body of knowledge in all fields. In addition, the system gives access to a vast array of videos, games, and simulations from all countries of the world.

A large number of workers spend their time putting the information into the system, usually working from their home-based workplace. There is little need to actually go anywhere. People can communicate with each other, regardless of where they are located. Shopping can be done by selecting and ordering from your home. Resources you need can be sent to you via a vast automated distribution system.

DESIGN ACTIVITIES

• Conduct the necessary research on how human/machine systems in the scenario developed above might be used for recreation or communication. Write out the series of events that would be needed for this scenario to occur.

• Using some of the concepts you have learned about biotechnological developments in agriculture, write a scenario that describes how food is grown and distributed in order to combat hunger and malnutrition in the world.

Figure 14.13A ■

(1) A future scenario starts like a future history by imagining a point sometime in the future. Such a scenario could also be built upon several related future histories. In all cases you try to envision a set of related events at a given time in the future. (2) You continue to build your future scenario by identifying more and more possible events connected with information machines, this time taking a global viewpoint. Questions about translation of information from one form to another will be important. Project these events for the same general time frame in the future. (3) After you have drawn together the many related events, try to describe how they are connected. You join them like links in a chain. Your description should provide some insight about what it will be like when people can communicate easily with people in almost any part of the world. (4) As more and more connections are made between the various events, your descriptions of the time slice will become more complete. That description is your future scenario. Remember that the future can play havoc with any projected description. This happens often with even the best made projections.

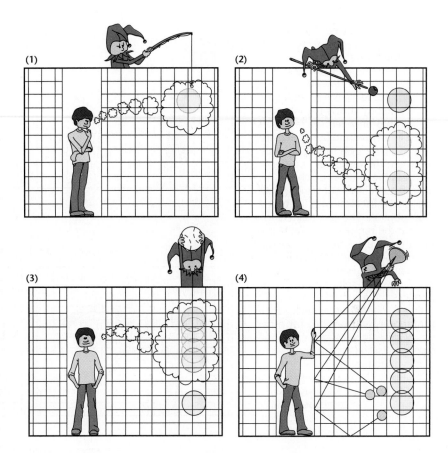

■ Figure 14.13B

In a relatively short, recent time, complex communication systems have improved our ability to communicate with large groups of people. What do you think will happen in the next century? *(© Ed Hornowitz/Tony Stone Images)*

Using Models

Modeling as a part of future studies is similar to the modeling work that you have done in earlier chapters. With modeling, you try to translate ideas into more concrete and visible forms. When enough information has been gathered over a period of time and when the data can be expressed in a mathematical form, the computer becomes a valuable tool for creating models. The models can be varied to show different alternatives.

Projecting into the future by modeling is possible only if an adequate and appropriate model is available. To be adequate, a model must (a) describe its component parts, (b) illustrate how those parts fit together, and (c) provide some of the instructions or rules for creating something from the model and its component parts.

Models do have their limitations, however. In earlier chapters, you have seen how structural models are used to create buildings, bridges, and towers. Structural models are appropriate for constructing but are less appropriate for transporting and are not appropriate at all for communicating. Similarly, a model can be very appropriate for studying and projecting some specific aspect of the future, such as the effects of an oil spill on a specific stretch of beach. That model would be inappropriate for determining the best crops or the most efficient means for transporting people in that area.

Figure 14.14A ■

Before we can model some aspect of the future, we must identify what parts there are and how these parts interrelate. (1) The model of the Resources of Technology shows the interrelationship between the resources used in any technological activity.

(2) Often we make trade-offs between the resources in order to meet the criteria to achieve the most effective outcome. (3) All models must deal with limits and constraints. The constraints indicate the range in which we can find a feasible solution. The IDEATE model helps clarify the limits and constraints and how they influence possible solutions. (4) To be successful, a model must help to find the optimum solution. Each criterion is weighed against other criteria before solutions and compromises are made in order to achieve the most goals at any one time. The costs of alternative solutions are weighed against expected outcomes so that the best quality can be achieved, given the available resources.

(1)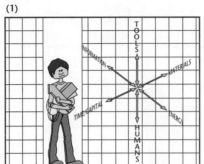

(2)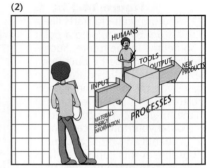

(3)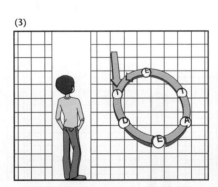

(4)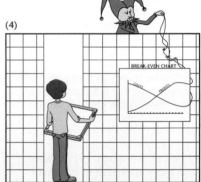

CONTEXT—A MODEL FOR DESIGN

The more complete the model, the better the projections will be. The IDEATE model is an approach you have used to think through a problem. The model indicates some of the thought processes and actions needed before you try out your best solution. By now, you have had a chance to use the IDEATE model several times. You may be so good at solving problems with the approach that you use the model as a reference and no longer go step-by-step. What you have developed is an approach—an attitude—that you can use as you start a design to create new products, environments, and systems.

DESIGN ACTIVITIES

• Conduct the necessary research that will help you develop a model for evacuating the school building in the event of a fire or accident. Your model should help show how changes to one part of the system or operation can create changes in other parts of the system. For example, what would happen if the fire started in one location versus another? At one time of day versus another?

■ Figure 14.14B
When the cost of a mistake is very high, more resources are needed in order to achieve only a small margin of error. This model of an airfield is used as part of the training for air traffic controllers. As students make decisions, the instructor can provide unexpected changes that simulate real dangers. (© Enrico Ferorelli)

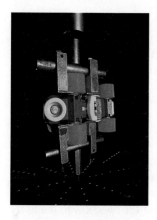

■ Figure 14.15
These roller skates endured testing with up to 278 pounds of tension before the front wheels came off. It was decided that this was a sufficient level of durability for the ordinary skater. (© 1989 Brian R. Wolff)

EXPANDING KNOWLEDGE FOR IMPROVING FUTURE INVESTIGATIONS

Throughout this book, we have seen how technology has developed. The products, environments, and systems that are developed at any time reflect the knowledge and attitudes of the time in which they were designed. Knowledge is accumulated as people try to improve their lives, learn more, and question old ideas. This ever-expanding body of information creates even more possibilities for the future.

For centuries, the expansion of knowledge came through the "trial and error" approach. Human curiosity would not be content with this approach alone, however. Accepted procedures for gathering and reporting new information and testing old ideas were developed. A scientific approach for research increases the likelihood that information is reliable and trustworthy. This approach is used today by scientists as they do pure research. Basic or **pure research** is research in which the primary aim of the investigator is for more complete knowledge or understanding of the subject under study. Little or no concern is given to how the knowledge will be used.

Most of the work that is done in technology can be called **applied research** and comes from using knowledge. New products, better processes, or improved systems and environments are the intended outcome. Applied research is also used to test and evaluate the prototype solution to a design process. Here, too, formal methods can increase the likelihood that the product has been given a fair test, and will perform as intended most of the time.

The procedures for both pure and applied research require attention to detail, attempts to guard against your own biases, and clear communication of results so that someone else can benefit from your work. As you follow each approach to investigating, you are likely to form an attitude toward your own work. You also may become aware that ideas and guesses are interesting and a good source of gossip. More reliably produced and interpreted information is critical, however, if you are to use it in your design and problem-solving work.

In the chapter on Humans, we described the way in which people view their reality—their own cognitive map. In order to organize the many bits and pieces of information that bombard us at each moment, we delete, distort, and generalize. The **scientific method** is used to correct for these natural tendencies. The investigator must record exactly what is done, and under what conditions. Procedures are precisely described so that another person can follow the same procedures and get the same results. When another investigator gets a different result, a flurry of new studies might be done to find out why. This information is published in order for other people to profit from the findings of other investigators. In this way, a body of knowledge grows and is shared.

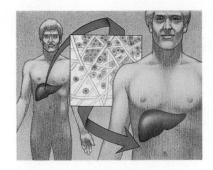

■ **Figure 14.16**
A major problem for people with liver disease is that there are so few organ donors. The development of a plastic scaffold may allow these people to grow their own livers. Liver cells are known for their ability to regenerate. The only problem with transplanting cells has been holding them close enough together to allow them to grow as a system. These cells, taken from living donors, are put into this plastic structure, designed to last just long enough to support the formation of a new organ.
(John Karapelou/Discover Magazine)

DEVELOPING—PUTTING KNOWLEDGE TO WORK

As we have discussed several times in this book, people use what they know to create what they want and need. Throughout history, as we develop more and more understanding, the next step has been to put that knowledge to work. The results are called **developments**. A study of human history shows many examples of research and development activities.

The first flights began in the early 1900s. Since that time, flight has developed to speeds faster than sound. Aircraft and spacecraft can now travel thousands of miles in an hour. Developments in technology continue to evolve at a faster and faster rate. Amazing things have already happened during your parents' lifetime. But even those will be overshadowed by many new technological developments during your lifetime.

Discovering, Inventing, and Innovating

Development is the planned use of knowledge from research and experience for the purpose of achieving a desired goal. The three parts of development are **discovery, invention,** and **innovation**. Each of these requires some knowledge and skills on the part of the person who creates the developments. Even if the change is the result of an accident, the person must have some prior understanding in order to interpret the accident and make use of it.

A discovery is new insight into the uses of something that already exists. An invention is something that a person makes that did not exist before. An innovation occurs when a person takes several things that already exist and brings them together to produce a new device or process.

If you found a piece of interesting rock, it would not be considered a technological discovery. But it would be a discovery if you are the first person to observe that small pieces of other materials were attracted to the rock and you used the rock to collect those pieces. You might call the rock a lodestone and the material, iron. It is an additional discovery if another person rubs a sliver of iron on the lodestone and the lodestone picks up other iron slivers. The lodestone can be called a magnet.

An iron magnet floating on a small piece of paper or wood in a bowl of water will always point the same way. The person who noticed this invented the basic compass. Someone later may have decided to shape the iron into a pointer so that it could turn on a pivot point rather than use a messy bowl of water. That person also decided to put a ring around the outside of the needle with marks for north, east, south, and west. These additional steps were innovations or improvements to the original invention.

Research and Development—The Search for Applications

Technological development has a place and function within a larger system of **research and development**. The research and development system uses existing knowledge to solve practical problems. This knowledge, in turn, can generate new knowledge about how practical problems might be solved more easily. It is a circular, continuous-loop process. Development is a system of using and generating knowledge that recycles almost continuously in search of better ways.

Suppose your mother makes the best piecrust around. One day, a friend takes one of her pies home to his mother. She is vice president in charge of product development for a national food industry. She thinks the pie is the best she has ever tasted. She wants to develop a new line of these pies.

The quality of the pies must be retained, even though they are mass-produced at the rate of 10,000 per month! A set of quality-control procedures must be developed. A crew of food technologists watches your mother's work. They analyze every material and process she uses. This crew gathers an accurate measure of her knowledge of pie baking. Observations of her behavior and statements that reflect her experience and intuition are included, along with measures of ingredients. The number of strokes she uses in mixing and the temperature of the ingredients are all recorded. They take the knowledge back to the factory laboratories. They can soon duplicate her pie in the lab.

The next step is planning. What processes are best done by machines and which ones are best done by people? What machines are needed? Where can large quantities of the ingredients be obtained? What is the effect of substituting one ingredient for another? What problems in production can the team anticipate? What is the best size and shape for the pie? What kinds of pie are already on the market? What kinds will sell best? The list will be long. This planning procedure gathers together the best knowledge available before the pies can go into production.

Several alternative procedures might look good. Tests will be conducted for each procedure, and results will be compared. Hours and hours of carefully controlled and recorded work will go into making pies in the laboratory. The pies will be carefully evaluated. Some may taste great, but fall apart easily. Others may stand up to being transported and sitting on a shelf, but taste like cardboard. Some taste great and can be delivered in good condition for sale, but a single pie will cost $20!

Another example will show development as a result of applied research. This is a special kind of controlled problem solving. In order for spaceships to be developed to carry people and animals into space, a reentry heat shield was needed. From previous flights, scientists and engineers learned that spacecraft entered the Earth's atmosphere at speeds of about 17,000 miles per hour. Because of the high speed, the craft was subjected to tremendous air friction. This friction could generate enough heat to melt the aluminum alloys and other structural metals then available. Special alloys were needed to withstand the high heat and strain. The heat generated by the air friction would be severe enough

■ **Figure 14.17**
The development of a new pie also includes the design of packaging and the development of marketing strategies. *(© 1993 John Olson/The Stock Market)*

■ **Figure 14.18**
New materials were custom-designed for these tiles used on space shuttles. The ability to absorb and release heat allows the shuttle to be launched and re-enter the Earth's atmosphere again and again. *(Courtesy of NASA)*

to raise the cabin temperature to dangerous levels—much too high for the tolerance limits of humans! Elaborate methods of insulating the cabin or carrying off the heat with cooling devices were tested. They all made the craft too heavy.

The space scientists and engineers faced a new research and development problem. They had to find a method of protecting the astronauts and capsule from heat during reentry. This method should not create excessive weight and should fit in with the structural requirements of the capsule.

The solution to the problem was the development of a new ablative material. It was lightweight and strong. When subjected to severe heat, the ablative material slowly burned away. As it disintegrated, it carried dangerous heat away from the spacecraft. This new, dynamic material changed its characteristics during use.

Although the first phase of development was complete, more problems remained. A method was needed to produce the new material at a reasonable cost. Techniques were developed, however, and the material was successfully used on dozens of manned space missions.

Later, the space shuttle was developed. It would be reusable. The same spacecraft would fly several times. Since earlier heat shield material burned away and could be used only once, an entirely new material was needed. Again, a solution was found. Engineers found that silicon tiles would do the job. They would be glued to the spacecraft. A new glue was required. Do you see how one technological development often leads to another?

Development may have several specific outcomes. The three general categories are new or revised products, processes, and systems. The most evident and often the most spectacular results of development are the end products. These may include new machines, tools, and toys, a more comfortable fabric, more efficient buildings, bigger spacecraft, smaller artificial hearts, or more powerful computers. The new products include those that are consumable, or used up in use, such as new foods, fuels, and materials. Other types of products are improved tools and devices. Often, innovations or improvements make the tools more efficient or better suited to a new want, or are developed to take advantage of a new material or fuel. New structures and methods of construction, production, and transportation are the result of other types of development work.

IMPROVING—MAKING THINGS BETTER

Ways of Thinking and Developing

For a few highly creative geniuses, the processes of discovery, invention, and innovation seem to come easily. Sometimes, the discovery or idea about an invention is made quickly. Often, the discovery is followed by the slow developmental work required to show others the idea. For most people, however, breakthroughs come only after a lot of hard work and frustration. The person with a unique idea may also have trouble finding any support for developing the idea. People may reject the idea simply because it is new. One method of presenting new ideas and preventing rejection is the use of prototypes.

■ Figure 14.19A
Prototype of the Wright brothers'
plane as a kite/glider. *(THE
BETTMAN ARCHIVE)*

■ Figure 14.19B
The Wright brothers' finished
plane. *(UPI/BETTMAN)*

Different production processes require different prototypes. An airplane is often custom-manufactured. Each aircraft is made individually with time given for extensive testing of each product. Televisions are most often mass-produced by intermittent or continuous methods. A fully functioning prototype is required to determine how to set up the manufacturing lines.

Prototypes are built to incorporate the best ideas found in the last stage of the modeling process. The work of the Wright brothers is a very early example of prototyping a product after extensive modeling for scientific and technological purposes. The brothers worked for years to build a flying machine. They found that developing an efficient wing shape was very important in making the airplane fly. To find that best shape, they built a wind tunnel. They made small versions of many different wing shapes. These models were placed in the wind tunnel and the lift upward, caused by the air flowing over the wing, was measured.

The Wright brothers could not be satisfied with only the results drawn from models in a wind tunnel. They proceeded to construct a full-size prototype of the aircraft they intended to fly. These prototypes were flown first as giant kites to determine if the full-size prototype glider could actually fly. (See Figure 14.19A.) Only after extensive work of this type did they trust the plane enough to actually ride in it. Even then, they tested unpowered aircraft before adding the engine. Carefully, the development of the airplane evolved. Their first successful flight was the culmination of years of developmental efforts. They also drew upon the research and development of several other people before them.

Optimization: Finding a Better Way

A developer is seldom lucky enough to find the best answer to a proposed problem and then use the solution without changes. Ordinarily, new procedures must be developed continually to improve the old ones. One improving activity is optimization.

Optimization is the process of seeking the most favorable condition or solution to a problem or a goal. Each time you tune your radio to a station, you are involved in the process of optimization. You move the dial back and forth until you get the best sound available. In this case, the goal or **criterion** is high quality of sound. The dial setting is the variable you manipulate to get the best sound possible. This process is called **maximizing the criterion**. The clarity of sound changes as the dial is changed. The optimum dial setting for the maximum sound is in a very narrow range. (Refer to Figure 14.20.)

Most problems of optimization are not this simple, however. Usually, there are two or more conflicting criteria. A television set provides an example. The sound and picture of your set, although controlled by the same dial, do not always match.

The best setting for the picture and the best setting for the sound may occur at different places on the dial. You must choose a setting that is a compromise between the two competing criteria. The setting you choose will depend on what is important to you about the program you are watching. If you are watching a musical show, you might place the dial closer to the optimum sound setting. If you are watching a documentary film on art, you may choose a setting for optimum picture clarity.

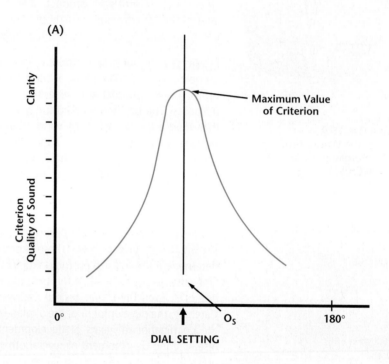

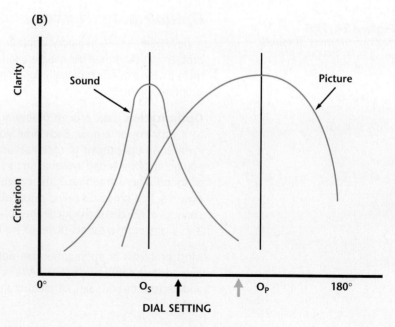

Figure 14.20A, B ■
(A) The optimum setting for sound (O$_S$) can be found by tuning the radio dial. (B) When tuning a television, the optimum setting for the picture (O$_P$) is different than O$_S$. The operator must decide which is the most important and how to achieve the best compromise.

Decision-making within technology is often similar to optimization. Many times there are competing goals and criteria. Sometimes short-term wishes compete with long-term wants. Decisions weigh one type of risk against another. For example, you know the research findings that show the risk of getting skin cancer from overexposure to the sun. You decide to take the risk because you think you will have more friends if you have a suntan. Or you may decide to use a sunblock and take your chances on attracting friends with your personality instead of your looks. You are faced with searching for the best balance or an optimum compromise. The necessary give-and-take required to reach a compromise is called the **trade-off process.** A choice between two behaviors, both of which are based on probability, requires that you decide what is an **acceptable risk.**

ASSESSMENT OF QUALITY OF PRODUCT DESIGN

The process of optimization can lead to high-quality products. Product designers have described the ideal in this way: An economical design is an optimized design—getting the most for the least. It is a design without excess, a design with every part exactly sized and shaped for its intended purpose. No material is used that isn't needed. The material is placed and shaped to function at the maximum allowable stress level throughout. Weight, number of parts, number of assemblies, amount of machinery, and complexity of shape are all minimized. The least expensive (but most appropriate) materials are selected and fabricated in the least expensive manner with the least amount of manual labor. In general, economy in design is the development of a product to do only and exactly what it is designed to do, and no more, with the least expenditure of materials, resources, production effort, and labor.

The following questions can help you move toward an optimized design:

1. Does it cost as little as possible (considering resources for production, maintenance, and replacement)?

2. Is the design as simple as possible, while still providing the required and wanted features?

3. How reliable and dependable is the product and how long will it last?

4. Is the product appropriately rugged and durable?

5. Does the product function as it is intended?

6. Is the product safe while in operation, as well as when stored?

7. Is the product attractive and pleasing to use?

8. Is the design appropriate for the ergonomics and developmental level of the intended user?

9. Can the product easily be taken apart to replace parts that break down, rather than having to replace the entire product?

■ **Figure 14.21**
Would you like a pair of these all-season skates? The blades can be inserted for ice skating. When you are ready to travel on pavement or sidewalks, just change the blades to rollers. *(Photo courtesy of DuPont)*

10. Does the design allow for recycling materials once the product is no longer wanted?

11. If failure of the product has costly consequences, does the design include ways to guard against breakdown or methods to protect the user should failure occur?

Iterating—Recycling Information for Improvement

Iterating is the process of developing and improving an idea, process, product, or system by sending it through a development stage more than once. The procedure is similar to the recycling and feedback used in the Delphi survey. Iterating is used in computers. Loops placed in the programs cause the computer to recycle through selected operations.

An idea is iterated by recycling it back through one or more of the steps of development. An idea may be returned "back to the drawing board" and moved again through the designing, implementing, and improving phases. Or an idea may be recycled through the optimization step only.

Iterating has sometimes been incorrectly labeled a "redesign." Iterating is actually a part of the normal design, development, and problem-solving cycle that attempts to find the best possible answer to a problem. Recycling an idea for its improvement through iteration techniques is part of the ongoing design and problem-solving process.

Valuable feedback can be acquired through the process of iterating. The failure of a product to perform in an idealized way provides valuable information about changes that could be made in the process and/or the product. The difference between the best possible operating product and the product as it now exists can be viewed as a mismatch between the desired goal and the system output. Automated or cybernetic machines correct their operation based on the degree of error that is sensed and sent back into the control system. The designer uses the mismatch or error as information to be sent back into the design process. This information is required for the improvement of the product. Eventually, the development team agrees that the error between the real and the ideal product is as small as possible. The iterating process is then stopped and the product is ready to be used, constructed, or mass-produced. The risk level for error is at an acceptable level.

SUMMARY
CHAPTER 14

Investigating and developing appear to be basic activities for humans. As they investigate, people need access to knowledge through such activities as researching and exploring. This knowledge is merged with creative practice to support the design of new developments in technology. For example, as people explore they often have a need to move about. This in turn requires technological devices for navigation. The devices help people to determine their direction of travel, to calculate their exact position, and to document where they have been.

Space exploration was initially limited to using receiving devices that collected energy that reached the Earth's surface. The rocket eventually allowed humans to transport themselves, as well as exploration devices, into space. Other types of space exploration include efforts to study and chart the ocean floors, the interior of the Earth, and the microspace of molecular and atomic matter.

Humans also explore within the dimension of time. Specific technological devices help us explore the history of events while other devices and procedures help us explore possible future events.

As the pace of change quickens, it becomes more important than ever to forecast and plan for the future. The new field called future studies has evolved. People who specialize in this field (futurists) have developed several methods of forecasting, including trend projection, future histories, analogies, scenarios, surveys, modeling, and networking. Each of the methods can be useful for studying the future and for helping people make better choices in the present.

Exploring is important to technology because it produces new knowledge through research. Practical knowledge comes from applied research, which is intended to solve specific problems. Pure research is conducted to produce more knowledge without concern for the immediate practical use of that knowledge.

Development is the planned use of knowledge from research and experience for the purpose of achieving desired goals. The general intent is to design new products or to improve existing products, environments, and systems. Products include consumables, tools, and structures.

People engage in a range of developing activities such as discovering, inventing, innovating, designing, implementing, and improving. These activities are supported by techniques that guide the work of designers and developers in generating models and prototypes. Other techniques and procedures are used in working toward the optimization of the products, processes, and systems that emerge from the efforts of developing.

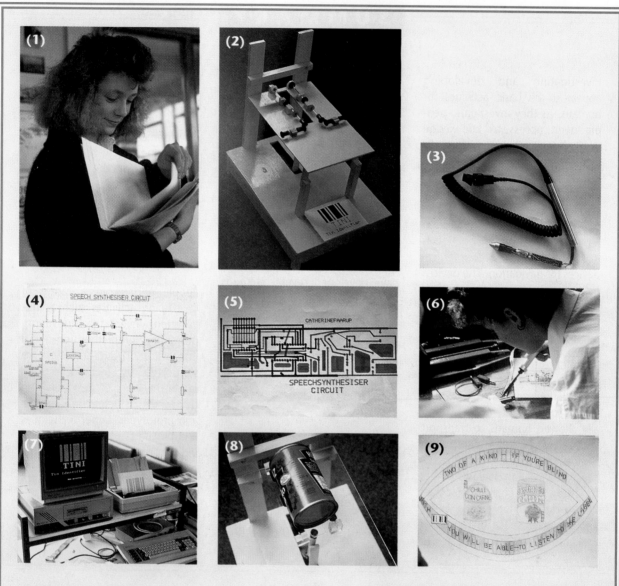

■ **Figure 14.22**
These photos show some of the work done by Catherine Faarup, a student who faced a problem similar to the one encountered by April and Enrique in Chapter 1. This device was designed to scan the bar codes on cans and then announce aloud the contents of the can.

The sequence of photos, from left to right, show (1) Catherine engaged in research and investigation, (2) an early model showing the scanning mechanisms, (3) a light pen provided by a supportive company, (4) Catherine's design for the speech-synthesizing circuit, (5) the circuit ready to be etched as a printed circuit board (PCB), (6) fabricating the PCB, (7) Catherine's voice-synthesizing program, (8) the scanning, synthesizing system in operation, and (9) Catherine's portfolio. *(Courtesy of Catherine Farrup, student; and Ted Gill, teacher)*

ENRICHMENT ACTIVITIES

- Choose a consumable material, tool, or system in which you are interested. Trace the development process as improvements were made to it over the last 50 years.

- Select a product that you use daily. Assess the design, using the criteria given in this chapter. Weigh the value of the criteria that the product meets readily against those criteria it misses. Compare these findings against a similar product, with another brand. Decide which is the best design overall.

MATH/SCIENCE/TECHNOLOGY LINK

- Design and conduct a research study to gain information on a topic of interest to you. Include all the steps of the scientific research method.

- Conduct applied research procedures to gather data for improving the design of the product analyzed above.

- Identify interest groups that regularly communicate their findings on a topic of interest to you. Use data banks and available information networks.

- Choose the best antitheft device and best automobile to use in your geographical area. Use all available data on theft of autos, effectiveness of various devices, insurance costs, etc. Decide on an acceptable level of risk. Identify the criteria, other than safety from theft, that you want the car to meet. Choose the best car and device, given all information and criteria. Write out the process, showing the weight you gave to each criteria. Illustrate your decision-making process.

ENRICHMENT ACTIVITIES

1. Name some inventions and innovations that increased people's abilities to tell the direction in which they were traveling—to determine their location on Earth. Describe how each device works.

2. What are the barriers to human travel in space? Under the sea?

3. Describe at least one method used to determine the age of artifacts from an ancient time.

4. Compare and contrast three different methods for investigating the future.

5. Distinguish between discoveries, inventions, and innovations. Give examples of each.

6. What is the difference between "pure" and "applied" research? Identify how each contributes to technological development.

7. Describe an instance in which you attempted to optimize the match between two or more variables. Discuss your decision-making process.

8. Cite at least five criteria for a "good design."

CHAPTER
15

Consequences and Decisions

Figure 15.1 ■
The major tool for understanding the future is the past. Humans have passed through many historical ages like doors that lead from one room to another. The next room has not yet been completed. We stand on the threshold of that new age.
(© 1989 Steven Hunt/ The Image Bank)

INTRODUCTION—DEALING WITH CHANGE, NOW AND BEYOND

Technology is changing our world. It will continue to do so in the future. With each generation of humans, the rate of change is quickening. Technological advances that we enjoy today would have amazed our great-grandparents. What do you suppose life will be like 50 or 100 years from now?

449

KEY TERMS

consequences
constraints
decision networking
decision tree
ecological consequences
emotional effects
external controls
first-order effects
hard technologies
if-then statements
internal controls
knowledge explosion
personal consequences
physical effects
results
second-order effects
social consequences
soft technologies

Technology has had many different effects on people, some of which have been desirable and intended, some of which have not. Some of the effects are immediate, while others are delayed. If technology is to be used to achieve our desired purposes, we must become better at predicting **consequences** before we act. Technology is a tool in itself. To use it effectively, humans must think things through and make difficult choices about possible outcomes.

In a democracy, citizens have the opportunity to participate in making decisions about technology. To make intelligent decisions, however, we must have knowledge. In this book, you have studied many aspects of technology and how it works. In the previous chapter, we discussed approaches used by futurists to forecast future consequences. This chapter explores some of the consequences of technological development and approaches to making decisions. With the tools you have learned, hopefully you will be prepared to contribute to making better decisions.

LOOKING TOWARD THE FUTURE

Developing Decision Networks

Making decisions from the projections for the future often is a complex process. One approach to using projections is **decision networking**. This is the identification of events that must happen if a goal is to be reached. For example, if people in the United States are to decrease their use of nonrenewable energy resources, then a number of prior events must take place. For one thing, gasoline consumption must be reduced. For that to happen, people must use more economical cars or drive fewer miles by car. At the same time, substitutes for heating oil and other petroleum products must be found. Nationwide efforts in Brazil have been devoted to reducing gasoline use. One method is to add or substitute alcohol, refined from sugar cane. In Denmark and other Scandinavian countries, experiments in using alternative energy sources range from the use of solar and wind energy, to producing methane from cow manure. The list of problems for each of these possibilities could go on and on. Each event requires that many other events also happen.

When all the necessary events are placed on a chart, the result is called a **decision tree**. Like a tree on its side, each branch (event) leads to other branches (events). When the total chart or picture of events is plotted, it becomes more clear what people will need to do to reach the goal.

■ **Figure 15.2**
Using information from the past, you can identify some potential events of the future. These projections vary in the probability that they will occur.

DESIGN ACTIVITIES

• Use the decision-networking approach to learn what is required to cut in half the amount of energy used in your home. Each energy-saving idea and its cost must be listed, along with the steps needed to achieve them. Then decide which steps must be completed first, second, third, and so on. Lay out the decisions in graphic form for the network.

- Work with a number of your classmates to determine how to reduce the energy costs for your school. Develop your decision tree as a part of a presentation that could be made to the school administration and the Board of Education.

Figure 15.3A ■
(1) We could work to become energy independent in our homes and in this country, not requiring energy from outside sources. (2) After we have a sharp image of the future event, we try to imagine the final decisions needed to reach that event. It can be helpful to picture the decisions as a tree, with each branch as a decision to be made. (3) Each decision becomes an event and each event requires earlier decisions. Your decisions spread out like a tree. In fact, you are developing a decision tree. (4) Finally, the limbs of our decision tree reach the present. This will allow us to see if specific key decisions have been made that will help us reach the desired event in the future.

■ **Figure 15.3B**
These solar collectors provide available, renewable energy to power a remote camera at an iron ore mine in Australia. (© 1985 Phillip Hayson/Photo Researchers, Inc.)

Technology Assessment

New products of technology are being developed each day. In some cases, new products and processes are being discovered faster than we can learn how to deal with possible consequences. Nuclear power plants, for example, are being developed even though we are not yet sure of their long-term effects on human health. Similarly, we are developing genetically engineered products with little understanding of the results of those efforts. Because of this, it is becoming more important that the possible effects of new technologies be evaluated before they are put to use. The process of systematically evaluating new technologies is called **technology assessment**. This is a study of the immediate and the delayed effects which result when an aspect of technology is introduced or changed. Decisions are made about acceptable risks regarding the intended/unintended and desired/undesired outcomes. A study of the immediate/delayed impacts of potential technological developments on individuals, society, and the environment is conducted.

There was a time when all new technologies were widely accepted without question. Today people are beginning to be more cautious. The United States Congress formed a neutral advisory group, the Office of Technology Assessment (OTA). This office is to investigate the possible effects of new technological development. The effects on people and the environment are carefully examined. Detailed reports are made to Congress. This information is used to help legislators make decisions related to the new developments. There will be even more concern about regulations and assessment of technological consequences as we trade with nations throughout the world. Some agreements about acceptable assessment procedures are complicated by differences in political systems, power struggles, and views about the value of life and acceptable changes. It will be far easier to make the products than to get people from different cultures to agree upon what is just, responsible, and right.

Figure 15.4A ■

(1) Determining the effects of decisions—made in the present to cause something in the future—is very difficult. The process is like trying to knock over some dominoes that we cannot see and cannot be sure are even there to be knocked over. (2) Determining the effects is difficult because we cannot cause the desired effects directly. It is necessary to plan many decisions and events that we hope will cause the effect we want. We can plan the events by asking "what if this and this were to happen?" The dominoes are possible events. (3) After we have made our plan of decisions, we can start the sequence by initiating the first event, knocking over the first domino. It is impossible to know exactly what will take place and what will interfere. (4) It is only after the events have run their course that we will know their effect. Some events we wanted to happen may not occur. Some effects of other events may not become apparent until years later.

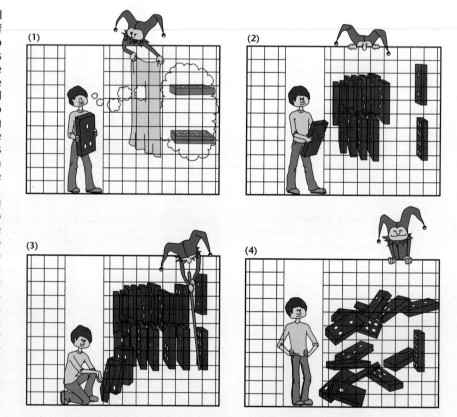

Special interest groups and consumer unions also help watch over new technologies that might be dangerous. Sometimes technology assessment slows down the process of implementing new technologies. For example, when a new fuel processing plant is proposed, it may require years of public study before construction can begin. During this time, numerous citizen groups and government agencies study and debate the good and the bad aspects of the plant.

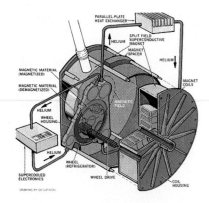

■ Figure 15.4B
Concern for the impact of refrigerants on the Earth's ozone layer has prompted a search for methods other that those that depend on CFCs (chloro-fluorocarbons). This design uses crystals that heat up when magnetized and cool down when they are demagnetized. Super-cold temperatures can be produced. *(Drawing by Ed Lipinski)*

CONTEXT—LEARNING FROM HISTORY

The need for technology assessment is not new. The Roman aqueducts, built nearly two thousand years ago, were constructed almost entirely with muscle power. Hoists, winches, and a variety of other mechanical devices were available but were underutilized. The Romans were more interested in getting the job done than in developing new machines. This kept large numbers of workers and slaves busy but did little to advance the level of technology. Water delivered to the city of Rome by the stone aqueduct system was distributed to most people through public fountains. The nobles, however, had their water piped into their homes. The pipes were made of lead, which leached a poison into the water. It appears the water system slowly killed the leaders of Rome through lead poisoning. Some historians believe this was one important cause of the fall of the Roman Empire. Lacking the tools and techniques of technology assessment and a scientific understanding of the effects of lead on humans, the Romans were unable to assess and avoid the consequences of using such a system of transporting and distributing water.

DESIGN BRIEF

• Using the model developed by the United States Office of Technology Assessment, given to you by your teacher, write a report on a study of the impact of the telephone. Write the report as though it was conducted very soon after the telephone became popular.

• Identify other problems of interest to people in your community regarding possible future developments. As a Technology Assessment team, conduct the necessary research to determine what changes one of the future developments might make in the setting or environment of the community.

Any of the techniques presented in this section could be used in the development of your portfolios. The future orientation of the techniques will help you turn attention to the future and how it will effect your plans and work. These techniques will also help you be more attentive to the consequences your products and approaches may have in the future.

CONSEQUENCES—RESULTS AND EFFECTS

In order to assess effects of technology, several aspects must be considered. Special attention can be given to projection of unintentional (accidental) consequences. The consequences of technology can be divided into three general categories. They include the following:

1. **Personal consequences**—the effects technology has on our bodies and minds.

2. **Social consequences**—the effects upon our homes, schools, workplaces, and other human organizations.

3. **Ecological consequences**—the effects of technology upon the physical world in which we live.

Personal Consequences

There is considerable evidence that early humans and even some animals used tools. Only humans learned to make tools for many specific purposes. Scientists believe that success in making and using tools changed our ancestors in specific and important ways.

The use of tools required skilled hands to make and use tools. The requirements for thought and communication increased as the complexity of the tools and tasks increased. Early humans who could use tools were more likely to survive in difficult environments, live long enough to reproduce, and nurture their children. As a consequence, more of the offspring of tool-users probably survived. These children could advance beyond their parents' development. Over many generations, humans developed better tools and improved technological environments.

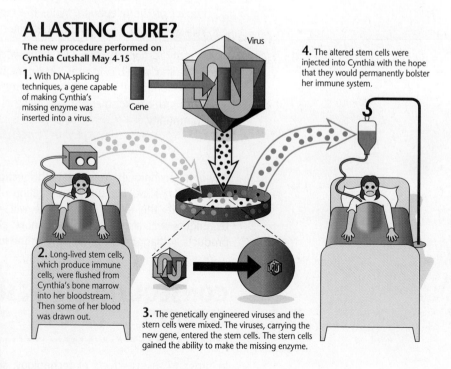

A LASTING CURE?

The new procedure performed on Cynthia Cutshall May 4-15

1. With DNA-splicing techniques, a gene capable of making Cynthia's missing enzyme was inserted into a virus.

Virus

Gene

4. The altered stem cells were injected into Cynthia with the hope that they would permanently bolster her immune system.

2. Long-lived stem cells, which produce immune cells, were flushed from Cynthia's bone marrow into her bloodstream. Then some of her blood was drawn out.

3. The genetically engineered viruses and the stem cells were mixed. The viruses, carrying the new gene, entered the stem cells. The stem cells gained the ability to make the missing enzyme.

Figure 15.5 ■
Advances in science and technology, such as this new procedure, can have profound personal consequences. *(Graphic by Joe Lertola/TIME Magazine)*

With the development of a reliable food supply and adequate stores to protect against emergencies, human efforts could be turned to tasks other than survival. People who had more time to develop new ideas and communicate with others could build a shared culture. Those who could communicate were able to share knowledge and that, in turn, helped them and their offspring survive.

Physical effects. In almost every job we do, we reach for a tool to help us. Tools extend our use of physical strength. They eliminate some of the hard work experienced by our ancestors. Much of our heavy lifting and hauling is now done by machine. One of the consequences of our easier life today is that our bodies can get "soft."

In countries where much of the economy is based on information and communication, physical fitness has become a significant problem. Thinking and communicating can be as tiring as hard physical labor. Jogging or doing bodybuilding exercises can be often seen as fun, but often is only done if there is time after work. Our great-grandparents would have found this strange! They looked forward to a chance to sit down and rest! By making work easier for us, tools can also leave us physically weaker.

Technology affects our physical health through foods we eat. With biotechnological changes, we have the capability of producing most of the food and fiber that the world needs. However, the problems of distributing food to everyone still remains. One approach to producing food in large quantities is the development of giant farms, thousands of acres in size. Each farm is devoted to a single or a few kinds of crops. These large-scale operations make it difficult for growers who have small farms and gardens to compete and stay in business.

Production of food on a large scale is quite capital-intensive. Huge equipment, large doses of chemicals, and only varieties of foods that will stand up to mechanical treatment are used. Large supplies of energy are needed for fuel, pesticides, and herbicides. Foods that are available have become more standardized, and many varieties are no longer planted on a regular basis. When the same crop is grown over and over on the same soil, trace nutrients that change nutritional value and the taste of the food are not replenished. Foods are picked while green so they can withstand the trip from the plant to the market. Additives may be used to mask the loss of flavor and increase the shelf-life of the product once it is in the store.

Without chemicals, the present system of food production could not exist. The system in the United States is one of the most efficient in the world. A relatively small portion of our total population is involved in producing food. Yet they can supply the needs of the nation and a large portion of the rest of the world. But what are the consequences of modern food technology? We depend on a large energy supply to keep the soil replenished. Frozen, canned, and dried foods, as well as fresh produce, are available year-round. The variety of foods increases the likelihood that you will get the nutrients you need. However, there is concern about the effects of additives and the residues of pesticides and other chemicals that we consume as we eat food and drink water.

■ **Figure 15.6**
In many parts of the world, high-quality frozen foods are available year-round. The potential for having tasty, nutritious fruits and vegetables is no longer limited to the season of the year. *(© 1995 Ned Matura)*

D E S I G N   B R I E F

Physical Impacts

Assess the effect that two technological products, the bicycle and computer games, have had on people.

What impact (physical effects) have they had on you?

What future impacts might these products have on you?

MATH/SCIENCE/TECHNOLOGY LINK

Make a collage of the ingredients of some of the products you use. Include cosmetics, foods, household cleaners, and the like. From the list of ingredients, stage a contest to see whether people in your school can tell what the product is. Find out what some of the ingredients are and what their function is in the product. Categorize the ingredients according to function:

color   thickness   smell   flavor   preservative

nutritional value   texture   extender   grease cutting

Have you read the play *Romeo and Juliet*, or other stories of people about your age, as they lived centuries ago? What the stories may not make clear to you is that young people of that time did not reach puberty until much later than teenagers do today. Some people say this is a physical effect of technology. Plants that are grown in a greenhouse, given the nutrients they need and protected from extreme conditions, mature faster than plants that must fend for themselves under natural conditions. Perhaps the same effect is occurring for people as they mature. Whatever the reason, we do know that people are reaching maturity sooner and, on average, living longer. How might you plan your life if you thought you would live for 40 years instead of your current life expectancy of almost 80 years?

As we look deeper into human history, we see some common patterns. Some of the same problems that our ancestors faced are still relevant. Technology may be able to produce enough resources to meet the physical needs of the whole world, yet human greed and hate easily destroy any progress that is achieved. Governments often go into debt to pay for technology to make war and kill. At the same time, they cannot find the resources to feed and shelter their own citizens, much less the needy in other parts of the world. Although people may plant crops for their food, armies may destroy them or take the products. Young boys and men may leave home to fight, leaving women to do all the work of providing food for the family.

It was once seen as inevitable that war, disease, and famine would keep world resources in balance. If we had too many people, one of these killers would keep population in control. This is not an acceptable idea for many people of today, however. Peace efforts have had some limited effects in containing war efforts. We have worked to alleviate some diseases. Producing and distributing more food, so that people will not starve or be malnourished, has been a goal of many technological efforts.

Most people believe death control is desireable. We have also attempted to control the birth process. Timing and controlling the birth of animals (other than humans) has met with little conflict. However, decisions about birth and death control for humans are complex. Major questions have arisen as to just how many people the planet can sustain. Yet varying ethical perspectives have produced heated debates, conflicts, and little action.

THE MATHEMATICS OF POPULATION AND CONSUMPTION

The Issue:

1. World population has doubled since 1950. In terms of new people, the planet gains the equivalent of New York City every month.

2. The rate of net increase of population is also increasing. In 1993, the increase was 87 million people; in 1950, it was 37 million.

3. Ninety-five percent of the world's babies are born into the poorest countries of the world. However, the population in the United States is expected to double in the next 60 years.

4. Because of our population size, energy and other resource use, and creation of waste and pollutants, the **United States has been called the most overpopulated nation in the world**.

5. In terms of yearly energy use, the consumption of three million United States citizens is equal to that of 90 million Indian citizens.

6. Considering the expected consumption of natural resources of a newborn over its expected lifetime, a baby born in the United States costs the planet as much as 30–40 babies born in developing countries.

Changes Needed to Foster Sustainable Development:

1. Reduce the demand for energy use.

2. Equalize the birth rate and death rate.

3. Diminish resource use and the creation of wastes by reducing the demand for consumer resources, reusing resources, and recycling materials.

4. Recognize that happiness and well-being do not come from the accumulation of possessions.

5. Recognize that "more" does not mean "better," and the fact that something *can* be done does not mean it *should* be done.

Source of data: *The Zero Population Growth Reporter*, Vol. 26, No. 1, February 1994.

The beginning of any new technology is sure to raise social and ethical issues. Along with the benefits offered by new technologies comes the possibility of risk. A good example of this was the development and use of nuclear energy. In the 1950s, it was believed that an unlimited supply of useful energy for consumers would be possible from splitting atoms and harnessing the energy released. But in using this new method of energy production, a whole host of unforeseen problems arose. Issues included how to control the dangers of radioactivity and what to do with radioactive waste. Every new technology brings with it the possibility of both benefits and risk. The benefits must be weighed against the risks. An informed public is needed to balance the positive and negative effects and to decide if the risks are worth the benefits offered. This must be continually done for the many potential advances being made in technology.

Emotional effects. Technology changes people emotionally, as well as physically. As we discussed in the chapter on Resource Management, tools for measuring time have changed our ideas about what is important. In an agricultural economy, people live their lives according to natural rhythms—sunrise/sunset, the seasons, tides, and lunar cycles.

With the development of clocks, charts, schedules, and plans for large segments of our time, most people in the United States now live in cycles and rhythms that are consequences of technology. Some say that we live in artificial rhythms. Since these rhythms are made by humans, they seem to be more arbitrary and changeable than natural rhythms. Who says that schools should not be in session year-round? Why not at night? Think of how your life is regulated by time and timekeeping devices.

What might happen if all the clocks and watches in the world mysteriously stopped running? Could our modern production system keep going? We saw in the chapter on Biotechnology that even agriculture is being changed by attempts to change the natural rhythms—to hurry up or slow down growth and maturation according to human wants and needs. One of the goals of many design and problem-solving efforts is that of efficiency. Efficiency may be defined as more goods produced with fewer resources.

One of those resources is time. We tend to believe that we can waste time, time is limited, and we should hurry things up. Processes that once took years, days, or hours to accomplish now take minutes. What shall we do with the "saved" time? Leisure, retirement, vacations—these are not concepts that concern people who must spend almost all of their time and labor providing for safety and physiological needs.

When people have many of their basic needs satisfied by the results of technological systems, they have time and energy to turn to issues of belonging, achievement, esteem, and higher-order motives. Their anxieties and their joys are different in nature from the anxieties and joys of people who are worried about physical and safety needs. But both groups of people have worries and satisfactions!

The Consequences of Knowing

Books have had a profound effect on people's lives. In the centuries before printing technology was developed, important thoughts and events were recorded in picture, song, and story, if at all. As each generation passed, there were a few people (scribes) who wrote down some of the things that happened. The handwritten books could be reproduced only if someone copied them by hand.

This was a very slow process. The results were available to only a privileged few. In Europe, monks isolated themselves in monasteries and spent their entire lives hand copying earlier religious writings. The few writings that existed were available only to leaders in the church and government. Most of the people were forced to make decisions based on what little they were told and what they experienced themselves. Most of the ideas and experiences of earlier generations were unknown to all but a few persons.

Johann Gutenberg developed a system of movable metal type in about 1440 A.D. Within a few years, books became available to many more people. The new printing process made it possible to reproduce, relatively cheaply, thousands of copies of any book. The Bible was one of the first to be produced. This made it possible for many people to have direct access to those religious ideas and teachings.

Reading became an important skill. Schools evolved so that people could learn to read and interpret the Bible. The consequences were tremendous. People began to question authority in new and different ways. They began to interpret for themselves, rather than relying on a few leaders for an interpretation. In addition, some people were not allowed to learn to read. Females and some racial or socioeconomic groups were not considered intelligent enough to learn to read, and laws were passed to restrict their access to education. (Laws such as this are still in place in some countries of the world.)

As more books on different subjects were produced, scholars in science, mathematics, philosophy, and religion were able to learn from other generations. They could share their ideas with colleagues and with people in other fields of study. In the previous century, a scholar could know most of the available information at the time.

Today, modern printing processes turn out books, magazines, and newspapers at a rate that would have amazed Gutenberg. Copying a page to share with someone is easy and you can even fax it to someone far away. Data banks of information around the world have changed the nature of libraries. Many people cannot find time to read the few magazines they find interesting. Can you imagine being assigned to read every book that was ever printed? Yet, around the world there are people who have *no* access to books or data banks. Just as once there were divisions between people based on who could and could not read and write, now there are divisions based on access to information sources. Those who have access to information have more power to decide for themselves and others.

Every few years the amount of information known to humans doubles. This rate of increase has been called the **knowledge explosion**. The pace of change can be expected to increase every year! Technology has made the knowledge

■ **Figure 15.7**
The availability of books allowed people to have the joy of learning new things. With this learning came new responsibilities of knowing. *(Courtesy of THE BETTMANN ARCHIVE)*

explosion possible, by providing faster and cheaper methods of communication. Access to this much information can be exciting, as well as overwhelming! Even in relatively precise areas of study, specialists may not be able to keep up with advances. It is becoming more and more difficult to know what is happening in more than one area of study.

Books and computers are tools for thinking. The development of such tools has had an effect on the development of the human intellect. Historians estimate that the average scholar two hundred years ago probably had a vocabulary equal to that of today's junior high school student. This doesn't necessarily mean that people of today are wiser, however. Having access to information does not mean that people always take advantage of what is available. It does indicate, however, one major consequence of modern communication systems. Information is all around us. As needed, new words are generated almost faster than they can be added to dictionaries. The challenge for critically analyzing information, judging its reliability, integrating it, and applying it to improvement of human life is greater now than ever before.

Social Consequences

Beginning with the very first invention, technology has caused social changes. A historical example of some developments in farming can illustrate these changes. Once people had developed technology for planting crops and domesticating animals, they had a more reliable food supply than when they hunted or foraged. Consequently, people began to form small villages instead of living as nomads. Animals provided food and did work. For thousands of years the social structures of tribes and villages changed very little.

When the wheeled plow was invented in Europe during the sixth century A.D., things began to change. Before the wheeled plow, each family farmed its own small plot of land. The wheeled plow required large oxen for power. Families began to join with others to buy teams of oxen. They also combined their small plots into larger farms.

Especially in Europe, this practice led to the development of a communal lifestyle. People in the village began to specialize. Some farmed, others produced clothing, while still others made and repaired tools. All these social changes were related to technological advances.

A similar pattern of change has occurred more recently on the plains of the United States. Owners of small farms formed "co-ops." This helped them purchase the large machinery and storage bins required for agribusiness. Together they stored grain and sold it whenever the price was favorable.

In both examples, a new set of social relationships was formed. People were required to give up some of their individual freedoms to benefit from group cooperation. A measure of independence was lost as people became more interdependent. Technology, you see, can have both positive and negative consequences.

Inventions do not have to be big to be important. Some small devices can have powerful consequences. The stirrup illustrated in Figure 15.8 is an example. Invented more than a thousand years ago, the stirrup changed the course of history. Before the stirrup, people rode bareback or with light saddles. When stirrups were attached to the saddle, the rider could more easily stay on the horse. The horse and rider could work together, thus using more of the horse's strength.

During the eighth century A.D. in Europe, kingdoms were warring among themselves for control of land and riches. A leader of the French, Charles Martel, observed that stirrups enabled his mounted troops to use long steel lances as battering rams. The stirrups gave them the power to overwhelm the soldiers on foot or the mounted warriors without stirrups.

To increase the number of mounted troops, Martel promised plots of land to those who would join his army. He soon had a vast force of mounted warriors. Martel and his son, Charlemagne, built an empire that lasted for a century. When the army disbanded, descendants of the original mounted troops claimed the land they had been promised. Each landowner ruled a mini-kingdom, complete with knights, serfs, and all the trappings of Camelot. And it all began with the stirrup.

Each of these examples—the clock, book, plow, and stirrup—has influenced the development of humans and their social structures. The development of humans and technology is closely connected.

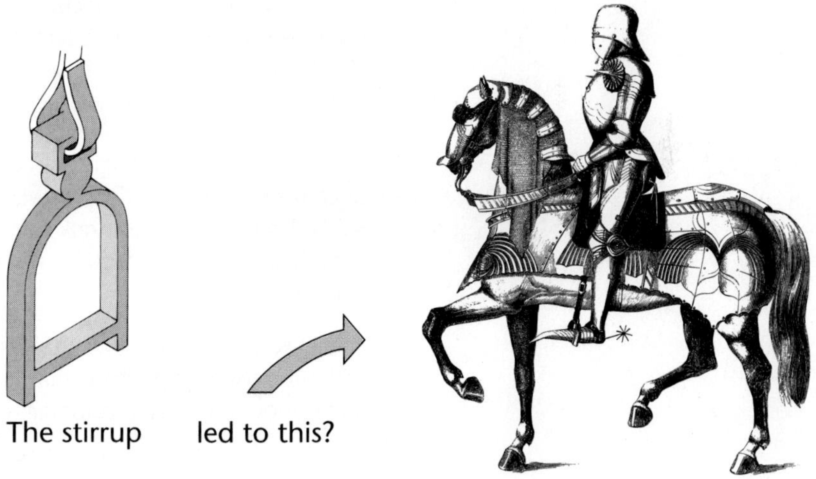

■ Figure 15.8

The stirrup was a very small technological innovation with major consequences. Because of this device:

- It was possible for warriors to fight more effectively from horseback. These warriors became the equivalent of the modern tank. Equipping these warriors with horse and arms was terribly expensive.
- To cover these expenses, monarchs took large tracts of land from the church. These tracts of land were given to those warriors who would fight for the monarch.
- These mounted warriors became knights.
- Strong castles were built for the safety of the people when other knights invaded their area.
- The castles helped the knights become more and more powerful. Finally the knights rebelled from the monarch and set up their own small kingdoms.
- The land, the social structure, and the government had gone through many important changes. The initial trigger for this change was the lowly stirrup.

(Courtesy of THE BETTMANN ARCHIVE)

The stirrup   led to this?

CONTEXT—CURRENT SOCIAL IMPACTS

- Several other human and ecological issues are involved in decisions about growing and harvesting food. Now, as in the past, land is owned by only a few people. Decisions about what to grow are made by these few owners, based on what will make the most profit for them. The vast majority of the people in some countries work on the land doing the labor of planting, harvesting, and processing crops. Most of the land is used for cash crops, such as coffee, tobacco, cacao, beef, poppies, and pineapple. These goods are sold to affluent people in other countries and bring a large profit to the landowner. Laborers usually are paid only a small wage. This situation would be difficult enough for the worker, but in addition, most of the fertile land of the country is devoted to cash crops. Consequently, space and resources for growing food to feed the laborers and their families is scarce. Many people suffer from malnutrition even though they live right next to a highly fertile field, planted with crops that provide luxuries for foreigners. Once again, the issue is not one of being able to feed a population, but of how decisions are made.

- Some methods for improving harvesting techniques have focused on changes in the tools and machines. Extensive systems of harvesting technology have been developed to plant, cultivate, harvest, and store grain crops. One example is the self-propelled harvesting machine, sometimes called a combine. This harvester combines the reaper, a machine that cuts grain, and the thresher, a machine that separates the grain from the rest of the plant. The combine separates the grain from the straw without damaging the grain. This bulky machine must move over relatively soft soil. It must operate efficiently while bouncing over uneven surfaces. The combine changed farming and the system of harvesting grain.

Before this invention, each farm family cut their own grain and gathered it into "sheaves" and "shocks" to let it dry. The owner of a threshing machine came from farm to farm to thresh the grain for each farm family. Usually, the crew of people who worked with the thresher relied on a "cookshack," a portable kitchen designed to feed the crew of workers. The entire community often became involved in harvesting the grain. Men, women, and children from each farm served as part of the thresher or cooking crew. They cooperated and harvested each field, in turn. They worked hard, but often took out time for gossip and socializing. They even held dances and other social events on weekends or when the harvesting was over.

Today, the combine allows each farm family to complete the process without outside help. The community cooperation is no long needed. If a farmer does not have a combine, he pays someone who does to harvest his crop. What advantages and disadvantages do you see in this change?

Farming machines are often specialized equipment that do only one task. Consequently, much money must be invested in different machines. When several specialized machines are needed, however, the cost of producing foods is high. Farming by using many machines is a capital-intensive process. Multipurpose machines could do several operations, reducing the overall expense.

Ecological Consequences

Environmentalists and ecologists are beginning to make us aware of what humans have been doing to our natural resources for centuries. They observe that for generations, people have been using raw materials, processing them, and pouring wastes in the rivers, oceans, air, and soil. They point out that nature holds a delicate balance. Humans, they charge, are destroying plants, animals, even the air we breathe and our protection from the sun's rays. Because it takes such a long time for ecological consequences to be discovered, it is difficult to know today how serious the situation will be in the future.

Some scientists point out that humans have always had an effect upon the environment. At the end of the last Ice Age, about ten thousand years ago, herds of wild horses on the North American continent disappeared. This happened in spite of an abundance of grass for food. The best guess is that the horses were slaughtered for food faster than they could reproduce. Finally, there were none left. More than nine thousand years later, Spaniards brought horses from Europe. The breeds of horses we now have in the United States are not native to North America, but are descendants of the Eastern Hemisphere.

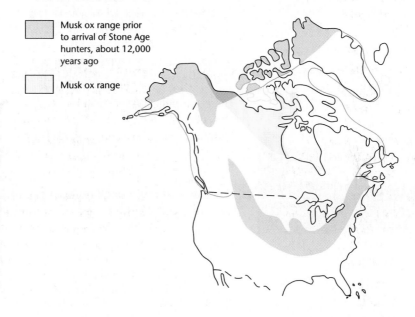

Musk ox range prior to arrival of Stone Age hunters, about 12,000 years ago

Musk ox range

Figure 15.9 ■
During the Ice Age, humans migrated into the western hemisphere from what is now Russia. Prior to their arrival, the musk ox were plentiful. Soon the musk ox disappeared except in places where the glaciers made it impossible for humans to reach the animals. The patterns of human migration follow closely the patterns of extinction of large animals. This suggests that humans may well have been a major cause of that extinction.

Investigations by archaeologists indicate that wherever humans have migrated, the balance of nature has been changed. About 12,000 years ago, humans moved across a land bridge that used to connect what is now Russia and Alaska. This seems to be the most plausible explanation for why Alaskan Eskimos have physical and cultural characteristics akin to Oriental people. Prior to that time, the musk ox lived on the north and the south sides of glacial ice covering most of Canada. Soon after these Stone Age humans migrated into North America, the musk ox began to disappear. Because the northern side of the glacier was inaccessible to these early people, the animal survived and prospered there for centuries.

Concern about what happens when a species dies out has become even more intense as we have developed the tools to study DNA. Genetic diversity is seen as a safeguard as we face issues that have not occurred in the past. Some of the consequences of technology on plants and animals are easily observed. There are also important effects that we cannot see. Consider the following chain of relationships.

The Earth is surrounded by a relatively thin layer of gases. This layer of gas, the biosphere, sustains all living things. The air we breathe consists mainly of three important gases: oxygen (21 percent), nitrogen (78 percent), and carbon dioxide (less than 1 percent). Bacteria and algae in the soil take nitrogen from the air and convert it to ammonia. Ammonia is poisonous to humans. Fortunately, other microorganisms convert the ammonia into nitrates. Nitrates are needed by green plants to produce proteins and oxygen. Animals, in turn, eat protein and breathe the oxygen. When animals and plants die, their protein is converted back to ammonia and the cycle continues. Life abounds! If ever there are not enough green plants left, humans will be in trouble.

Surely we recognize the importance of nature's life cycle in the biosphere. Or do we? For many years, vast amounts of a chemical called 2-4D were used in highly developed countries to kill weeds. During the Vietnam War, 2-4D ("Agent Orange") was sprayed from planes to wipe out entire forests. Few questions were raised until years later, after the damage was done. Not only were green plants destroyed, but 2-4D turned out to be extremely dangerous for humans, as well. Years later, veterans are suffering from illnesses and disabilities.

There are difficulties in determining the possible consequences of technology. One difficulty is that the effects may not become apparent until long after a harmful chain of events has started. Once that sequence has begun, it takes a long time for it to be stopped, if it can be stopped at all. This factor reemphasizes the importance of using all the skills we have to try to predict consequences.

Nuclear testing is another example of an activity which has long-term consequences. A large number of nuclear devices were exploded both underground and in the atmosphere between 1948 and 1964. Halfway through that period, some scientists indicated that we need not worry about the level of radioactivity in the atmosphere. In fact, people were regularly exposed to levels of radiation that we now know are damaging. By the 1960s, scientists found that radioactivity was poisoning our atmosphere, our soil, our food, and even the milk we drank. They realized that we had reached a crisis. They stopped the cause of the crisis, nuclear testing in the open air. They could not stop the chain of events that had been started, however. For some years to come, we may be affected by the consequences. The same will be true of disasters such as those that occurred in such places as Hiroshima, Chernobyl, and Three Mile Island. It also means that international agreements on such testing are vital.

We are still waiting to see the effects of radioactivity from nuclear wastes. We already know that currently the storage containers are made of materials that will wear out and leak in less than 50 years. Yet nuclear wastes remain radioactive and potentially harmful for several centuries!

■ **Figure 15.10**
Tires are certainly necessary to run the modern automobile. But what to do with them when they are worn out? This recovery center extracts reusable materials from the tires. *(Waste Recovery, Inc., Courtesy of Compressed Air Magazine)*

Many of the by-products of technological activities are materials that do not occur naturally in the environment. Toxic chemicals that do not naturally break down into safe products have potentially high economic and ecological costs. Fortunately, people are beginning to become more aware of environmental issues. We often hear of acid rain, greenhouse effects, global warming, depletion of ozone, and the like. Each of these issues is complex and will need attention to both prevention of further damage and efforts to repair what has been done.

DESIGN BRIEF

The World Watch is an organization that regularly assesses what is happening in most of the world's ecological systems. Choose a system that has been studied by this group and gather information on the most recent indicators of the health of that system. Develop a plan for action that can be accomplished by students in your school that will affect that system. Develop a list of ways people can make changes locally, yet have global consequences. Create a video or program for adults in your community to report your results and indicate what they can do to make a change.

MATH/SCIENCE/TECHNOLOGY LINK

Form discussion and work groups to consider some of the ecological consequences that change the balance between the living and the nonliving things in the biosphere. Select any of a range of problems that have important consequences for the environment. Several possible topics are listed below:

• Excess burning of fossil fuels

• Dumping of waste materials and garbage into rivers, lakes, and soil

• Nuclear testing and its resulting atomic fallout

• Use of pesticides and weed killers

• Oil leaks from shipwrecks or offshore drilling

• Storage of nuclear waste

• Deforestation of the world's woodlands

• Acid rain

• Greenhouse effect

Determine what some of the possible effects have been on humans and animals that have been omitted from the above list.

Consequences and Control

We do not have to be caught unprepared for the consequences of technological developments. Technology assessment, described previously, can help us plan for future consequences. We can learn to use technology assessment to try to foresee the immediate and the delayed effects of a proposed technological event.

These delayed effects are called **second-order effects**. They often occur in a sequence. They appear to interact like a row of dominoes. The dominoes sometimes fall very slowly. They may topple each other in strange and unusual patterns. A **first-order effect** (the desired effect of the action) may cause second-order effects. These, in turn, may cause others, and so on. Technology assessment is used to study and, hopefully, project strings of possible effects. We would like to foresee positive and negative effects before the events happen.

DECISIONS—MAKING TOUGH CHOICES

Many major concepts and ideas about technology have been presented throughout this book. This section shows how your knowledge can be used to make decisions about technology as it affects your life and the lives of others.

You have learned that tools are an important part of technology. Many of the tools that have been discussed are considered **hard technologies**, such as machines and instruments. Decision-making skills are **soft technologies**. They are tools that can be used to change society. There are a number of decision-making tools. How well they are used may determine if the technology of the future will result in the consequences we desire.

Kinds of Decisions

What is a decision? You make them every minute, every day. A decision is a choice among alternatives in the face of uncertainty. In making a decision, one course of action is chosen instead of others.

If you decide to buy a new ten-speed bike rather than a skateboard, hopefully you make your decision based on the best information available. You consider how much long-distance travel you do. You might consider how much adjustment and maintenance a bike will require and how to store it safely. In the long run, your decision is an educated guess about meeting current and future needs.

There is always an element of uncertainty in making decisions. Most people dislike being uncertain about things. Ordinarily, we try to find answers that are clear-cut. We want the "right answer" or the "truth"—right now! We want to choose either this or that, not one choice from several alternatives.

In reality, the world is not that simple. Most decisions in our technological world are selected from dozens of possible choices. Making decisions leads to accepting responsibility. When people make a decision, they must accept the consequences. Dealing with uncertain consequences can be a challenge to the decision-maker. The only alternative is to let others make the decisions. If you do that, you may have to live with consequences you will not like and didn't choose.

■ **Figure 15.11**
We make many decisions every day about how to spend our time and energy. (©1992 Robert Cerri/The Stock Market)

Several desirable alternatives. One type of decision is a choice among several alternatives, all of which seem to have desirable consequences. When this is the case, a person must decide which choice is the "most right." The process of making a choice is usually more complicated than flipping a coin. You will want to know as much as possible about each of the consequences and the likelihood (probability) that each will happen.

Usually, the more you know about the options, the easier it is to make a choice. It still may be difficult to make a decision of this type. You may want to go roller skating *and* buy a new T-shirt *and* put some money in your bank, even though you only have enough money for one of these choices. To decide, you must weigh the consequences (effects) of each possible choice against the others.

Desirable versus undesirable alternatives. A second type of decision is that made when one alternative is what you clearly want, and the other is something you wish to avoid. This kind of decision is easiest to make, but only if your goals are clear.

Instead of buying a new bike, for example, you might decide to continue riding your old one. The choice is easier to make if you know that, in order to buy the new bike, you must go without lunches, sell your baseball card collection, and give up the money you were saving for a new CD player. You might also remember that other people with great new bikes have had them damaged or stolen.

However, life is seldom really this easy. You have learned from your design work that some criteria are more important than others. Not all alternatives are equal in weight. Also, sometimes we fool ourselves by believing that we can avoid the undesirable consequences. People make judgments about just how much of a risk they are willing to take. For example, the scientific research is very clear about the link between smoking and certain diseases and decreased life expectancy. However, these are the *probabilities* that we must weigh, not certainties. It is easy to say to ourselves, "Sure, nine out of ten will get this disease, but I will be the one who stays healthy." Usually, there are several variables at work at once. For example, people who depend on the lumber industry for employment often agree that saving the spotted owl species is important for their grandchildren. However, they may not even have any grandchildren if they cannot take care of their children, presently. Keeping their jobs, feeding their children, and staying in their homes are more important for now. They may just want the future to take care of itself, as they struggle right now.

Choices from undesirable alternatives. A third type of decision is the kind that people must make when all the choices seem to lead to undesired consequences. This is the most difficult of decisions because it requires you to decide which alternative has the least negative consequences. Making choices from several undesired alternatives places added importance on techniques such as risk assessment as a means of helping make the best decision possible.

An example of this kind of decision is when you make a choice between doing the dishes or cleaning your room; or between going to the dentist or writing your term paper; or telling someone that your friend is in trouble and needs help, when you think your friend might hate you for telling. Anxiety and conflict may

■ **Figure 15.12**
As the demand for land use and development becomes greater, making decisions to set aside some land for the use of everyone becomes even more critical. Through such decisions these lands can be preserved in their natural state for the enjoyment of future generations. *(Photo provided by Alabama Power Company)*

■ **Figure 15.13**
Around the world, the supply of safe, pure drinking water is threatened. Something as basic as water is in danger as a consequence of technology. Technological solutions to reverse this trend may provide an answer. *(© Lawrence Migdale/Photo Researchers, Inc.)*

■ **Figure 15.14**
The personal computer provides access to a range of services. This vast array of products, services, and information makes the decision-making process easier, at the same time it becomes more difficult. *(Courtesy of Hewlett-Packard Company)*

be evident. You may try to postpone the decision as long as possible while searching for other alternatives. Usually, however, the decision must be made sooner or later.

When we make any decision, we make our best fit—we try to optimize the probability of achieving the outcomes we want. The more we learn about the probable consequences, the easier it is to make a decision. More than likely, you have learned by now that there are no guarantees that even the best choice will work out as you hope. And, because we are human, each of us sees the situation and predicts the consequences differently. That is why group decisions, such as those about building a waste incinerator, are so difficult to make. Different people see different effects and place different values on them. Often they want other people to take the negative consequences, not themselves.

Often the decision-making process is even more difficult when we realize that by not deciding, we have in effect made a decision. As a result of not deciding, we leave our future either to chance or to the decisions made by others. If you choose not to act when someone needs your help, for example, your decision may have consequences. It seems that, as difficult as it sometimes is, we cannot avoid making decisions. In technology, our decisions often have personal, social, and ecological consequences.

In the face of uncertainty and change, we must make decisions based on the best information we can get. We will make some mistakes, but we can learn from past errors and avoid repeating them. We can make even better decisions if we develop our skills. One way to do so is to explore the different methods people use to make decisions.

How Do We Make Choices?

The four general bases for decision-making are authority, feelings, logic, and experience. Each of these are considered in turn.

Authority. In the first approach, people make choices based on the advice of others. You may have decided to try a new brand of shampoo because a classmate recommended it. Manufacturers often pay famous people to recommend their product, hoping that many customers will make a decision based on the star's "authority."

On more serious matters, we often rely on the advice of a physician, a religious leader, parents, or a trusted friend. Before purchasing a product, some people consult a consumers' magazine to see how it and similar products have been rated. Decisions based on authority usually are made when we do not have the technical knowledge or experience to judge the probable consequences ourselves. As technology becomes more complex, we often must rely on specialists in a given field. It is imperative that we check out the reliability of the experts and assess any biases that may influence their interpretation of the information. We also need to keep an open mind, so that new information can be considered as it is gathered.

Feelings. A second approach involves making a choice based on feelings. In this case, people choose not to rely on authority, but upon what pleases them.

Such decisions are often the ones that "feel right." Often, they might as well say, "Don't confuse me with facts."

This method involves no particular information. It is a kind of personal sense of what is best. Young children often make these kinds of decisions because they have not yet gained the knowledge required to make choices based on facts.

Sometimes, a feeling or sense of what is right is helpful. A decision to invite a new classmate to a party might be based on your intuition or feeling that it is the right thing to do. There may be no reliable facts available, and no authority to consult. The results can be positive. At other times, however, decisions based on feeling are made only according to what pleases the individual at that moment. Retail stores are aware of shoppers' tendency to buy products on impulse. They often construct flashy displays and packaging to catch their attention. People often find that they have spent resources they need elsewhere because they did not use both their mind and their whim.

As you learned in the chapter on Humans, thoughts and emotions are related. Strong emotions are signals that expectations and rules, developed in the past, are being used to perceive the current situation. If you do feel strongly about something, it is useful to tune into what you are thinking at the time. Rather than deny the emotion or act on it—first, think about it. Why do you feel this way? In some cases, the emotion is a good indication that the decision will be right for you. In other cases, you have deleted important information, distorted the facts, and refused to see how what is happening now differs from what happened before. You may be responding as you did when you were a child. Then, that was all you could do! But now, you have experiences and abilities for thinking things through. You can consider probabilities and handle more information at once. You can consider other people's feelings as well as your own.

Consequences. As a third way to decide, you can rationally consider the consequences of the decision. Using this method, people logically weigh the probable effects of a choice and judge the worth of each effect. The process usually involves **if-then statements**. If you buy the cheaper shoes, then they will probably wear out faster. However, they may not be in fashion any longer than that anyway. If you install a wood-burning stove in your home, then someone must cut or buy the firewood. Rational consequence decisions require that people know how and why things work. If you know how a furnace thermostat works, you can better decide if it will save money to set it manually at a lower setting each night. If you know that you tend to forget, getting a thermostat that you can program to automatically change at a given time may be worth the investment.

Experience. A fourth method for making decisions is to use personal experience. People sometimes make choices this way by remembering an experience in similar circumstances. Professional baseball pitchers keep a notebook to remind them of the pitches that have worked against opposing batters. The actions and past consequences are used as a guide for new decisions.

Experience can also be used to indicate if a new choice is necessary. You might choose an old tool to tighten a loose bolt on your bike. If it doesn't work, you may look for a new tool. In general, the old idea, "If it ain't broke, don't fix it," applies here. Only when we sense a problem do we look around for new ways of responding. As you have found out in this book, this is the source of most technological efforts.

Each of these methods of choosing may be used to make decisions about technology, whether on a personal, national, or global scale. In some instances, people make choices because a respected authority indicates a given decision is best. In other cases, choices may be made on a feeling about the situation. Old hatred and past resentments may block out the ability to think through the alternatives.

The decision can be based on the probable consequences. We usually make choices based on what worked in the past. Some of these consequences may now take on a global or environmental perspective that was not considered before, however. Each method of decision-making has certain advantages and disadvantages.

What Questions Must Be Answered?

Decision-making requires a consideration of goals, resources, actions, and results. As you know from your design work, each involves thinking through the problem and making a plan.

What is the purpose? Goal-setting requires that you have an indication of what you wish to achieve. It includes an awareness of what short-term and long-term outcomes you want to achieve and what undesirable consequences you wish to avoid. Some of the goals are at a practical, personal level. Others may be based on moral and ethical standards. When you consider your action in light of its effects on other people, you are concerned about justice, fairness, freedom, and responsibility. What is right must meet a set of ethical and moral criteria.

What needs to be done? Decisions also require that necessary actions be considered. It would be foolish to set a goal to clean up the neighborhood baseball field, without recognizing the work and resources needed. Technology decisions often require a lot of dedication and effort. The series of manned explorations of the moon, for example, required the cooperation of thousands of scientists, engineers, politicians, and technicians.

What resources will be needed? The general resources that support technological efforts include space, labor, capital, knowledge, and time. These resources provide the needed support for the tools, materials, processes, energy, information, and humans that are required for a technological act to happen.

The resources we choose will be directed by the desired outputs, availability, and the trade-offs between resources. In many cases, our best choice is the one that uses the fewest resources. This often happens when a needed resource is in short supply. Car buyers, for example, are beginning to choose more fuel-efficient models. This tendency comes from increased awareness that gasoline is a limited, nonrenewable resource. The current cost of the resource is also part of the consideration.

At other times, decisions may be made on the basis of an oversupply of resources. During the 1930s in the eastern part of the United States, a large number of chestnut trees were killed by disease. The wood was quite valuable and beautiful. Many home builders used chestnut lumber extensively for finished trim in the houses they were building. By the time the last trees were cut, worms had bored into the wood, leaving traces of their miniature tunnels. Houses built in this later period were trimmed in what was called "wormy chestnut."

In these cases, choices were made to take advantage of the available resources. Today, people are beginning to look at garbage as a resource, not just as a problem. The costs of disposing of garbage have become great and a new perspective is needed. Systems of reclaiming and reusing are needed. New designs that consider what happens when a product is discarded are needed. Consumer behavior that encourages a decrease of waste is needed.

Results.   What might happen? Every possible choice has some **results**. This may be the most difficult component in the decision-making process. It is not always possible to predict clearly the consequences of each option.

Who can be sure what will happen if your school decides to have your lunches prepared by a fast-food restaurant? Perhaps the consequences would be better than building a new cafeteria in the school, if that is the alternative. As with all decision components, however, it helps to have the necessary information.

It is often helpful to consider a sequence of effects of your actions. These are called first-order effects, second-order, and so on. Consider, for example, having a choice between using herbicides (plant-killing chemicals) on a vacant lot or removing the weeds by hand. The immediate results may be the same (fewer weeds). There may be second-order effects, however. You might save time by using the herbicide, but the chemical may wash off into your water supply or into a local stream or river. Both the human and fish populations may be harmed.

"Okay," you are probably thinking, "but I am only one person. How can my spraying chemicals on one tiny patch of dirt hurt the fish population, hundreds of miles away?" The answer to your question comes if you recognize that every other person could make the same decision you did. Then the effects would be multiplied and the consequences may occur.

If you remove the weeds by hand, you may avoid those unwanted second-order effects. You may also work off some excess energy, achieve some satisfaction, and get a suntan in the process. But you may have to give up the time you would prefer to spend on sports and you may get blisters. Many decisions involve second-order effects.

One way to make decisions is to chart the possible goals, actions, resources, and their results. This approach is similar to the design approaches discussed earlier. Decision components can be listed in order to show the interrelationships among actions and consequences. This may even cause you to rethink your original goals.

■ **Figure 15.15**
With all of the newspapers and other print material produced every day, it has become imperative that the paper be recycled. (© David Woodfall/Tony Stone Images)

The decision chart may help you focus on the necessary information, but it will not make a decision for you. Decisions are rarely made in a straight line sequence of goals-actions-resources-events. More often, decisions are reached by a continuous cycling of the process. After going through the process several times, the best choice may become evident.

GOALS	ACTIONS	RESOURCES	RESULTS
Produce fresh vegetables, get exercise	Gardening, planting (mulching or hoeing)	Seed, tools, fertilizer, time, growing season	Fresh vegetables, sunburn, sore muscles, blisters
Produce fresh vegetables, save money and time	Pest control (spraying or bug-eaters)	Tools (sprayers), chemicals (pesticides)	Larger yields, reduced costs, killing bees and birds
Produce fresh vegetables, get suntan	Cultivation (no till or weeding)	Tools (hand, power), chemicals (nitrogen, herbicides)	Increased yield, good suntan, health risk from chemicals and sun
Produce fresh vegetables, stay out of sun	Gardening (greenhouse or hydroponics)	Greenhouse, equipment, chemicals	Vegetables year-round, increased costs, work, knowledge, health risks from chemicals but not from sun

Figure 15.16 ■
This decision chart can help compare alternatives for choosing the most appropriate approach to gardening.

Limitations on Decision-Making

Controls and constraints. Decisions are not made in a vacuum. Normally, there are personal, social, and ecological factors involved. Each factor may bring certain limitations to the decision-making process. Often, there are limits to our knowledge of the options, our resources, and our ability to foresee consequences. We cannot dwell on the past but must decide in the present, from present options. (Refer to Figure 15.17.)

Sometimes our choices are limited by factors we cannot change. These are called **external controls** or **constraints**. For example, you had no choice about when and where you were born. To some extent, you may have little choice about where you now live and where you go to school. You may have little control over the actions of other people who are involved in the decision. Choices you make each day are constrained by your current age, the circumstances under which you do live, and your surrounding environment. Many decisions involve external factors such as these.

Internal controls are those things which you are able to change, at least in theory. They include personal attributes such as the skills and stamina needed to bring about the desired consequence. If you are able, for example, you might choose to ride your bike or walk to school rather than to be transported by bus or car. Others who have physical handicaps may not have that choice. They are limited by internal (personal) circumstances. You might consider buying soda in returnable bottles to save energy and materials. That choice is possible if there are no external controls; i.e., if returnable bottles are available. The internal control in this case is your own willingness to make an extra effort.

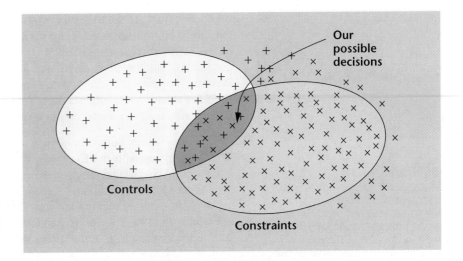

Figure 15.17 ■
Controls and constraints rule out some of our possible decisions. The decisions that are open to us may not bring about the goals and effects we want.

This leads us to consider the relationship between values and decisions. Throughout this book you have seen how people use technology to alter the environment. In the decision-making process, technology can be used to obtain information that will help make better choices. Computer simulations, for example, determine the possible consequences of building a new superhighway around a town. The fact remains that humans decide which factors to consider. The ultimate decisions are human ones.

For example, years ago, the United States constructed a system of underground missiles. There were some people who thought that technical devices should be designed to fire the missiles automatically in the event of attack. Others believed that humans should remain in control. The issue was one of values—which judgment should be trusted, human or machine? The system was built to require a series of human decisions, including the president's, before the missiles could be fired.

Making the "right" decisions. In our society, we often speak of what is "right." The term can mean many things, including what is ethical and moral, and what is technically correct (true). You have probably been taught many "shoulds" (how the world should be and how people should act). These "shoulds" reflect our values. Unfortunately, decisions that people make do not always fit what they say they value. Actions do not always correspond to words.

Centuries ago, the world population was small. Materials, food, and energy resources seemed plentiful. People defined their society as those people who were like them or who belonged to their group. This feeling of community did not include people who spoke another language, who were from another racial group, or who had different religious beliefs. Decisions were made in terms of the immediate effects upon the local group.

In modern societies, decisions about technology are not so easily made. We are learning more about the total human population on the Earth. We are becoming more aware of how far-reaching our decisions actually are. Personal decisions often have social consequences, and vice versa. Decisions made in one part of the world may have profound effects on other parts of the world. For many of us, our community involves all people.

■ **Figure 15.18**
Many of our technological decisions should consider what is fair to us, to other people in the world, and to those who will live here in the future. Decisions about resource use and distribution are critical. Ethical and moral dilemmas can be expected to get even more complex and vital in the future. (© *Sean Sprague/Impact Visual*)

We are beginning to see that many of the Earth's resources are limited. We realize that decisions must be faced about sharing these resources with billions of other people. Who should have access to the Earth's resources? What about future generations? In the past, there was little concern for the long-term future. Recently, however, decision-makers have begun to examine possible effects centuries ahead. These futurists urge us to consider all of our decisions in terms of what effects there may be on our children's children.

All this may make our technological choices even more difficult. The only alternative, however, is not to decide. Would you like to see the world continue as it is now? What will be the consequences for you and the rest of the world, for now and for future generations? Perhaps, with the insights you have gained, you are ready to participate in making decisions that reflect the best knowledge that is available right now. An understanding of how we are both independent and interdependent can help us make choices that are more likely to be fair and just.

SUMMARY
CHAPTER 15

Technology is changing our world. Like a snowball rolling downhill, technological systems are growing faster and faster. Each new invention or process makes other developments possible. The new developments in technology bring change. Change is sometimes difficult for people and societies to accept.

Technology can have a significant impact on each of us, on our social institutions, and on our environment. Some of the consequences are helpful and some are harmful. It is becoming increasingly important for us to plan ahead. We need to investigate all new technological developments for their potential effects.

Technology assessment procedures can help us understand and project the consequences of new proposed technologies. When we understand the consequences, we can decide whether or not to use each new technological development. The benefits should outweigh the problems. If we do not learn all we can about new technologies, then we are in danger of losing the choice about using them. We can find ourselves having to live with their effects for generations. The effects may extend to innocent people who had no say in the decisions that affect them.

The "soft" technology of decision-making is serious business. Even our small, day-to-day choices may have far-reaching effects. You base

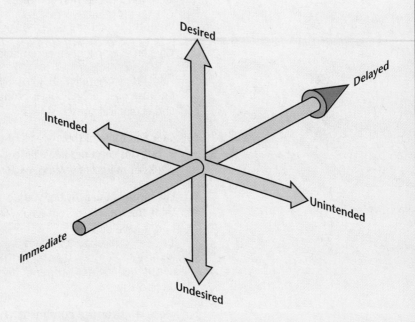

■ **Figure 15.19**
All technological decisions have consequences—some immediate, some delayed. The consequences may take four different dimensions. They may be intended or unintended, desired or undesired. The most dangerous consequences are those with unintended, undesired, and delayed effects.

some of your decisions on advice from authorities. Other decisions are based upon your own feelings and beliefs. If you know the potential consequences of your decisions, logic may help you make choices. Finally, you may fall back on your experience from similar situations to help you decide. In all cases, you need to acquire as much information as possible. That way, your decisions will be the best possible under the circumstances.

Decisions require that we consider at least four factors: goals, actions, resources, and consequences.

Often, there are constraints or limitations on each of these factors. Choices must be made accordingly. A decision may require that we cycle several times through the decision-making process.

Ultimately, the "right" or "best" decision is the result of human judgment. The process is becoming more difficult all the time as we become aware of the interconnectedness of our decisions. The decisions we make about technology may have serious consequences for the rest of the world and for future generations.

ENRICHMENT ACTIVITIES

1. Describe a decision tree. Indicate how you would construct one.

2. What is the purpose of technology assessment? How is it done?

3. Give an example of an acceptable risk.

4. Cite three examples of personal consequences that have occurred because of a technological product or event. Do the same for social and ecological consequences.

5. Describe some of the impacts of the development of print on the lives of humans centuries later. Describe your life if there were no television, books or magazines, radio, or automobiles.

6. List some of the activities you do alone and with a group. How do you think this list differs from what your ancestors, two hundred years ago, might have done?

7. Identify some of the concerns that people have about future consequences for current decisions and behaviors related to technology.

8. Make a list of the constraints and controls that place limits on the decisions that are possible for you to do.

9. Describe the types of alternatives that are involved in choices people make.

10. What are the major sources that people consult when they decide?

11. Give some examples of situations in which ethical and moral values must be considered in making a decision.

Epilogue

Many people currently enjoy the fruits of technology. And yet there is increasing concern about where technology is taking us and whether we are becoming dependent on technology. As technology becomes more complex, it is increasingly difficult to make informed decisions. Some people throw up their hands and say an individual cannot do anything about it. Many times we buy and use what we want and do not consider long-term consequences to ourselves, other people, and the future.

We hope, now that you have read this book, that you are ready to use your head rather than throw up your hands. We hope you have increased your knowledge and willingness to work so that technology adds to the well-being of people and the environment, for both now and in the future. You now understand that technology is a very human endeavor, and that you can have a part in creating products, environments, and systems. This occurs by your own design and by the choices you make as you use the outcomes from the technological activities of others. Technology is like any other tool; it can cause damage if the user is unaware, unskilled, or unwilling to exercise caution.

Yet, you have also seen how technology benefits people, if used with understanding and enlightened concern. Control of technology takes the willingness to make difficult decisions and to constantly reevaluate old information and approaches. The dynamic nature of such a process can be both exciting and overwhelming.

Making decisions about technology is a complicated process. It would be easy to sit back and let others make the choices. It is easy to be overwhelmed and wonder what one person can do. We would like to suggest several things you can do to help.

(1) Continue to develop your technological skills and understanding to the highest degree possible, given the resources you have. You are among the most privileged group in the world. You have better access to the information, tools, materials, and other resources of technology than do most people. You live in the most highly developed technological society in the world. You have the chance to become a leader by being a smart producer and user of technology.

Continue to look for ways of making improvements in satisfying human wants and needs. Use your design and problem-solving skills to make products and more effective ways to improve the lives of yourself, your family, and people in your neighborhood.

(2) Develop your survival skills and understanding. It is easier to make difficult decisions if you know that you can live with the choices. Suppose all electricity was turned off and all fossil fuels were unavailable for two weeks. Maybe there was a power failure and the energy networks closed down. If you are technologically prepared, you can use your problem-solving abilities to seek out alternatives as you cook your food, stay warm, get where you want to go, and get along until the power again becomes available.

Suppose there was a major storm in your area and the food supply system was cut off. Stores run out of food. Could you go into the country and find edible plants and catch animals for food? Do you know how to take care of yourself if the transportation system was to break down? Transportation moves food and fuels, as well as people. Hopefully, you will not have to test these skills, but you have begun to develop a way of looking for information and solving problems that should be of help.

Survival skills can help people to live in their environment with limited technological resources. This kind of training can help you avoid the "technology trap." That is the trap of being helpless if the technology systems break down for a time. We hope that you are better equipped now than you were when you started this book. There are many opportunities to learn more by reading, watching television specials and videos, taking more technology courses, and participating in clubs and educational organizations.

(3) Interpret technology to others. We have indicated how rapidly information is growing and how technology changes people's lives. The pace of change may make young people the experts on technology. Older people may be required to unlearn old ideas before they can grasp the new. Be patient as you help others learn. You may be in a better position to ask pertinent questions about technology than are your parents, grandparents, and teachers.

(4) Consider personal, social, and ecological consequences when you go about your life every day. The phrase used often is "Think globally; act locally." Evaluate your behaviors and choices until concern for the consequences becomes a habit. You might ask yourself such questions as the following: Is this made of nonrenewable materials? Can I conserve or recycle it? What is the energy used for making and operating this tool or machine? If it is from a nonrenewable source, can I find another way of reaching my goal? If not, how can I conserve that energy?

In your choice of products, you may give preference to those made locally, so that the need for shipping is less. You may wear clothes designed to keep you warmer in winter, so you can be comfortable in buildings that are minimally heated. You may help insulate your home and take advantage of solar heating techniques. You may choose to use mass transportation, walk, bike, or form a car pool. Many technological decisions have ecological consequences. As you continually remind yourself of these consequences, you may change some of your behavior and urge others to change theirs also.

Industries are sometimes reluctant to spend the money to prevent pollution and clean up their waste products because the money spent means reduced profits or increased prices. Yet several industries have responded to documented evidence of problems they cause. Groups of concerned students and adults can show the company the importance of being a good community member.

Vacant lots might be turned into playgrounds or community gardens. You might plant a small garden just to find out how it is done. You also decrease your technological dependency. Grass clippings, leaves, and hair from a barbershop or beauty shop can be recycled to nourish the soil and to control weeds. You can learn to use organic bug and pest controls. Check out the special plants, fertilizers, herbicides, and insecticides produced by modern biotechnology.

Locally, you can ask questions about your environment and help improve it. With your friends, perhaps you can hold a car wash or bake sale to earn money. You can spend it on paint to cover graffiti or on other improvements in your area. You might urge local business owners to contribute the materials while you help with the labor. You can organize a litter cleanup day or ask your elected officials to promote research about garbage disposal.

You are best able to decide where to put your individual energy and resources. Look at several possibilities before you decide what action you wish to take.

(5) Lend your support to larger groups that are organized to have impact on industry and government. Many citizens' groups have shaped local, state, national, and worldwide decisions about technology. Some examples of positive actions include stopping the construction of a dangerous intersection, lobbying for a better sewage treatment center, and boycotting a company that produces unsafe products. You may fail to reach your goal the first time, but the process of applying pressure and asking questions often changes similar decisions in the future.

Consumer groups acting together on a local, state, national, and international basis have been effective in influencing the decisions of industries, businesses, and government. Consumer groups can be powerful, partly because they represent a large number of people. These large groups are customers and voters, so decision-makers tend to listen. Political decision-making is very important. It can set directions and decide actions about technology. It is important to elect representatives who are knowledgeable about technology and its potential consequences.

It is not enough simply to elect individuals to make technology work for us. We must also observe continually. You can keep track of the decisions of your political representatives. This way people can know whether their wishes are being observed. Groups have organized to serve this purpose. Even if you are too young to vote, politicians and groups that watch them are often glad to have your help. You could volunteer to distribute literature or do other kinds of work for a candidate or organization you support. Your questions at home and in your community can be valuable to alert politicians that people are concerned about the consequences of technological decision-making.

(6) Remember that even if you cannot cast a political vote, you vote every time you spend money. As you decide what to purchase, you tell companies what you

do and do not want. Whether you join or form groups, or focus on developing your decision skills, you do have the potential to make a difference.

You can be action-oriented and be involved in the decisions of industry, government, or other groups. You can also be an advocate and work to promote what you believe in. You can spread information. Your choice of what effort to join requires a working knowledge of technology, and its resources, systems, controls and consequences. We hope that this book has laid the foundation upon which you can build.

The first decision that must be made, however, is whether or not you will get involved. If you decide not to engage in the scuffle, you automatically lose control of an important aspect of your life. A decision not to act has many consequences, too. The choice is yours.

Glossary ....

absorption – to soak up or take in.

acceptable risk – choosing between possible outcomes, based on the degree to which error and unknown, undesirable outcomes can be tolerated.

accumulator – a device to store air or fluids. Accumulators can be used to create a time delay in pneumatic systems.

achievement – a result brought about by resolve or persistence.

acoustical processes – employ the use of sound to create internal changes in materials.

action – the operation or movement of a mechanism.

addition process – adding materials together to form a new material or product that combines the characteristics of each of the ingredients.

additives – a substance added to another in relatively small amounts to impart or improve desirable properties or suppress undesirable properties.

adhesion – the connection formed between materials in a surface bond in which the atoms and molecules of the two surfaces remain separate and the adhesive material is "sandwiched" in between. Adhesive processes include gluing soldering, brazing, printing, or painting.

adhesives – materials used to adhere and assemble parts, such as building a model or assembling a product. Examples of adhesives include rubber cement, contact cement, polystyrene cement, acrylic cement, or hot glue.

adjusting – to change slightly to a more satisfactory state, process, or result.

adsorption – the process by which a material is made to adhere to the surface of a material (solid or liquid) without combining with it chemically.

aesthetics – attractive or pleasant qualities of a design.

air motor – mechanisms that produce rotary motion from the force of compressed air.

algae – aquatic, nonvascular plants such as seaweed or pond scum.

allergens – substances that cause allergic reactions such as sneezing, skin rashes, itching, and respiratory problems.

alloying – the process used to change materials into new materials and into products.

alloys – a metal mixed with another sometimes more valuable metal to give durability or some other desired quality.

alternatives – the opportunity to decide between two or more courses of action.

amorphous materials – lack any specific structure, organization, or form and are not organized as in a crystal or in a string-like molecule. Amorphous materials are nondescript and disorganized in their form and may take a variety of forms.

analog devices – tools that do continuous sensing and measuring of something that is happening. Examples of analog devices include the gas gauge in a car; a clock with hands; a pressure gauge in a boiler; a speedometer in an automobile.

analogies – things and events that have similarities between them. Because they are similar, whatever happens in one event may also happen in the other.

anthropometrics – (See ergonomics) the study of human size and the science of measuring people.

antibiotics – medicine that kills bacterial growth.

appearance model – presentations of what an object will look like when and if it is produced; used to show the final form of products, buildings, and environments.

applied research – using the results of previous research and knowledge to develop and test new products, systems, and environments.

architectural model – presentations commonly used to show the buyers or financiers of a project what a building or development will look like once it is built; representations are also used to study what changes a new building might make in air movement and patterns of sunlight in surrounding areas.

artifacts – the remains of previous events and living events and living beings that are used to give clues about the conditions of the time.

assessing – determining the importance, size, or value of something.

assets – valuable, tangible items such as gold, minerals, gems, land, raw materials, inventory, and other salable products.

asymmetrical – objects that are different on each side of an imaginary center line.

atomic energy – the energy released from an atomic nucleus during fission or fusion.

atoms – a unit of matter, the smallest unit of an element.

attraction – a mutual force between particles of matter drawing them together and resisting separation.

auditory representation – ideas communicated through sound.

auditory – relating to or experienced through hearing.

automation – a process by which machine or control mechanisms are designed so that a process will follow self-regulating, predetermined sequences of operations or respond to encoded instructions.

autonomy – the quality or state of being self-directing.

bacteria – any of a class of microscopic plants having round, rodlike, or filamentous single celled or noncellular bodies, often aggregated into colonies that live in the soil, water, organic matter or bodies of plants and animals.

balance – in design, stability created when the same design treatment is used on both sides of the object. Balance is also created by working around a center point.

basic materials – natural materials that have been changed to make them more useful for certain products; necessary for the production of most products.

bell crank lever – a Class 1 lever that allows force and motion to be transmitted through a right angle.

belonging – a sense of being an accepted part of a group, in harmony with.

belts – device used to link two or more pulleys that are some distance apart.

bevel gears – gears that can be used to transmit motion through a right angle. Bevel gears always have different sizes for the driving and driven gears. Bevel gears are used when the rotary speeds of the input and output shafts are to be different.

binary code – a signal system in which the data is encoded through the use of bits of 0 or 1, where 0 represents "off" and 1 represents "on".

binary – a system of signals using only two signs, 0 and 1.

biocatalysts – enzymes or cells used in bioconversion to speed up the processing.

bioconversion – changing the state, form, or structure of a material through the use of a living organism or part of an organism.

bioethics – concerned with understanding all legal, social, cultural, and ethical and economic implications of conducting biotechnology research and implementation.

biological energy – source of energy from plants and animals that are still alive.

bioprocessing – changing materials through biological means.

bioreactors – a vessel in which cell, cell extracts, and enzymes carry out a biological reaction. A growth chamber used to provide optimal growing conditions for a microorganism.

biosensors – combine a micro-electric component with a biological component.

biosorbants – biological materials that are used to absorb undesirable components (pollutants) so that they can be separated out or made inert.

biotechnology – using living organisms or a part of an organism to make some product or perform some service.

bit – the smallest unit of information recognized by a computer.

black box – a conceptual model that provides the meas of discussing the inputs and outputs of a process without attending to the actual process.

bond energy – the energy of attraction and repulsion of the bonds between molecules.

Boolean Logic – refers to the combinations of 0's and 1's produced by the different logic gates. The term *Boolean* comes from the name George Boole who first studied this digital logic. Boolean logic provided the theory that describes how computers and control systems function.

byte – grouping of adjacent binary digits operated on by the computer as a unit.

callus – a small lump of cells grown for the purpose of cloning or tissue culture. By using this process, a number of identical, complete plants can be grown, all from a single cell of one parent plant.

cam shaft – an axle upon which one or more cams are attached.

cam – an unevenly shaped mechanism that, when caused to move, can produce variable or reciprocating motion in other parts.

capital intensive – technological enterprises in which economic resources are spent to obtain machines, materials, and energy; a decrease in the reliance upon humans as the source of energy and guidance.

capital – an accumulation of economic resources that can be used for trade, sale, or investment.

carrier – the vehicle that transmits information. It is the link between sender and receiver through or on which a message can travel. Examples include radio waves, paper, tape, cable, and laser beams.

casting – the process of forming a desired shape by liquefying a material and pouring it into a mold. When the material cools or sets, it retains the shape of the mold.

cells – the basic building blocks of plants and animals. All living things are made up of one or more cells.

ceramics – of or relating to the manufacture of any product made from a nonmetallic mineral such as clay by firing at a high temperature.

chain of command – the management of personnel which communicates patterns of authority and responsibility.

chain – a series of links or rings connected to or fitted into one another and used for various purposes such as support, restraint, transmission of mechanical power, or measurement.

chains of molecules – the basic building units of materials connected in long chains held together by molecular forces of attraction.

change – to make different, to transform or modify.

characteristics – distinctive traits, qualities, or properties.

chemical bonding – process in which enzymes can be linked chemically to other molecules.

chemical energy – energy released by breaking high energy bonds in molecules and reforming lower energy bonds.

chemical processes – a series of actions or functions that change the structure, properties, or reactions of matter especially of atomic and molecular systems.

chip – In communication, a small, thin, silicon wafer on which electronic components are deposited in the form of integrated circuits. Chips perform various functions such as mathematical calculations, serving as a computer's memory, or controlling other chips.

chromosomes – cell nuclei of plants and animals, containing the DNA responsible for the determination and transmission of hereditary characteristics.

clone – an exact replica of a single parent source, created through genetic engineering processes.

closed-loop process – communication system in which there is return contact between the receiver and sender thus closing the communication circle.

clutch – a common device for switching mechanical energy on and off.

coating – addition process used to cover a base material for protection or decoration.

code – a system of symbols used to represent assigned meanings.

cognitive templates – a conceptual model used to describe how information is processed by humans. (See sensory, cultural, and individual templates.) (Also see template.)

cohesion – a joining process activated by heat, pressure, or both causing the materials to bond through attraction of molecules.

color wheel – a model which explains how pigments or light sources can be mixed in various combinations.

color – the appearance of objects or light sources caused by differing qualities of light reflected or emitted by them.

combustion – an energy conversion process in which a chemical reaction releases energy from molecules containing carbon. The initial heat released allows those molecules to break some molecular bonds and combine rapidly with oxygen.

comfortable – elements of a design that makes the user feel relaxed, content, and free from stress or tension.

communication medium – the method by which a message is sent and received. Examples include the voice, television, print, and pictures.

communication system – the network of interrelated technological enterprises that support the transmission of information.

communication vehicles – (See carrier.)

communication – the processing of information from the context, the words, and the nonverbal expressions.

compatibility – capable of performing in harmonious or agreeable combination with another or other.

composite – complex materials in which two or more distinct substances are combined to produce structural or functional properties not present in any individual substance.

compound – combinations of two or more different elements in definite proportions and

usually having properties unlike those of its constituent elements.

compression – the state of being pushed or squeezed together; also the force on a material or structure from pushing or squeezing.

compressor – device which provides the source of compressed air required to operate pneumatic devices and systems. Compressors are pumps that push air into a storage tank. The air in the tank is then available as needed. The force of the compressed air can be used to cause different pneumatic devices to move.

computer model – a program that uses a mathematical formula to describe a process or event, such as describing the flow of traffic through a tunnel under different circumstances; a computer-generated image that is rendered electronically to add realism to the object.

concept – an abstract idea generalized for particular instances.

conceptual map – a graphic presentation showing one or more important relationships between ideas, events, or objects.

conceptual model – a graphic presentation that shows ideas as they interact and influence each other.

conceptual skills – the ability to formulate ideas, visions, and plans that provide meaning and direction to proposed actions. In management, the ability to view the organization as a whole, as

well as in context with other organizations. A characteristic that becomes increasingly important among top executives in an organization.

consequence – the relation of a result to its cause.

constraints – restrictions or limitations.

constraints – restrictions, limitations, and regulations.

containing structures – constructed enclosures that hold themselves up and something else in. Examples include corals to restrain animals, containers to store food, etc.

contrast – a striking dissimilarity between things or characteristics, such as light and dark, smooth and rough, etc.

contrasting colors – colors which are opposite from each other on the color wheel. For example, red contrasts with green.

control – the process of operating, regulating, or guiding a machine, vehicle, or system.

control device – a tool used to sense, direct, and guide the operation of a machine or system.

control system – an organized set of elements that include switches or sensors used to provide input to a process unit that in turn activates the output device(s). (See open loop and closed loop systems.)

conversion – the act of changing into another form.

convert – to change or adapt from one form, substance, state, or product to another; to transform.

coordination – the act of combining various aspects together to form a harmonious whole.

cosmic energy – radiation from extraterrestrial sources; energy at the extreme high frequency end of the known energy spectrum.

crank – a device that uses a lever to create a rotary motion and mechanical advantage. Cranks are commonly used in a windlass and in a bell crank lever.

crating – a drawing technique in which a transparent box of lightly drawn lines is used to visualize and construct the image of an object being drawn. Crating allows the illustrator to sketch out the object before it is actually drawn.

criteria – a standard on which a judgment or decision may be based.

criterion – a standard, rule, or test on which a judgment or decision can be based.

crowding – a perception that the number of people or materials is too great in relationship to the amount of space.

crystal – a three-dimensional atomic or molecular structure consisting of periodically repeated, identically constituted unit cells.

cultural templates – the expectations that a person brings to an experience used in processing information. Based on the rules, definitions, values, and expectations formed as part of living in a given cultural group.

custom manufacturing – producing products that are designed for specific purposes or users, usually made one at a time.

cutaway drawings – an illustration showing portions separated or "cut away" to highlight specific areas.

cybernetic – the theoretical study of control processes in electronic, mechanical, or biological systems. The automatic control of a machine through systems that are self-regulating.

cylinder – a device that creates linear movement and force from controlled air pressure; most common of the pneumatic components used for producing movement.

data – factual information used as the basis of reasoning, discussion, or calculation.

decision networking – the identification of the choices that must happen if a goal is to be reached.

decision tree – a map of the sequence and relationships between the choices that are needed in order to lead to a desired end.

decode – to translate or convert into intelligible, understandable language.

dedicated devices – computer-based device that performs only one specific function.

deficit spending – excess spending beyond the amount of capital that one has.

delete – to eliminate by erasing ignoring, cutting, or otherwise leaving out. One method by which humans are able to make sense out of a large amount of information. (See also, distort and generalize.)

Delphi survey – a method of projecting the future in which the opinions of a group of experts in a given field are questioned. The process is repeated until an acceptable level of consensus is reached.

design brief – a short statement that outlines a problem to be solved.

design limits – considering the practical possibilities and limitations of a design.

design loop (See IDEATE) – a form of technical problem-solving that includes the term IDEATE (**I**dentify and define the problem; **D**evelop the design brief; **E**xplore possible alternatives; **A**ccumulate and assess the alternatives; **T**ry out the best solution; and **E**valuate the results).

design process – a systematic, yet creative process involved in turning ideas into real objects, products, systems, and environments.

designing – translating ideas into things and thought into action; the translation of ideas into something tangible.

desired outcomes – expecting positive, predictable results.

development – the process of creating new ideas, products, or systems based on previous knowledge; term used in design and technology to describe how a three-dimensional form is made from two-dimensional (flat) materials.

development – the process of creating new ideas, products, or systems based on previous knowledge. Making a three-dimensional model from a two dimensional construction.

directing structures – constructed enclosures that control the flow of materials and humans going from one place to another. Examples include pipelines, canals, mazes, tunnels, etc.

discovery – a new insight into the uses of something that already exists.

dissonance – lack of agreement, consistence, or harmony; for example, distraction resulting when objects are out of proportion to what we expect.

distort – to twist or change the meaning or form. In human information processing, forcing information to fit expectations or current wants or needs. (See also, delete and generalize.)

DNA – (deoxyribonucleic acid) any of various nucleic acids that are localized especially in cell nuclei that are the molecular basis of heredity in many organisms. They are constructed of a double helix held together by hydrogen bonds between purine and pyrimidine bases which project inward from two

chains containing alternate links of deoxyribose and phosphate.

documenting – (See portfolios) an important part of the design and technology process which involves recording any plans, ideas, notes, drawings in journals, logs, reports, and portfolios etc.

driven gear – the gear that transmits motion and force from another gear.

driving gear – the input gear of a gear train or system.

dwell – a programmed time delay, such as is provided by the shape of a cam.

dynamic load – the force created on material or a structure by mass that is in motion.

dynamic materials – materials that change characteristics or form during use.

E. coli – a specific species of one-celled bacteria, often used in genetic engineering.

ecological consequences – the effects of technology upon the physical world in which we live.

economic resources – resources that are scarce, transferable, measurable, and usable for production purposes

effectiveness – producing a desired effect in the design of a product.

efficiency – creating or producing effectively with a minimum of waste, expense, or unnecessary effort.

electrical processes – cause changes in materials by passing a

stream of electrons through the materials. The flow of electrons can cause materials to change internally and vary their properties.

electromagnetic energy – range of radiation from extremely high frequencies to zero frequency, including in descending order of frequency: cosmic ray photons, gamma rays, X-rays, ultraviolet radiation, visible light, infrared radiation, microwaves, radio waves, heat, and electric currents.

electromagnetic – a core of magnetic material surrounded by a coil of wire through which an electric current is passed to magnetize the core.

electronic modeling – An approach used to show how electrical and electronic parts will be assembled and tested. Circuits can be modeled using real components on a solderless breadboard. (See electronic prototyping board.)

electronic prototyping board – (See electronic modeling)

elements – a substance composed of identical atoms of one kind.

emotional effects – outcomes that are shown in mood, attitudes, and behavioral tendencies.

energy conversion – the change of one energy form into one or more different energy forms.

energy differential – the difference between high and low energy states such as the difference between the positive and negative poles on a battery or the difference between water at higher and lower levels.

energy spectrum – the range of energy forms from atomic and molecular to cosmic.

enlightened self-interest – a recognition that in caring about the welfare of others and the ecology for both now and in the future, one is also caring for the self.

entrapment – a process in which a cell or enzyme is held inside a porous gelatin-like substance. Because the gelatin is porous, it allows the chemicals to reach the cell or enzyme inside the gelatin that is surrounding them. It also makes it possible to use the same cells or enzymes repeatedly.

entropy – the measure of the disorder in a system; the loss of energy or information due to disorder.

environment – the circumstances, objects, or conditions by which something or someone is surrounded.

enzymes – any of numerous, complex proteins that are produced by living cells catalyzing specific biochemical reactions to body temperatures.

ergonomics – the study of ways in which objects, systems, and environments can be safe, effective, efficient and of appropriate size and strength requirements for the proposed users.

evaluation – to determine the significance or worth of

something usually by careful appraisal, study, or testing.

expandable materials – are of two types: those that flex and those that foam. Both types of materials change more rapidly than in materials where composition or structural changes occur.

expert power – the ability to influence based on the experience, knowledge, and skills of an expert.

exploded view drawing – an illustration or diagram of an object that shows its parts separately but in positions that indicate their proper relationship to the whole.

external controls – outside restraints that limit available choices. The means or devices for operating, regulating, or guiding a machine, vehicle, or system.

external program – the signals that are encoded into the computer by the operator to do unique, specific tasks.

extracting – the process of taking natural materials and necessary energy for the Earth, air, and sea to produce products.

extruding – forcing a material through an opening (a die) to control the cross-sectional area and the shape.

fabrication – constructing or creating a model, product, or structure.

families of materials – materials that have similar traits and can be considered related. Examples include ceramics, foods, textiles, and woods.

family work – the tasks required in order to keep the family group functioning effectively. Usually done through cooperation and sharing and seldom seen as a source of economic resources. Often overlooked as necessary and valuable work when measurement of labor and productivity are calculated and reported.

fastening – the process of joining parts or components through mechanical means. Examples include bolting, nailing, pinning, sewing, and riveting.

feedback – the return of a portion of the output of a process of a system to the input, especially to maintain the performance or to control a system or process.

fermentation – a process that uses yeast to digest sugar and create alcohol and carbon dioxide. Also, the growth of cells or microorganisms in specialized vessels.

ferrous metals – materials derived from or related to iron, such as steel and wrought iron.

fiber optics – a very thin, transparent fiber of glass or plastic which employs light to transmit large quantities of data at very high speeds.

filtration – the process of passing through a filter.

firing – a process usually referring to the heating of ceramics to temperatures that cause a non-reversible change in the materials. (See vitrification.)

first-order effects – the outcome of actions which lead to further related outcomes.

fission – nuclear energy resulting from tearing the nucleus of the atom apart.

floating structures – transporting structures that float such as boats.

flow control regulator – device used to adjust the amount of air entering or escaping a cylinder or other air operated mechanism.

fluidics – the use of logic systems comprised of pneumatics and hydraulics devices; the general use of pneumatics and hydraulics.

flying structure – transporting structures that fly such as airships or planes.

force – the effort that is applied to the lever.

forecasting – attempts to project future events, using technological tools and information currently available.

form – three-dimensional drawings.

forming – a process which shapes materials by impact or pressure.

forming – processes that are used to change the form of materials without necessarily removing any of the material.

fossil fuels – a hydrocarbon fuel such as petroleum coal or peat derived from the decayed remains of prehistoric plants and animals.

fractional distillation – a process for separating material, by heating and removing and capturing constituent parts (fractions) as they boil out at different temperatures.

free form – outline with the absence of geometrical shape.

frequency – the number of waves per second produced by a vibrating body of matter (See Hertz.)

fructose – a naturally occurring sugar from fruit and related sources, that can be chemically altered to greatly increase its sweetening potential. (See high fructose corn syrup.)

fulcrum – the pivot or the point around which the lever turns.

functional materials – change back and forth from one state to another as an outside force is applied and then removed.

functional model – the fabrication and arrangement of parts to see if an idea or device will work; useful in investigating an idea or concept that might lead to an innovation or invention.

functional – practical design developed chiefly from the point of view of use.

functions – the actions or roles for which materials, objects, machines, systems or people are particularly well suited.

fusion – combining the nucleus of one atom with the nucleus of another.

future histories – a means of studying the future. The process begins by pretending to be alive at a time far in the future. By using their imaginations, futurists can then look back, in a way, and write a "history" of events that must occur if the given future event is to happen.

futurists – individuals who specialize in studying the future.

gas – state of matter distinguised from solid or liquid states by very low viscosity and density.

gear reduction – the meshing of two or more gears in order to decrease or increase the output speed. Gear reductions that decrease output speed provide an increase in mechanical advantage. Gear reductions that increase speed decrease mechanical advantage.

gear train – a series of two or more meshed gears that operate together as a system.

gear – toothed wheel, cylinder, or other machine element that meshes with another toothed element to transmit motion or to change direction, speed, or mechanical advantage.

gene splicing – a process by which the DNA strand is broken and a string of genes from another organism is inserted, or a segment of the DNA is removed.

generalize – to use information and conclusions from previous or related experiences in order to make a judgement in the present. One of the methods used by humans to make sense and order from information. (See also, distort and delete.)

generativity – a stage in human development in which the person is motivated to care for the next generation and to create something that has a meaningful and lasting impact.

generator – device that converts mechanical motion and magnetic energy to produce electrical energy.

genes – an element that controls transmission of a hereditary character by specifying the structure of a particular protein or by controlling the function of other genetic material and that consists of a specific sequence of purine and pyrimidine bases usually in DNA.

genetic engineering – a change in the genetic structure created by inserting or removing a new segment, thus forming a new DNA molecule.

geometrical – of or relating to geometric shapes such as lines, squares, or circles.

geothermal energy – a vast reservoir of heat that comes from molten rock

glucose – the basic energy ingredient of the foods we eat.

grain – the arrangement, direction or pattern of fibrous or string-like materials such as wood, cloth, and plastic; also a state of fine crystallization as in metals.

graphic communication – refers to the visual forms of sharing information or ideas which includes printing technologies or visual

communication such as sign language.

gravitational energy – energy available or resulting from the action of gravity such as falling water or falling weight.

green – a dried but unfired state of ceramics that allows the materials to be reversed and changed back to pliable and usable forms.

growth chamber – a vessel used to grow cells, cell extracts, or enzymes under controlled growing conditions.

growth medium – an especially prepared nutrient mixture.

gyroscopic action – the process that uses the information provided by a spinning mass to maintain control of a vehicle or system.

half-life – the period of time required for half of the atoms of a radioactive substance to disintegrate.

handling – a process that occurs when materials are transported to, through, and from one site to another, or to, through, and from a machine.

hard technologies – machines or instruments, as compared to the techniques and tools, for the decision making process of technology.

harmonious colors – colors that are close to each other on the color wheel.

harmony – pleasing combinations of the elements that form a whole.

harvesting – the act or process of gathering in a crop.

heavier-than-air-vehicles – aircraft that can be made to fly by moving rapidly enough to create a low pressure (vacuum) above its wings. The suction lifts the aircraft and holds it up.

Hertz – a unit of frequency equal to one cycle per second.

hierarchies – an order in which things are arranged in order of importance.

high fructose corn syrup – a very sweet form of sugar, made by changing the molecular structure of fructose.

highlights – show different shades and light reflections important for portraying form.

hormone – a chemical substance formed in one organ of he body that, when carried to another organ or tissue, stimulates a specific function or effect.

human capital – the resources that a person has that can be exchanged for economic resources. Examples include talents, understandings, muscle power, and skills that can be improved through investment of time and effort.

human control subsystem – includes the nervous system and all of the senses (sight, sound, smell, taste, and touch).

human energy transmission system – includes muscles, tendons, and cartilage.

Human Genome Project – a project whose goal is to identify and map the entire complement of genetic material of human beings.

human motivation – the need or want that drives the human to act. Varies with age and the person's level of need satisfaction or deprivation.

human skill – the ability to communicate, motivate, and otherwise work with other people individually and in groups. In management, a necessary skill at all levels of the organization.

human structure and cover system – includes the skeleton and skin.

hydraulics – the study and use of liquids under pressure.

hydrocarbon – organic compound containing only carbon and hydrogen and often occuring in natural gas, coal, and bitumens.

idea generation – includes brainstorming, developing mind maps and question webs, or approaches which help to develop ideas more fully.

IDEATE – an acronym used for the design loop which includes the following: identifying and defining a problem, developing a design brief, exploring possible alternatives, accumulating and assessing the alternatives, trying out the best solutions, and evaluating the results.

if-then statements – the practical form of hypothesis development or hypothetical statements. For example, *if* you walk in the rain with no umbrella or rainwear, *then* you will probably get wet.

immobilization – a method for separating out the microorganism and growth medium from the product, through chemical bonding, adsorption, or entrapment.

immune – having a high degree of resistance to disease and change.

impact structure – a structure that is designed to be capable of dissipating or absorbing the energy from collision with a stationary or moving mass.

improvements – to enhance in quality or value.

individual template – includes an individual's traits, past experiences, and the interpretation of those experiences.

information conversion – the changing of symbols and signals into new forms.

information conversion – the process of changing information into a different form or forms.

information – knowledge gathered for investigation, study or instruction.

information – knowledge gathered from investigation, study, or instruction.

infrared – electromagnetic radiation of wave lengths greater than visible light and shorter than those of microwaves.

innovation – occurs when an individual takes several things that already exist and brings them together to produce a new device or process.

input sensor – a device that receives or responds to a signal or the presence of a specific form of energy.

insulation – materials that do not conduct electricity or heat.

insulin – a chemical that is made by animal cells. Insulin is necessary to metabolize sugar in the blood stream.

integrate – to make into a whole by bringing together the various parts, components, or ideas.

integration – combining different resources to form a unified system. In control, this involves combining hardware and software in order to develop a coordinated set of processes.

integrity – a stage of human development in which the individual accepts his/her own life experiences as having meaning and purpose.

intellectual property – the product or idea that belongs to someone because they produced it from the workings of their mind. Often protected by patents and registration.

intended outcomes – planned results.

interest – the fee charged for loaning money.

interlacing – to lace together, to interweave, to intertwine.

intermittent manufacturing – a production process that is done in "lots" or "batches" in order to make a quantity of material (such as paint) or many items (such as cookies) that are alike.

internal controls – inside means of guidance. In human decision making, the alternatives through which we can make choices.

internal program – program which controls the operation of the computer and is determined when the computer is built.

interval – a level of measurement that requires equal distance or amount between each data unit.

intimacy – very close association or warm friendship.

invention – something that an individual makes that did not exist before.

investigating – to study or observe by close observation.

investment – the use of resources in such a way that they can be expected to increase in value over time.

isometric – a drawing technique in which three equal angles are used to create an image. Similar to two-point perspective drawings, the vertical lines remain vertical and all horizontal lines are drawn at 30 degree angles.

iterating – the process of developing and improving an idea, process, product, or system by sending it through a development stage more than once.

iteration – the process of going through the same process another time in trying to improve the product, results, or outcomes.

joining – adhesion, cohesion, and mechanical processes used to fasten materials together for structural or decorative purposes.

kiln – an oven-like device capable of reaching temperatures of 2,000 degrees Fahrenheit and more. Often used for firing ceramic materials.

kinesthetic modality – a body sense that is developed from movement; changes in internal state and information received by the body.

kinetic energy – energy in motions such as energy in falling water.

knitting – to form by interlacing yarn or thread in a series of connected loops with needles.

knowledge explosion – the exponential growth of information available for study and use, especially when contrasted with what was known in the previous century.

labor intensive – enterprises that use many people to do the work instead of investing in costly equipment and processes to do the job.

language systems – a set of symbols, that when put together in orderly arrangements to follow grammatical rules, can communicate highly complex sets of ideas. In communication between humans and computers, this involves the use of symbols that are familiar to humans.

layering – building up a "material sandwich" by adding multiple coats or thicknesses.

legitimate power – the ability to exercise influence over the behavior of others because of position or titles. May or may not be accompanied by expert knowledge and skills.

levels of measurement – a means of describing the degree of precision of the information that has been organized for a presentation or report.

lever – a simple machine consisting of a rigid body, typically a metal bar, pivoted on a fixed point (fulcrum).

lighter-than-air-vehicles – aircraft that are held aloft by the difference in the weight of the craft and the weight of the air that it displaces. This ratio can be changed by changing the weight or volume of the craft.

line – a thin, continuous mark as made by pen, pencil, or brush applied to a surface.

linear motor – a device that converts energy into straight line power and movement.

linear – to arrange in a line; chronological; to arrange according to a time line.

linkage – two or more linked levers used to transmit force and linear motion from a force applied as an input.

liquid – the state of matter in which a substance, relatively high in compressibility, exhibits a readiness to flow and little or no tendency to disperse.

load – a weight or mass that is carried, lifted, or supported. (See dynamic and static load.)

lobe – the raised portion on a cam.

location – a site. The position and boundaries of a place.

logic gates – a computer switching circuit that performs an arithmetic logic function.

logic – the formal principles of reasoning and thought. The use of electronics and circuits for making decisions with the output of the circuit being dependent on the input circuit.

logo – the name, symbol, or trademark of a company.

magma – molten rock found under the Earth's mantle or crust.

magnetic North – the direction that corresponds to the Earth's magnetic pole. The direction that a needle of a compass will point.

magnetic poles – the two limited regions at which the magnetic force is the most intense; described as north or south poles.

magnetic processes – the changing of energy or materials through the use of magnets or magnetism.

maintenance – processes done in order to counteract the effects of entropy, wear, and disorganization that occurs with use and time.

making – producing or building something that represents a solution to a problem.

manufacturing – the activities through which useful and salable products are made.

mass production – a process of manufacturing by which very large numbers of an identical product or component parts of a product are made.

master – of or relating to the original from which duplicates are made.

material layers – "material sandwich" composed of different layers and coats.

material mixtures – a composition made of two or more diverse ingredients produced by combining or blending.

material systems – two or more materials combined into a system. The ingredients are combined for a specific function or to have a desired characteristic.

materials conversion – the process of changing materials into other forms or into energy or information.

maximizing the criterion – a process of trade-offs between established standards of output, in order to reach the highest level possible for several possible options.

mechanical advantage – the relationship of the load moved compared to the input effort; mechanical advantage equals load divided by effort.

mechanical energy – energy available from or produced by physical forces, as in tools and machines.

mechanical fastening – materials joined together with fasteners such as screws, bolts, zippers, rivets, staples, etc.

mechanical model – shows how mechanical parts fit and work together. May employ linkages, gears, pulleys, cams, pneumatic cylinders and other devices to produce desired movements.

mechanical processes – employ shock to cause a change in materials. Mechanical processes include hammering, peening, or explosive-forming.

mechanisms – a system of parts that operate or interact like those of a machine.

media – (See communication medium.)

memory – the capability of organizing, storing, retrieving, and using data.

message – information sent by a sender to a receiver, transmitted through a given form of media.

metal – selected elements or different alloys of two or more metallic elements that are ductile, malleable, and conductive.

metamorphic materials – change slowly from their original characteristics to something different as they are used.

microbes – complete organisms made up of only a single cell. These tiny beings do not depend on other cells for their survival. Each cell is able to perform all the functions necessary for its life and reproduction.

microorganisms – tiny living beings that cannot be seen with the unaided eye. Usually single-celled plants or animals. (See microbes.)

microscope – a device used to investigate micro (small) space.

mitre gears – gears used when it is necessary to transmit motion through a right angle. Mitre gears are always the same size and when used make no change in the rotary speed.

mixing – to combine or blend into one mass or mixture so that the constituent ingredients are indistinguishable.

modality – the system of the human senses in which data are perceived, stored, and retrieved for communication. (See visual, auditory, and kinesthetic.)

model – simplified version of real objects, events, or systems; often a smaller, simpler version of a real thing.

modeling – an attempt to translate ideas into concrete and visible or tangible form. In future studies, using existing ideas and tools to project future events and circumstances.

modifications – to make changes in a design.

mold – superficial, often woolly growth produced in damp or decaying organic matter or on living organisms caused by fungus.

molding – combination of extrusion and casting in which material is forced into a mold (casting) under pressure (extrusion).

molecular level – pertaining to, consisting of, caused by, or existing between molecules.

molecules – simplest structural unit that displays the characteristic physical and chemical properties of a compound.

moment – the rotation produced in a body when a force is applied; the product of a quantity and its perpendicular distance from a reference point. (See torque.)

monetary system – an exchange system that allows the trade of goods and services between groups of people and from one time to another.

monolithic materials – materials made of only one ingredient, element, or compound that do not change during use.

monomer (mer) – the basic element of a polymer. (See polymer.)

Morse code – a signal system, consisting of dots, dashes, or long and short sounds, developed by Samuel Morse for use on the telegraph. Messages are transmitted by audible or visual signals.

motif – a pattern or central theme.

mutation – a change in one or more characteristic of the offspring of an organism, caused by an alteration in the genetic material and transmitted during reproduction. These changes may be naturally occurring or may be genetically engineered.

mycoprotein – produced by protein-rich microorganisms, often grown for good. Also called SCP, single-cell protein.

natural – shape or form based on objects and features found in nature.

natural materials – materials produced or existing in nature.

need – relating to design, a lack of something necessary, desirable, or useful.

noise – interference during transmission.

nominal – the least precise level of measurement that only requires that data units be given names.

non-economic resources – resources consumed for purposes other than the production of goods and services.

norm – a standard, model, or pattern regarded as typical of a specific group; also a mode or average.

nucleus – a complex, usually spherical body of protoplasm that exists within the cell of each living organism. The nucleus contains the cell's hereditary material and controls its metabolism, growth, and reproduction.

nucleus – the positively charged central region of an atom.

nutrient agar – a gelatinous material prepared from certain marine algae and used to support bacterial culture development.

nutrients – the chemicals that are needed in order to support the growth, development, and maitenance of living cells.

nutritional supplements – additives designed to provide more than the nutrients that are naturally found in food.

oblique drawing – a drawing technique which employs slanted lines to create a picture of an object. Similar to one-point perspective, the lines do not converge and are drawn at a 45-degree angle.

open-control loop – a communication system that does not incorporate feedback and therefore the sender or processor receives no information about the effectiveness of the transmission

optical processes – refer to the change of materials that is caused by light.

optimization – the process of seeking the most favorable condition or solution to a goal by balancing and trading-off results on more than one criterion.

ordinal – the level of measurement that requires that data units can be placed in order, e.g. from most to least important.

organization chart – hierarchy within an organization; a chart which shows who is in charge of each division and identifies various levels within the organization.

orthographic projection – drawings that include a front view, a top view, and at least one side view of an object intended for development of construction

or production of an object.

output device – a mechanism or apparatus that is operated by a control system to create a final, desired outcome; especially important in a control system.

ownership – to possess or to have and hold as property.

oxidation – the combining of a substance with oxygen. Examples of rapid oxidation include burning and exploding; examples of slow oxidation include decaying and rusting.

patent – exclusive rights to make, use, and sell an invention or product for a specified period of time.

pathways – the course or path vehicles travel.

patterns of growth and development – the expected changes that occur as a result of time and experience over the entire life span.

patterns – repeating lines, textures, colors and other elements.

penetration – to cause to permeate into or diffuse throughout a material; also to pierce.

personal consequences – the effects technology has on our bodies, minds, and emotions.

personal recognition – one form of reward that many people will work for; to be noticed for individual accomplishments.

perspective – pictoral drawing technique or process of representing on a plane or curved surface the spatial

relationship of objects as they might appear to the eye.

pesticides – a chemical agent used to destroy pests, such as insects and rodents.

pharmaceuticals – a general term that includes all drugs and medicines.

photosynthesis – the process by which chlorophyll containing cells in green plants convert light to chemical energy and synthesize organic compounds (especially carbohydrates) from inorganic compounds.

physical effects – changes in materials, systems, and environments that are tangible. Examples include, increase or decrease in strength, color, or size.

physical model – three-dimensional representations of something real. Physical models fall into three general categories (functional models, appearance models, and production models).

pictogram – the logo for a company or organization in which a picture is used to represent the organization.

plane – a flat surface that is identified by two dimensions.

planning – the development of a detailed scheme, program, or method for accomplishing an action or objective.

platelets – small disk-like components of selected materials, such as blood, wool, and clay.

pliability – easily bent or shaped, flexible.

pneumatics – the study and use of air and fluids under pressure.

point – a mark made by a sharp or tapered object; also a dimensionless geometric object having no property but location.

pollutants – undesired material or energy, may be a waste material found in soil, air, or water or may be a form of noise or interference.

polymer – materials composed of long molecules of natural or synthetic origin. Plastics is the most common family of synthetic ploymer.

polymerization process – a chemical reaction in which two or more small molecules combine to form larger molecusles that contain repeating, structural units of the original molecules.

portfolio – is a compilation of the various records maintained throughout the design and problem solving process of a project. These records include materials outlining the process as well as the outcome of the project.

potential difference – voltage; the voltage drop or reading across any component in an electrical circuit.

potential energy – energy that has potential for doing work but is basically at rest.

prediction – the ability to guess future events.

pressing – shaping sheet material by squeezing it between two dies or textured surfaces that force the material to take on a three-dimensional shape.

primary color – of or pertaining to the basic colors from which all other colors can be derived.

primary energy sources – large, usable reservoirs of energy from which other secondary energy sources derive their energy.

primary source – an original source of data and information, usually the person or people who conducted the research or developed the idea or object.

principle of supply and demand – an economic exchange theory that indicates that when the supply of goods and services is greater than the demand, the price of the goods and services will be low. The opposite can also be expected.

privacy – personal; not known or intended to be known publicly; intended for the use of a particular person, group or class.

private property – resources owned by a particular person or group.

problem – a situation, question, or challenge that presents uncertainty, perplexity, or difficulty.

process – a series of actions or operations that lead to a particular result.

processor – an information conversion device used in computers and programmers; a tool or machine that is designed to carry out the necessary processes on the material in order to produce the desired outcome.

production model – a finished pattern, plan, or object from which an item or items will be produced.

products – objects, structures, or environments produced by human or mechanical effort.

program – a sequence of instructions that direct the process for achieving a desired outcome; a sequence of coded instructions for a computer, telling it how to process data.

program – the series of commands that identify the steps and sequence of actions to be taken in order to achieve the desired outcome.

promissory note – a written promise to pay a sum of money to an individual or organization.

properties – the characteristics of materials.

proportion – an appropriate relationship of objects in terms of their size and quantity.

protein – a group of complex organic compounds that contain amino acids that occur in all living matter and are essential for the growth and repair of animal tissue.

prototype – advanced form of model-making in which a full-size version of a product, system, or environment represents a solution to a problem; prototypes actually work and can be tested under realistic conditions.

psycho-social stages – Erik Erikson's model which explains human motivation, the stages in human development, and which needs are dominant at different ages in life (refer to Figure 8.19B).

pulley – a simple machine, based on the wheel, used to change the direction and intensity of a pulling force.

punishment power – the ability to influence behavior through the threat of pain or deprivation.

pure research – research in which the primary aim of the investigator is for more complete knowledge or understanding of the subject under study. Little or no concern is given to how the knowledge will be used.

purpose – an object or result to be obtained.

quality circles – a management plan in which representatives from each stage of the production and distribution of a product are included in the decision-making process.

rack and pinion gears – a geared device that can produce rotary or linear motion depending on which element (rack or pinion) is the driving and which is the driven gear.

radial balance – arrangement in a uniform manner in which the design radiates out from a center point.

radiant energy – emmitting heat or light; consisting of or emmited as radiation.

radio telescope – a tool developed to see into the distance, operated by radio energy.

radioactive carbon – an element that is used to date artifacts, based on its established half-life.

rapid deposition modeling – computer-controlled modeling in which an object is built up from layers that are sprayed on one at a time. (Compare to stereolighography.)

ratio – the level of measurement that, in addition to being precise enough to have reached the interval level, also requires that there be an absolute zero. (Absolute zero means that there cannot be less than "none" such as a negative balance in a checking account.)

ratio – the relative size of two quantities expressed as the quotient of one divided by the other. The ratio of 5 to 1 is written 5:1 or 5/1.

read only memory – the capability of a computer to access but not change information, usually accomplished through an integrated chip that is programmed at the time of its manufacture and cannot be reprogrammed by the user.

receiver – the participant in the communication process for which the message is developed and sent, for the purpose of sharing information that is previously unknown to the receiver.

recombinant DNA (rDNA) – a new form of DNA prepared through laboratory manipulation in which genes from one species of an organism are transplanted or spliced to another organism.

recycling – reusing resources in order to regain material for human use.

reduction – refining procedure in which excess oxygen is driven out to reduce metal to a more pure form.

redundancy – designing so that more than is needed to achieve the desired outcome is built into the process. Done in order to increase the probability that a communication will be received accurately or that a product will continue to perform when some elements wear out or fail. Examples include an excess number or volume of transmitted signals or symbols and excess material or back-up systems in a product.

referent power – the ability to influence behavior because of one's association or connection with someone who has even greater authority.

refining – a separation process used to remove impurities and unwanted parts of a material.

reflector – a surface that returns energy, such as light or radiation, in a straight line, back to its source.

refractor – a surface that deflects or scatters waves from a straight line.

reiteration – going through the same process several times which increases the level of

496

experience and familiarity with the project and enables the researcher to ask better questions related to the problem.

relay – an electromechanical switching device that is magnetically activated by a small current or voltage change.

replication – the ability to duplicate or reproduce a product according to the specifications of the original product.

report – a formal oral or written presentation with documentation outlining the strengths and weaknesses of designs and proposed solutions.

representation – a likeness or image or something.

repulsion – to push away; to repel.

research and development – the process that utilizes information and previously developed resources to produce new and innovative products, systems and environments. (See IDEATE.)

research – searching for new information and knowledge related to a problem which may take the form of reading materials related to the problem or interviewing and listening to people who have experience related to the problem.

reservoir – a receptacle or chamber for storing a fluid (gas or liquid).

resource management – the process designed to achieve the desired outcome while using the minimum amount of resources.

resources – the tools, materials, energy, information, humans, time, space, and capital required to implement a technological process.

restriction enzymes – enzymes that weaken the bonds between specific strands of DNA; used to separate the targeted portion from the rest of its DNA.

result – to occur or exist as a consequence of a particular action, process, or outcome.

result – to occur or exist as a consequence of a particular action.

reward power – the ability to influence behavior by promising or actually supplying desirable goods and services.

rolling structure – transporting structure in which the load is focused on very small places of contact between the wheels and the road or other surface such as in an auto or on a bike.

ROM (read only memory) – (see read only memory.)

rotarios – historic maps that showed the directions and listed the directions, distances, and hazards of a particular course.

rules – established conventions that provide structure, security, and knowledge of what is expected.

sampling – the process of selecting a small portion of the whole, often in a controlled manner so that the sample represents the entire group or population.

sawing – separation process requiring the use of a tool that is pointed with wedge-shaped teeth evenly spaced along the edge of the blade.

scale model – a representation of something else in a different size. Scale models are usually smaller than the object being modeled, but in some cases may be larger.

scenarios – imaginative "pictures" of what could happen in the future.

scientific method – a process used for carefully testing and expanding knowledge. Prescribed ways for controlling conditions under which the researcher makes observations, records and analyzes data, and reports findings, so that the process can be replicated by other researchers.

second-order effects – delayed outcomes that occur in a sequence, and related to an original action.

secondary colors – colors made by mixing equal parts of two primary colors. For example, with pigments red and blue yields violet, red and yellow produces orange, and blue and yellow produces green.

secondary energy source – sources of energy that are derived from primary sources of energy, such as wind energy from the action of the sun and the spinning of the Earth.

secondary sources – a source of data and information other than the original, usually a person or persons who write about the work of others.

sectional drawings – presentations of figures and drawings that show an object as though it had been sliced by a plane or planes; also to shade or crosshatch to show a section.

sedimentation – matter that settles to the bottom of a liquid.

selective breeding – technique in which organisms exhibiting beneficial traits are selected and crossed to produce offspring that display a blend of those traits.

semiconductor – any of a class of solids whose electrical conductivity is between that of a conductor and that of an insulator.

sender – the participant in the communication process that originates the message for the purpose of sharing information that is known by the sender.

sensors – devices that use a small sample of energy and change that energy into information (signals) for communicating with or controlling a machine.

sensory templates – the guides used for perception, storage, and retrieval of information obtained through the human senses during an experience.

separation processes – the act of removing material; dividing materials into components or parts.

sequence – a related or continuous series; a succession of events.

shading – a technique that helps to make a pictorial drawing of an object look more like the real thing. Shading reveals the way light is reflected from the object where it is lighter and darker.

shape – the use of a line to enclose space. Shapes are two-dimensional in that they have no thickness. Shapes fall into three categories: natural, geometrical, and free-form.

shaping – to modify, form, or create for a particular use or purpose.

shearing – separation process in which one or two smooth blades are used to separate materials without producing waste.

sheltering structures – a container that is designed to hold itself up and keep something else out. Examples include houses, castles, forts, etc.

shelters – products designed to protect protect individuals, materials, and equipment from outside forces and to maintain the desired environment.

shortcuts – development of processes that take less time to produce a desired result.

signal entropy – the loss of usable information due to disorder caused as information is changed from one form to another. (See entropy.)

signal – an impulse or pattern of transmitted energy that represents coded information. Used for communication among machines and between humans and machines.

single cell protein (SCP) – (see mycoprotein.)

slow oxidation – the gradual process materials combining with oxygen as when carbon in decaying materials combines with oxygen to form carbon dioxide gas.

sludge digester – assortment of organisms introduced to break down the residue of sewage treatment.

social consequences – effects upon the interaction of groups of people shown in their cultural practices and expectations.

soft technologies – decision-making skills; tools that can be used to change society.

solar energy – energy radiated from the sun. Solar energy includes visible light, infrared waves, microwaves, ultraviolet, and several other energy wave forms as well as atomic particles.

solenoid – a device that operates by converting electrical energy that passes through a coil of wire to magnetic energy. This in turn causes the metal core of the solenoid to move, as in the starting system of an automobile.

solid – a substance that is neither liquid or gaseous.

source of light – a point of illumination identified in pictorial drawings to help determine intensity of shading of surfaces and the orientation of shadows.

specifications – a clarifying statement that accompanies a design brief which explain the "specific" details of the design. This detailed description indicated what must be accomplished by the completed design. (Often referred to as "specs.")

sprocket – a wheel rimmed with tooth-like projections used to engage the links of a chain, as in a bicycle drive system.

spur gear – gears with teeth cut on the outside edge of the gear that resemble cowboy's spears. Spur gears are set up so the teeth of the gears mesh in order that the rotary motion of one shaft is transmitted to a parallel shaft.

stability – resistant to sudden change, the capacity to maintain a steady process or motion and limit fluctuation.

stamina – the ability to work at something over a period of time.

standardization – the setting of size and shape of products or parts to be manufactured; also the establishing of accepted procedures and processes and expected levels of performance.

static load – weight that does not move or change.

steady state – a series of adjustments that continue until the machine reaches the desired state of operation.

stereolithography – a process of fabricating parts using computer-driven lasers that trace out the part in a bath of liquid plastic (polymer). As the plastic hardens layer by layer, a 3-D model emerges from the bath.

stick chart – an early map used in the South Pacific to fix the position and distance between the islands.

stress – applied force that tends to strain or deform an object or material.

structural model – representation that shows how the parts of a structure fit together; used to provide the framework to support mechanical and electrical components when creating dynamic models.

structure and cover system – subsystem of a machine that supports and maintains the position of its parts and protects the parts and the machine operator.

structure – something made up of a number of parts that are held together in a particular way; also the way in which parts are arranged and put together to form a whole.

subsystem – a smaller set of interrelated elements that form a part of a system.

sucrose – a form of sugar commonly used for sweetening foods.

support structure – an object built to hold itself and other things up. Examples include bridges, lighthouses, table, etc.

symbiotic relationship – the condition in which a mutually beneficial relationship is developed between two living beings, each of which depends on the other for comfort or for continued existence.

symbol – a sign image, sound, or gesture that conveys inferred meaning. Used for communication among humans.

symmetrical – (see asymmetrical) when one side of the object is the mirror image of the other.

synchronous satellite – a transmitter/reporter device for transmission of information that is placed in an orbit whose speed matches the speed of the turning earth.

synthesizer – a person or machine that combines separate elements or substances to form a coherent whole. Often done with elements that are not of natural origin. Examples include a music synthesizer.

synthetic hormones – chemical substances manufactured to simulate the action of naturally occurring hormones.

synthetic skin – material, such as clothing or other means, for insulating and ventilating the body to protect the body and skin from temperature extremes and potential injury.

synthetic – that which is produced artificially through chemical or biochemical synthesis.

systems model – illustrates how subsystems are to be assembled and demonstrates or suggests how the system will operate. Used to build a smaller part of a large system to see if that part or subsystem actually operates. Systems models are also very useful in helping to communicate to others what each system involves and how the parts of the system relate and work together.

systems – a group of interacting, interrelated, or interdependent elements forming a complex whole.

systems – a set of interacting, interrelated or interdependent elements forming a complex whole.

tactile – the ability to feel surface finish through the sense of touch.

technical skill – the ability to actually do a task or make a product. Especially important among supervisors of production and less significant as a management skill at higher levels of the organizational hierarchy.

telecommunication – to transmit long distance communications.

telegraph – a mechanism of communication using coded signals at a distance by electronic transmission over wire.

template – a pre-set guide that provides limits and direction. In production, the template is used to guide the material or machine to increase the accuracy and to standardize the final product. In humans, a model used to describe how information may be perceived, processed, and stored.

tension – the state of being stretched or pulled apart; also, the force on materials or structures from pulling or stretching.

territory – a geographical area belonging to or under the jurisdiction of a government authority.

tertiary color – the colors produced by mixing two secondary colors.

tessellations – elements of shape or form that fit together in an interlocking manner.

testing – the critical analysis, examination, or comparison of how well a design, process, or product performs.

textile – fabric that is woven or knitted.

texture – tactile and visual surface characteristics of an object.

theoretical model – a model based on theory or abstract ideas.

theory – systematically organized knowledge that applies to a relatively wide variety of circumstances; abstract idea not necessarily linked to experience or practice.

thermal energy – pertaining to, using, producing, or caused by heat.

thermal processes – producing changes in materials that are dependent upon a change in temperature. Examples include baking, boiling, drying, and freezing.

thermocouple – a device used to measure temperature differences by converting part of the available heat into an electrical signal.

three-dimensional – giving the illusion of depth or varying distances.

tidal energy – draws energy from the gravitational forces of the sun and moon. The gravity of the sun and moon serve to make oceans pulse in long, steady counts

timetable – a schedule that determines how much time is needed to complete a project.

tissue culturing – a genetic engineering technique by which an entire plant can be grown from as few as one cell from the original plant.

tolerance limit – the concept that human operation is affected by the physical variables of the environment. Beyond these limits, great discomfort, physical harm, and even death can take place.

tonal range – the range of the shades of a color.

tone – a quality or value of a color; with pigments, the degree to which a color is mixed with white (tint) or black (shade); with project light, the degree of illumination (light versus dark); a slight modification of a particular color, such as three tones of blue.

tooling up – preparing machinery for the production of a product or a part of a product.

toothed belt and pulley system – a system that replaces the standard flat belt with one with teeth formed in the belt. When combined with toothed pulleys, higher speed and torque is possible. Such systems are used in riding mowers, washing machines and some hand power equipment.

torque – the tendency of a force to produce torsion and

rotation about an axis; the moment of a force.

toxins – poisonous substances.

trade-off process – the necessary give-and-take required to reach a compromise.

transducer – a device that converts input energy of one form into output energy of another form.

transferring – the carry-over or generalization of something familiar from one situation to another.

transmit – to send information in the form or a code or message to a receiver.

transporting structures – containers that provide support and move themselves and something else. Examples include vehicles that fly , hover, float, etc.

transporting units – separate containers for moving goods that provide support for containing goods, materials, or people. For example, railroad cars and modular containers loaded on the bed of a truck.

trend analysis – studying cycles or trends in order to make predictions.

true North – the point on the Earth's surface that coincides with the axis of the Earth's rotation.

truss – a special kind of triangular brace that takes advantage of this stable geometric form.

trust – assured reliance on the character, ability, strength, or trust from someone or something.

truth table – used with logic devices to describe the state of the output, based on the state of the input.

ultrasonics – pertaining to acoustic frequencies above the range audible to the human ear (above 20,000 cycles per second); also, the technology that uses ultrasonic processes as in medical treatment and materials testing.

undesired outcomes – unexpected, unwanted, unpredictable results.

unintended outcomes – unplanned and unanticipated results.

vaccines – a chemical substance that is administered to produce or artificially increase immunity to a particular disease.

valve – device used to regulate the flow of gases, liquids, or loose materials through directing structures, such as tubes and pipes.

variable resistor – an electric device that can be used to change or vary electrical resistance.

vehicle – a means of carrying or transporting something.

velocity ratio – the relationship of the speed of the input effort and the speed of the load; velocity ratio equals the velocity of effort divided by the velocity of the load.

velocity – speed of movement; velocity is often expressed as a vector (a line graphically representing both magnitude and direction) when the speed represents the magnitude of the vector and the direction representing the orientation of the vector.

virtual reality – combination of video and computer in which an individual can get close to the sensation of "being there."

virus – any of a large group of ultramicroscopic organisms that can only grow and reproduce within the cells of a living host and cause various diseases and disorders in humans, animals and plants.

visible light waves – waves that have a higher frequency than radio waves. Visible light vibrates at frequencies up to a million times faster than AM radio waves. Visible light waves can be seen with the unaided eye.

visual design – the aesthetic and emotional appeal of ideas, objects, and experiences to an observer.

visual modality – a preference for visual data for receiving and processing information.

visual representation – portraying ideas, objects, and experiences to others through sight.

vitrification – the point at which platelets melt and fuse together, crystals form, and a material begins to look like stone.

vulnerable – open to attack or damage; capable of being physically wounded.

want – of or relating to a strong desire or need for something.

waterhead – the difference in depth or height of a liquid at two different points; the fall of stored water from its greatest height to its lowest.

wavelengths – the number of times light waves vibrate per second. Messages are carried on the wave length from one vibration to another. The more vibration per second, the more information that can be carried.

ways – the paths that direct and limit the movement of transproting structures, as they move from one place to another. Ways include roadways, highways, seaways, railways, waterways, and airways.

weaving – to construct by interlacing; interweaving string-like elements to produce a fabric-like material.

wedge – a simple machine of triangular shape used for splitting, tightening, securing, or levering.

windlass – any of a variety of lifting or hauling machines consisting essentially of a drum or cylinder wound with a rope or cable and turned by a crank.

wood – the hard, fibrous substance beneath bark which makes up the stems and branches of trees or shrubs.

worm gear – a gear consisting of a threaded shaft and a wheel with teeth that mesh with it.

x rays – relatively high energy photons used for their penetrating power to record on film the internal structure of opaque objects.

yeast – microscopic fungus cells that are used in fermentation to break down starch or sugar into alcohol and carbon dioxide. Used in making wine, beer, and baked products.

zero force – processes that do not require mechanical force. For example, bleaching, etching, dry cleaning, laser cutting, and so forth.

Index ....

A

AC motors, 95
acceptable risks, 443
accumulating alternatives, 8-10
accumulators (reservoirs), 94
acoustical processes, 134
actions (conversion process),
    124-126
addition processes
  creation of molecular bonds in,
      136
  described, 126-129
adhesion between platelets, 114
adhesion processes, 127
adhesives
  assembling models using, 62-64,
      66
  biotechnology and, 331
adjustments
  in closed-loop systems, 398
  control system output, 390
  in open-loop systems, 397-398
  steady state operation and, 400
adsorption, immobilization by, 347
aesthetics/aesthetic appeal, 25
agriculture. See also foods
  food from microbes, 353-356
  growing and harvesting, 312-314
  plant and animal biotechnology,
      349-352
  social consequences of advances
      in, 460, 461-462
  tissue culturing and cloning,
      352-353
air, compressed. See compressors
air cylinders. See pneumatic
      cylinders
air motors, 91-92
air pressure, 170-172. See also
      compressors
aircraft
  as flying structures, 299-301
  gyroscopic action and, 398-400
  lighter- vs. heavier-than-air,
      305-306
  steady state operation in, 400
Alaskan pipeline, 298
algae, overgrowth of, 350
allergens, 336
alloying process, 125

alternatives
  accumulating and assessing
      alternatives, 8-10
  choices from undesirable,
      467-468
  decisions as choices among, 466
  desirable vs. undesirable, 467
  exploring, 7-8, 18
  guidelines for assessing, 19
  multiple desirable, 467
aluminum, 315
American Standard Code for
      Information Interchange
      (ASCII), 365
ammeters, 174
amorphous materials, 119
amperage. See current, electrical
amperes, 151. See also current
analog devices, 408
analysis
  of signals by humans, 205
  task, 271-273
  trend, 426
AND logic gates, 192
  described, 199-200
  in "smart" machines, 402-403
animal biotechnology, 349-352
animals, increasing productivity
      of, 351-352
anthropometric chart, 33
anthropometrics, 33
antibiotics, 336
appearance (design consideration),
      25
appearance models. See also
      models
  architectural, 61
  described, 60-61
  materials for, 64-65
  production, 61
applied research
  defined, 437
  developments resulting from,
      439-440
appropriateness of product size, 34
archaeological exploration,
      424-425
arches, 294
architectural models, 61. See also
      models
artifacts, archaeological, 424

ASCII (American Standard Code
      for Information Interchange),
      365
assessing
  design alternatives, 8-10, 19
  design process, 19
  outcomes, 4
assets, 265
assymmetrical objects, 31
astrolabe, 422
atomic energy
  in energy spectrum, 152
atoms. See also molecules
  in adhesion processes, 127
  in chemical elements, 109
  connections among, 111
  decay of radioactive carbon, 424
  energy derived from
      fission/fusion of, 146
  in magnetic processes, 134
  nuclei of, 146, 153
  overview, 108-109
attraction, force of
  in atomic structure, 108-109
  magnetic poles, 94
  in molecular chains, 112
  vs. repulsive force in bonds, 153
audio in portfolio presentations,
      377
auditory modalities, 238
auditory representations, 24
authority, decision-making and,
      468
automation
  integration of elements of, 410
  in production, 406
automobiles
  as containing and propelling
      vehicles, 307
  as rolling structures, 299
autonomy, development of, 244
axles. See wheel and axle

B

bacteria. See also microorganisms
  agricultural uses, 349-352
  algae-eating, 350
  controlling animal diseases using,
      351-352

E. coli, 344
nitrogen-collecting, 350
potential for waste disposal, 331
balance (sense of)
ergonomic design and, 35
inner ear and, 239
balance (visual)
described, 31
as visual design principle, 30
bar codes, 364-365, 409
bar graphs, 210
basic materials
defined, 118
types of, 119
batteries
chemical energy and, 154
as energy storage systems, 160
invention of, 372
measuring voltage across, 175
behavior, human, 240-241
bell crank levers, 81-82. *See also*
levers
belonging, social need for, 244-245
belts, 85-86. *See also* pulleys
bending processes, 132
bevel gears, 87. *See also* gears
bicycles
as machines, 97-99
as rolling structures, 299
binary code, Morse code as, 372
binary system. *See also* logic
as digital system, 408
symbols used, 189-190
biocatalysts, 346
bioconversion
described, 346
supporting technologies required,
342
bioethics, 356-357
biological energy, 147-148
biomass, 148. *See also* biological
energy
bioprocessing
bioconversion, 346
fermentation, 335, 346-348
immobilization, 347-349
overview, 342
bioreactors
defined, 340
as growth chambers, 343
biosensors, 348
biosorbants, 355-356

biotechnology
in agriculture, 349-356
applications of, 329-332
bioethics, 356-357
bioprocessing, 342-349
cells and genetic engineering,
338-342
early history of, 332-333
pollution control, 355-356
stage one applications, 333-335
stage three applications, 337-338
stage two applications, 336
bits, memory, 408
black boxes, 197
boats. *See* ships; submarines
bond energy, 153. *See also*
molecular bonds
bonds. *See* material conversion
processes; molecular bonds
Boyle's Law, 170-171
brain, human. *See* nervous system
breeding, selective, 340
building models. *See* model-
making
buzzers, 192
bytes, memory, 408

C

callus, 352
calories
amounts produced from
nutrients, 148
calculating number generated,
152
cam shafts, 89
cams
as control devices, 193
described, 89
in mechanical models, 60
cantilever structures, 294
capital, economic. *See also* human
capital; human resource
management
described, 263-264
exchanging human capital for,
268-270
worth of money, 264-265
capital management, 270-271.
See also human resource
management
capital-intensive enterprises, 269

carbon
amorphous forms of, 119
decay of radioactive, 424
fuels derived from, 142, 147
carbon dioxide, 335
carrier vehicles
described, 366-367
multiple messages and differences
among, 368
casting processes. *See also* contour
change processes
described, 131, 132
forging vs., 136
in molding processes, 133
cells
cloning plant, 352-353
immobilization methods,
346-348
in multicellular vs. single-cell
organisms, 338
structure of, 338
ceramics
described, 114-115
as family of materials, 122
reduction and refining, 315
chain of command
described, 279
in organization charts, 276
chains. *See also* sprocket and chain
assemblies
chains of molecules, 112-114
change (conversion process),
124-126
characteristics of materials,
106-108
Charlemagne, 461
charts
advantages of using, 380-381
communicating performance of
solutions using, 19
navigational, 423
organization, 276
plotting trends on, 429
sharing information using,
214-216
cheese making
in factories, 336
microorganisms and, 335
chemical bonding, immobilization
by, 347
chemical compounds. *See*
compounds, chemical

chemical elements. *See* elements, chemical

chemical energy
  biological energy and, 147
  described, 153-154
  in energy spectrum, 152
  in fossil fuels, 147
  release of, 153

chemical processes
  described, 134
  safety precaution, 154

chitosan, 355

chromosomes
  cell division and, 339
  Mendel's discoveries, 335

circles
  in isometric drawings, 43
  in oblique drawings, 42

circuits
  designing control system, 392-397
  measurements in, 174-175
  planning layout of, 395-396
  processor, 197-198
  switches in, 189

clay. *See also* ceramics
  molecular structure of, 114-115
  platelets in, 114-115

clones
  bacteria, 340
  plant tissue, 352
  process creating, 127

closed-loop control systems, 398

closed-loop feedback processes, 371

clothing, 230

clutches, 193-194

coating processes, 128-129. *See also* addition processes

codes
  interpreting symbolic, 204-206
  signals and, 193
  for symbol sequences in messages, 206

cognition, designing and, 237-240

cognitive templates, 234, 235-237

cohesion processes, 127

cohesion within platelets, 114

cold, human limits of tolerance of, 35

color codes, resistor, 179-180

color markers, 48-49

color media, 49

color pencils, 48-49

color wheel
  described, 27-29
  harmonious colors, 29
  mixing pigments vs. combining light, 28

colors. *See also* color wheel
  as element of visual design, 25
  importance in visual design, 27
  as symbols, 207

combustion. *See also* oxidation
  defined, 125
  release of energy from fuels during, 142-143

communicating. *See also* documenting; information; signals; symbols
  advantages of IDEATE Design Loop for, 4
  as human skill, 275-276
  ideas using models, 55
  three-dimensional forms of, 382-384
  visual/auditory representations in, 24

communication
  machine, 363-366
  by signals, 372-373
  in well-functioning groups, 279

communication systems
  basic components of, 191-192
  feedback, 371-374
  interactive, 379-381
  knowledge explosion and, 459-460
  media, 366-371
  overview, 361-362
  telecommunication, 372
  types of, 373-374

communications medium/media, 366-369

compasses, magnetic, 421

compatibility in communications, 362-363

compounds, chemical, 109-112

compression, force of
  described, 155-157
  in structures, 291-293, 294

compression testing of materials, 118

compressors
  described, 91
  energy conversion in, 170-172
  safety precautions, 91

computer models. *See also* models
  described, 69
  as a means of forecasting, 435

computers. *See also* memory, computer; programs, computer
  analog vs. digital, 408
  binary system and, 190
  communications and, 365-366
  controlling machines, 410
  extending human abilities, 78
  input data, 375
  in integrated systems, 410-413
  interactive communication systems on, 379
  outputs from, 377-378
  weather forecasting and reports, 379-380

concept, 56

conceptual map, 57-58

conceptual models. *See also* models
  described, 56-58
  types of, 59

conceptual skills, 276

concrete, reinforced, 295

cones, shading, 47

consequences
  as decision-making basis, 469
  ecological, 463-465
  of knowledge, 459-460
  overview, 450
  personal, 454-458
  planning for delayed effects, 466
  social, 460-462
  types of, 454

conservation of materials, 322-324. *See also* ecological consequences of technology

constraints
  considering in planning process, 19
  on decision-making, 472-473

construction
  defined, 290
  earthquake-resistant, 296
  model. *See* model-making

construction kits, model-making, 67

construction systems. *See also* structures
  overview, 289-290
  transporting systems and, 303-304
consumers, humans as, 221-251
containing structures. *See also* transporting structures
  defined, 291
  described, 298
  overview, 290
containing vehicles, 307
continuous manufacturing. *See* mass production
contour change processes. *See also* material conversion processes
  described, 131-133
  molecular bonds in, 136
contrast
  described, 31
  role in design, 30
  as visual design principle, 29
contrasting colors, 29
control devices
  cams and, 193
  signals and, 190-191
control systems
  analog and digital devices, 407
  aspects of, 389
  automation of, 406, 410
  basic components of, 191-192
  bicycle, 97-98
  closed-loop systems, 398
  computers and, 408-413
  design and development of, 391-397
  electricity as, 194
  feedback in, 400-402
  fermentor, 343-344
  human, 224
  inputs, 389-390
  integrated systems, 410-413
  modeling circuit design, 395
  open-loop, 397-398
  outputs, 391
  overview, 387-389
  process aspect of, 390-391
  self-regulating, 398-400
  "smart" machines, 402-407
  types of, 397-400
controls (on decision-making), 472-473

converters (energy), 161
coolants, 155
coordination, human movement and, 224
cosmic energy
  described, 158-159
  in energy spectrum, 152
cover systems
  bicycle, 97-98
  human, 222-223
cranks
  calculating velocity ratio for, 166
  turning forces and, 165
  as wheel and axle, 83
  worm gears and, 87
crating technique, 43
criteria. *See also* specifications
  in design brief descriptions, 18
  for desired outcomes, 280
  developing list of, 8
  in optimization process, 441-443
crop production
  genetic engineering and, 350-352
  plant cell cloning and culturing, 352
cross staff, 421-422
crowding, perception of, 259
crystals
  in ceramic firing process, 115
  described, 115-116
  molecular structure of, 115-116
cubes, shading, 46
cultural templates
  described, 236
  sense of privacy and, 259
current, electrical
  Ohm's Law and, 176-177
  vs. watts and joules, 151
curves
  in isometric drawings, 43
  in oblique drawings, 42
custom manufacturing, 319
cutaway drawings, 45
cybernetic devices
  described, 398-400
  logic in, use of, 402
cylinders. *See* cylinders (shape); pneumatic cylinders
cylinders (shape)
  shading, 47
  as standard form in drawings, 46

D

dams
  as containing structures, 298
  and generating electricity, 145-146
data
  charts, graphs, and maps, 214-216
  levels of measurement of, 209-213
  representing by using symbols, 208
  as units of information, 188-189
DC (direct current) motors, 95
decision networks, 450-451
decision trees, 450
decisions
  making "correct", 473-474
  types of, 466-468
  values and, 473
decision-making. *See also* consequences
  bases for, 468-470
  decision networks, 450-451
  human, 248
  limitations on, 472-474
  machine, 199-201
  overview, 466
  questions to be answered, 470-472
  recommendations, 477-480
  technology assessment, 451-453
  types of decisions, 466-468
  on use of resources, 280
decoding messages, 204-205
dedicated devices, 410
deficit spending, 264-265
defining problems, 5-6
deleting information, 234
Delphi survey, 430-431
deoxyribonucleic acid (DNA). *See* DNA (deoxyribonucleic acid)
design brief
  describing, 18
  developing, 6-7
  specifications with, 8
design considerations
  aesthetics and appearance, 25
  control system outputs, 391
  elements of visual design, 25-27
  ergonomics, 32-37
  principles of visual design, 30-35

design limits, 8
Design Loop. *See* IDEATE Design
  Loop
design process
  accumulating and assessing
    alternatives, 8-10
  compromises in, 443
  developing design briefs, 6-7, 18
  evaluating results of, 13-15, 19
  exploring possible alternatives,
    7-8, 18
  iteration in, 444
  overview, 3-4
  problem identification and
    definition, 5-6, 17-18
  reiterating, 16
  role in technology, 23-24
  symbols in, using, 208-216
  trying out best solution, 11-13,
    189
designers, humans as, 221-251
designing. *See also* design
    considerations; design process
  control systems, 391-397
  defined, 3-4
  described, 23-24
  human cognition and, 237-240
  for human users, 224-225.
    See also ergonomics
  product packages, 382-383
  structures for loads, 293-295
designs
  effects of colors in, 29
  requirements for success of, 24
desired outcomes
  in design process, 7-8, 24
  developing criteria for, 280
development, human. *See also*
    psycho-social stages in human
    development
  of cognitive templates, 235-237
  growth and, 226
  initiative, 244-245
developments (model-making),
    66-67
developments, technological
  discovering, inventing,
    innovating, 438
  iterating, 444
  methods of thinking and
    developing, 440-441
  optimization, 441-444
  research and development,
    439-440

digital devices, 408
direct current (DC) motors, 95
directing structures
  defined, 291
  described, 298
  overview, 290
direction-finding devices, 420-421
discovery, in development process,
    438
dissonance in design, 30
distorting information, 234
DNA (deoxyribonucleic acid)
  discovery of structure of, 337
  in gene splicing, 340-341
  Human Genome project and, 341
  recombinant, 339-340
documenting
  advantages of IDEATE Design
    Loop for, 4
  defined, 37
  formats for, 37-38
  graphics in, 49-50
  importance of, 37
  ownership of intellectual
    property, 270
  recording ideas, 38-39
double-action cylinders, 91
drawing techniques. *See also*
    presentation techniques
  creating perspective.
    See perspective drawings
  model-making developments
    and, 66
  outlining, 48
  pencil rendering, 46-47
  shading, 46
  sketching, 39-40
drawings. *See also* pictorial
    drawings
  clarifying ideas using, 24
  as communication, 380-381
  design and problem solving
    process requirement for, 11
  nonpictorial, flat, 44-45
  recording ideas using, 38
drilling processes
  described, 312
  materials extracted by, 310-311
driving and driven gears, 87
drop forging process, 137
drugs, pharmaceutical, 336, 344
dwell, 193
dynamic loads, 293
dynamic materials, 121-122

E

E. coli bacterium, 344
ecological consequences of
    technology, 463-465
economic resources, 263. *See also*
    resources
efficiency
  defined, 458
  variations among biological
    energy sources, 148
Einstein's Law of Relativity, 153
electric air compressors, 91
electric generators
  bicycle headlight, 100
  described, 96
  driven by falling water, 145
electric motors, 95
electrical energy, 194. *See also*
    electricity; electromagnetic
    energy
electrical power, 175-178
electrical processes, 135
electricity. *See also* electromagnetic
    energy; electromechanical
    mechanisms
  as control system, 194
  generated from water movement,
    145-146, 160-161
  generated by wind energy,
    144-145
  measuring properties of, 174-175
  produced from chemical energy,
    154
electromagnetic energy
  described, 157-158
  in energy spectrum, 152
  sensing, 194
electromagnetic spectrum
  frequencies within, 157-158
  vs. mechanical bands, 157
  transmitting information using,
    368-369
electromechanical mechanisms
  electric generators, 96
  electric motors, 94-95
  measurements, electrical, 174-175
  overview, 94
  power, electrical, 175-178
  resistance values and tolerances,
    179-180
  resistors in circuits, 178-179
  safety precaution, 94
  solenoids, 95-96

electromotive force (EMF).
See voltage
electronic models, 60.
See also models
electronic prototyping boards, 60
elements, chemical
atoms forming, 109
defined, 109
in Periodic Table of the Elements, 110
elements of visual design
form, 26-27
line, 26, 40
overview, 25
point, 26
shape, 26
texture, 27
EMF (electromotive force).
See voltage
emotions, human
as decision-making basis, 469
described, 240-241
energy
atomic, 152, 153
chemical. *See* chemical energy
control system inputs as, 389
as driving force of technology, 141-142
effect on functional materials, 122
electromagnetic, 157-158
forms of, 150-159. *See also* energy sources
mechanical. *See* mechanical energy
radiant and cosmic, 158-159
safety precautions, 152, 158
sensors linking to information, 191
sources of. *See* energy sources
thermal energy, 155
use by humans, 232-233
energy conversion
chemical energy and rate of, 154
control system outputs, 391
creating thermal energy from materials, 125
defined, 160
electromechanical mechanisms, 174-180
energy production, 316-317
fluid mechanisms, 170-174

machines, 100
mechanical energy. *See* mechanical energy
for storage, 160-161
for use, 161-162
energy differentials, 144
energy sources
fossil fuels as, 147
geothermal energy, 144
life forms as, 147-148
movement of water as, 145-146
nuclear, 146
overview, 142-143
renewable, 324
solar energy, 143
wind as, 144
energy spectrum
described, 152
electromagnetic vs. mechanical bands, 157
frequencies within, 157-158
sensor detection of portions of, 191
energy transmission systems
bicycle, 97-98
human, 223
engines, internal combustion, 161
English Channel tunnel, 304
enlightened self-interest, 248
entrapment, immobilization by, 347
entropy
lost energy as, 150
passage of time and, 255-256
signal, 363, 370
environment, technological impact on, 463-465
environmental tolerance zones for humans, 36, 225
enzymes
in bioconversion processes, 346
cellular production of, 339
in cheese making, 335
in detergents, 331, 346
in immobilization bioprocesses, 346-348
restriction, 340-341
equilibrium
of compression and tension forces, 291
described, 163
of pulley systems, 168

equipment
model-making, 65-66
modeling center, 63
ergonomics
balance (sense of), 35
human senses, 35-37, 238-240
movement, 34
overview, 32
size, 32-33
strength and stamina, 35
Erikson, Erik, 241-242
evaluating design results, 13-15, 19
Evans, Oliver, 406
evaporation
energy from moving water and, 145
latent heat of, 155
exhibits, 383
expandable materials, 121. *See also* dynamic materials
experience, decision-making and, 469-470
expert power, 274, 275
exploded view drawings, 45
exploration
of alternatives, 7-8, 18
archaeological and historic, 424-425
below ocean surface, 423-424
direction-finding, 420-421
Guide for Design Questions, 17
microscopic, 424
position-finding devices, 421-423
of space, 423
external controls. *See* constraints
external programs, 409
extracting processes
drilling and pumping, 312
minerals, 310-311
extruding processes, 132. *See also* contour change processes

F

fabrication of models, 12-13, 67.
See also model-making
families of materials, 122-123
family work, 265-267
farming. *See* agriculture; foods
fasteners, 128
fastening processes. *See* addition processes; mechanical fastening processes

feedback
  in control systems, 400-402
  from iterating process, 444
  open vs. closed-loop processes,
    371
feelings, decision-making and,
    468-469
fermentation
  described, 335, 342-345
  producing insulin by, 344
fermentors, 343-344
ferrous metals, 123
fertilizers
  nitrogen-collecting bacteria, 350
  from purified sewage, 355
fiber optics, 368
filtration, 346
finishes for appearance models, 65
firing ceramics
  changes occuring during, 115
  defined, 114
first-order effects, 466
fish factories, floating, 314
fission, nuclear, 146, 153
five-way transport, 305
flat belts, 86. See also belts; pulleys
flat drawings, 44-45
flexible materials, 121
floating structures
  described, 298-299
  fish factories, 314
flowcharts, 410-411
flow control regulators, 92-94
flow restrictors, 92-94
fluidic mechanisms
  air motors, 91-92
  compressors as, 91. See also
    compressors
  flow control regulators, 93
  overview, 90-91
  reservoirs (accumulators), 94
  valves, 92
fluidics, 90-91
flying bicycles, 99
flying buttresses, 294
flywheels, 161
foaming materials, 121-122
followers, cam, 89
foods. See also agriculture;
    nutrients
  biotechnology and, 331
  effects of large-scale production
    of, 455

fermentation, 335, 343-344
growing and harvesting, 312-315
immobilization techniques,
    346-348
from microorganisms, 353-356
microorganisms and, 333-334
processed, 229
refining, 316
sense of taste and, 239
force. See also attraction, force of;
    moments of force; pressure;
    torque; turning forces
  binding atomic nuclei together,
    146
  compression, 155-157, 291-293,
    294
  electromotive. See voltage
  levers and, 80, 162
  magnification by pulleys, 90
  pistons, generated by, 172-173
  processes involving zero, 129
  repulsion, 94, 108-109
  separation processes and
    mechanical, 129, 136
  tension, 155-157, 291-293
  transmission of, between shafts,
    86
  transmission of, by linkages, 82
forecasting
  analogies, 428-429
  conducting surveys, 430-431
  described, 426-427
  future histories, developing,
    431-433
  future scenarios, developing,
    433-435
  modeling, 435-436
  projecting trends, 427-428
forging processes
  vs. casting processes, 136
  drop, 137
  overview, 132
formats, documentation, 37-38
forming processes. See also contour
    change processes
  described, 132
  in model-making, 68
forms
  creating drawings using standard,
    46
  drawing in two-point perspective,
    41-42
  as element of visual design, 25

in ergonomic design, 35
in patterns, 32
sketching and drawing, 40
types of, 26-27
formulas, theoretical models as, 58
fossil fuels. See also fuels
  demand vs. supply of, 312
  described, 147
  as secondary energy sources, 143,
    147
four-way transport, 304
fractional distillation, 316
free-form shapes and forms, 27, 40
frequencies
  defined, 157
  light vs. radio wave, 368
friction, number of pulleys vs., 90
fuel cells, 319
fuels
  as energy sources, 142-143
  fossil, 143
  refining, 316
fulcrum
  defined, 80, 162
  equilibrium and, 163
Fuller, Buckminster, 222
"functional" foods, 354
functional materials, 122. See also
    dynamic materials
functional models. See also models
  described, 58-60
  materials for, 65
functional objects, size of, 32
fusion, nuclear, 146, 153
future histories, developing,
    431-433
future, predicting. See forecasting
future scenarios, developing,
    433-435
futurists, 427

G

gamma rays, 158
Gantt timeline, 39
gases
  Boyle's Law, 170-171
  conversion by microbes to
    nitrates, 464
  molecular structure of, 111
  transporting structures for, 298

509

gear reduction
  defined, 87
  of worm gears, 167
gear trains, 86
gears. *See also* rack and pinion
    gears
  bevel and mitre, 87
  described, 86
  in mechanical models, 60
  spur, 87
  types of, 87
  velocity ratios, 166-167
  worm, 87
gene-splicing process. *See also*
    genetic engineering
  described, 340-341
  producing insulin from E. coli,
    344
generalizations, 234
generativity, 247
generators. *See* electric generators
genes, 339. *See also* chromosomes;
    genetic engineering; genetics
genetic diversity, 349
genetic engineering
  bioethics and, 356-357
  cells and, 338-342
  crop production and, 350-352
  defined, 339
  discovery of DNA and, 337
  gene splicing, 127, 340-341
  increasing animal hormone
      production, 351
  increasing enzyme efficiency, 331
genetic information, 339
genetics
  Mendel's discoveries, 335
  mutations, 336
geometric shapes and forms, 26,
    27, 40
geothermal energy
  described, 144
  limited use of, 317
glucose, 147-148
glues. *See* adhesives
goals
  clarifying, 270-271
  consideration in decision-
      making, 470-472
Gossamer Albatross, 99
grain of materials, 113-114
graphic communication, 375

graphics
  animation of, 377
  inputting into computers, 375
  as presentation technique, 49-50
graphs. *See also* charts
  advantages of using, 380-381
  bar, 210
  communicating performance of
      solutions using, 19
  sharing information using,
      214-216
gravitational energy, 145
green ceramics, 114-115
groups
  assessing health of, 279-280
  working in, 273-274
growing and harvesting, 312-314.
    See also agriculture
growth chamber, 343
growth and development, human,
    226
growth medium, 342-343
Guide for Design Questions, 17-19
Gutenberg, Johann, 459
gyroscopes, 398-400
gyroscopic action, 398

H

half-life, 424
handling, materials, 406
harmonious colors, 29
harmony
  described, 31
  role in design, 29
  as visual design principle, 30
harvesting, 312-314. *See also*
    agriculture
HB pencil, 40
hearing, sense of, 35
heat. *See also* thermal energy
  effect on materials containing
      platelets, 114
  energy storage and, 161
  human limits of tolerance of, 35
  in metal refining, 314-315
  produced by electrical power, 177
  in rapid oxidation, 142
  sensing by thermocouples, 191
heat pumps, 161
heavier-than-air vehicles, 305-306
height, anthropometric chart of,
    33

Hertz (Hz), 157
HFCS (high fructose corn syrup),
    331
hierarchies, management, 276
high fructose corn syrup (HFCS),
    331
highlights, 47-48. *See also* shading
highways, 296-297
historical exploration, 424-425
hormones
  in animal agriculture, 351-352
  applications of, 341
  synthetic, 351
human body
  compared to machine parts, 225
  control subsystem, 224
  energy transmission system, 223
  materials protecting, 230-232
  movement of, 34, 224
  strength and stamina, 35
  structure and cover system,
      222-223
  study of size and measurement
      of, 34
human capital. *See also* capital,
    economic; human resource
    management
  brains and brawn, 267-268
  described, 265-268
  exchanging for economic capital,
      268-270
human cognition, 237-240
Human Genome Project
  described, 341
  patents and, 356
human motivation. *See* motivation
human resource management. *See
    also* capital management;
    human capital
  groups, assessing health of,
      279-280
  motivation to work, 278-279
  overview, 270-271
  power in relationships, 274-275
  skills needed, 275-278
  task analysis and scheduling,
      271-273
  working in groups, 273-274
human skills, 275-276
humans. *See also* ergonomics
  Buckminster Fuller's definition of,
      222
  as decision-makers, 248

designing for, 224-225. *See also* ergonomics
emotions and behaviors, 240-241
energy use, 232-233
environmental tolerance zones for, 36, 225
impact on environment, 463-465
information input through senses, 187-188. *See also* symbols
as information processors, 233-237
interaction with objects through senses, 35-37
machines vs. labor of, 268-270
as material users, 227-228
nutrients and, 228-230
technology and, 221-222
tools extending abilities of, 77-79
types of energy detectable by, 152
hydraulics, 90-91. *See also* fluidic mechanisms
hydrocarbons, 316
hydrogen gas, 160
hydroponics, 313
hydropower, 144-145, 161
Hz (Hertz), 157

ICs (integrated circuits)
AND gates, 402-403
NOT gates, 404-405
OR gates, 403-404
ideas
generating, 18
methods of recording, 38-39
translating into objects, 55-56
IDEATE Design Loop
advantages of, 4
Guide for Design Questions, 17-20
manufacturing, 321-322
modeling future using, 436
symbols in, using, 208
identifying problems
in design process, 5-6
guidelines for, 17-18
identity, sense of, 245-246
if-then statements, 469
images
animation of, 377

combining with text for printing, 377
inputting into computers, 375
immobilization technique, 346-348
immune system, viruses and, 343-344
impact forming processes, 132
impact structures, 301
improvements
to designs, 440-444
to future investigations, 437
identifying possible, 19
individual templates, 236-237
industry, sense of, 245-246
inflatable structures, 297
information. *See also* data
in assessing alternatives, 18
as capital, 265
charts, graphs, and maps, 19, 214-216
converted into code, 362-363
for design work, 214
genetic, 339
human processing of, 233-237
human sensory input of, 187-188
for humans. *See* symbols
increase in rate of available, 459-460
levels of measurement, 209-213
lost through entropy and noise, 370-371
machine. *See* signals
methods of collecting, 18
positive and negative effects of, 248
required on packages, 382-383
stored in DNA, 338
uses of, 188-190
information conversion machines, 100
information conversion processes
creating information from materials, 126
sensors and, 191-193
information highways
as communication systems, 373
defined, 369
infrared waves
generating, 157
as solar energy component, 143
initiative, development of, 244-245

innovation, in development process, 438
input devices, selecting control system, 391
input sensors, 192
circuit operation, 395
in planning control system design, 389-390
input signals
changing to outputs, 390-391
into control systems, 390-391
from sensors, 191-193
types of, 192
insulation for human body heat, 230
integrated systems, 410-413
integrating signals, by computers and microprocessors, 191
integrity, 247-248
intellectual property
documenting ownership of, 270
patents of genetically engineered organisms, 356
intended outcomes, 8
interactive communication systems, 379-381
interest, loan, 264-265
interlacing processes, 129. *See also* addition processes
intermittent manufacturing, 319
internal change processes
acoustical processes, 134
chemical processes, 134
described, 137
electrical, 135
magnetic processes, 134
mechanical, 133
optical processes, 134
thermal, 133
internal combustion engines, 161
internal controls on decision-making, 472-473
internal programs, 409
interval level of measurement
defined, 210
vs. other levels of measurement, 209
intimacy, 246-247
inventions
in development process, 438
impact of small, 461

investigating
  expanding knowledge for
    improving, 437
  problems in design process, 6
  through exploring, 420-426
  through forecasting, 426-436
investing resources
  in human capital, 267-268
  supply and demand governing,
    264
iron ore, 315
isometric drawings, 42-43
iterating process, 444
iteration, 16

J

jobs
  humans vs. machines
    performing, 268-270
  management of complex,
    271-273
joining processes. *See also* addition
    processes
  described, 127
  model parts, 66
joules
  electrical potential energy in, 175
  vs. watts, 151, 177
journals, 38

K

kilns, 115
kinesthetic modalities, 238
kinetic energy
  in falling water, 145
  thermal energy and molecular,
    155
knitting, 129
knowlege explosion, 459-460

L

labor
  human vs. machine, 268-270
  performed by family members,
    265-267
labor-intensive enterprises, 268
language systems, 365-366
lasers
  defined, 158
  in optical processes, 134

latent heat, of evaporation and
    fusion, 155
Law of the Lever. *See* Lever, Law of
layering, coating by, 129
layers, material, 120-121
LDRs (light-dependent resistors)
  in circuit operation, 395
  as input signal sources, 192, 196
LEDs (light-emitting diodes), as
    output devices, 192
legitimate power, 274-275
lettering for appearance models, 64
levels of measurement
  described, 209-213
Lever, Law of
  described, 79-81, 164
  turning forces, 163-165
levers
  bell crank, 81-82
  described, 79-81
  in equilibrium, 163
  principle of. *See* Lever, Law of
light
  as carrier, 368
  as energy, 232
  in optical processes, 134
  sensing energy of, 195
  as solar energy component, 143
  spectrum of visible, 157
light-dependent resistors (LDRs)
  in circuit operation, 395
  as input signal sources, 192, 196
light-emitting diodes (LEDs), 192
light sensors, 192
light sources, 46
lighter-than-air vehicles, 305
linear motion, 89. *See also* rotary
    motion
linear motors, 95
lines
  as element of visual design, 25
  in isometric drawings, 42-43
  in oblique drawings, 42
  in perspective drawings, 40-41
  in plan drawings, 44-45
  sketching and drawing, 40
  uses of, 26
linkages
  described, 82
  in mechanical models, 60
liquids
  Boyle's Law and pressurized, 171
  molecular structure of, 111
Lister, Joseph, 334-335

loads
  compression and tension in
    structural, 291-293
  described, 163
  designing structures for, 293-295
  in equilibrium, 163
  lever and, 80, 162
  limits for transport, 307-308
  materials to withstand, selecting,
    295-296
  static and dynamic, 293
lobe, cam, 89
location
  devices determining current,
    421-423
  maps and charts, 423
log, as documentation format, 38
logic
  in cybernetic machines, 402
  described, 197-198
logic gates
  AND gates, 192, 199-200,
    402-403
  described, 199-200
  NOT gates, 404-405
  OR gates, 192, 200, 403-404
  as signal processors, 192
logos, 206

M

machine communication
  described, 363-366
  feedback in, 371-372
machine logic. *See* cybernetic
    devices; logic
machines. *See also* tools
  belts in early, 86
  communicating with. *See* signals
  computer control of, 410
  decision-making. *See* logic
  as extensions of human abilities,
    77
  function of, 100
  history of, 96-97
  humans vs. labor of, 268-270
  lever as basis for, 79
  role in communications, 362
  "smart", 402-407
  subsystems of, 97-99
mag-lev systems, 302-303
magma, 144
magnetic compasses, 421

magnetic energy, 194. *See also* electrical energy; electromagnetic energy

magnetic fields
  generating electrical current from, 94
  levitating vehicles on, 302-303
  solenoid operation, 95-96

magnetic North, 421

magnetic poles, 94

magnetic processes, 134

maintenance, 256

management. *See* capital management; human resource management; resource management

manufacturing processes
  automation of, 406, 410
  custom, 319
  intermittent, 319
  mass production, 61-62, 319-320, 405
  overview, 318-320
  robots, use of, 410

maps
  development of, 423
  of ocean floor and currents, 424
  sharing information using, 214-216

markers, color, 48-49

Martel, Charles, 461

Maslow, Abraham, 241-242. *See also* psycho-social stages in human development

mass production
  control systems and, 405-407
  described, 319-320
  prototypes and, 61-62

masters, production models as, 61

materials conversion processes
  addition processes, 126-129
  contour change processes, 131-133
  described, 124-126
  to energy, 232-233
  internal change processes, 133-135
  refining, 314-316
  separation processes, 129-130
  similarities in, 136-137
  types of, 126

material layers, 120-121

material mixtures, 120

material systems, 119-121

material units. *See* molecules

materials. *See also* adhesives; material systems
  amorphous, 119
  for appearance models, 64-65
  atomic and molecular structure of, 108-109
  characteristics of, 106-108
  common to many models, 62-64
  conservation and reuse of, 322-324
  constructing models from basic, 67-68
  constructing models from two-dimensional, 66-67
  conversion of. *See* material conversion processes
  crystals, 115-116
  dynamic, 121-122
  elements and compounds forming, 109-112
  families of, 122-123
  for floating structures, 299
  forming, 68
  functional categories of, 118-122
  in functional models, 59
  as fundamental to technology, 105-106
  humans as users of, 227-228
  loads, selecting to withstand, 295-296
  molecular chains, 112-114
  monolithic, 118-119
  platelets, 114-115
  protecting human body, 230-232
  in prototypes, 61
  refining, 314-316
  shaping, 68
  in structural models, 60
  tools for shaping model, 66

materials conversion machines, 100

maximizing the criterion process, 441

measurements
  electrical, 174-175
  levels of, 209-213
  of position of sun and stars, 421-422
  of time, 203-204
  units of, 214

mechanical advantage
  of cranks, 166
  provided by levers, 80, 164
  of pulleys, 169-170
  in windlasses, 83

mechanical energy
  calculating MA (mechanical advantage), 162
  described, 155-157
  in energy spectrum, 152
  flywheels and, 161
  pulleys, 168-170
  sensing, 193-194
  in separation processes, 129, 136
  turning forces, 163-165
  variations on wheel, 166-168

mechanical fastening processes, 128

mechanical models, 60

mechanical processes, 133

mechanisms. *See also* fluidic mechanisms; gears; machines; pneumatic mechanisms; tools
  as extensions of human abilities, 77
  machines as combinations of, 97
  safety precaution, 85
  vs. tools, 78-79

media. *See* communications medium/media

medicine
  bioethics and biotechnology, 357
  biosensors and, 348
  diabetes and biotechnology, 344
  diagnoses using enzymes, 346
  early explanations for illness, 334
  microscopic exploration in, 424
  penicillin production, 336

memory, computer, 198, 408

Mendel, Gregor, 335

Mercator projection, 423

mers. *See* polymers

messages, 188
  conveyed by collections of symbols, 207
  emotions and, 240
  interpreting symbolic code in, 204-206
  vs. media and carriers, 367
  requirements for understanding received, 362-363

metals
  crystals in, 115
  as family of materials, 122-123
  molecular structure of, 115-116
  refining, 314-316
metamorphic materials, 121. *See also* dynamic materials
microbes. *See* microorganisms
microorganisms. *See also* fermentation
  controlling pollution using, 355-356
  food from plant or animal, 353-356
  foods and, 333-334
  growing and harvesting, 313
  in history of biotechnology, 333
  human survival and, 227-228
  manufacturing of, 336
  patenting new, 356
  potential for water purification, 331
  producing chemicals for plastics, 355
  rapid reproduction of, 338
  in spread of illness, 334-335
microprocessors, 410
microscopes
  development of, 424
  types compared, 425
microscopic exploration, 424
microwaves, cooking process using, 157-158
migrations, human, 463
mind-hand interaction model, 24
mining, 310-311
mitre gears, 87. *See also* gears
mixing processes, 127. *See also* addition processes
mixtures, material, 120
model-making
  construction techniques, 66-68
  materials for, 62-65
  tools and equipment, 65-66
modeling
  circuit, 395
  as forecasting method, 435-436
models. *See also* model-making; scale models
  appearance, 60-61
  computer, 69, 435
  conceptual, 56-58

fabrication of, 12-13, 67. *See also* model-making
  physical. *See* physical models
  prototype. *See* prototypes
  testing best solution using, 19
  theoretical, 58
  as three-dimensional communication, 383
  translating ideas into objects using, 55-56
  types of, 60
modifications, identifying possible, 19
molding processes. *See also* contour change processes
  described, 133
molds, 333. *See also* microorganisms
  in cheese making, 335
  as food sources, 354
  production of penicillin from, 336
molecular bonds
  in addition processes, 127, 136
  in contour change processes, 136
  effect of thermal energy on, 155
  as energy source, 142-143, 153
  in internal change processes, 137
  in materials conversion processes, 136
molecular chains, 112-114
molecular structures
  in ceramic firing process, 115
  materials compared, 111-112
molecules
  in adhesion processes, 127
  atomic structure forming, 108-109
  chains of, 112-114
  in crystals, 115-116
  in elements and compounds, 109-112
  in platelets, 114
moments of force
  levers and, 80
  turning forces and, 164, 165
monetary system, 264
money. *See also* capital, economic
  as form of communication, 374
  worth of, 264-265
monolithic materials, 118-119
monomers, 112
Morse code, 204-205, 372

Morse, Samuel
  recording instrument invented by, 205
  telegraph and code inventions, 204-205, 372
motif, 32
motion. *See also* rotary motion
motivation
  as human skill, 275-276
  power relationships and, 274-275
  in psycho-social development, 241-242
  types of power influencing, 274-275
  to work, 278-279
motors
  air, 91-92
  electric, 95
  as output devices, 192
  stepper, 95
movement. *See also* linear motion; rotary motion; transport systems
  effect of linkages on, 82
  human, 34, 224
  Newton's Laws of Motion governing, 155-156
  sensing, 193-194
  of water, 145-146, 160-161
mutations, genetic, 336
mycoproteins, 354

N

natural materials
  defined, 118
  types of, 119
natural shapes and forms, 26, 40
navigation
  development of maps and charts for, 423
  technological advances in, 421-423
needs, human
  physical, 243-244
  psychological models of, 241-242
  role of design work in satisfying, 23-24
nervous system
  growth and development of, 226
  as human control subsystem, 224
Newton's Laws of Motion, 155-156

Newton's Second Law, as theoretical model, 58

noise
  human limits of tolerance of, 36
  loss of information due to, 370-371

nominal level of measurement
  defined, 209
  vs. other levels of measurement, 209

non-economic resources, 263

nonpictorial drawings, 44-45

norm, human variations from, 33

NOT logic gates, 404-405

notes, recording ideas using, 38-39

nuclear energy, 146, 317. *See also* atomic energy

nuclear testing, consequences of, 464

nucleus
  atomic, 146, 153
  cellular, 338

nutrient agar, 352

nutrients. *See also* foods
  amounts of calories produced from, 148
  biotechnology and, 336
  human requirements for, 228-230

nutritional supplements, 229

O

oblique drawings, 42

oceans, exploration below surface of, 423-424

Ohm's Law, 176-177

oil
  cleaning up spills of, 355-356
  drilling and pumping, 312
  refining, 316

one-celled animals. *See* microorganisms

one-point perspective drawings, 41

one-way transport, 304

opamps (operational amplifiers), 192

open pit mining, 311

open-loop control systems, 397-398

open-loop feedback processes, 371

operational amplifiers (opamps), 191

optical processes, 134

optimization, 441-444

OR logic gates
  described, 200
  as signal processors, 192
  in smart circuits, 403-404

ordinal level of measurement
  defined, 210
  vs. other levels of measurement, 209

organization chart
  chain of command in, 276
  purpose of, 276

orthographic drawings
  described, 44
  in model-making, 66

OTA (Office of Technology Assessment), 452

outcomes of design process, 7-8

outlining objects, 48

output devices, 191
  driven by control signals, 192, 391
  selecting control system, 391

ownership, sense of
  capital and, 263-264
  cultural differences in, 260
  intellectual property, 270

oxidation. *See also* combustion
  copper, 111
  rapid vs. slow, 142
  slow, 142, 147

P

packaging
  creating models of, 66-67
  as three-dimensional communication, 382-384

paints. *See also* coating processes
  for appearance models, 64-65
  molecular bonds formed when applied, 136

parts
  replacement, for humans, 223
  scale model, 65

passageways, 298

Pasteur, Louis, 333

patenting products, 16

patents
  advantages of IDEATE Design Loop for obtaining, 4
  for microorganisms, 356

pathways, transportation
  freedom vs. control, 306-307
  types of, 304-307

patterns
  of human growth and development, 226
  as visual design principle, 30, 32

PCBs (printed circuit boards), 395-396. *See also* electronic prototyping boards

PD (potential difference), 175

pencil rendering, 46-47

pencils
  color, 48-49
  freehand drawing and sketching, 40

penetration, coating by, 129

perceptions, sensory modality and, 238-240

Periodic Table of the Elements, 110

personal consequences
  emotional effects, 458
  history of, 454-455
  physical effects, 455-458

personal recognition, 279

perspective drawings
  common features of, 40-41
  computer-generated, 69
  one-point perspective, 41
  two-point perspective, 41-42

pesticides, reducing quantity of, 350

petroleum. *See* oil

pharmaceutical drugs, 336, 344

photography in optical processes, 134

photoresistors, 195

photosynthesis
  glucose production from, 147
  solar energy and, 143

photovoltaic cells, 195

physical models. *See also* models
  computer-generated, 69
  functional, 59-60
  types of, 58

physical systems
  overview, 289-290
  production, 309-324
  transporting, 302-309

pictograms
  described, 206
  universal symbols, 207

pictorial drawings
  crating technique, 43
  elements of perspective drawings, 40-41
  isometric, 42-43
  oblique, 42
  one-point perspective, 41
  two-point perspective, 41-42
  types of, 40
pigments, producing color by mixing, 28
pinion gear. *See* rack and pinion gears
pipelines, 298, 307
pistons, 172-173. *See also* pneumatic cylinders
plan drawings, 44. *See also* orthographic drawings
plane, requirement for living on level, 258-259
planning
  circuit layout, 395-396
  goals clarification and, 270-271
  resource, 12
  time required for design, 257
planning webs, 57-58
plans, recording ideas using, 39
plant biotechnology, 341, 349-352
plastics
  molecular structure of, 112
  as polymers, 112-113
  produced by biotechnology, 355
  in RDM and stereolithography processes, 69
  reinforcing, 295-296
platelets, 114-115
pliability of platelets, 114
pneumatic cylinders. *See also* pistons
  air flow rate vs. speed of motion of, 92-94
  Boyle's Law and, 170
  described, 92
  in mechanical models, 60
  reservoirs and, 94
  valves in, 92
pneumatic mechanisms
  air motors, 91-92
  compressors, 90-91
  flow control regulators, 92-94
  reservoirs (accumulators), 94
  safety precautions, 91
  valves, 92

pneumatics, 90-91. *See also* fluidic mechanisms
point (design element), 26
polymerization process, 316
polymers, 112
population vs. consumption, 457
portfolio
  audio and video, inputting, 377
  as communication system, 374
  design brief exercise, 50
  developing presentation, 51
  as documentation format, 38
  images, inputting, 375-377
  presentation, 50-51
  printing and presentations, 377-378
  processing text and images, 377
  purposes of, 15-16
  selecting media for, 375
  text, inputting, 375
post and lintel structures, 293-294
potential difference (PD), 175
potential energy, 145
power, electrical, 175-178
power, managerial, 274-275
predictions, 426
presentation portfolio, 50-51
presentation techniques
  color techniques, 49-49
  drawing techniques, 45-48
  graphics, 49-50
presentations
  adding audio and video to, 377
  graphics as technique for, 49-50
  as portfolio output, 378
  software for, 377
pressing processes, 132. *See also* contour change processes
pressure
  compressed air, 90-91
  energy conversion and air, 170-172
primary colors
  described, 27
  harmonious colors and, 29
primary energy sources
  defined, 143
  solar energy as, 143
primary information sources, 18
printed circuit boards (PCBs), 395-396

printing
  combining images with text for, 377
  as communication system output, 377
  invention of movable type for, 459
prisms, shading, 47
privacy, 259
problem solving. *See also* solutions
  information for, gathering and using, 214
problems, identifying, 5-6, 17-18
process devices, 191
processed food, 229
processes
  automation of, 406, 410
  vs. input signals and output devices, 191
  material conversion. *See* material conversion process
  resource management, 254
  time management, 257
processors, 192, 197-198
production
  automation of, 407, 410
  control systems and, 405-407
  defined, 310
production models, 61. *See also* models
production systems
  conservation and reuse, 322-324
  energy production, 316-317
  extracting minerals, 310-311
  growing and harvesting, 312-314
  manufacturing, 318-322
  overview, 289-290, 309-310
  refining metals, 314-316
products
  from addition processes, 126-129
  assessing design quality of, 443-444
  designs as, 4
  factors determining materials used for, 106-107
  from materials conversion processes, 125
  as outcome of IDEATE Design Loop, 4
  packaging, 66-67, 382-384
  prototypes for mass-produced, 61
  from recycled materials, 323

programming codes into machines, 193
programming languages, 365-366
programs, computer, 408-410
projecting trends, 427-428
projection drawings. *See* orthographic drawings
projection, Mercator, 423
promissory note, 264
propelling vehicles, 307
properties
  of crystalline materials, 115-116
  dependency on molecular structure, 115-116
  of materials, 108
  of plastics, 112-113
  pliability of platelets, 114
proportion, 30-31
proteins. *See also* enzymes
  mycoproteins, 354
  single cell, 354
prototypes
  defined, 13
  in design process, 13
  in development process, 441
  of electronic models, 60, 395
  testing best solution using, 19
prototyping
  boards for electronic, 60
  defined, 61
psycho-social stages in human development
  achievement, 245-246
  autonomy, 244
  initiative, 244-245
  intimacy, generativity, and integrity, 246-248
  learning to trust, 243-244
  overview, 241-242
pulleys. *See also* belts
  described, 90
  in mechanical models, 60
  velocity ratio, 166
pumping processes
  described, 312
  materials extracted by, 311-312
punishment, power to confer, 274
pure research, 437
purification, 346
pyramids, shading, 47

Q

quadrant (navigational aid), 422
quality
  automation and control of, 407
  of product design, assessing, 443-444
quality circles, 277
question webs, 57-58

R

rack and pinion gears. *See also* gears
  described, 89
  velocity ratios, 167
radiactive carbon atoms, 424
radiactive wastes, 464
radial balance, 31
radiant energy
  described, 158-159
  in energy spectrum, 152
railroads, 304
rapid deposition modeling (RDM), 69
ratio level of measurement
  described, 210-211
  vs. other levels of measurement, 209
raw materials, 118
RDM (rapid deposition modeling), 69
read only memory (ROM), 198
receivers, 362-363
reclaiming materials, 322-324
recombinant DNA (rDNA), 339-340
recording information
  activities for time management, 256
  quantity and value, 203
  sharing information by, 204-206
  time measurements, 203-204
recycling materials, 322-324
reduction refining, 314-315
redundancy in information transmission, 370
reed switches, 194
referent power, 274-275
refining
  foods, 316
  fuels, 316
  metals, 314-316

reforming process, 316
refrigeration
  latent heat of evaporation and fusion, 155
  steady state operation in, 400
reiteration, 16
Relativity, Einstein's Law of, 153
replication (of messages), 367-368
report
  of design strengths and weakenesses, 14
  as documentation format, 38
representations, models as, 56
repulsion, force of
  between atoms, 108-109
  magnetic poles, 94
research
  forms of, 7
  pure vs. applied, 437
research and development, 428, 439-440
reservoirs (accumulators), 94
resistance
  circuits operating on "change of", 197-198
  vs. current flow, 174
  defined, 175
  Ohm's Law and, 176-177
  values of and tolerances, 179-180
resistors
  calculating required size of, 176
  described, 178-179
  determining resistance of, 179-180
  variable, 192
resource management
  capital, 263-270
  human. *See* human resource management
  overview, 253-254
  space, 258-263
  time, 254-258
resources
  conserving and reusing, 322-324
  consideration in decision-making, 470-472
  gathering from outside groups, 280
  interrelationship of, 280
  planning which are required, 12
restriction enzymes, 340-341
results (conversion process), 124-126

results (decision-making), 471-472
rewards, power to confer, 274
risks
  acceptable, 443
  vs. benefits of new technologies, 458
roads
  construction of, 296-297
  historical development of, 305
  rebuilding, 323
robots
  biosensors and, 348
  mimicking humans, 223
rolling structures, 299
ROM (read only memory), 198
rotarios, 423
rotary motion
  cams, 89
  gears and, 87
  right angles, transmitting through, 87
  shafts, transmitting between, 86
rpms (revolutions per minute), 91
rules, 279

S

safety, developing sense of, 244
safety precautions
  chemical processes, 154
  compressors, 91
  electromechanical mechanisms, 94
  energy, 152, 158
  hot-wire cutters, 68
  mechanisms, 85
  pneumatic mechanisms, 91
sampling energy, 191
satellites, synchronous, 373
sawing, 129
scale model parts, 65
scale models, 67
scanning images, 375-377
scenarios, 433-434
scheduling tasks, 271-273
scientific method, 437
SCPs (single cell proteins), 354
second-order effects, 466
secondary colors, 27-28
secondary energy sources, 143
secondary sources, 18
sectional drawings, 45
sedimentation tanks, 355

selective breeding, 340
self-regulating control systems, 398-400
semiconducting materials, 195
senders
  coding information to be transmitted, 362
  compatibility with receivers, 362-363
  feedback from receivers, 371-372
senses, human
  in ergonomic design, 35-37, 238-240
  as human control subsystem components, 224
  information input through, 187-188, 233-235
sensing
  electrical energy, 194
  light energy, 195
  magnetic energy, 194
  mechanical energy, 193-194
  sonic energy, 195
  thermal energy, 195
sensors, 191-193
sensory modalities, 238
sensory templates, 235
separation processes
  described, 129-130
  immobilization method, 346-348
sequence (conversion process), 124-126
sewage processing, 355
sextant, 422
shading. See also highlights
  described, 46
  in pencil rendering technique, 46
shafts, transmission of force between, 86
shapes
  described, 26
  as element of visual design, 25
  in ergonomic design, 35
  in patterns, 32
  sketching and drawing, 40
  as symbols, 207
shaping materials, 68. See also forging processes; pressing processes
shearing, 129
sheltering structures
  defined, 291
  described, 297
  overview, 290

shelters, human, 230
ships
  determining at-sea postion of, 421-423
  exploration of the oceans using, 420-421
  as floating structures, 298-299
  as three-way transports, 304
shortcuts, 257
sight, sense of, 35
signal entropy
  contribution to garbled information, 363
  described, 370-371
signals
  code common to senders and receivers, 362-363
  in computer communications, 365-366
  conversion into symbols, 205-206
  digital information devices for sending, 189
  electrical energy, sensing, 194
  human analysis of, 205
  as information for machines, 190-193
  light energy, sensing, 195
  magnetic energy, sensing, 194
  mechanical energy, sensing, 193-194
  processors, codes, and logic, 197-198
  purpose of, 188
  sensors providing input, 191-193
  sonic energy, sensing, 195
  symbols vs., 188-189
  thermal energy, sensing, 195
  vs. media and carriers, 367
signs as symbols, 207
simulations, computer, 379-380
single cell proteins (SCPs), 354
single-action cylinders
  described, 92
  valve control of, 92
sintering, 137
six-way transport, 306
size in ergonomic design, human, 32-33
size (of objects)
  in ergonomic design, 35
  object functionality and comfortable use vs., 32

skeleton, human
  growth of, 226
  as human structure, 222-223
sketching techniques, 39-40
skills, management, 275-278
skin, human, 222-223
slow oxidation, 142, 147
sludge digesters, 355
"smart" machines
  AND gates, 402-403
  logic and decisions, 402
  NOT gates, 404-405
  OR gates, 403-404
  production and controls, 405-407
smell, sense of, 35
social consequences of technology, 460-462
software, presentation, 377
solar energy
  creation of fossil fuels and, 147
  described, 143
  photosynthesis and, 147
  wind energy and, 144
solenoids, 95-96
solutions. See also problem solving
  developing design of best, 11-13, 18
  experimenting and developing, 19
  searching for, 18
solvents, 114
sonic energy, 195
sound. See also auditory
    modalities; sonic energy;
    ultrasonics
  in acoustic processes, 134
  as energy, 232
space
  familiar, 260-261
  requirement for orientation in, 258-259
  territorial, 259-260
space exploration, 423
space management. See also
    resource management
  techniques, 261-262
spacecraft
  as flying structures, 301
  as six-way transport vehicles, 306
speakers as output devices, 192
specifications. See also criteria
  defined, 8
  determining which solution will
    meet, 10

in generating design briefs, 18, 19
  identifying success in fulfilling, 19
  testing for compliance with, 13
spectrum. See electromagnetic
    spectrum; energy spectrum
speech synthesizers, 366
spheres
  shading, 47
  as standard form in drawings, 46
splicing genes. See gene-splicing
    process
sprocket and chain assemblies.
    See also belts; pulleys
  described, 86
spur gears
  described, 87
  worm gears and, 87
stability, human sense of, 35
stamina, human, in ergonomics, 35
standardization, defined, 33
static loads, 293
stature, anthropometric chart of, 33
steady state operation, 400
stepper motors, 95
stereolithography, 69
stick chart, 423
strength, human
  in ergonomics, 35
  growth and development vs., 226
  of hand, 225
strengths, design
  assessing alternatives to identify, 19
  assessing in final evaluation, 19
stress, mechanical, 155
stretching materials, 156. See also
    contour change processes
strip mining, 311
structural models, 60. See also
    models
structures
  bicycle, 97-98
  compression and tension forces, 291-293
  containing, 290, 291, 298
  designing for loads, 293-295
  directing, 290, 291, 298
  human, 222-223
  of materials, 108-118
  materials to withstand loads, 295-296

sheltering, 230, 290, 291, 297
static and dynamic loads, 293
supporting, 291, 296-297
transporting systems and, 303-304
transporting, 290, 291, 298-299
types of, 290-291
submarines
  as five-way transport vehicles, 305
  six-way transport and, 306
subsystems
  of machines, 97-99
  in systems models, 60
sun. See also solar energy
  direction-finding devices and, 420-421
  gravitational energy, 145
  nuclear fusion within, 153
supply and demand principle, 263-264
supporting structures
  defined, 291
  described, 296-297
  overview, 290
surface finish, 36
switches
  in digital systems, 408
  as input signal sources, 192
  mechanical, 193-194
  reed, 194
  sending signals using, 189
  sensing magnetism, 194
symbiotic relationships, 227
symbolic communication, 374
symbols, 217
  assembly into messages, 206
  binary, 189-190
  code common to senders and receivers, 362-363
  collections of, 207
  described, 189
  for elements and compounds, 110
  as ergonomic consideration, 37
  evoking emotions using, 240
  exchanging information using, 204-206
  human use of, 188, 202-203
  in IDEATE model, 208
  logos and pictograms, 206
  loss of meaning over time, 370
  representing numerical values, 203

systems of, 207
vs. media and carriers, 367
symmetrical images, 31
synchronous satellites, 373
synthesizers, 366
synthetic hormones, 351. *See also*
hormones
synthetic materials
for construction, 295
development of, 106
synthetic skin, 230
systems. *See also* entries for specific
systems
complexity of, 60
machine subsystems, 97-99
as outcome of IDEATE Design
Loop, 4
systems models, 60. *See also*
models

T

tactile aspect of texture, 27
tasks, analyzing and scheduling,
271-273
taste, sense of, 35
teams
in flattened hierarchies, 277
working in, 273-274
technical skill
defined, 275
persons requiring, 276
technology
conceptual models of, 56
consequences and decisions,
449-476
energy as driving force of,
141-142
hard vs. soft, 466
humans as creators and users of,
221-222
materials as fundamental to,
105-106
predicting consequences of, 450
recommendations, 477-480
risks vs. benefits of new, 458
role of designing in, 23-24
symbols in, using, 208-216
technology assessment, 451-453
telecommunication, 372
telegraph
defined, 372
invention of, 204, 372

telemetry, defined, 372
telephone, defined, 373
telescope, defined, 373
teleshopping, defined, 373
teletext, defined, 373
television
as communication medium, 368
defined, 373
temperature. *See also* heat; thermal
energy
of objects handled by humans,
35
in thermal processes, 133
temperature sensors, 192
templates
cognitive, 234, 235-237
for machine response to signals,
205
tensile testing of materials, 118
tension, force of
described, 155-157
in structures, 291-293
territory
sense of, 259-260
staking claim to, 260-261
tertiary colors, 28
tessellations, 32
testing designs, 13, 19
text
combining with images for
printing, 377
inputting into computers, 375
textile family of materials, 122
textile systems, 230
texture
described, 27
as element of visual design, 25
food, 239
theoretical models. *See also* models
described, 58
types of, 59
theory, 58
thermal energy
derived from mechanical energy,
155
described, 155
producing, 125
sensing, 195
thermistors, 195
thermocouples
as input sensors, 191
as thermal energy sensors, 195

three-dimensional communication
forms and, 26
packaging, 382-383
three-way transport, 304
tidal energy, 145
time. *See also* resource
management; time
management
emotional effects of changed
sense of, 458
entropy and, 255-256
and speed of transportation, 308
symbols and, 203-204
time management. *See also*
resource management
overview, 254-255
techniques, 256
timers, 192
timetable, 11
tissue culturing and cloning,
352-353
Tokamak nuclear fusion reactors,
146
tolerance limits, human, 36, 225
tolerances, resistor, 179-180
tonal range, 29
tones
creating in pencil renderings, 46
mixing pigments to change, 29
tooling up process, 61
tools. *See also* gears; linkages
development of principles of
operation, 79
as extensions of human abilities,
77-79
human evolution and use of,
454-455
lever principle, 79-82
model-making, 65-66
results of replacing physical
effort, 455
wheels and axles, 83-84
toothed belt and pulley, 86. *See
also* belts; pulleys
torque, defined, 86
touch, sense of, 35
toxins, 336
trade, symbols and, 203
trade-off process, 443
transistors
in control system circuit, 395,
404
as signal processors, 192

transmitting information, 366-371
transporting systems. *See also* vehicles
  described, 303
  load limits, 307-308
  mag-lev, 302-303
  overview, 289-290
transport units, 307
transportation
  purposes of, 308
  wind energy and ships, 144-145
transporting structures
  defined, 291
  described, 298-299
  overview, 290
trend analysis, 426
trends
  analysis of, 426
  projecting, 427-428
true North, 421
truss structures
  defined, 293
  towers, 297
trust, learning to, 243-244
truth tables, 200
turning forces, 163-165
two-point perspective drawings, 41-42
two-way transport, 304

U

ultrasonics, 157
undesired outcomes, 7-8
unintended outcomes, 8
Universal Product Code (UPC), 364

V

vaccines, 345
values, decisions and, 473
valves
  flow regulators as, 92-94
  pneumatic, 92
variable resistors
  in control system circuit, 395
  as input signal sources, 192
  photovoltaic cells and photoresistors as, 195
vehicles. *See also* transporting structures
  containing and propelling, 307
  heavier-than-air, 305-306

as impact structures, 301-302
  lighter-than-air, 305
  mag-lev, 302-303
velocity ratio
  of gears, 87, 166-167
  wheel and axle, 166
VHST (Very High Speed Transit) vehicles, 302
vibration
  electromagnetic vs. mechanical energy bands, 157
  human limits of tolerance of, 36
  mechanical energy and, 157
video in portfolio presentations, 377, 379
virtual reality
  as interactive communication, 379
  overview, 239-240
  in presentations, 377
viruses, 333, 344-345. *See also* microorganisms
visible light waves, 368
visual aspect of texture, 27
visual communication
  enhancing, 24
  packaging as, 382-383
visual design
  color as element of, 27-30
  contrast as principle of, 31
  as ergonomic consideration, 37
  form as element of, 26
  harmony as principle of, 31
  line as element of, 26
  overview of elements of, 25
  overview of principles of, 30-35
  pattern as principle of, 32
  shape as element of, 25
  texture as element of, 27
visual modalities, 238
vitrification, 115
voice recognition and synthesis, 195
Volta, Alessandro, 372
voltage
  measuring, 174
  Ohm's Law and, 176-177
  power rule and calculating, 178
voltmeters, 175-176
volume, fluid, Boyle's Law and, 170-171

vulnerability, human
  development of technology and, 225
  growth and development vs., 226
  need for belonging and, 245

W

wants, human
  psychological models of, 241-242
  role of design work in satisfying, 23-24
waste materials
  microbes eating, 355
  radioactivity from nuclear, 464
  recycling, 324
water
  energy from movement of, 145-146, 160-161
  energy from stored, 160-161
  purification of, 355
waterhead, 145-146
Watt, James, 388
watts
  vs. joules, 151, 177
  measuring power in, 175-177
wavelength, 368
ways, transportation, 304-305
weaknesses
  design, assessing alternatives to identify, 19
  identifying design, 19
  of materials along grain, 113-114
weather forecasting and reports, 380
weaving processes, 129. *See also* addition processes
wedges, 129
wheel and axle
  described, 83-84
  turning forces and, 165
  variations on wheel, 166-168
  velocity ratios, 166
wind energy, 144-145
wind farms, 145
windlass
  described, 83
  turning forces and, 165
wood
  conserving, 324
  as family of materials, 122
  growing and harvesting trees for, 313-314

molecular structure of, 112
as polymer, 112
workers. *See* human resource
    management
worm gears
  described, 87
  velocity ratios, 167

X

X rays, 158-159

Y

yeasts, 333, 335, 343. *See also*
    microorganisms

Z

zero
  in ratio level of measurement,
    210-211
  as symbol, 203
zero force processes, 129